Electrical Biosignals in Biomedical Engineering

Peter Husar · Gabriel Gašpar

Electrical Biosignals in Biomedical Engineering

Medical Sensors, Measurement Technology and Signal Processing

Springer

Peter Husar
Institut für Biomedizinische Technik und Informatik
TU Ilmenau
Ilmenau, Germany

University of Žilina
Žilina, Slovakia

Gabriel Gašpar
University of Žilina
Žilina, Slovakia

ISBN 978-3-662-68000-1 ISBN 978-3-662-67998-2 (eBook)
https://doi.org/10.1007/978-3-662-67998-2

This Springer imprint is published by the registered company Springer-Verlag GmbH, DE, part of Springer Nature.
The registered company address is: Heidelberger Platz 3, 14197 Berlin, Germany

Paper in this product is recyclable.

Foreword

Biomedical engineering or biomedical technology is an interesting field of science. It is truly multi-engineering and multi-scientific field crossing also over several natural sciences including physics, chemistry, biology, and clinical sciences. Combining all of these biomedical technology areas or even a certain subset like topics around biosignals in a book is a formidable challenge. However, I think Peter Husar has made a very nice job on compiling his book for the introduction of the area of bioelectric biosignals, their origin, measurements, and processing.

I am a professor of Biomedical Engineering at Tampere University, and during my professional career I have given courses several different subjects on the field from bioelectric phenomena, biosensors to instrumentation, biosignal and systems, and physiological system modeling. Further acting as a president of the Finnish and the European Societies on Biomedical Engineering Sciences provided me a broader view on the biomedical engineering, and especially its core contents. What I can see is happening is the integration of topics inside the fields of biomedical engineering and integration of biological or clinical fields into engineering core concepts.

I have known Peter now for over 10 years as we have had our joint EU FET-OPEN project that I was coordinating. That project aimed to provide novel measurements for emerging 3D in-vitro neuronal cultures—a work well ahead of its time as FET projects often are. In a way that project provides a mirror image of this book Peter has written. Our project had an engineering challenge that was to develop 3D electrode arrays for the stimulation and recording of the bioelectric signals for in-vitro human stem cell-derived neuronal 3D cultures—a quest that still today is only partially solved. In that project the key point was to understand the neuronal system biology and electrophysiological signaling in order to develop novel 3D measurement system. In the project, as almost always when novel sensor system is developed and thus new biosignals are recoded, it is also of importance to develop the tools to assess the acquired signals and further to develop methods to analyze and model the resulting information to gain better understanding on the underlying biological processes, e.g., for assessing drug effects or disease models.

Even the EU FET-OPEN project was back in time, however, still the same engineering principles of the bioelectric signal acquisition, measurement, and analysis system design are fully valid today. In order to have the ability to design novel bioinstrumentation and analysis tools biomedical technology students should have the basic understanding of all the signal processing components and scientific principles in their toolbox—and have at least basic understanding on the underlying biology and how the signals are translated from the biological sources through the sensors and electronics to data in computer. And this book does a very good job in introducing all these concepts.

In his book on *Electrical Biosignals in Medical Technology - Medical Sensors, Measurement Technology and Signal Processing* Peter has brought together the core of electrical engineering principles of biomedical engineering. This book starts with an introduction of the biological origin of signals using the nervous system as an example. Then the origin and the theory of the bioelectric fields are covered including the electric excitation of cells and projection of the resulting electric fields to the measurement system and sensors. The book provides a deeper look at the sensors introducing the principles of various sensor types and then dives into introducing typical neuronal biosignals and their properties. The next chapters provide a good overview on signal acquisition from amplifiers, digitalization, and AD conversion. The signal processing covers the principles of analog and digital filtering of the signals including the time, frequency, and compound modes. Rest of the book provides more comprehensive introduction to the principles of assessment of the derived parameters with fundamentals of the statistical analytics and tests and finally using the statistical approaches on signal analysis by considering the biosignals as stochastic processes introducing statistical tools for time series, signal detection, and decomposition techniques.

All this provides a very good overview for the students of bioelectric signals from the sources to the estimation of the importance of the acquired signals and their time or frequency-based components and parameters. The book includes rather extensive literature review and exercises to students especially regarding the instrumentation, acquisition, signal analysis, and statistical methods, thus proving good basis for developing introductory courses for biotechnology and biomedical engineering students providing comprehensive view on engineering principles needed to design instrumentations, signal acquisition, and their processing, parametrization and assessing the outcomes. The book can easily serve as a great introductory book on these topics for bachelor level, or also as a basis to teach biosignal processing in M.Sc. or even for Ph.D. students or developers entering this area. I would like to conclude to congratulate Peter for the achievement putting together this nice and comprehensive book.

Tampere, Finland
August 2023

Jari Hyttinen

Reviewers

Our great thanks also go to the peer reviewers who, in their demanding work, have found time for constructive criticism of our forthcoming book. Thanks to:

- *Prof. Ing. Ladislav Janoušek, Ph.D.*, University of Žilina, Faculty of Electrical Engineering and Information Technology, Žilina, Slovak republic.
- *Prof. Ing. Dušan Maga, Ph.D.*, Czech Technical University in Prague, Faculty of Electrical Engineering, Prague, Czech Republic.
- *Prof. Dr. Andreas Voss*, University of Applied Sciences Jena, Jena Germany.

Contents

Part I

Origin, Acquisition, Analog Processing, and Digitization of Biosignals

Origin and Detection of Bioelectric Signals 1

1.1 The Neuron

The neuron (nerve cell) is the basic building block of all sensory (visual, auditory, somatic, gustatory, olfactory) and motor systems (motor end plate) as well as of the central (CNS) and peripheral nervous system (PNS). Its basic structure and connections to other neurons are shown in Fig. 1.1. From a signal analysis point of view, and the neuron has several signal inputs (the synapses) and a signal output (the axon). So that from the systemic point of view, it can be interpreted as a MISO system (Multi-Input–Single-Output).

From the signal processing technology point of view, a neuron is a hybrid, analog, and digital mixed system (MSP, Mixed Signal Processor). All input signals at the synapses are given temporally and spatially variable analog weights, and the information weighted this way is integrated temporally and spatially in the cell. The integration result is compared with a potential threshold, similar to the discriminator: If the integral is above the threshold, an action potential (AP) is emitted via the axon (output). If the integral is below the threshold, nothing happens at the axon. Since an AP has a constant form at the axon, the output signal can be simplified as a quasi-random binary sequence. The intensity of the temporally and spatially integrated input signals is encoded at the output into the AP frequency (temporal density). In a sense, this is the first natural conversion of analog signals into a binary sequence, whereby this inherently has a nonlinear, dynamic, and spatially variable transfer function.

At rest, the concentrations of the ions inside and outside are different. So that a potential difference of 50–100 mV forms across the membrane, inside negative

Supplementary Information The online version contains supplementary material available at https://doi.org/10.1007/978-3-662-67998-2_1.

P. Husar and G. Gašpar, *Electrical Biosignals in Biomedical Engineering*, https://doi.org/10.1007/978-3-662-67998-2_1

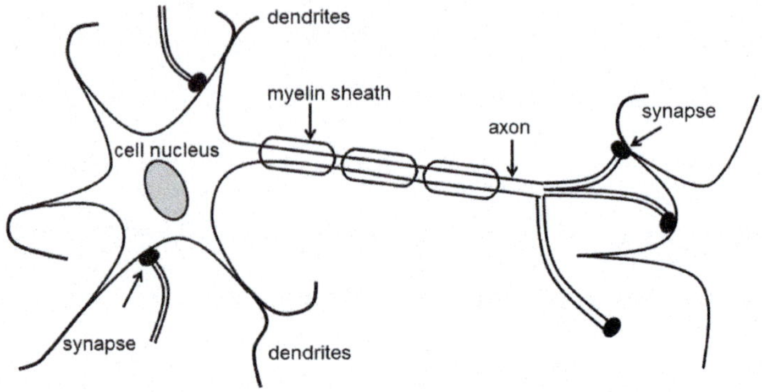

Fig. 1.1 The neuron is the basic building block of the sensory and motor systems and the CNS and PNS

against outside (Figs. 1.2 and 1.3). This potential difference is maintained with the help of the so-called Na–K pumps, and a dynamic equilibrium is formed. Influences from the outside (via the synapses) shift the equilibrium until it is no longer sustainable above a threshold, and the Na–K channels open the way for ion flows. The flow dynamics consist of three phases (Figs. 1.2 and 1.4):

- *Depolarisation*: The synapse signals accumulate in time and space and exceed the threshold. The Na channels open, and Na^+ ions flow like an avalanche (positive current in Fig. 1.4). Afterwards, the Na channels close and remain closed for a specific time (*absolute refractory period*). No further AP can be triggered during this time.
- *Repolarisation*: Na channels are closed (refractory phase), and K channels open (negative current in Fig. 1.4).

Fig. 1.2 Time course of an action potential. The membrane conductivity for $Na^+(y_{Na})$ increases faster than for $K^+(y_K)$. As a result, the cell interior becomes more positive (depolarisation), then more negative (repolarisation). Until equilibrium is restored, a third phase of hyperpolarisation follows. The AP (action potential) has an average width of 1.5 ms

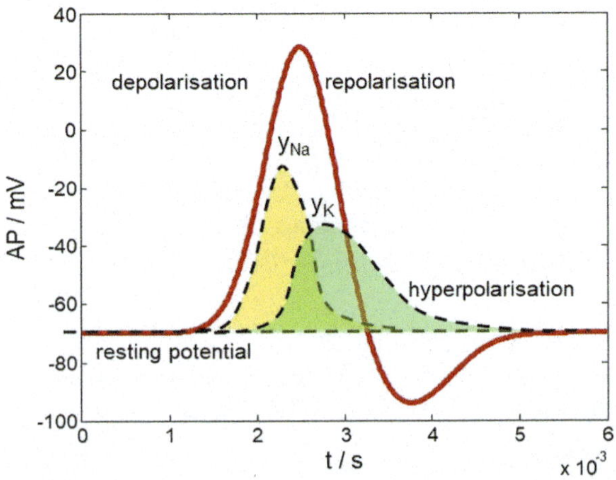

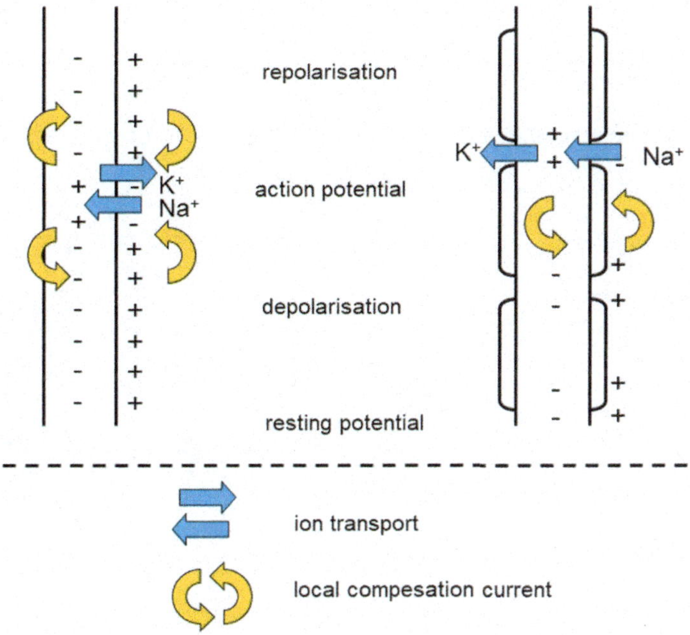

Fig. 1.3 Course of the action potential along a membrane. On the left, electrotonic propagation (continuous). On the right, saltatory propagation across the myelin sheath (spatially erratic)

- *Hyperpolarisation:* K^+ ions flow outwards, and K channels close with a delay compared to Na channels so that overshooting occurs, and then the equilibrium is re-established. In this phase, the threshold is raised so that higher intensities from outside are necessary to trigger the next AP. Therefore, this phase is called the *relative refractory period.*

This three-phase course has a constant signal shape called an action potential (AP). It has an average width of 1.5 ms (cardiac cells up to 300 ms) and an amplitude of up to 100 mV (across the membrane, potential difference inside-outside). The AP continuously continues along the membrane in electrotonic propagation (Fig. 1.3). In saltatory propagation in spatial jumps between Ranvier's lacing rings, reaching 1–100 m/s.

The Na^+ /K^+ ion fluxes through the membrane channels are a charge shift from an electrical point of view. This results in an actively impressed electric current (from the Na–K channels) across the membrane—the *membrane current* (Figs. 1.3, 1.4 and 1.5). From the time course of the voltage across the membrane and the current through the membrane (Fig. 1.4) the membrane acts like a capacitor. The signal passes through a high-pass filter or a differentiator from the primary source (AP of the cell) to the outside (body surface).

The conduction of an AP presents itself to the outside, like spatial and temporal local negativity (negative charge). The local potential drop at the outer wall

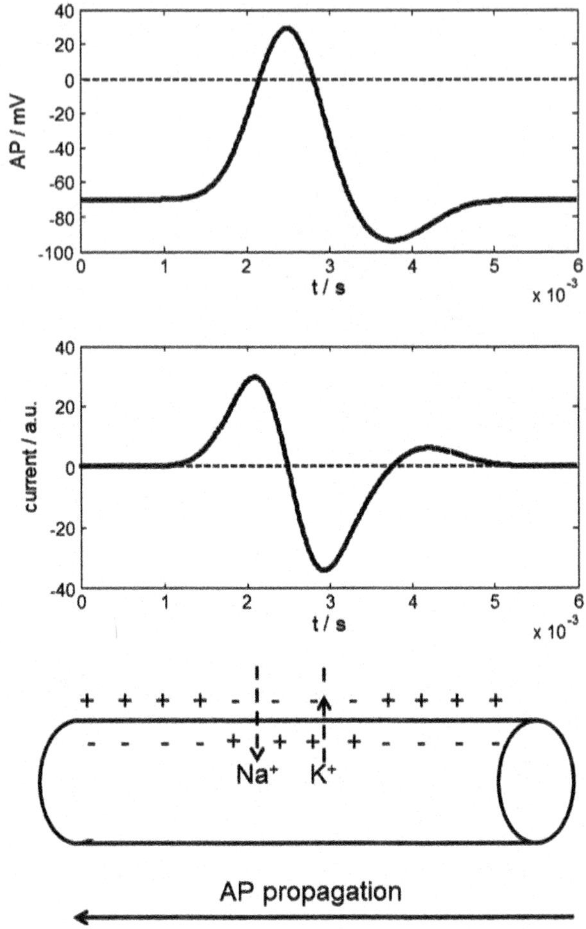

Fig. 1.4 Time course of the action potential (potential difference across the membrane, top) and the electric current through the membrane (bottom). It should be noted that the current course results qualitatively from the first time-derivative of the potential difference (the membrane acts electrically like a capacitor)

creates a current sink, which can be interpreted as two dipoles (tripoles) directed in opposite directions (Fig. 1.5). As a result of the external local current sink, extracellular, passive currents i_A flow. A spatially and temporally local positivity inside the cell arises as a mirror image of the external current sink. This positivity, considered active by nature, results in intracellular currents i_I.

The effect of the externally observed local negativity in the propagation of an AP provides, on a macroscopic scale (surface derivation in humans), an explanation for why in neurology, contrary to the usual conventions in electrical engineering/signal analysis, negative electrical activity is shown to be directed upwards on the potential axis over time. A high negative potential (potential difference) indicates high neuronal activity for neurologists.

Another source of measurable potentials is the synapse or its contact point to other neurons (Fig. 1.6). In the presynaptic axon terminal, the AP triggers a Ca^{2+} influx, which results in neurotransmitter release. The neurotransmitters open ion

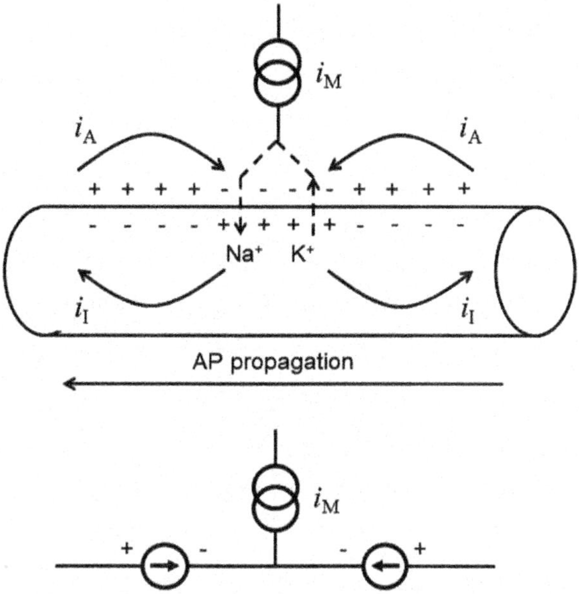

Fig. 1.5 Currents and potential differences during AP continuation along a membrane, electrical interpretation of AP continuation: the primary signal source is the Na–K ion currents that generate the electrical membrane current i_M. It can be represented as a current source across the membrane. The outer local negativity of the cell presents itself as two oppositely directed dipoles or a tripole

channels at dendrites, which change the membrane potential depending on the ion current. Since the membrane potential is changed behind the synaptic cleft, it is called the postsynaptic potential (PSP). Depending on the function of the synapse, a distinction is made between excitatory (EPSP, excitatory PSP) and inhibitory (IPSP, inhibitory PSP). The PSP directly determines the weight of the signal received via the synapse in the total sum of all synapses. It is then compared with the threshold for triggering an AP.

1.2 Electrical Excitation Conduction and Projection

For research purposes or precise medical questions, activity measurements of individual neurons are carried out. Typically, one examines the conduction of stimuli at macrostructures, such as nerves, cortical centers, or motor units. For this purpose, combining the activity of many neurons (up to 100,000) in space and time into manageable electrical sources is beneficial. A common, albeit very simplistic, interpretation of neuroelectric activity is modeling using an electric dipole. For this purpose, some basic approaches will be considered first.

Suppose a dipole source is located in an infinitely large, homogeneous, and isotropic (conductivity equal in all directions) medium. In that case, the current lines are distributed rotationally symmetrically along, and the equipotential lines transverse to the dipole axis (Fig. 1.7).

In practice, however, one must reckon with natural, spatially limited volume conductors. This situation is shown in Fig. 1.8. Due to the finiteness of the body, compressions of the electric fields occur. As a result, projections can arise on the

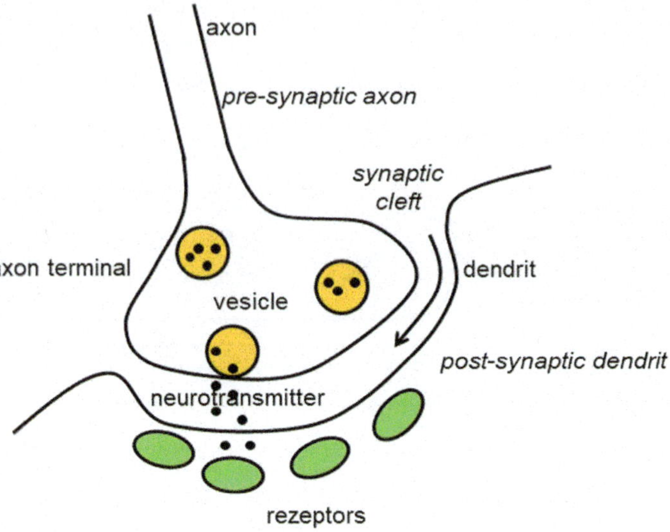

Fig. 1.6 The synapse and its connection. The neuron with its synapses is, to a large extent, a dynamic and nonlinear system. Individual synapse weights (PSP) change over time, causing the neuron to react to changing conditions and information. The spatially and temporally integrated information from the synapses is fed to a threshold via a nonlinear transfer function. If the threshold is exceeded, an AP is generated. The process described is naturally used to store information patterns, compare them and generalize similar patterns. This behavior is used in artificial neural networks (ANN) to model dynamic, nonlinear, and fuzzy problems of classification, recognition, and generalization and solve them at least suboptimally

Fig. 1.7 Currents $i(r, t)$ (solid lines) and equipotentials (r, t) (dashed lines) of the electric field of a dipole source in an infinitely extended, homogeneous, and isotropic medium (r is the radius, t the time)

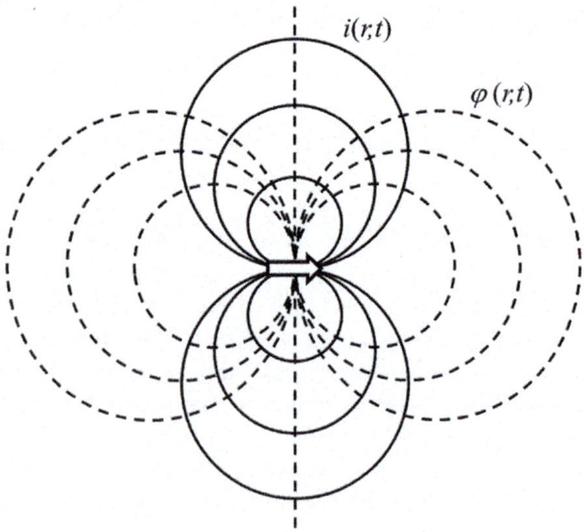

body's surface that is identical for different sources and spatial positions. Mathematically expressed, the potential differences on the body surface can be described as follows (Eq. 1.1.):

$$\Delta\phi(y, i) = f(\phi_i(i, r, t, x, y)) \tag{1.1}$$

The relationship from Eq. 1.1 connects a source's depth and intensity and the potential differences derivable at the surface. If one keeps the coordinate x and the time t constant in the relationship according to Eq. 1.1—which corresponds to the comparison of two sources at different depths y or with different intensities *i—one* finds the following: The potential difference measured at the body surface at the points (x_i, x_j) can be identical for different depths y of the sources and different intensities *i.* One can derive the potential difference at the body surface alone *fr*om the potential difference m*ea*sured at the points (x_i, x_j). It follows that one cannot derive any conclusion about the source depth from the potential difference at the body surface alone. Mathematically, the relationship, according to Eq. 1.1 can be derived with a two-dimensional uncertainty factor (Eq. 1.2):

$$\phi(y, i) = c_y + c_i + \int f(\phi(i, r, t, x, y))\mathrm{d}y\mathrm{d}i \tag{1.2}$$

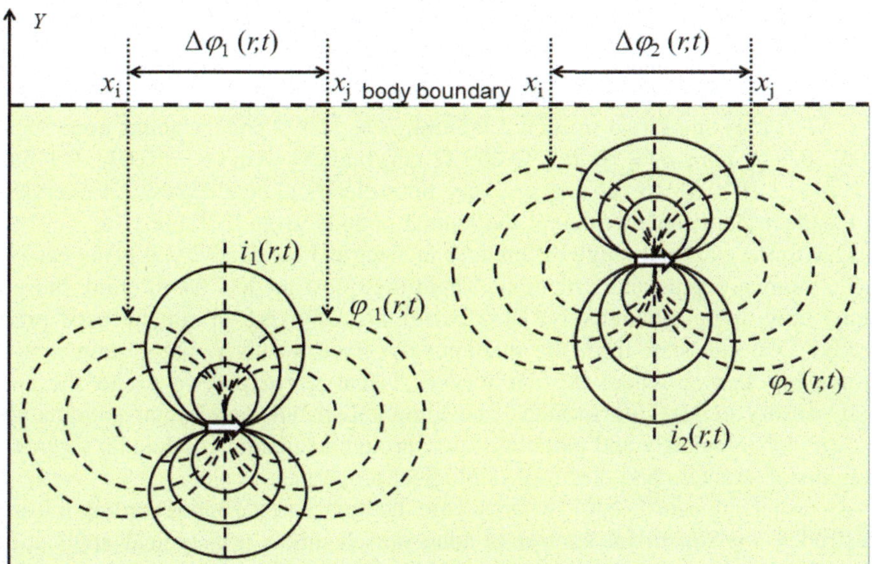

Fig. 1.8 Currents $i(r,t)$ (solid lines) and equipotentials s(r,t) (dashed lines) of the electric fields of two dipole sources of equal strength in a finite spatial, homogeneous and isotropic conductor (r is the radius, t the time). The body boundaries mark the volume conductor. The spatial dimension is reduced to a surface representation

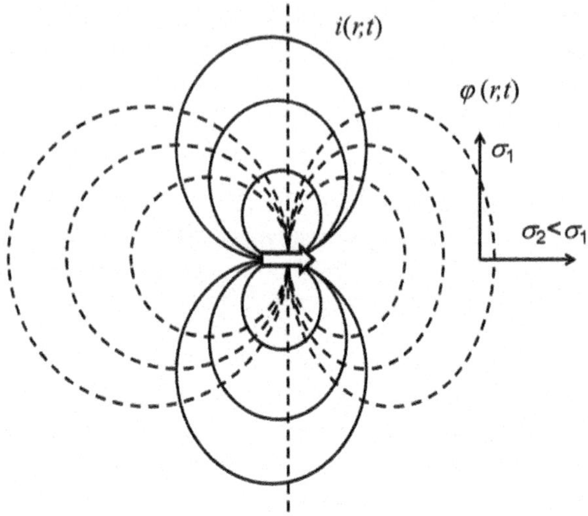

Fig. 1.9 Deformation of the electric field due to anisotropy of the volume conductor. The electrical conductivity σ_2 is horizontally lower than σ_1 vertically in the right part of the body. This leads to the compression of the potential lines

This consideration follows for biosignals' areal or spatial analysis: An identical areal or spatial projection of neuroelectric activities does not necessarily conclude that the sources are equally deep or intense in the tissue.

Another essential property of natural bodies that influences the electric fields is anisotropy or the spatial inhomogeneity of the electrical conductivity (Fig. 1.9). The differences in conductivity in the human body are at least three decades. The highest conductivity or lowest resistivity is found in body fluid ($\sigma = 15$ mS/cm, $\rho = 67\ \Omega$ cm) and blood ($\sigma = 6.7$ mS/cm, $\rho = 150\ \Omega$ cm), whereas bone ($\sigma = 0.05 \ldots 0.2$ mS/cm, $\rho = 5000 \ldots 20.000\ \Omega$ cm) and dry skin ($\sigma = 0.03 \ldots 0.1$ mS/cm, $\rho = 1.10^4 \ldots 3.10^4\ \Omega$ cm) are very poor electrical conductors. An overview of the electrical conductivities of the human body is given in Table 1.1.

Due to the extremely high differences in electrical conductivity, it is not easy to model human organs and structures. Therefore, most models (skull/brain, thorax/heart) only apply qualitatively. How can this relativizing insight be used practically? One can start from the qualitatively given anatomy. For example, it is sufficiently known where the auditory or visual nerve pathways and the cortical sensory centers are located. The temporal-spatial distribution patterns are known physiologically and pathologically through excitation conduction's general anatomical structure and electrophysiological properties.

By projecting neuroelectric activity onto the body surface and recording it using electrodes, the functional control of long muscle fibers is a typical application (Fig. 1.10). Since long muscles (e.g., in the extremities) are essentially controlled neuronally only one-dimensionally (along the fibers), a sensor series along the examined muscle is sufficient for diagnostics. This way, the conduction speed and possible interruption can be determined relatively precisely. An areal recording of the projection may be necessary when dealing with organs (heart) or flat muscles

Table 1.1 Specific conductivity and specific resistance of selected substances, organs, and tissues of human body (Silny, 2009)

Tissue, fabric, liquid	Spec. conductivity mS/cm	Spec. resistance ρ/Ω cm
Physiological saline solution	20	50
Body fluid	15	67
Blood	6.7	150
Heart muscle	1…2	500…1000
Brain	1.7	590
Kidney	1.6	600
Skeletal muscle	0.8…2.5	400…1250
Lungs	0.7…2.5	400…1430
Fat	0.2…1.0	1000…5000
Bones	0.06…0.2	500…16,000
Skin	0.03…0.1	$1.10^4 \ldots 3.10^4$

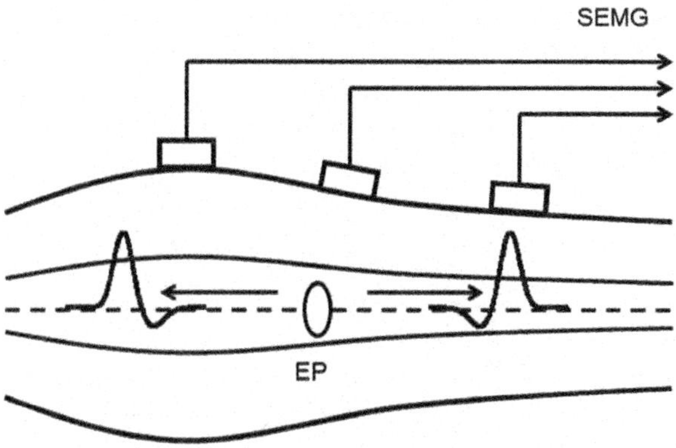

Fig. 1.10 One-dimensional derivation of a surface EMG (SEMG = surface EMG). The motor end plate (EP) in the muscle fiber triggers action potentials propagating along the muscle fiber and are recorded at the body surface via electrodes

(chest, back) (Fig. 1.11). Often the potentials between the sensors are interpolated to create an image matrix, which is then called an activity map.

The brain is the most challenging projection, derivation, and interpretation structure. The neuronal excitations propagate in turns and furrows of the cortical tissue (Fig. 1.12) and form signal sources that can generally assume any spatial position and current direction. From the point of view of signal analysis, one classifies in a simplified way between radial or tangential sources (Fig. 1.13). This binary separation does not exist; all cortical sources always form a mixture of both types.

With an EEG recording system, no more than one-third of the head surface can be recorded. Therefore, especially in the radial sources (Fig. 1.13 left), they only appear as monopolar distributions since the other pole cannot be recorded at all (in this case, in the face). Analyzing and interpreting diverse sources is particularly

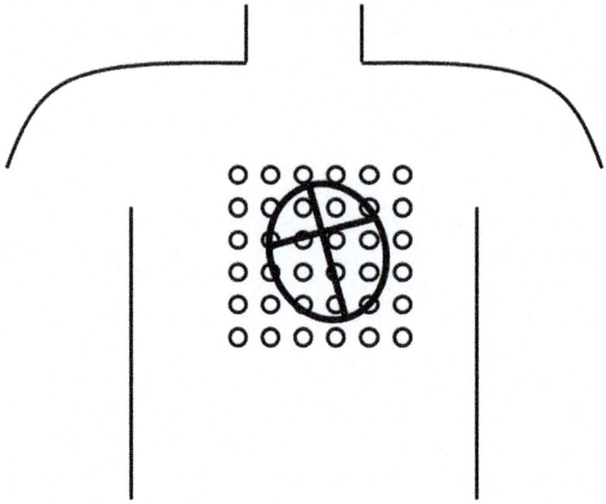

Fig. 1.11 Two-dimensional recording of a surface ECG. If the density of electrodes is sufficient, disturbances in the conduction of excitation in the heart can be localized relatively precisely. One interpolates between the sensors and creates an image matrix representing the activity. It is called ECG mapping

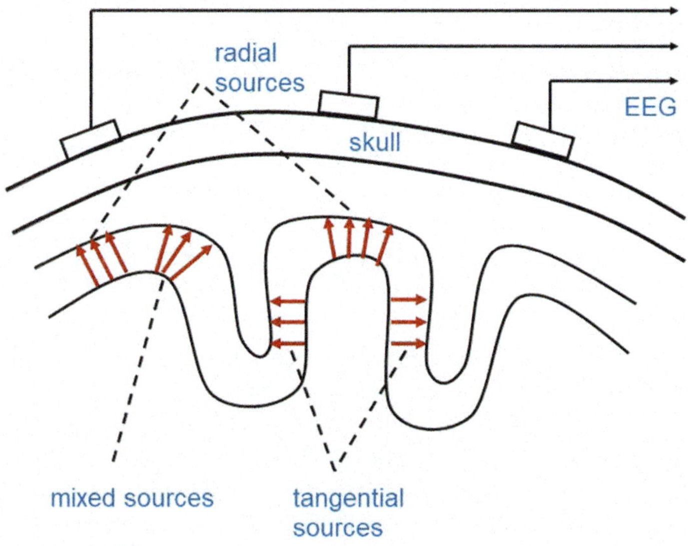

Fig. 1.12 Three-dimensional recording of a surface EEG. The electrodes are located on an approximately spherical surface. The excitation spreads along the turns and furrows and forms signal sources that occur radially, tangentially, and mixed

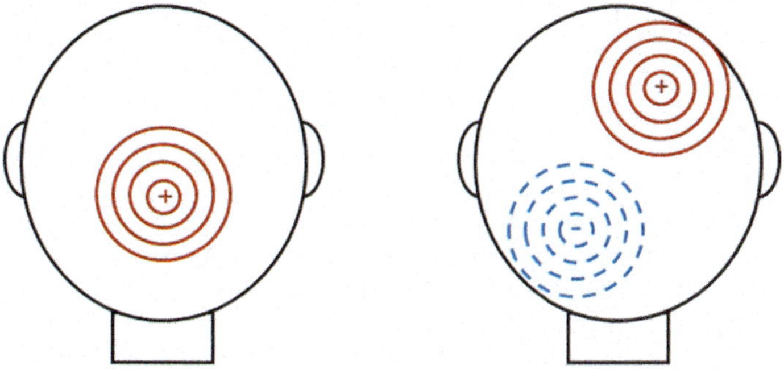

Fig. 1.13 Projections of a radial (left) and a tangential (right) source, such as those produced during the derivations of visual evoked potentials over the visual cortex

difficult, mainly if they are located in the arcs between turns and furrows during excitation propagation (Fig. 1.12). From the point of view of signal analysis, this is a moving source that also rotates during movement (translation). Such sources produce ambiguous head surface projections, i.e., one cannot clearly distinguish between translation and rotation.

1.3 Galvanic Sensors

1.3.1 Basics

By galvanic sensors, we mean electrodes that have direct (conductive) contact with the tissue. It is initially irrelevant whether the contact is between metal and tissue or metal salt/electrolyte and tissue. In recent years, so-called "dry electrodes" have been presented more often. This term implies that previously there were only "wet electrodes," i.e., those that required an electrolyte (physiological NaCl solution) or conductive gels or electrode paste (Ag/AgCl electrodes) to function. This classification does not apply, for example, to the precious metal electrodes (Pt, W, Au, In) that have been in use for a very long time, although they do not require a contact medium. At this point, the classification of the electrodes is carried out under a different aspect: For sensor and measurement technology, it is primarily not crucial whether an electrode is wet or dry. What is essential are its electrical or signal-analytical properties. It is known from bio- and electrochemistry that the electrical properties of an electrode—offset voltage, impedance, stability—depend decisively on the material and the construction of the electrodes. Figure 1.14 shows in a very simplified way which charges are effective at the boundary layer or how a space charge is formed: Basically, two different processes can be observed (also simplified).

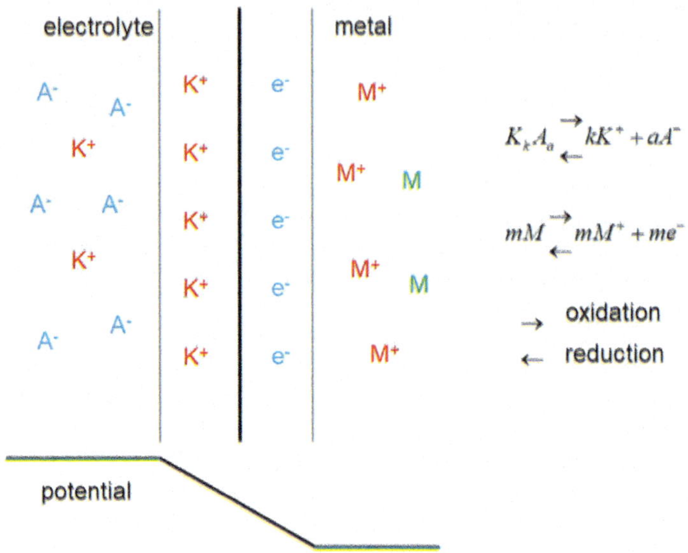

Fig. 1.14 Simplified situation at an interface between metal (conductor of the first type, electron conductor) and electrolyte (conductor of the second type, ion conductor). At the interface, a space charge zone forms from the free charge carriers of the metal (electrons) and oppositely polarised free charge carriers of the electrolyte (cations). The electrical potential is, therefore, always higher in the electrolyte than in the metal. With fully polarisable electrodes, no chemical reactions occur, so no current flows, and therefore the offset voltage is relatively high. With non-polarisable electrodes, charge transport occurs through the interface (current flows), so the offset voltage is much lower. *M* stands for metal, *K* for cation, *A* for anion, *e* for the electron, *k*, *m*, and stands for proportions or numbers

- Formation of a space charge zone through the attraction of electrons in the metal and cations in the tissue
- Charge migration through the contact point due to chemical reactions (oxidation, reduction).

A general electrode's metrologically and signal-analytically important properties are analyzed in the following.

1.3.2 Offset Voltage

In a first approximation, the offset voltage is a temporally constant voltage difference at the contact point, set according to the position of the metals involved in the electrode or the electrolyte and the tissue in the electrochemical series. With common electrodes, this voltage varies between approx. 220 mV (Ag/AgCl) and 1.2 V (platinum). These significant differences result from the different characters of the materials. Ag/AgCl is an almost non-polarisable electrode, i.e., one in which the charge current through the contact point is relatively strong, and thus the offset voltage remains low. No chemical reaction occurs with platinum, i.e., no charge transport through the contact point. This means this electrode is completely polarisable, and the offset voltage level corresponds to this.

The question arises as to which of the two borderline electrode types is better suited for measurement technology concerning the offset voltage or for which signals. If we do not compare the level of the offset voltages but their stability over time, we find that this is at least one order of magnitude worse with Ag/AgCl than with platinum. This is because the charge transport through the contact point depends on many factors that cannot be influenced, e.g., the temperature of the contact point, the concentration of the ions, geometric conditions on site, mechanical pressure, and weld formation. Notably, the offset voltage of non-polarisable electrodes is relatively low but fluctuates strongly over time. It overlaps with the low-frequency components of the biosignals, from which it can no longer be removed. Nevertheless, such electrodes are preferred in practice for several, not necessarily signal-analytical, reasons.

For measurement technology/signal processing, the question arises of controlling such offset voltage fluctuations. Their time constants are not infrequently in the one to ten seconds range. Some metrological standards specify strict limits for the corner frequencies. A maximum lower cut-off frequency of 0.05 Hz must be observed for the diagnostic ECG. This corresponds to a time constant of approx. 3.2 s. It can be realized electronically with analog high-pass filters. However, filters with such time constants react to every disturbance with a compensation process of the length of three times the time constant, i.e., at least about 10 s. In EEG analysis, this can mean that after a simple cable artifact, one has to wait almost half a minute for the measurement system to reach an average level. With today's technology, analog filters with such time constants are no longer necessary. However, this requires ADC (analog–digital-converter) must be used, which intercepts the high, but unwanted DC component by their large bit width (20…24 bit). After digitization, artifact reduction in the system clock (a few milliseconds) is possible.

From the point of view of signal processing, polarisable electrodes are preferred. They have a high offset voltage (about six times more than Ag/AgCl). On the other hand, this voltage is much more stable. These electrodes have another considerable advantage: since there are practically no chemical reactions at the contact point, there is also no stochastic charge movement, which is effective in signal processing as (disturbing) noise. In summary, polarisable electrodes are more suitable than other alternatives from the point of view of measurement technology and signal processing in terms of offset voltage.

1.3.3 Impedance

Since with non-polarisable electrodes (here Ag/AgCl), a charge transport takes place at the contact point, i.e., a measuring current flows, the real component of the impedance is relatively low (1…10 kΩ) and the capacitance relatively high (an imaginary component of the impedance low, frequency-dependent, 100 Ω…10 kΩ). This results in a low cut-off frequency of the electrode impedance for the measuring arrangement, which is very important for biosignals. The spectrum of biosignals has been proven to start at 0 Hz, so this electrode type is the better spectrum.

The impedance behavior of polarisable electrodes (here platinum) is inverse to that of non-polarisable electrodes: The real part of the impedance is very high (1 MΩ…100 MΩ), up to four decades higher than with non-polarisable electrodes (1 kΩ…10 kΩ). On the other hand, the imaginary part (capacitive) is up to four decades smaller. It results in a cut-off frequency for polarizable electrodes up to four orders of magnitude higher than for non-polarisable electrodes. The conclusion is that polarisable electrodes are suitable for high-frequency biosignals, e.g., EMG, where they are preferably used in practice.

1.4 Capacitive Sensors

1.4.1 Sensor Technology

In many situations in medical measurement technology, a galvanic connection of the sensor to the tissue/neurons is not wanted or does not make sense. For example, wiring with electrodes is practically impossible if one has to monitor the vital functions of safety-relevant persons (bus drivers, pilots, and air traffic controllers). There are also areas where galvanic contact is harmful.

In recent years, multi-electrode arrays (MEA) have also been offered for the scientific field. The prospect of a sensor array that measures almost at single neurons (sensor diameter 50 μm) is up-and-coming. However, these constructions should be analyzed intensively from offset voltage (see above), especially from a biological point of view. Galvanic sensors build up an offset voltage of well over 100 mV up to 1 V when they come into contact with tissue. A living nerve cell can only tolerate a few millivolts of external voltage on its body in the long term. Because of the high contact voltage, the neurons die in a few weeks or days. Capacitive sensors offer a contactless alternative that does not require a galvanic connection to the tissue (Fig. 1.15).

The way capacitive sensors work can be compared to the principle of an electroscope: One can effectively isolate the previous (galvanically coupled) electrode from the tissue (thin layer of glass) so that a micro-electroscope is created. The level of the collected charge can then be read out using electronics (MOSFET with free-floating gate). In this way, one has contact-free information about the (also static) charges on the neuron, even at its synapses and at the neurons.

In practice, sensor matrices are available that have up to 4000 sensors and can accordingly record the activity of small neuronal structures. It makes it possible to measure or monitor (also stimulate) biological neuronal networks individually at the level of their true dimension without being disturbed by an offset voltage (Fig. 1.23).

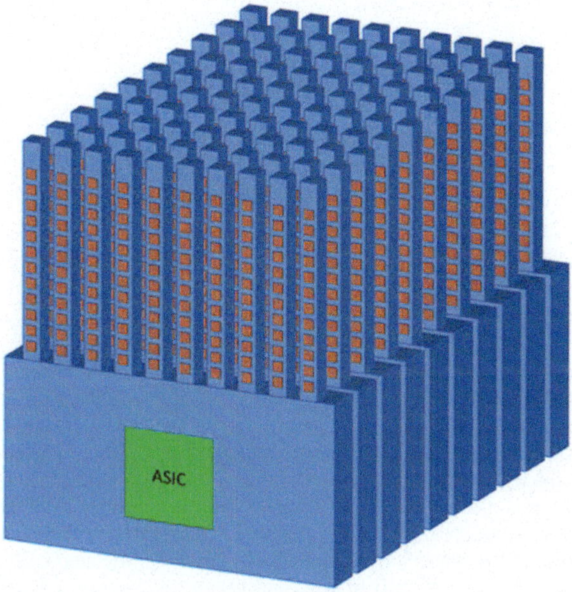

Fig. 1.15 3D sensor array consisting of a matrix with 10 × 10 × 8 passivated (SiO_2) metallic surfaces (capacitive sensors). The living and growing neurons are located in the gel in the space between the combs of the spatial sensor matrix and are supplied with nutrient solution from below. A comb consists of a matrix of 10 × 8 sensors and an ASIC (see front view). The edge length of the sensor surfaces is approx. 50 μm so that measurements can be taken at the cellular level. *Source* 3dneuron.eu research project in FETopen, project documents of the TU Ilmenau, Biosignal Processing Group in cooperation with the Institute for Microelectronic and Mechatronic Systems, IMMS, Ilmenau

1.4.2 Metrology

A capacitive sensor (ideally an insulated free-floating conductive surface, the gate of a MOSFET) has the decisive advantage over galvanic sensors that it does not need direct contact with the tissue, i.e., it is, to a certain extent, independent of the signal source. However, it also has disadvantages compared to galvanic contact. The major issue arises from the nature of the electric field detected by the capacitive sensor: Not only are the desired electric fields of the biological signal source detected, but also all interfering fields from the environment. In the case of galvanic contact, these electric fields are also present, but their effect is reduced mainly by the electronics (differential amplifier). Nevertheless, the capacitive measuring circuit can be optimized constructively, electrically, and signal-analytically so that the desired signal remains evaluable. Thus, at the cellular level, one can place the entire measuring circuit in a Faraday cage. On a macroscopic level (monitoring of vital human functions), large capacitive plates can be built into the backrest of an armchair, or conductive threads can be woven/knitted into textiles.

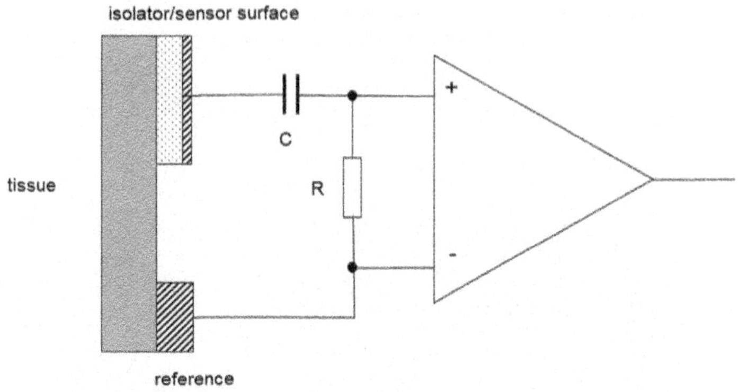

Fig. 1.16 Principle circuit diagram of an arrangement for capacitive detection of the electric field of neuronal activity in tissue. If the sensor surface is passivated, i.e., galvanically separated from the tissue, the sensor behaves like a capacitance modeled with C. The capacitance is the same as the capacitance in the tissue. Regardless of whether the derivative reference is galvanically connected to the tissue or is also galvanically separated from it, the capacitive sensor forms a high pass with the modeled input resistance of the amplifier. The high-pass filter's cut-off frequency is generally unknown since the capacitance is individually and spatially/temporally variable. One can try electronically and signal-analytically to determine the cut-off frequency and compensate for the high-pass behavior inversely (by low-pass). In this way, it is theoretically possible to measure the initial response despite the galvanic isolation of the capacitive sensor

Constructing a combined capacitive-inductive sensor is possible: If a large-area coil (spiral, diameter approx. 30 cm) is formed, it can function simultaneously as an inductive and capacitive sensor. Using known relationships between the electric and magnetic fields, the reliability of the signal detection can be increased considerably.

For measurement technology, the challenge with capacitive sensors is that potential differences between discharge points of an electric field are measured and their time derivative or high-pass filtered components. The circuit diagram in Fig. 1.16 shows this highly simplified structure. If one connects a capacitive sensor directly to the input of a measuring amplifier, it behaves like a high-pass filter with a generally unknown cut-off frequency. Suppose one can dimension or configure the measuring circuit. In that case, the high-pass behavior can be partially corrected inversely, and one obtains the undistorted original despite the lack of galvanic coupling.

1.5 Experimental Data

The selection of experimental data from the abundance of electrical biosignals in this publication is made from the point of view of signal-processing methodology. Therefore, the focus is on which biosignal is best suited for demonstrating the performance and effectiveness of signal processing methods and, at the same time, robustness against interference. Under these two aspects, the ECG is the most suitable test signal:

- The ECG has the most robust transient character (temporal and spectral dynamics). Slow waves (P- and T-wave) alternate with a three-phase implus-like process (QRS complex), while the pauses between the waves or the QRS complexes represent cardiological rest. From the point of view of signal analysis, we are dealing here with extreme non-stationarity, in which phases of rest, impulse character (QRS), DC (true DC in the biosignal, S-T stretch), and relatively slow processes alternate in the millisecond range (respiration in the ECG, variability of the cardiac period, migration of the baseline). Due to this strong non-stationarity, the ECG is the best test signal for established and newly developed methods.
- Usually, the ECG has a qualitatively constant shape. Therefore, when analyzing the signal, one knows relatively precisely which course or which graph element of the signal belongs to the desired part and which is to be qualified as a disturbance.
- The disturbances we must reckon with in the ECG (technical and biological) have a very variable shape. It is, therefore, advantageous that one can distinguish relatively clearly between the desired signal (ECG) and the undesired signal (all disturbances) based on the signal shape alone.

In order to demonstrate the applicability of the methods to other electrical biosignals and to evaluate their advantages and disadvantages, other biosignals are also compared and analyzed at appropriate points (EMG, EEG, EP, AP).

1.5.1 Action Potentials of Natural Neurons

Since the neuron forms the signal-analytically elementary signal source, its AP (action potential) is the elementary signal course of electrical neuronal activity. APe are not only of research interest. The clinical practice already uses them in diagnostics, e.g., using needle electrodes to test the local EMG in muscles (motor end plates). The AP has a constant signal shape and levels at the source (see Sect. 1.1). However, the levels and signal shape change depending on the position or distance of the electrode or capacitive sensor from the neuron to be examined or from a neuron population. Due to the limited spatial resolution of sensor matrices, differentiation of individual neurons is hardly possible but also diagnostically not useful. Using signal processing and statistical methods to separate individual sources by classifying the signal shape is possible. Figure 1.17 shows an image of several thousand individual APs of a cultured neuron population (Just, 2013) divided into three clusters. These cluster courses can also be used, among other things, for investigations of functional (causal) relationships. At this point, it should be noted that all AP courses have an identical signal form at the neuronal

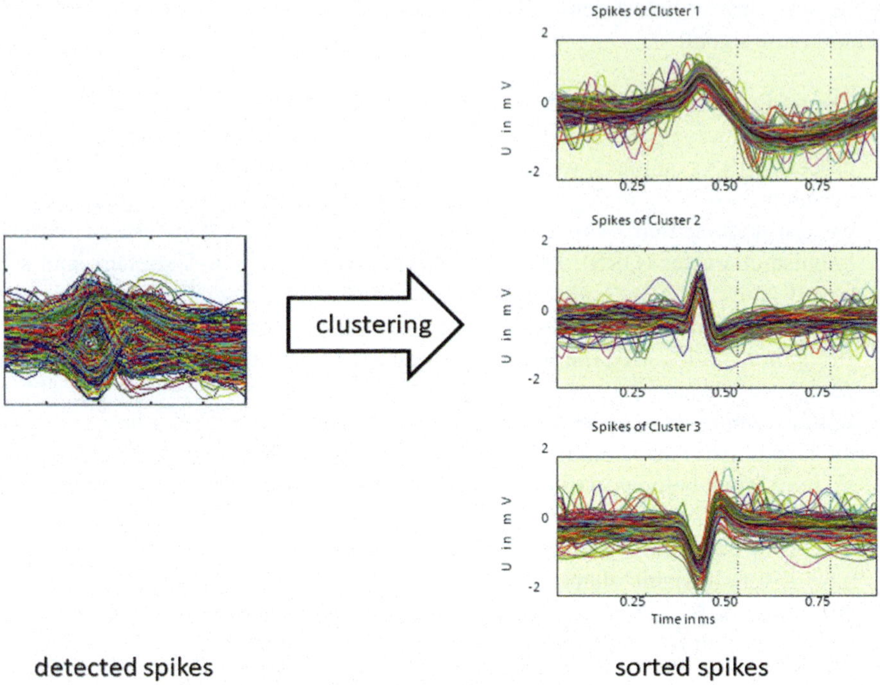

Fig. 1.17 The action potentials measured in the derivative system shown in Fig. 1.15 were classified according to similar shapes using a new clustering algorithm, PARAFAC2. Weis (2015) to classify the action potentials according to similar shapes. It should be noted that all action potentials always have the same shape at the point of origin and only change their shape through spatial and temporal integration (convolution with transfer function). Courses were recorded at one of the sensors and divided into three clusters, see graphic. The result is three partly completely different shape curves of identical signal sources. Thickly printed curves present the mean value of the respective class (Just, 2013)

signal source and only change the signal form through their spatial and temporal propagation in the tissue.

1.5.2 EEG, Sensory System

While conventional EEG is used for the global assessment of brain activity, it can be used analytically, anatomically, and physiologically related to humans' sensory or motor systems (see Sect. 1.5.3). In this macroscopic projection, however, it is no longer possible to measure individual APs (Sect. 1.5.1) since the activity of neuron associations ($10^4 \ldots 10^6$ neurons) is presented at the surface electrodes as the result of a spatial and temporal sum. It is known that the EEG changes depending on the physiological or mental state and in which spectral ranges the related activities present themselves (δ-, θ-waves for sleep, α-waves for visual relaxation, β-

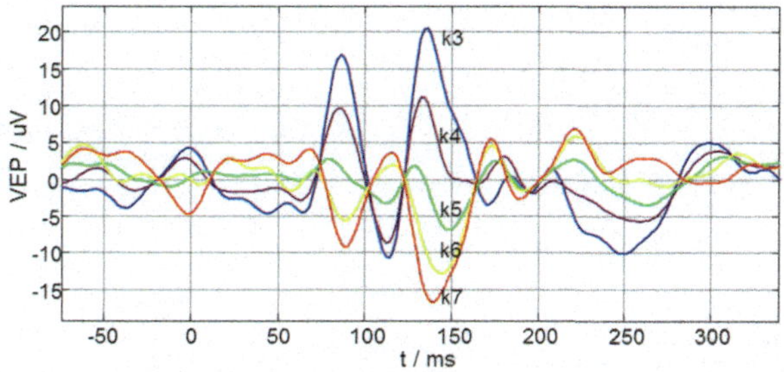

Fig. 1.18 After a local optical stimulus (white flash of 20 ms duration) at time t = 0 ms, the EEG was recorded with five occipitally arranged Ag/AgCl electrodes, referenced to a virtual reference (CAR) and averaged over 64 individual responses (synchronous averaging). The typical (reproducible) sensory stimulus responses ("waves") can be reliably detected at 90, 110, 140, and 250 ms. These form the methodological measurement basis for neurologists to make a diagnosis

and γ-waves for concentration). Independently of this, the sensory centers in the brain (visual, auditory, sensorimotor, olfactory, gustatory) produce form-specific neuronal responses to adequate (defined) sensory stimuli. For example, after stimulating the eye with a brief flash of light, a typical neuronal response of the visual cortical center (occipital, "back of the head") can be observed (Fig. 1.18). This course of stimulus–response is qualitatively identical in all human sensory systems. It is also unsurprising since the neuronal structures of the cortical sensory centers behind the sensory layers (rods/cones in the retina and hair cells in the cochlea) are mainly identical or structurally similar.

1.5.3 Needle and Surface EMG

The same technology and methodology can also be used to investigate neuronal control's (active) motor side: the EMG is the measurable electrical activity of muscle activity. From a diagnostic point of view, it is essential to measure with sensors locally in the muscle—in the vicinity of the so-called motor end plates: structurally, the motor end plate resembles a neuronal synapse but transmits the activity into a muscle cell. Tiny needle electrodes made of precious metal (fully polarizable electrode) are used for this. With such needle electrodes, it is possible to measure and diagnose the electrical activity of motor end plates. However, such a measurement is painful (puncture in the muscle), and during an activity examination (muscle tension), the local pain increases many times over. Therefore, one tries to follow the dictate of a non-invasive examination using surface EMG (Ag/AgCl-unpolarisable electrode). The surface EMG is (as with the surface EEG) the result of the spatial and temporal integration of the individual potentials of the motor end plates. As a result, one loses an order of magnitude of spatial or depth resolution in such a measurement. New methods of biosignal processing attempt to compensate for this deficit; see chapter on space–time processing.

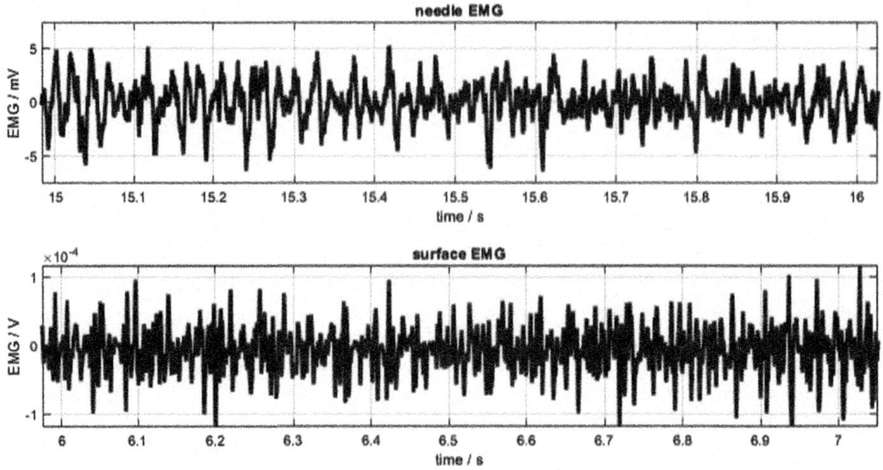

Fig. 1.19 Simultaneous recording from the biceps during maximum tension: the surface EMG (top) was recorded with Ag/AgCl electrodes from the upper arm surface and the needle EMG (bottom) with precious metal needle electrodes in the muscle. Heidelberg University Hospital, Clinic for Paraplegiology, Section Experimental Neurorehabilitation, kindly provided the measurement data for this publication

Figure 1.19 shows simultaneously recorded time courses (tense biceps) of the two EMG types for comparison; Fig. 1.20 shows the corresponding spectra. It can be seen that the signal shape (time domain) shows few differences. It is mainly because the EMG typically has a stochastic signal character; therefore, no distinctive waveform forms. However, the difference is visible in the spectrum. The surface EMG shows a much lower level across the spectrum, especially at higher frequencies. This low-pass character results from the spatial and temporal integration of the tissue between the motor end plates and the surface electrodes.

The data in Figs. 1.19 and 1.20 are based on empirical data obtained simultaneously with needle electrodes in the biceps and with surface electrodes on the upper arm during a contraction experiment: In the time range 0–8 s, the biceps was relaxed, which was followed by tension up to approx. 14 s. After a resting phase up to approx. 18 s, a more muscular tension followed up to approx. 23 s, then resting again. Figure 1.19 shows the time course during the most muscular tension. Differences can already be seen visually: The needle EMG shows narrower pulses and larger amplitudes. It has corresponding spectral effects, which can be observed in the time–frequency representation in Fig. 1.20. The spectrum (SPWD, Smoothed Pseudo Wigner Distribution) at this time is significantly broader and more temporally resolved in the needle EMG than in the surface EMG. Fundamental differences can be observed, mainly based on the TFD (Time–Frequency Distribution):

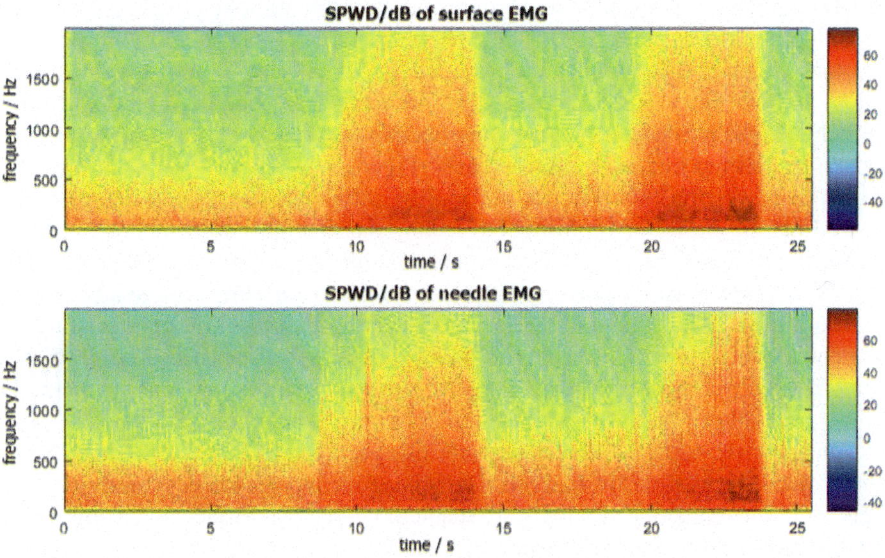

Fig. 1.20 Time–frequency distribution (SPWD) of the complete recording from Fig. 1.19. The signal energy is colored logarithmically (in dB) with significant derivative differences with up to 20 dB spacing (minimum energy). The dynamic difference is up to approx. 30 dB favoring the needle EMG (bottom graph) compared to the surface EMG (upper graph)

- Needle EMG has a broader spectrum at all stages of muscle loading.
- The temporal resolution of the needle EMG is an order of magnitude better than the surface EMG.

Both effects can be traced back to a common cause: In the case of surface EMG, the low-pass effect of the perfused and water-containing tissue between the muscle fiber and the surface comes into play. This spatial and temporal low-pass simultaneously results in a temporal smoothing effect in addition to spectral smoothing, as can be observed in the comparison of the TFD in Fig. 1.20.

1.5.4 Stress ECG

The stress ECG is one of the most important diagnostic tools for cardiology to assess the current stress capacity of the cardiovascular system. As mentioned above, the signal form of a normal ECG is known. Based on this, one can recognize the external disturbances (technical) and the biological artifacts (muscle tremors, respiration, movement) and reduce them if possible. However, it is complicated to separate the disturbances in the stress ECG: due to the periodic movements on the bicycle ergometer as well as on the treadmill, the disturbances are significantly stronger than the biological signal (10…100 times stronger, SNR $= -20\ldots-$

40 dB), so that simple methods to improve the SNR are out of the question. In addition, the interferences overlap entirely with the ECG biosignal in the spectrum and time domain. Therefore, they are neither separable by simple spectral filters nor by simple time series analysis. This problem poses one of the most significant challenges to medical measurement technology today and the technology for separating desired and interfering signals. Figure 1.21 shows the ECGs of a test person at rest before a stress test and during the stress test. It can be seen very clearly that the signal curve during the resting phase almost corresponds to a sample ECG, whereas the same ECG is no longer subjectively recognizable during the stress test. This situation poses a diagnostic problem for clinical practice. Detecting the critical cardiac actions in the movement artifacts is no longer possible. It is a challenge for the BSP, which is being tackled with various approaches (see time–frequency analysis, TFA). Figure 1.22 shows partial results of our methods, which gradually or partially improve the diagnostic reliability.

Conventional Ag/AgCl electrodes are also used for recording the stress ECG. Therefore, this area hardly has room for maneuvering to reduce movement artifacts. In clinical practice, attempts are made to reduce the artifacts by changing the positioning of the electrodes (depending on empirical values) in contrast to standardized systems. In the case of success, the artifacts are less pronounced, but there

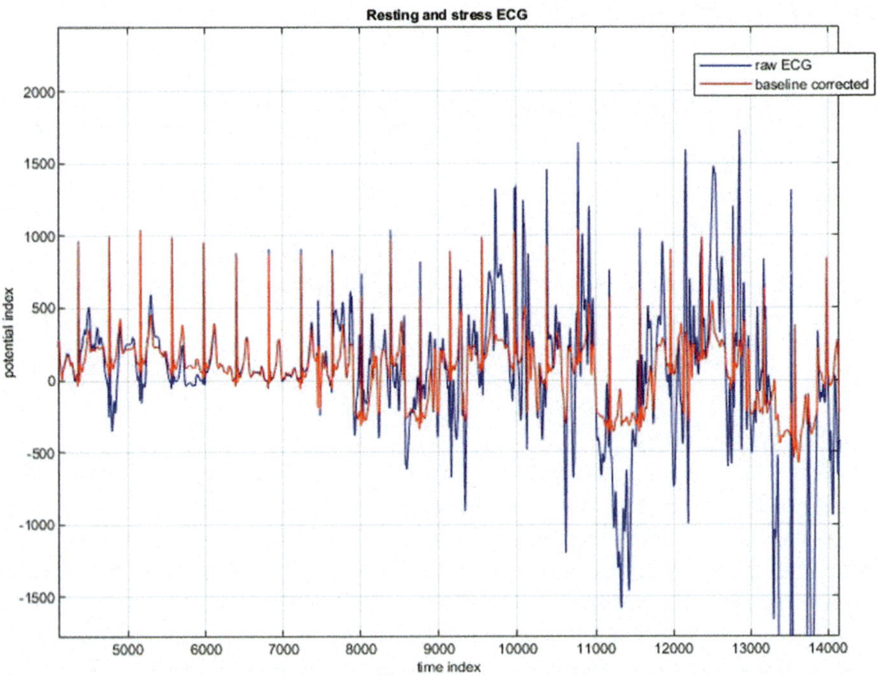

Fig. 1.21 Resting ECG (left, blue) and exercise ECG (right, blue) recorded while cycling on a cycle ergometer. The red curve shows the effect of a single-stage adaptive baseline filter

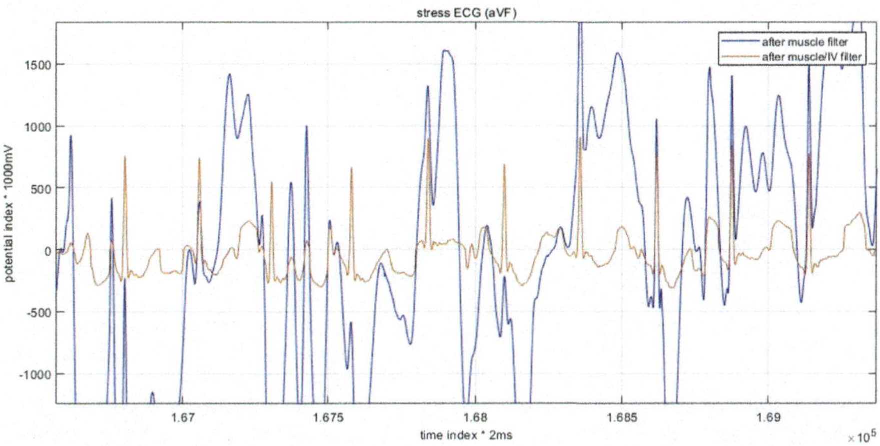

Fig. 1.22 Stress ECG heavily disturbed by motion artifacts; the original ECG is no longer recognizable (blue). With a combination of own methods (adaptive correlation detector, linear time-varying filter, dynamic compression of the integral vector in the vector ECG), the ECG can be identified (red). The QRS complexes are reliably detectable even after a locally extremely low SNR of − 40 dB or less. However, this is at the expense of the justified demand for linear-phase processing. Therefore, as shown here, the ECG tracings are only suitable for rhythmology (monitoring ECG) rather than form-based diagnostics (diagnostic ECG)

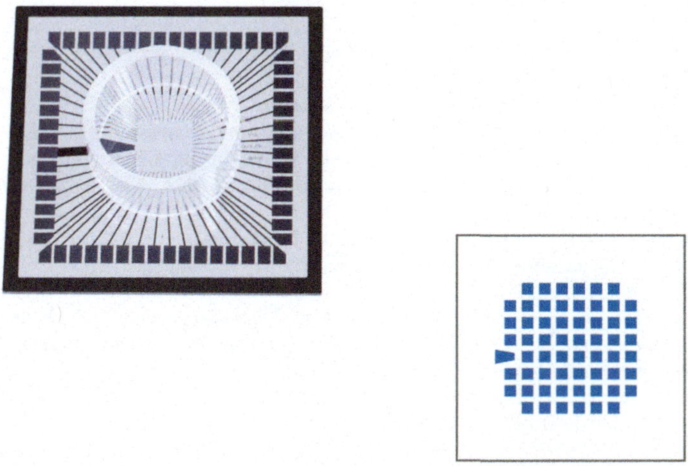

Fig. 1.23 Microelectrode matrix of size 8×8 for recording the electrical activity of neuronal cultures (published with the permission of Multi Channel Systems MCS GmbH)

is no basis for comparative studies for the biosignal because of the changed electrode positions. The consequences or causes are more in the medical-methodical area, so they are not considered in more detail here from the point of view of measurement technology and signal processing.

References

Baraniuk, R., & Jones, D. (January 1994). A signal-dependent time-frequency representation: Fast algorithm for optimal kernel design. *IEEE Transactions on Signal Processing*, 134–146.

Berger, R., Akselrod, S., Gordon, D., & Cohen, R. (September 1986). An efficient algorithm for spectral analysis of heart rate variability. *IEEE Transactions on Biomedical Engineering*, 900–904.

Bergh, J., Ekstedt, F., & Lindberg, M. (1999). *Wavelets with applications in signal and image processing*. Springer.

Boashash, B. (2003). *Time frequency signal analysis and processing*. Elsevier.

De Lauthauwer, L., De Moor, B., & Vandewalle, J. (April 2000). A multilinear singular value decomposition. *SIAM Journal on Matrix Analysis and Applications*, S. 1253–1278.

Debnath, L. (2002). *Wavelet transforms and their applications*. Birkhäuser.

Engelhardt, A. (2019). *Crash course in statistics*. Retrieved from http://www.crashkurs-statistik.de

Haardt, M., Roemer, F., & Del Galdo, G. (July 2008). Higher-order SVD based subspace estimation to improve the parameter estimation accuracy in multi-dimensional harmonic retrieval problems. *IEEE Transactions on Signal Processing*, 3198–3213.

Heisenberg, W. (1927). On the descriptive content of quantum-theoretical kinematics and mechanics. *Zeitschrift Für Physik, 43*, 172–198.

Henning, G., Hoenecke, O., Husar, P., & Schellhorn, K. (March 1996). Time-frequency analysis in objective perimetry. *Applied Signal Processing*, 95–103.

Husar, P. (2010). *Biosignal processing*. Springer.

Hyvärinen, A. K. (2001). *Independent component analysis*. Wiley.

Ingle, V. K., & Proakis, J. G. (2007). *Digital signal processing using MATLAB*. Thomson.

Just, T. K. (November 2013). Neuronal cell spike sorting using signal features extracted by PARAFAC. In *Proceedings of 6th Annual International IEEE EMBS Conference on Neural Engineering* (pp. 472–475).

Liavas, A., Moustakides, G., Henning, G., Psarakis, E., & Husar, P. (1998). Periodogram-based method for the detection of steady-state visually evoked potentials. *IEEE Transactions on Biomedical Engineering*, 242–248.

Manolakis, D., & Ingle, V. (2011). *Applied digital signal processing*. Cambridge University Press.

Nikias, C. L., & Petropulu, A. P. (1993). *Higher-order spectra analysis; A nonlinear signal processing framework* (p. 07632). PTR Prentice-Hall, Inc.

Pan, J., & Tompkins, W. J. (March 1985). A real-time QRS detection algorithm. *IEEE Transactions on Biomedical Engineering*, 230–236.

Renn, J., & Sauer, T. (September 1996). Einstein's Zurich notebook. *Physikalische Blätter* , 865–872.

Silny, J. (2009). *Electrical, dielectric and magnetic fabric properties*. Aachen.

Stewart, G. W. (1993). On the early history of the singular value decomposition. *SIAM Review*, 551–566.

Weis, M. (2015). *Multi-dimensional signal decomposition techniques for the analysis of EEG data*. Ilmenau.

Weiss, C. (1999). *Basic knowledge of medical statistics*. Springer.

Widrow, B., Glover, J., McCool, J., Kaunitz, J., Williams, C., Hearn, R., … Goodlin, R. (December 1975). Adaptive noise cancelling: principles and applications. *Proceedings of IEEE*, 1692–1716.

Amplification and Analog Filtering in Medical Measurement Technology

2

This chapter deals with *metrologically* relevant, characteristic properties of biosignals and disturbances in the time and frequency domain. Interference coupling mechanisms into the measurement circuit are analyzed, and measures for their reduction are proposed and discussed. Based on the knowledge of the formation mechanism of biosignals and depending on the externally acting disturbances, requirements for medical measurement amplifiers and the measurement arrangement can be formulated. The basis of the engineering consideration is a detailed analysis and simulation of the differential amplifier as a fundamental building block of biomedical measurement technology. More complex design stages (instrumentation amplifiers, isolation amplifiers, guarding technology, and active electrodes) are evaluated concerning their circuit advantages and disadvantages. Analog spectral filters are required for signal conditioning, and this chapter deals with their design theoretically and practically. The demand for a linear phase frequency response of the filters and the entire measurement chain is justified, and possible realizations of this demand are discussed. Evaluated specific examples of technical design and their practical implementation are the focus of interest.

Supplementary Information The online version contains supplementary material available at https://doi.org/10.1007/978-3-662-67998-2_2.

P. Husar and G. Gašpar, *Electrical Biosignals in Biomedical Engineering*,
https://doi.org/10.1007/978-3-662-67998-2_2

2.1 Properties of Biosignals and Disturbances

Many sources of interference surround the biomedical measurement system. Most can be attributed to media, energy supply, industry, traffic, and communications engineering.

For biosignal processing, periodic (power supply network, monitors) and quasiperiodic (rotating machines, streetcar) disturbances are still a comparatively minor problem for biosignal processing because they can be explicitly suppressed with spectral filters in the analog measurement chain (Sect. 2.3) or digitally (Sect. 5.2) after analog-to-digital conversion (Sect. 3.3). The situation is much more difficult if transient disturbances are present. These have an unknown and non-reproducible course. If a temporary disturbance does not impair signal acquisition by overdriving the measurement amplifier and deviates from the shape of the biosignal, it can be detected and possibly eliminated. However, this is usually not possible. This section focuses on the characteristics of biosignals and interference from the point of view of medical measurement electronics. For the technical component of biosignal processing (sensing, amplification, filtering, AD conversion), biosignals are first classified into the two domains of time and frequency. These two domains are decisive for deriving the requirements for the technology and for the circuit solutions to be developed.

2.1.1 Properties of Biosignals and Disturbances Over Time

2.1.1.1 Periodic Course

The signal traces are classified according to the objectives and tasks of biosignal processing (Table 2.1).

A *harmonic waveform* is given when the signal has a sinusoidal character, where one or more integer multiples of the fundamental may be present. For example, the power supply network produces a fundamental wave with a frequency of 50 Hz and the second harmonic (third harmonic) at 150 Hz. Harmonic progressions in the medical environment are of technical origin.

Table 2.1 Periodic course

Course	Example
Harmonic	Power supply network
Non-harmonic	Pulse curve, tube monitor
Impulsive	ECG, switching power supply
Rectangular	Data transmission
Quasiperiodic	EEG, high-frequency transmitters
Stochastic	Noise, random signals

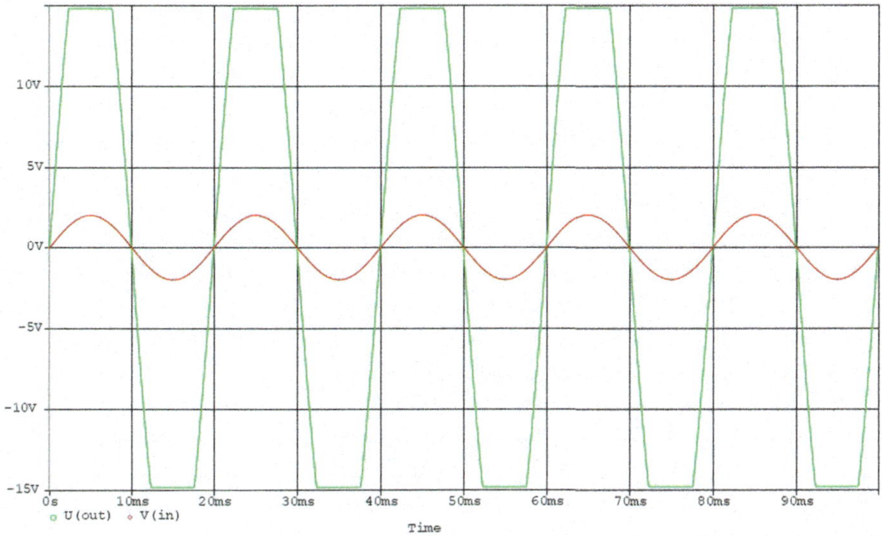

Fig. 2.1 If the interference voltage (green) at the amplifier input is too high (in this case, a mains interference), the measuring amplifier is overdriven and goes out of the operating range into the limitation (red) so that further processing of the biosignal is no longer possible (PSpice simulation BSV_2.1)

With correspondingly strong coupling and intensity, they can overdrive the measuring amplifier to the limit and impair the measurement (Figs. 2.1 and 2.2, Exercise 2.1) so that a non-harmonic response is present at the amplifier output. The leaving of the operating range or the occurred limitation should be monitored by circuitry as close as possible to the input (signal monitor) because it cannot be detected later after the always existing analog spectral filters (antialiasing low pass). It can lead to the complete defacement of the biosignals, so undetected incorrect data are evaluated.

A *non-harmonic curve* is present when periodicity is present, but the waveform is not sinusoidal. A typical representative of this class is the line deflection voltage in tube monitors, which has a sawtooth shape and is therefore strongly affected by harmonics (Fig. 2.3). The plethysmographic pulse curve (time course of a volume change) of the cardiovascular system or the triangular currents used for electrical stimulation therapy are also such signals.

Pulse-like progression occurs when high-energy, short-time courses occur against a relatively calm background. A typical technical disturbance of this category is the electromagnetic field emitted by switching power supplies. Several biosignals also have impulsive periodic character: the electrocardiogram (ECG) in Fig. 2.4 or the electromyogram (EMG) in Fig. 1.19.

The rectangular course is typical for communications technology in data transmission. It occurs in wired (data networks, ISDN, DSL) and wireless (radiotelephone, RFID) communication. Communications technology is one of the

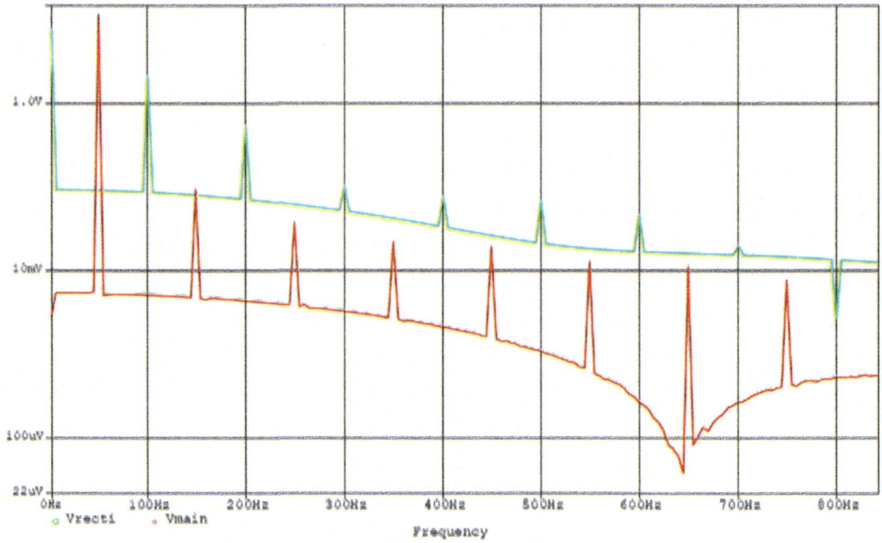

Fig. 2.2 Spectrum after a two-way rectification of the mains voltage at the rectifier or smoothing capacitor (dotted line) and its feedback effect on the mains due to the finite internal resistance (full line). Note that the nonlinearity of the rectifier produces odd harmonics in the grid, of which one usually evaluates the first harmonic (50 Hz) and the third harmonic (150 Hz) (PSpice simulation BSV_2.2). The rectified voltage contains even harmonics, as shown in the spectrum. The rectification has halved the period

Fig. 2.3 Periodic, sawtooth-shaped course, for example, in horizontal deflection in tube monitors or symmetrically (triangle) for electrical muscle stimulation

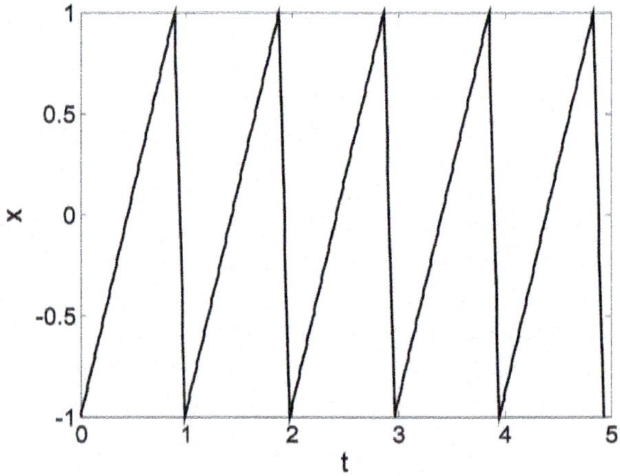

most potent sources of interference in the medical field. However, the frequency range is several decades higher than biosignals (radio telephones operate at 900 and 1800 MHz, and biosignals range up to 10 kHz). High-frequency interference

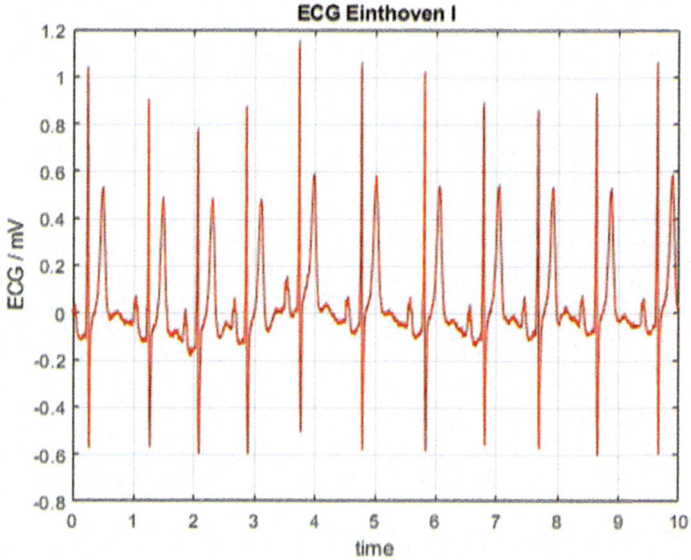

Fig. 2.4 A pulse waveform of the ventricular complex QRS of an electrocardiogram (ECG). The amplitude of the R-spike is modulated by respiration

is caused by the unwanted amplitude demodulation of the measurement amplifiers (see Sect. 2.1.3).

Quasi-periodic behavior is when the periodicity prevails only in a limited time frame. This behavior is typical for a series of biosignals; the electroencephalogram (EEG) represents this. The EEG presents itself in limited spectral ranges, which are characteristic of the mental and physical state of the human being.

2.1.1.2 Stochastic Course

Stochastic is the term used to describe time courses that (from the point of view of the observer) have a random character. The structure of stochastic progressions ranges from actual, purely random processes, such as thermal or semiconductor noise, to partially deterministic processes, such as the EMG or EEG. These represent a mixture of apparent randomness and determinism. Especially biosignals are not random processes but complicated and hardly manageable dynamic and nonlinear process couplings. For the analysis, they are often called stochastic processes after a substantial simplification.

2.1.1.3 Transient Time Course

Transient processes are challenging to classify and differentiate due to their non-reproducible course. Therefore, we form the following classes based on experience (empirical classification in Table 2.2).

Table 2.2 Transient course

Course	Example
Movement	Patient movement, cable
Clearing transaction	Capacitor discharge
Tips and jumps	Sparks, mechanical contacts
Switching operations	Switching machines off and on

Motion artifacts are the most common transient disturbance in the medical environment. Sensitive medical measurement technology is connected to the patient via sensors. This contact point does not form a rigid connection but is flexible, such as surface electrodes for recording EEG, ECG, or EMG. Relative movements between the patient and the measuring circuit (the patient himself or the cables of the measuring amplifier move) change the coupling properties at the connection point of the sensor. Thus, a movement artifact occurs in the signal.

Motion artifacts can also occur if there is no relative motion between the patient and the measurement setup. Interference sources from the environment, most frequently from the power supply network (see Sect. 2.1.3.2 Capacitively coupled interference), always affect the measurement circuit via existing stray capacitances. Even simple movements of medical personnel in the vicinity of sensitive measuring equipment can lead to the formation of artifacts. Such motion artifacts have an unknown time course, are not reproducible, hardly detectable, and thus represent a complex signal analysis problem.

Compensation processes are transient disturbances that are triggered by short disturbance pulses or steps but which can have a prolonged after-effect due to the extensive time constants of the analog filters in the measuring circuit (see Sect. 2.3) (Fig. 2.5). In medical amplifiers, analog high-pass filters are used to eliminate the high electrode voltages (100 mV to 1 V). Because of the shallow cut-off frequency, these filters have time constants of up to 30 s. As a result, even a short voltage pulse with a few milliseconds (switching peaks) or a jump causes an equalization process with a duration of up to one minute. If they do not overdrive the measuring amplifier, equalization processes can be partially eliminated.

Spikes and jumps are of technical origin and are usually clearly distinguishable from the biosignal. They are also easily recognizable and detectable (Fig. 2.5). If they do not result in an equalization process, they only affect the biosignal for a short time. They thus do not represent a signal analysis problem. Like most transient disturbances, they cannot be corrected. Still, due to their short duration, the affected signal sections can be excluded from further analysis without much loss of information.

Switching operations are, according to their origin, transient technical disturbances. High-energy disturbances are generated when switching electrical equipment on and off, propagating mainly via the power supply or magnetically as induction peaks. The concrete signal form depends on the technical and technological conditions (connected equipment, start-up and shutdown times, and secondary interference effects via capacitive and inductive coupling). Switching processes

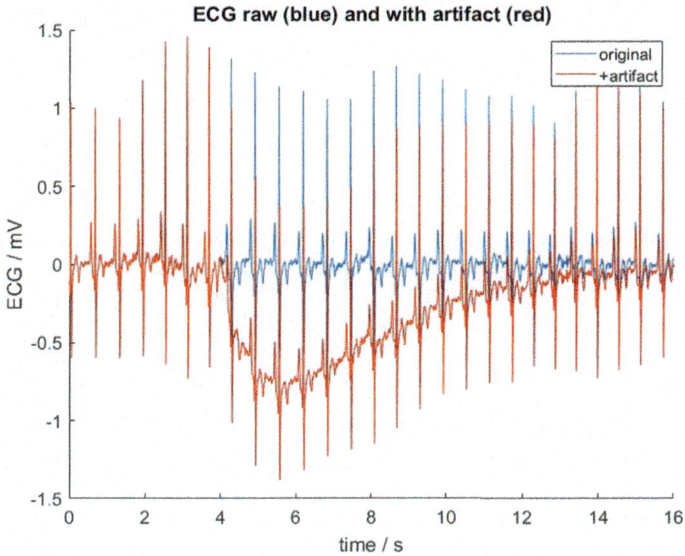

Fig. 2.5 Transient disturbance in ECG: voltage jump at t = 4 s caused by a motion artifact, leading to the compensation process (here with a time constant of approximately 2 s). The transient course of the disturbance is not reproducible

cover the range of transient disturbances between voltage peaks and jump up to equalization processes. Therefore, they are hardly detectable and cannot be eliminated.

2.1.2 Properties of Biosignals and Interference in the Spectrum

The information about the spectral composition of biosignals in the literature is very different and differs strongly, especially at the lower cutoff frequency. The reason for these significant differences is technical: Before electronic technology developed ADC (Analog–Digital Converter) with sufficient bit width (18…24 bit; 108…144 dB), one had to rely on a relatively low dynamic range (12…14 bit; 72…84 dB). If a biosignal is derived using electrodes, it contains strong DC components caused by the electrode voltages (see Chap. 1.3). The DC components are crucial for weak biosignals (EEG, EMG, AEP, VEP), which are many times larger than the biosignals, so the available dynamic range is insufficient for the desired signal. Therefore, attempts have been made to reduce the DC components using high-pass filters. Because developers and designers used different cut-off frequencies of the high-pass filters, an empirical spectral measurement showed different lower cut-off frequencies. Today, it is sufficiently proven that biosignal spectra start at 0 Hz or that the signal characteristics have DC or extremely low-frequency components.

Biosignals generally have a spectrum between 0 Hz and about 10 kHz, with *maximum energy* between 1 and 100 Hz.

For biosignal processing, this spectral position is very unfavorable since most technical interferences are either directly in this range (power supply network, tube monitors, rotating machines, streetcars, airplanes) or occur in this range after the undesired amplitude demodulation (radiotelephone, RFID, computer networks, transmitters with amplitude modulation, high-frequency surgery).

The *line spectrum* occurs only in the technical field, typically as spectral needles of the harmonics of the mains frequency or the image frequency in monitors or fundamental frequencies in transport (railroads, tramways, power electronics, see Fig. 2.2).

Since their frequency is constant and known, harmonic interferences can be effectively suppressed by spectral filters (band rejection filters), provided the interferences have not overdriven the measurement amplifier before filtering. Signal analytically correctly constructed band rejection filters affect the biosignal only insignificantly, although they are in the middle of the biosignal spectrum (Fig. 2.6; Table 2.3).

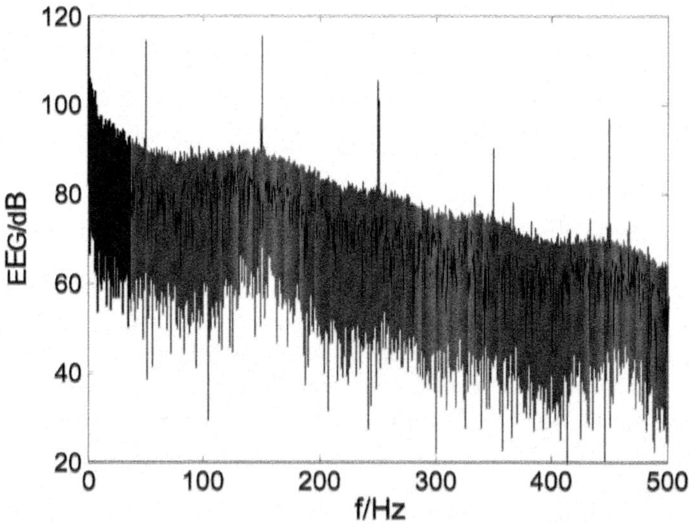

Fig. 2.6 An EEG recording superimposed with five harmonics of the mains frequency and the amplifier noise. The mains produce the prominent line spectrum, while the EEG and the amplifier noise show a smoothly decreasing continuous spectrum (sampling rate 1000 sps, no analog filter before AD conversion)

Table 2.3 Classification of signals and interference in the spectrum

Spectrum	Example
Lines	Mains with harmonics, tube monitors
Continuous	EEG, motion artifact
Narrowband	Pulse curve, radiotelephone
Broadband	EMG, impulsive disturbances
Combined	ECG, multiple disturbances

Continuous spectra are typical for quasi-periodic signals like EEG (between 0 Hz and about 500 Hz) or motion artifacts. The continuous spectra of disturbances and biosignals overlap entirely; therefore, separation using spectral filters is impossible.

Narrowband spectra (relative bandwidth about 0.01–0.1 of the total width) occur with periodic signals whose shape is non-harmonic or changes in time, such as the plethysmographic pulse curve. However, some technical interferences are also narrowband, the most common being the envelope of radio telephones caused by amplitude demodulation. As long as the interfering spectrum is sufficiently distant from the biosignal spectrum, it can be suppressed with spectral filters. In practice, however, this is only sometimes possible.

Broadband spectra (bandwidth above 0.1 of the total width) in medical measurement technology primarily result from impulse-like biosignals or disturbances. The EMG (electromyogram) has the broadest spectrum of biosignals since it is a quasi-random sequence of very short action potentials with a length of a few milliseconds and extends into the kilohertz range. Impulse interferences of technical origin are very broadband and extend well beyond the range of biosignals. They cannot be controlled with spectral filters. The only way to eliminate them is to exclude the disturbed signal section from further processing.

Combined spectra occur when the shape of the biosignal or disturbance contains several of the above characteristics. The electrophysiological signal ECG comprises very narrow impulse-like QRS complexes, which produce a broad line spectrum (over 30 harmonics of the fundamental frequency of the heart rate). The slower P- and T-waves as quasi-periodic signals have a continuous spectrum on which the line spectrum is superimposed, resulting in a combined spectrum in the sum.

2.1.3 Coupling of Disturbances into the Measuring Order

Interference can enter the measurement order in different ways: Galvanically, capacitively, inductively, or as a high-frequency (above about 50 kHz) electromagnetic wave. In the following, the ways of interference coupling are analyzed, and possibilities for their elimination or reduction are discussed.

2.1.3.1 Galvanic Interference Coupling

The galvanic coupling means this occurs on a DC path via a real ohmic and insulation resistance. The real resistance's significance depends on the concrete conditions in the measuring circuit (insulation resistance, humidity, condensation, creepage distances). The disturbance can be coupled on the signal path (Fig. 2.7) or the ground. For simplicity of presentation, only one signal line via R_{e2} has been considered. Of course, the network affects all lines via their respective insulation resistances. The total effect then adds up according to the *superposition principle.* This procedure is also continued in the following analyses:

> An exemplary action path is shown, which can be extended to all other branches and components according to the superposition principle.

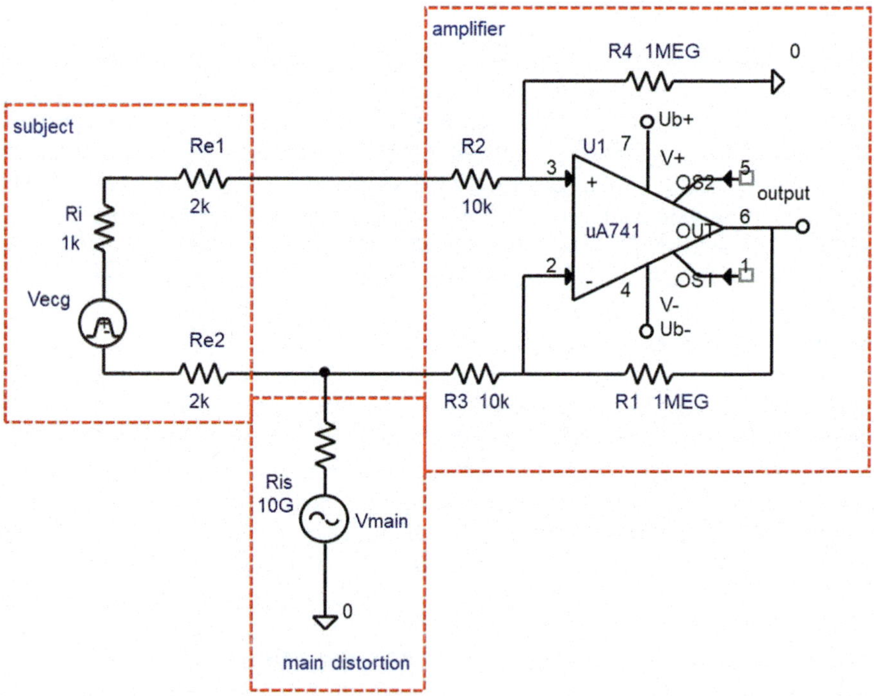

Fig. 2.7 The object to be measured (patient) is connected to the measuring amplifier via electrodes (R_{e1} and R_{e2}). A disturbance is galvanically coupled into the signal path from the mains via the insulation resistors R_{is1}, R_{is2}, and R_{is3} (PSpice simulation BSV_2.6). It must be ensured that the mains are galvanically isolated from the patient section. See symbols for ground and operating ground

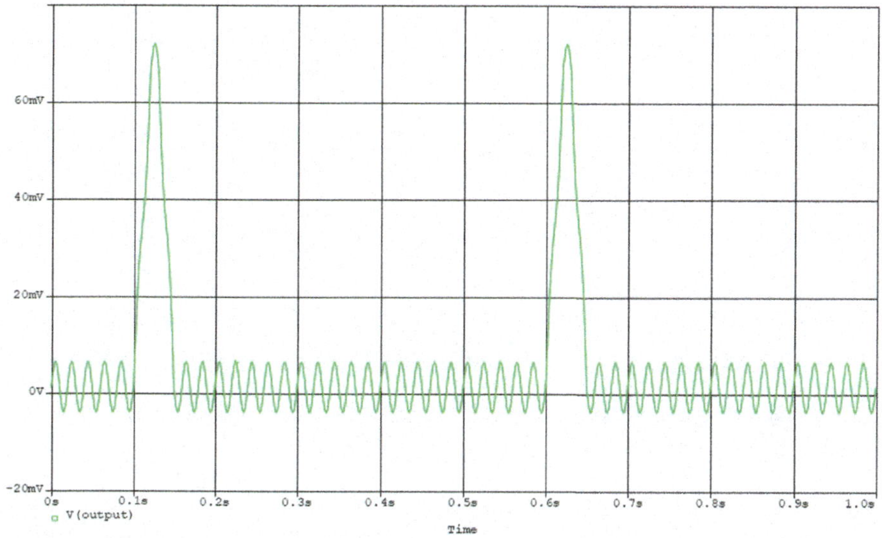

Fig. 2.8 Simulated ECG with a galvanically coupled mains disturbance according to Fig. 2.7 (PSpice simulation BSV_2.6)

As Fig. 2.8 shows, the interference galvanically coupled via the insulation resistance is relatively strong and impairs the biosignal. Therefore, it must be effectively reduced. The measures are constructive and must be implemented during device development: double or reinforced insulation of all parts carrying mains voltage, including compact encapsulation of the mains transformer or the power supply unit (exercise 2.3).

A frequent but difficult-to-identify interference coupling occurs via the ground line of the measuring instrument. In technical language, it is also called "ground loop" and is schematically shown in Fig. 2.9.

The voltage characteristics are qualitatively the same as in Fig. 2.8. Since several devices are generally connected to the patient, the operating current of one medical device may flow through the ground line of another device (current through R_g, in simulation BSV_2.7, is about 5 A, Fig. 2.9). This generates an additional voltage drop U_{loop} across the ohmic resistance of the operating conductor of the measuring device. According to the mesh equation (hence the ground loop), this voltage acts directly on the input of the measuring amplifier as if it were a patient-generated voltage. Detecting a ground loop in practical measurement, for example, when there are about 20 devices connected to a patient in the intensive care unit, is very difficult. The most effective way to avoid ground loops is to connect all the devices used on the shortest path in a power supply's single point (power distributor). If a ground loop is suspected, carrying out the installation or connection assignment anew and from the beginning, according to the principle mentioned (Exercise 2.4), is advisable.

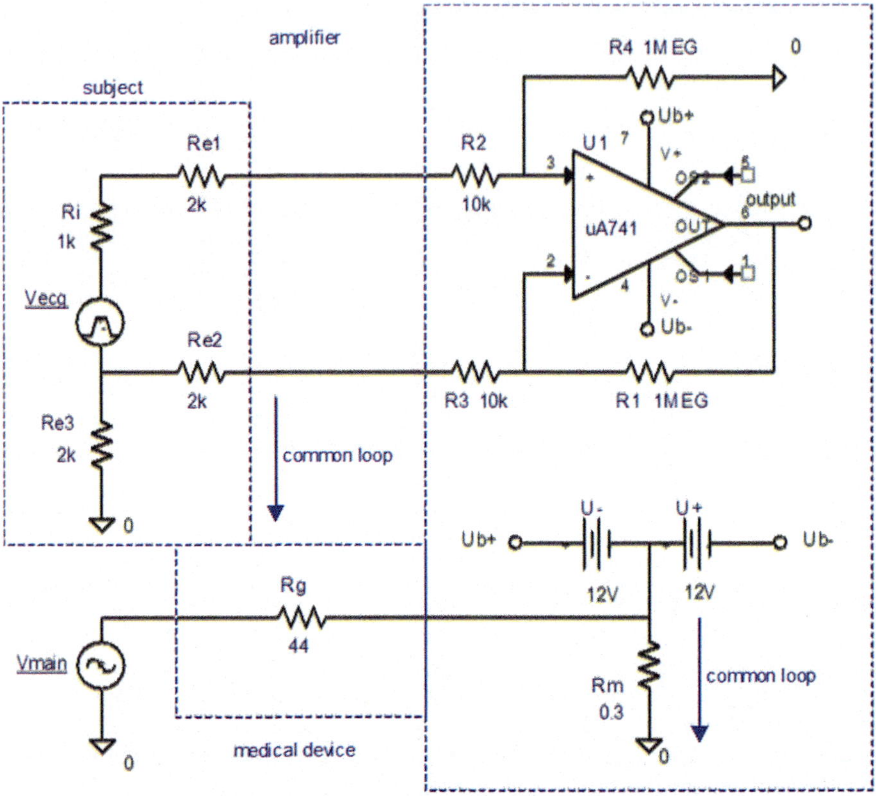

Fig. 2.9 Interference coupling via the ground connection, formation of a ground loop. The patient is connected to the current ground via the electrode R_{e3}. The ground connection of a device always has a finite real resistance R_m. Under unfavorable circumstances, the operating current of another device flows via R_m, which generates a voltage drop U_{loop} via R_m. It is directly carried over to the patient (PSpice simulation BSV_2.7)

2.1.3.2 Capacitively Coupled Interference

It is known from electrical engineering that two conductive parts (also liquids and electrolytes) form a capacitance. The capacitance magnitude depends on the effective area of the parts, their distance, and the substance (dielectric) between the parts, according to Eq. 2.1.

$$C = \varepsilon \frac{A}{d} \tag{2.1}$$

In Eq. 2.1, C is the capacitance between two conductive parts, ε is the dielectric permittivity between components, material constant (lowest value exists for vacuum), A is the effective (projected) parts area, and d is the distance between components.

If an electrical biosignal is to be derived in medical practice, this is done in rooms with the main installation. The patient forms a stray capacitance with the outer conductor (phase) and the neutral conductor (neutral). This situation is shown schematically in Fig. 2.10. For simplicity, it is assumed that the stray capacitances C_1 and C_2 are equal, and the patient is neither grounded nor connected to the operational ground of the measuring amplifier (the resistance R_m is not present). In this case, the patient is at about half the potential of the supply voltage, i.e., 115 V. At first; this seems dangerous for the patient. However, on closer examination, the following becomes clear: the danger to the human being arises when a potential difference greater than 24 V is generated externally between two points on his body and has a frequency of 10–1 kHz within the biological spectrum. Of course, the current path also plays a significant role—the path leading through the heart along its anatomical axis is the most dangerous. If a potential difference is present, it can only become dangerous if the circuit is closed. It is when the current can flow from the source through the patient back to the source. It is irrelevant whether there is a connection to the electrical ground or the earth. A widespread misconception is that the current always flows to the ground or earth; in a sense, they are interpreted as ultimate current sinks. The only decisive factor for the potentially dangerous effect of the electric current is whether the patient closes the circuit. It is independent of whether it is connected to the ground or earth intentionally or unintentionally. If this condition is not satisfied, then—as in this case—even high potential is harmless.

Let us consider the loop concerning the line voltage. The capacitances C_1 and C_2 form a voltage divider with the internal resistance of the body R_i according to Eq. 2.2. Thereby, voltages of 115 V drop across the capacitances whose impedance magnitude reaches 3.18 Γ at the line frequency. A voltage difference of 36 V is formed across the body impedance $R_i = 1$ κ. This is completely harmless to humans. However, this low voltage can significantly impair the measurement (see Fig. 2.11).

$$u_{LH,LF}\big|_{(R_m \to \infty)} = u_{L1,N} - u_{C1} - u_{C2} = u_{Netz} \frac{R_i}{R_i + |Z_{C1}| + |Z_{C2}|} \tag{2.2}$$

$$|Z_c| = \frac{1}{\omega C} \tag{2.3}$$

If the patient is connected to the ground (via the resistor R_m, which simulates the ground electrode with 5 κ), the conditions change significantly from a measurement point of view. The interference voltage at the left foot (LF) becomes much smaller than in the ungrounded case due to the ground connection R_m, which is only about 0.36 mV. However, the signal-to-noise ratio (SNR; for an ECG measurement, see Eq. (2.5)) will deteriorate compared to the measurement order on the ungrounded patient. Although the ECG amplitude remains constant at about 1 mV, the interference voltage falling across the patient increases to twice (72 V) (see Eq. (2.4)) since the following applies:

$$R_m << Z_{C2} \Rightarrow R_m \| Z_{C2} \approx R_m.$$

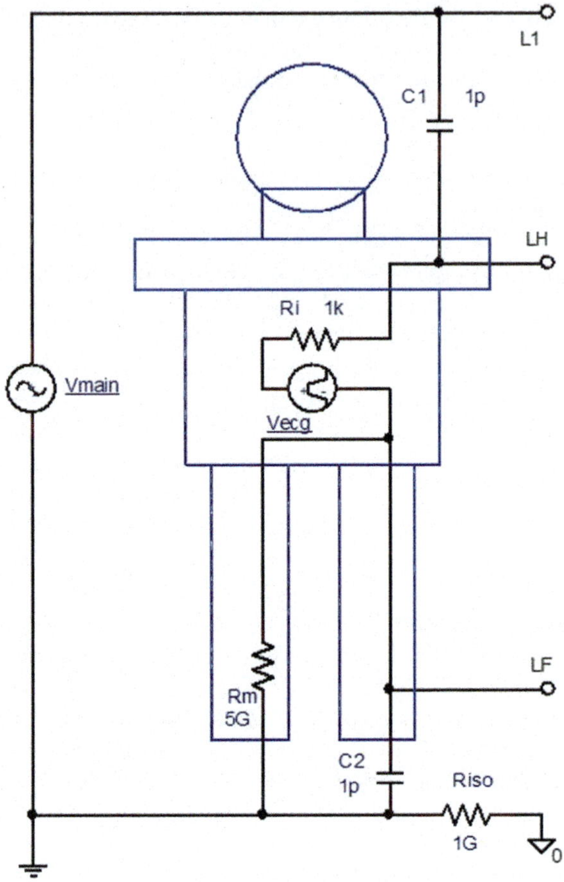

Fig. 2.10 Patient in the electric field of the power supply (mains voltage) during an ECG lead, where the potential difference between LH (left hand) and LF (left foot) generated by the heart is to be measured. L1 is the outer conductor of the power supply ("phase"), N is the neutral conductor ("zero"), and the capacitances C_1 and C_{gPE} are modeled stray capacitances of the patient's body to the conductors of the power supply. The resistor R_m represents the ground electrode. If this is not present, its resistance must be preserved in the PSpice simulation and set to a sufficiently high value (e.g., 5 G) (without R_m, U_{ekg} would have no conductive connection to ground, and this is not allowed in PSpice; PSpice simulation BSV_2.8)

$$u_{LH,LF}\Big|_{(R_m<<Z_{C2})} \approx u_{L1,N} - u_{C1} - u_{Rm} = u_{Netz}\frac{R_i}{R_i + |Z_{C1}| + R_m} \tag{2.4}$$

$$SNR_{EKG} = \frac{\hat{u}_{EKG}}{\hat{u}_{Ri(Netz)}} \tag{2.5}$$

$$SNR_{EKG}/dB = 20 \cdot \log_{10} SNR_{EKG} \tag{2.6}$$

Thus, the SNR deteriorates to half or decreases by 6 dB (see Eq. 2.6). The interference increased because the stray capacitance C_2 was bridged at low impedance by the ground electrode R_m. Since the resistors R_i and R_m are negligible against the impedances, it follows from comparing Eqs. 2.2 and 2.4 that the SNR drops to half, from 29 to 23 dB.

The measures to reduce capacitively coupled interference are straightforward in principle. However, they pose considerable challenges when applied in a practical setting: First, all parts carrying mains voltage (cables and devices) that are not

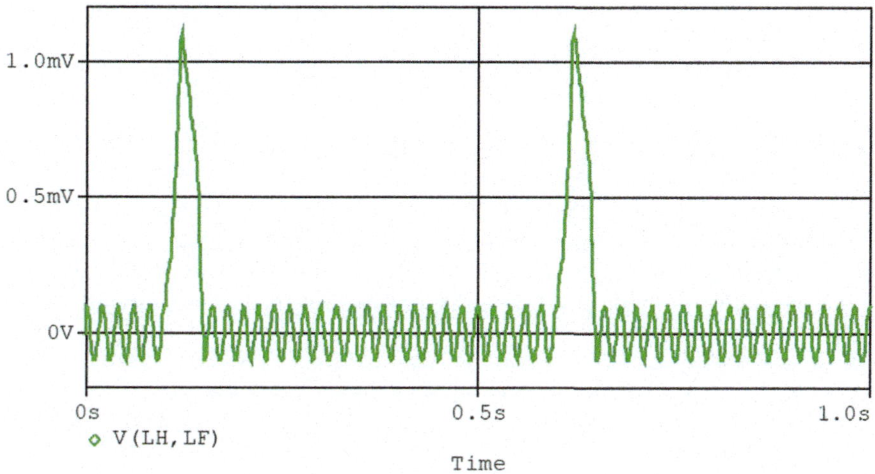

Fig. 2.11 ECG leakage (difference voltage between LH and LF) at the patient with capacitive interference from the installation of the mains and missing ground electrode R_m (PSpice simulation BSV_2.8, there $R_m > 5$ G)

necessary for measurement must be removed from the patient's vicinity. Switching them off is insufficient since a static electric field is present even when no current flows. It increases the distance that determines the stray capacitances. In case of extreme requirements on the measuring arrangement (medical research, experimental investigations, especially in the neurological field), the shielding of the mains installation or the measuring circuit (Faraday cage) may be necessary. Capacitively coupled interferences can be controlled by circuitry, as shown in Sect. 2.2.2.

2.1.3.3 Inductively Coupled Interference

While capacitively coupled interference is also present statically (electric field), inductively coupled interference occurs only when the interfering source produces a current flow or an electromagnetic wave. This situation is illustrated in Fig. 2.12: The mains supply the load R_l (medical device) with an operating current of magnitude 1 A. The patient is electrically conductive so that it forms a conductor loop with the connecting cables of the measuring amplifier on the left hand and foot. The magnetic field generated by the circuit of the mains flows through the surface of the patient measuring circuit. It induces an interference voltage at the input resistance of the measuring amplifier. According to Eq. 2.7 on the occurrence of an induced voltage, it can occur both due to the time change of the magnetic induction and the movement of the area flowing through. The temporal change of the magnetic induction is already given by the fact that in the installation of the power supply flows, a current harmonically varies with the mains frequency. In the medical measuring arrangement, it can happen that the area through which the magnetic flux also flows changes, e.g., due to the movement of the patient or the cables.

$$u_{ind} = -\frac{d\Phi}{dt}, \quad \Phi = A \cdot B \tag{2.7}$$

In Eq. 2.7, u_{ind} is the induced voltage in an open conductor loop, magnetic flux, B magnetic induction, and A area of the conductor loop.

The magnetic coupling between the power supply and the measuring circuit is simulated with the transformer T in Fig. 2.12. The voltages induced in the measuring circuit can reach a few millivolts and are thus very small compared to the capacitively interspersed disturbances. However, they act directly on the

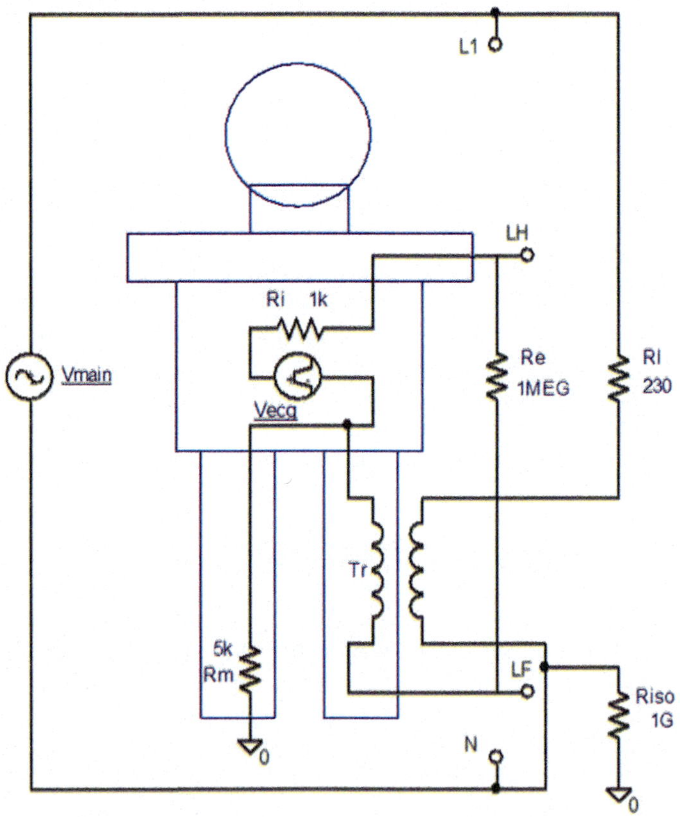

Fig. 2.12 Patient in the magnetic field of the power supply during an ECG lead, where the potential difference between LH (left hand) and LF (left foot) generated by cardiac activity is to be measured. L1 is the outer conductor of the power supply ("phase"), N is the neutral conductor ("zero"), R_l is the load resistance of a device on the mains, R_e is the input resistance of a medical measuring amplifier with which the ECG lead between the left hand and the left foot is to be measured. The transformer Tr models the inductive coupling between the current-carrying mains installation and the patient measurement circuit (PSpice simulation BSV_2.9)

input of the measuring amplifier as differential voltages. The desired ECG and the inductively coupled interference overlap additively in a first approximation.

The difficulties with inductively coupled interference result from their operating principle: They enter the amplifier input on the same path as the desired signals—from a differential signal source. They cannot be separated from each other in terms of circuitry. The possibilities for reducing induced interference are limited: The induced voltage depends on the size of the area magnetically flown through, the current in the installation, and the distance to the interfering source. In practice, it is possible to influence the area. It can be reduced by twisting the measurement cables between the patient and the amplifier. However, this measure can only be implemented in part and does not significantly improve the measurement situation because the patient area is always present and the necessary flexibility of the connecting cables. Shielding is not justifiable in frequencies up to 60 Hz for magnetic fields. A thickness of 50 cm of iron and more metallic walls would be necessary for effective suppression. Shielding with special alloys is reserved for a few unique designs, such as magnetically shielded chambers for biomagnetic measurements. However, effective measures can be taken before the construction of a hospital during project planning: One ensures during the planning stage that rooms with sensitive measurement technology are sufficiently far away from powerful equipment and its power supply, such as CT, MRI, radiology, and HF surgery.

2.1.3.4 Electromagnetically Coupled Interference

Electromagnetic coupling of high-frequency interferences take place on the same path on which wanted radio signals are received: The connecting cables of the measuring arrangement, which have a length of one to two meters, function like rod or frame antennas. The received frequency range extends over the entire spectrum of radio communication, from 50 kHz (long waves) to the gigahertz range (WLAN at 2.4 and 5 GHz). At such high frequencies, the question arises as to why such oscillations can interfere with biosignals far in the low-frequency range, up to 10 kHz. One explanation is the unwanted and undesired amplitude demodulation at the amplifier input of the measuring system. However, why does an amplifier input demodulate high-frequency signals when intended for the lowest frequencies? Starting with explanations of the basics of AM (amplitude modulation), this problem is explained in circuit terms.

To wirelessly transmit low-frequency signals (speech, images, biosignals) over long distances, imposing these signals on a high-frequency carrier's parameter (amplitude, frequency, phase) is necessary. AM is the oldest and most straightforward type of modulation. Mathematically, AM, according to Eq. 2.8, can be realized by multiplying the LF signal (low frequency, U_{NF}) by the RF carrier (high frequency, $U_{\mathrm{HF}} > > U_{\mathrm{NF}}$). This possibility was also chosen in the simulation in Fig. 2.13.

$$u_{AM} = \left(\widehat{U}_{HF} + \widehat{U}_{NF}\cos(\omega_{NF}t)\right) \cdot \cos(\omega_{HF}t) \tag{2.8}$$

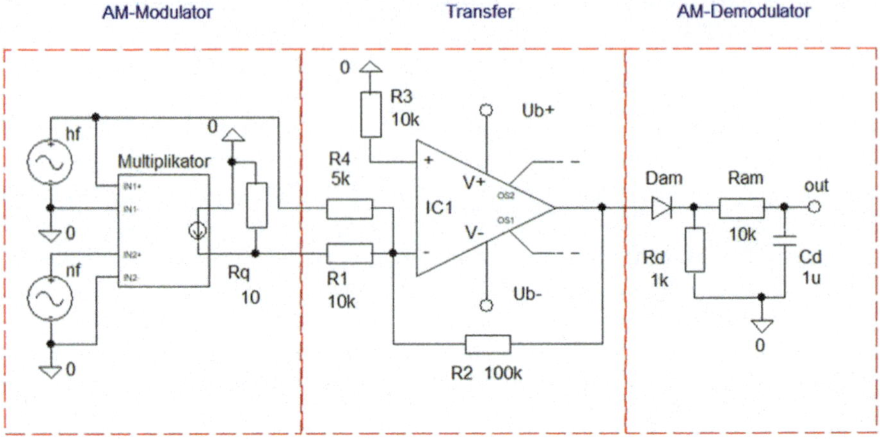

Fig. 2.13 Principle of amplitude modulation and demodulation. An LF signal modulates the RF signal. The transfer block represents the transmission path (propagation in free space as a radio wave or, for example, in the acoustic channel of an audio card of PC communication). Theoretically, the demodulator's three passive elements (D, R, C) are sufficient to recover the AF signal. (PSpice simulations BSV_2.10, a, b)

After multiplying both signals in the AM modulator, an AM signal is produced with the RF carrier's frequency and the AF signal's envelope, as shown in the lower graph of Fig. 2.14. The transfer path (amplification, antenna, transmission medium) was modeled here with a simple amplifier. The task of an AM demodulator is to recover the envelope from the AM-modulated RF carrier. Since the RF signal is zero-balanced, it must first be rectified; otherwise, the envelope would cancel out after the intended low-pass filtering. The diode D_{am} in the demodulator in Fig. 2.13 serves this purpose. After rectification, it is necessary to bridge the short periods of the RF oscillation, most simply with a low-pass filter. It is realized with the RC combination of R_d and C_d. In circuit terms, the demodulator arrangement is an envelope detector with a suitable time constant (large enough to bridge the RF period but small enough to follow the AF signal). The resulting waveform is shown in the upper graph of Fig. 2.14. This straightforward circuit solution was one reason for the spread of AM in the early days of communications and broadcasting.

In the following, the question arises as to which relationship exists between the AM-demodulator and a medical measuring amplifier: At the input of each amplifier, there are transistors as active components. The base-emitter path (gate-source path of a FET) is a PN junction, on the one hand, a rectifying element; on the other hand, the PN junction in the forward direction is a relatively large capacitance (up to some nF). The transistors of the input stage are equipped with resistors to define their operating parameters. Thus, the input stage of each measuring amplifier contains all necessary components for AM demodulation (rectifier, resistor, capacitance). It functions like an unwanted AM demodulator for high-frequency

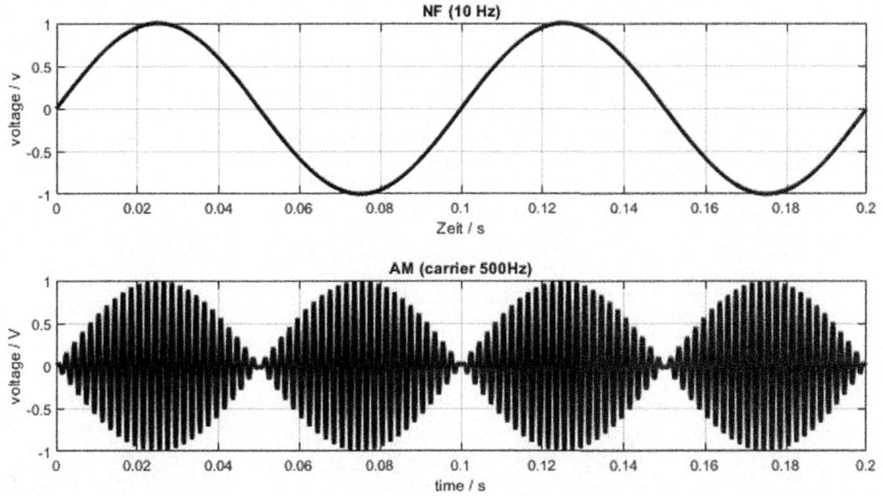

Fig. 2.14 Time courses for amplitude modulation and demodulation. The upper graph shows the AM-demodulated signal at the demodulator output, and the lower graph shows the AM RF signal at the demodulator input (Matlab simulation for circuit BSV_2.10 in PSpice). Note that the AM demodulator works as an envelope detector and reproduces each amplitude, not only the desired AM-modulated RF oscillation (RFID, mobile radio, Bluetooth, and others)

signals with variable envelopes. The interspersed RF signals do not have to originate from an original AM radio signal. It is sufficient if the amplitude of any RF signal (FM broadcasting) changes of its own accord simply due to changing propagation conditions (mobile radio while driving). It can happen due to movement in the terrain in the case of an FM signal or a movement-induced variable amplitude of the RF oscillation in the operating room in the case of RF surgery. This analysis concludes that any RF oscillation with *variable* amplitude can affect the measurement of biosignals.

Effective measures against electromagnetic interference are challenging to implement. In terms of circuitry, it would be conceivable to equip the inputs of the measuring amplifier with low-pass filters, as is common in technical measurement technology or telecommunications (Fig. 2.15). However, such filters would significantly reduce the required high input impedance of the amplifier, which is unacceptable in the medical field. Measures that can be implemented are organizational: the medical room must be set up far away from commercial transmitters, and mobile radio equipment (radiotelephone, RFID, remote control) must not be operated in such rooms.

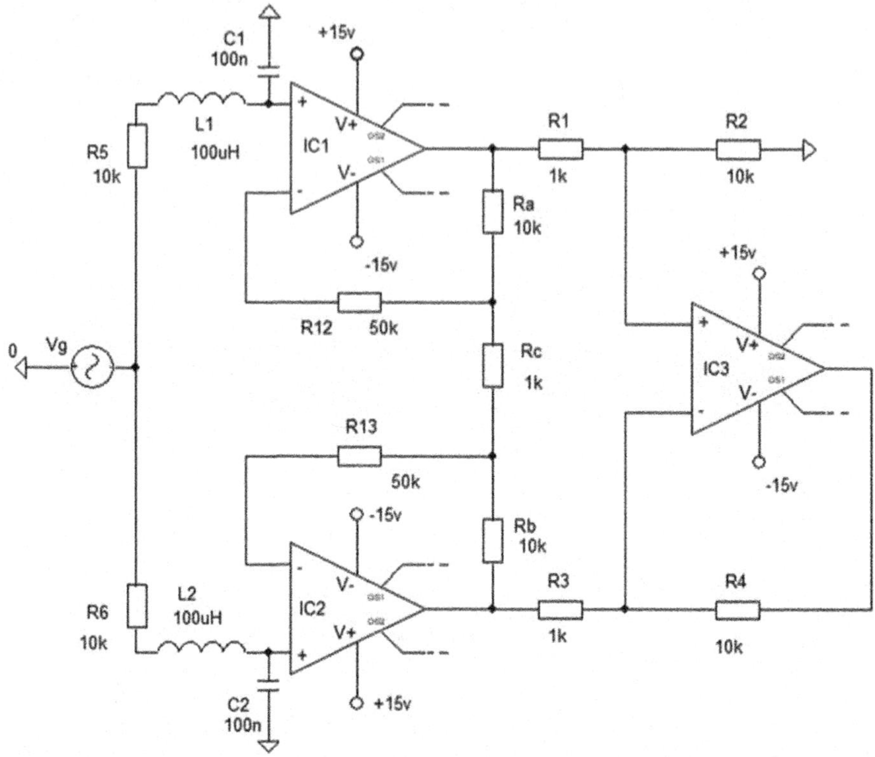

Fig. 2.15 Medical measurement amplifier as instrumentation amplifier with low-pass filters at the input to suppress high-frequency interference. The low-pass filters reduce the amplifier's input resistance from 600 M to 10 kR_5, R_6 at their cut-off frequency (PSpice simulation BSV_2.11)

2.2 Medical Measuring Amplifiers

In medical measurement technology, the task is to detect and amplify biosignals whose levels are in a few microvolts to millivolts and occupy a very low-frequency range between 0 and 10 kHz. These initial properties do not pose any problems in the technical field and would be manageable with a typical measurement amplifier.

The specificity of medical measurement technology results from the subject to be measured—the human being—with many endogenous biosignals and the sources of interference surrounding him.

2.2.1 Specifics of the Medical Measurement Technology

One of the essential principles of medical measurement technology dictates that we measure as non-invasively as possible for minimal patient exposure. The consequence of this requirement is that usually, none of the biosignal sources to be measured can be reached directly.

2.2.1.1 Accessibility of the Biosignal Source

The vital functions like heart activity (ECG), brain activity (EEG), and the activity of the nervous system as well as the motor potentials EMG, can only be measured as sum potentials from the body surface: respiration and the associated blood gases. (pH, pO_2, pCO_2 but one or more related auxiliary quantities are measured. A typical example is blood pressure measurement on the upper arm: the air pressure of the cuff is measured under the assumption that it is equal to the arterial pressure under certain conditions. An example of the cardiological examination of conduction at the AV node illustrates the point. All other electrical activities outside the structure under consideration have a disturbing effect. Because the measurement alone cannot be performed on the heart (cardiac catheter) but from the chest surface, all heart's electrical activities are inevitably recorded. Further interfering signals arise during this measurement in the biological and technical environment. The situation becomes even more complicated if the biosignal to be recorded is not an electrical activity, so a transducer is still required to convert the investigated quantity into an electrical signal (invasive blood pressure measurement, respiratory volumes, blood gas concentration).

2.2.1.2 The Complexity of the Biological Target

In the technical field, it is possible and customary to take assemblies of a system out of the system for function control and to check whether they are intact as a functional unit specifically. In medicine, this is not feasible for the patient, so the measurement of biosignals must always occur on the fully functioning measurement object. It is one of the most fundamental measurement methodological problems: One cannot switch off or interrupt the human organism's dynamically nonlinear coupled and nested control circuits. So, the measurement results can show significant errors under unfavorable circumstances and lead to false conclusions. Elaborate modeling methods are currently used to describe the individual control loops so that the undesirable influences on the measurement can be eliminated. However, these approaches are still far away from possible solutions. Nevertheless, some simple methods have become established. For example, it is feasible to eliminate blink and eye movements (EOG) from a measured EEG today.

2.2.1.3 The Reaction of the Biological Measuring Object

From the point of view of measurement methodology, it is essential to avoid influencing the object being measured and to minimize any possible feedback effects. In technology, this requirement can be implemented without further ado. In medical measurement technology, two problem areas need to meet this requirement. On the one hand, most sensors and transducers for measuring non-electrical quantities are, in principle, not feedback-free. Typical of this is blood pressure measurement with cuff methods on the upper arm. The artery is occluded under pressure, which is problematic from a measurement method point of view, apart from other measurement errors. The circulation is subjected to a short-term flow resistance by interrupting the blood flow, which inevitably influences the measurement. On the other hand, another source of error is compelling about.

2.2.1.4 Coupling of the Measurement System to the Biological Measurement Object

The focus of biosignal acquisition and processing in this publication is on electrophysiological signals, particularly EEG, ECG, EMG, and EOG. Using this selection, It shall demonstrate which fundamental problems must be overcome on the way from the sensor via amplification, filtering, AD-conversion, and digital signal processing. Of course, the principles and approaches discussed here also apply to non-electrical quantities such as blood pressure, oxygen saturation, or respiratory volumes. However, the size transformation will complicate the measurement and evaluation of these quantities. The most crucial difference to electrical measurement objects is that the biological measurement object—the human being—has no *ground* or *earth connection in* the electrical sense. From an electrical point of view, the human being is a closed, three-dimensional volume conductor, an electrolyte vessel. In this volume conductor, there are electrically conductive pathways (nerves) that conduct the electrical signals (action potentials, neuronal excitations) from and to the control centers (brain, spinal cord). To make statements about the type and local limitation of the electrical activity in this three-dimensional electrolyte, at least two sensors—in this case, two electrodes or two capacitive sensors—must be attached, and the electrical activity between the electrodes has to be measured. Considering an electrical engineering perspective, one always measures a potential difference in humans and thus—mathematically speaking—the first spatial derivative or the first spatial gradient.

The electrodes are entirely equivalent in terms of the electrical activity they detect. No location on the human body would have a preferential position in the sense of reference. Because the human body consists of living cells, all processes in the cells (metabolism, charge carrier movement) are accompanied by electrical effects (local current flows). Against this background, there is no reason to consider any part of the body as electrically inactive and to provide it for the attachment of indifferent electrodes or as a reference point—as is common in technology. In multi-channel measurement systems (ECG, EEG, EMG), it is necessary to establish a reference electrode or a reference point, but this does not change the principle mentioned above of spatial difference formation.

If only the levels of the signals to be amplified and their spectra are assumed, an amplifier commonly used in technical measurement technology (e.g., a microphone amplifier) is sufficient. Such an arrangement is shown in Fig. 2.16. Two electrodes are attached to the patient's hands, which derive the biosignal (ECG) from being amplified from the body surface and pass it on to the amplifier input via cables. The electrodes are complicated electrochemical structures as interfaces between the ion conductor human and the electron conductor metallic cable. Their properties will not be discussed in detail here (see Chap. 1.3). For this analysis; it shall suffice to state that the electrodes have an impedance whose magnitude can be simulated with about 2 kΩ. That simulated signal courses of the simplified ECG derivation without external disturbances are shown in Fig. 2.17.

In medical practice, however, a mains interference will always occur (even with galvanic isolation from the mains or battery operation), which can initially be regarded as a capacitively coupled interference. This situation is simplified with two stray capacitances acting on the patient's arms, as shown in Fig. 2.18. The simulated measurement setup in Fig. 2.18 is ideal because the electrode impedances Rer and Rel and the stray capacitances C1 and C2 acting on the arms are each equal. From the point of view of the patient connections "left hand" and "ground", there is an entirely symmetrical arrangement concerning the interference source. Therefore, it can be assumed that the mains interference has the same amplitude

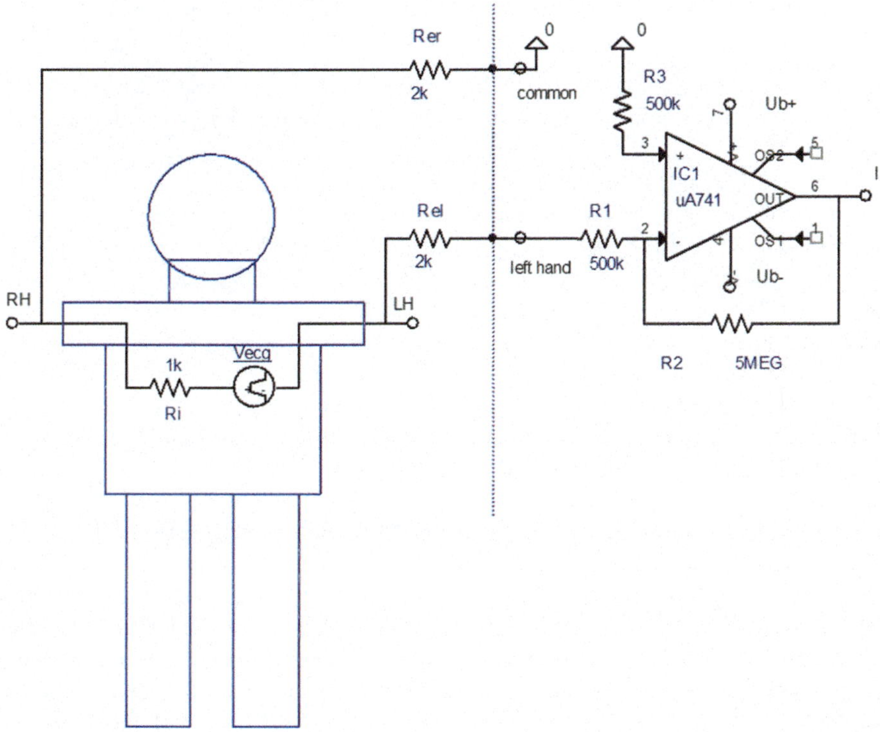

Fig. 2.16 Electrical equivalent circuit of a single-channel ECG lead: The internal resistance R_i of the signal source U_{ekg} and the electrode junction impedances (here simplified to resistances) R_{er} and R_{el} correspond approximately to reality. The ECG is amplified with a single-channel ground-referenced amplifier—as it is common in technical measurement technology (PSpice simulation BSV_2.3)

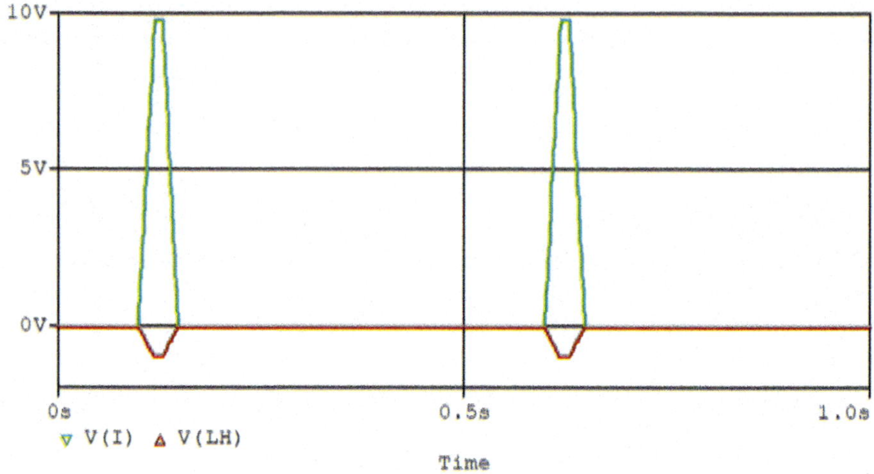

Fig. 2.17 Simulated ECG as a single-channel signal at the electrode of the left hand (red) and the output of a single-channel amplifier (green) according to the circuit in Fig. 2.16

and phase at both connections; its difference would consequently be zero without the connected amplifier and ground. However, the interference coupled to the right hand is connected to the ground. That to the left hand is amplified, as it is the intended ECG. The simulated course is shown in Fig. 2.19.

From this analysis of the desired signal (ECG) and the unwanted disturbance (mains), it is concluded that it would be helpful to implement a circuit measure that would amplify the *biosignal as a differential signal* between the two electrodes. At the same time, the *interference* (common mode) *applied identically to* both electrodes would be eliminated. Thus, a circuit is needed that amplifies only the potential difference at both terminals (left and right hand, LH and RH) in a ground-free manner and ignores the identically applied ground-related interference voltage. For this case, analog circuit technology offers the *differential amplifier* shown in its simplest form (Fig. 2.20 or 2.26).

2.2.1.5 Input Resistances of Medical Measuring Amplifiers

In measurement technology, a very high input resistance is generally required at the input of the measuring amplifier concerning the internal resistance of the signal source, not to falsify the voltage to be measured. This requirement depends mainly on the accuracy class of the measurement system. In medical measurement technology, it is commonly required that the input resistance of the measurement amplifier is at least one hundred times the internal resistance of the biosignal source. To quantify this requirement, an undisturbed measurement circuit must first be assumed (Fig. 2.20).

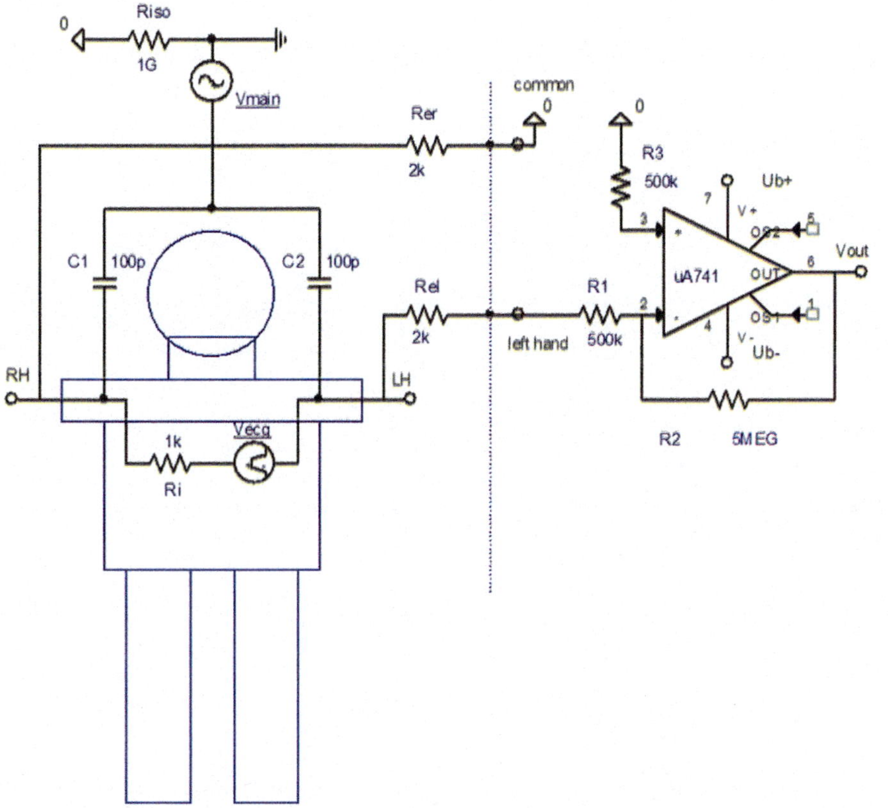

Fig. 2.18 Via the stray capacitances of the arms to the network installation, a network disturbance is transmitted to the patient, which is also processed in the single-channel amplifier (PSpice simulation BSV_2.4)

From the point of view of the measurement amplifier, the electrode resistors R_{el} and R_{er} must be added to the internal resistance R_i of the biosignal source, resulting in a global internal resistance of 5 κ. The input resistance of the differential amplifier is about 1 MΩ (the series connection of the two common mode resistors $R_{g+,-}$ is in parallel with the differential input resistance, so it is insignificantly less than 1 M). With this value of the differential input resistance, the general requirement for at least 100 times higher input resistance is well fulfilled and accepted in practice (Note. The catalog values of today's amplifiers specify an input resistance in the range 1–10 GΩ; however, it must be considered that the input stages also have capacitors that become effective with AC voltage and reduce the input impedance to a fraction of the otherwise high input resistance). Now the question arises about the necessary size of the common mode resistors $R_{g+,-}$. For this purpose, the measuring circuit is examined under a common mode disturbance on the line side, shown schematically in Fig. 2.21.

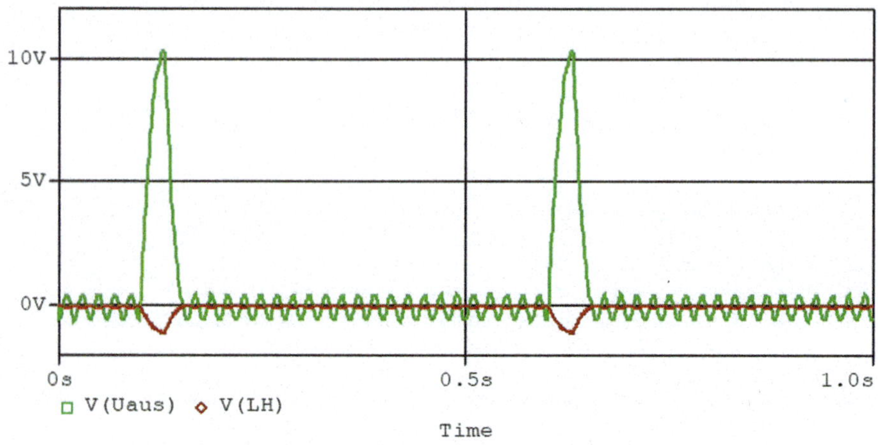

Fig. 2.19 Due to the capacitive interference of the network disturbance, it is amplified with the desired signal. The interference can become much stronger than the signal itself. Only the stray capacitances at the patient's arms were considered (LH red, output green)

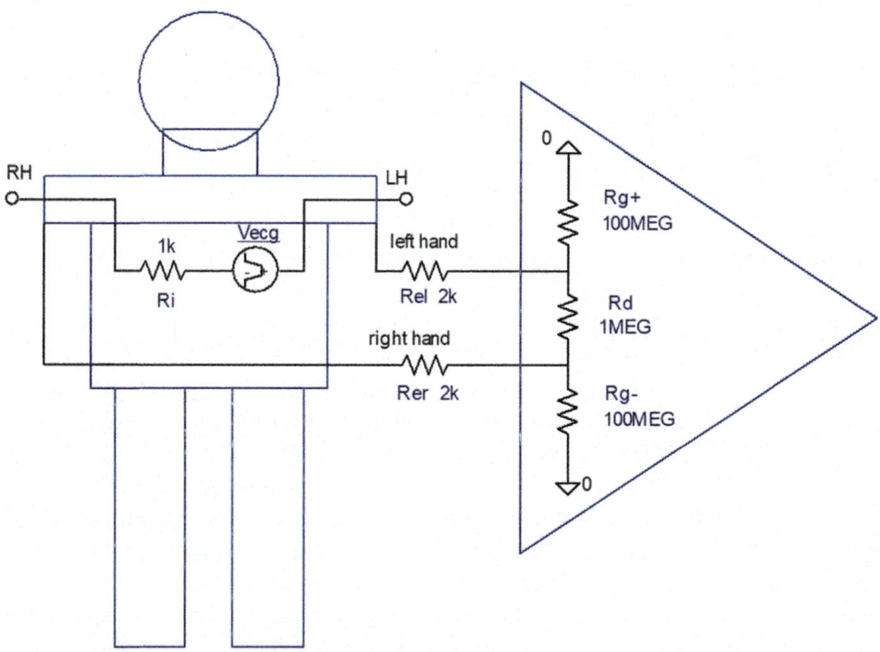

Fig. 2.20 Electrical equivalent circuit of a single-channel ECG lead and gain with a differential amplifier. The input circuit of the differential amplifier is modeled with three resistors: R_d for the ground-free differential voltage and $R_{g+,-}$ for the ground-referenced common-mode voltage at the respective input (details in the differential amplifier chapter). The modeled resistors correspond to the real values (PSpice simulation BSV_2.15)

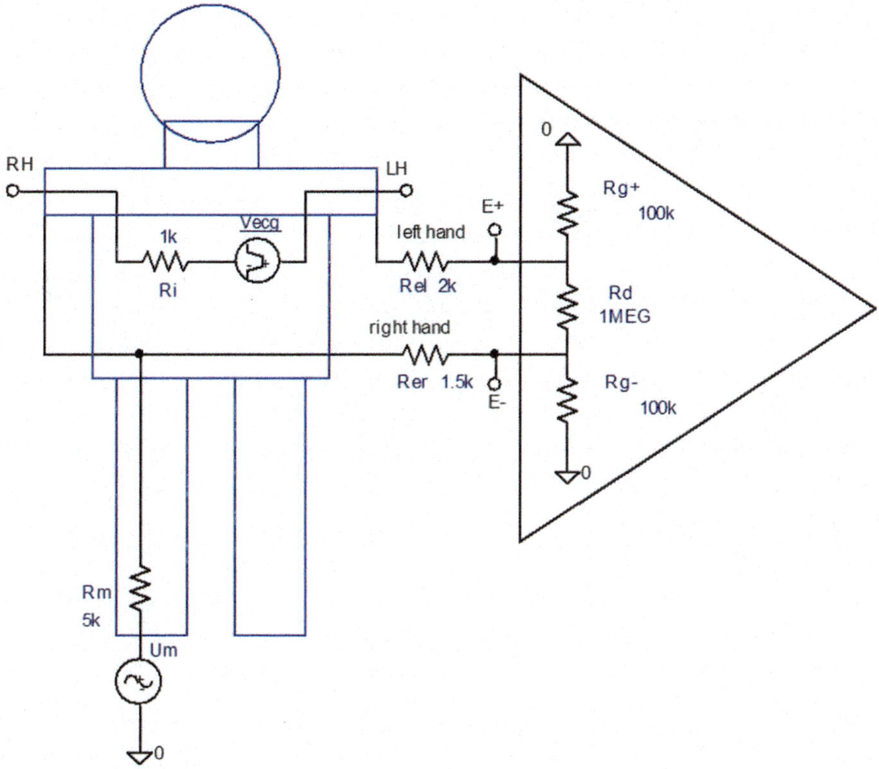

Fig. 2.21 Electrical equivalent circuit of a single-channel ECG lead processed with a differential amplifier. The voltage source U_m simulates the voltage drop across the ground electrode R_m, formed by capacitive interference from the power supply network, as shown in Fig. 2.10 (PSpice simulation BSV_2.16)

As shown in Fig. 2.10, a voltage drop in the lower millivolt range occurs across the ground electrode R_m, simulated with a voltage source U_m in Fig. 2.21. This spurious voltage acts on both inputs of the differential amplifier: on the inverting input $E-$ via the right-hand branch, invoking a voltage drop across the common mode resistor R_{g-}; on the non-inverting input $E+$ via the left-hand branch, invoking a voltage drop across the common mode resistor R_{g+}. Since U_m is a common-mode interference voltage (acting identically on both inputs), the difference between the inputs $E+$ and $E-$ caused by the interference voltage should be as close to zero as possible. In theoretical analysis, this is not a problem since one can make both branches completely identical. In practical measurement, however, the electrode resistances R_{el} and R_{er} will always differ. It arises because there are no two similar electrodes, and the contact points on the body always have different contact properties, humidity, and temperature. As a result, differences between electrode impedances in more than one hundred percent range are possible. The question arises about how the requirement for minimum differential voltage between the $E+$ and $E-$ inputs could be met in practice. Mathematically, the differential voltage caused by the common mode can be described according to Eqs. (2.9) and (2.10). Provided that the two common mode resistors R_{g+} and R_{g-} are identical, which is fulfilled in most amplifiers, holds:

$$u_{E+} \approx u_m \frac{R_g \| (R_g + R_d)}{R_m + R_i + R_{el} + R_g \| (R_g + R_d)} \tag{2.9}$$

$$u_{E-} \approx u_m \frac{R_g \| (R_g + R_d)}{R_m + R_{er} + R_g \| (R_g + R_d)} \tag{2.10}$$

$$u_d = u_{E+} - u_{E-} \to 0 \Rightarrow R_g \to \infty \tag{2.11}$$

The qualitative relations, according to Eqs. 2.9 and 2.10 were derived using the superposition principle and the approximation calculation. Precise nodal stress analysis leads to the following result (the numerators of the polynomial quotients were analyzed because the denominators are identical. Therefore, the quotient in this relation is to be interpreted purely numerically):

$$u_d \propto u_m \frac{1/(R_{el} + R_i) - 1/R_{er}}{R_m R_g}.$$

The goal of a circuit solution should be that the differential voltage u_d caused by the common mode voltage u_m at the input becomes minimal. It follows from this relationship that the common-mode resistance (impedance) should be as large as possible, theoretically infinitely large.

According to Eq. (2.11), the interfering differential voltage can be calculated and minimized. It theoretically results in several solutions, e.g., complete symmetry of the amplifier and the interference coupling. However, in terms of circuitry, only the resistors R_g and R_d can be influenced. Therefore, the resistor R_g must be maximized to minimize the differential voltage. A compromise between the theoretical maximum requirement and electronic feasibility is about 100 MΩ impedance per common mode input resistor. The finding that the interfering differential voltage is determined mainly by the common mode input resistance seems contradictory initially. However, through the circuit interpretation of Eq. 2.9 to Eq. 2.11, it becomes clear that the two common mode input resistors form two unwanted voltage dividers with the electrodes and the patient. Between the voltage dividers, the voltage U_d should become minimal, and consequently, there must be practically no voltage division at all, i.e., $U_{E+} = U_{E-}$. However, this is only possible if the common mode input resistors are infinitely large, practically by about 100 M.

2.2.1.6 Common Mode Rejection of Medical Measurement Amplifier

Biosignal sources are to be interpreted as differential signal sources because of the metrological necessity to derive their electrical activity as a potential difference. However, neuronal activity primarily generates local negativity (See Chap. 1.1). Mathematically, this can be formulated by forming the first spatial derivative.

Many disturbances, especially capacitively coupled, arrive at the measuring circuit in first approximation with the same level, frequency, and phase—in the so-called common mode as shown in Fig. 2.18. The task of a medical measurement amplifier is to amplify the desired differential signals (Eq. 2.12) and suppress the unwanted common mode signals (Eq. 2.13). In the case of a single discrete differential amplifier (Fig. 2.26) with a floating output voltage, the common mode problem would be solved by ideal symmetry and the difference between the ground-related output voltages (Eq. 2.13).

$$u_{as} = V_d \cdot u_{es} = V_d\left(u_{g1} - u_{g2}\right) \tag{2.12}$$

$$\left.\begin{array}{r} R_{d1} = R_{d2} \\ T_1 = T_2 \\ U_{g1} = U_{g2} \end{array}\right\} U_{d1} = U_{d2} \Rightarrow U_{as} = 0 \tag{2.13}$$

In Eq. 2.12 and Eq. 2.13, the indices *d* stand for difference, *g for* gate, *e for* capture, *a* for output, *s* for "symmetric", here massless, $d_{1,2}$ for drain.

In practice, the two approaches mentioned above are not feasible. Perfect symmetry is not achievable even in integrated electronics, so the common mode signal is always converted to a certain extent into a differential signal (Eq. 2.14).

$$\left.\begin{array}{r} R_{d1} \neq R_{d2} \\ T_1 \neq T_2 \\ U_{g1} = U_{g2} \end{array}\right\} U_{d1} \neq U_{d2} \Rightarrow U_{as} \neq 0 \tag{2.14}$$

Since the output voltage is processed in further stages of the measurement chain (AD conversion, display), a transition from the ground-free output voltage of a differential amplifier stage to a ground-referenced voltage is necessary for the last amplifier stage. Therefore, reducing the common mode by mere difference formation is no longer possible. Other circuitry measures Eq. (2.15) must effectively reduce the common mode.

$$\left.\begin{array}{r} u_{es} = u_{g1} - u_{g2} \\ u_{eg} = \left(u_{g1} + u_{g2}\right)/2 \end{array}\right\} u_a = V_d u_{es} + V_g u_{eg} \left\{\begin{array}{l} V_d \gg 1 \\ V_g \to 0 \end{array}\right. \tag{2.15}$$

When evaluating the differential (V_d) and common mode gain (V_g) of a differential amplifier, it is not the absolute values of the gains that are of interest but their quotient—the common mode rejection (Eq. 2.16).

$$CMRR = \frac{V_d}{V_g},\ CMRR/dB = 20 \cdot \log_{10} CMRR \tag{2.16}$$

In Eq. 2.16, *CMRR* for common-mode rejection ratio, V_d for differential gain, V_g for common-mode gain.

This formulation results from the view of the relations between signals and disturbances: In signal processing, it is not the absolute levels of signals that are of interest across the board. More important is how well a signal is distinguished from a disturbance. The measure for the quality of the distinction is the signal-to-noise ratio, also referred to as *SNR* (signal-to-noise ratio). In its general form, it is formed from the quotient of the rms values or average powers of signal and noise (Eq. 2.17). This general form is not applicable in medical signal processing because biosignals are usually highly transient. Accordingly (Eq. 2.17), the temporal averaging reflects the characteristics of the biosignals only smoothed in the level or power, which can be attributed to the low-pass effect of the temporal integration of the definition. Therefore, it makes sense to adapt the general definition 2.17 to the concrete problem as, e.g., with the ECG measurement according to the equations Eqs. 2.5 and 2.6. Also, with other biosignals, a problem-specific definition of the SNR is meaningful, e.g., with the EMG (impulse-like, stochastic character) or with the EEG/EP around the energy of individual waves or wave trains to judge.

$$SNR = \frac{s_{eff}}{n_{eff}}, \; SNR/dB = 20 \cdot \log_{10} \frac{s_{eff}}{n_{eff}} = 10 \cdot \log_{10} \frac{\overline{s^2}}{\overline{n^2}} \tag{2.17}$$

A CMRR of 100–120 dB is required for the medical sector. For the amplifications, this means that the differential amplification V_d must be higher than the common mode amplification V_g by a factor of 10^5–10^6. Achieving such a high-value circuit-wise is a design challenge. With today's technology, these values can only be achieved with reasonable effort using integrated electronics.

How valuable the (logarithmic) dB representation of the *SNR* and *CMRR* is, demonstrates the following example: In an ECG measurement, the amplitude of the R-wave is around 1 mV, the common-mode interference coming from the mains causes a voltage of 100 mV at the amplifier terminals so that the *SNR* at the input reaches −40 dB ($SNR_E = -40$ dB). If the *CMRR* is 120 dB, the ECG as a differential signal is amplified 10^6 -times more than the disturbance (the differential amplification value is 10^3–10^4), and the SNRA = SNRE + CMRR = 80 dB output is reached. The ECG is 104 -times stronger at the output than the common mode interference, significantly improving the ratios at the input. It is, of course, a purely mathematical improvement of the *SNR, which is* never achieved in practice because of additional asymmetries (see Sect. 2.2.2). It should also be noted that the *CMRR* is frequency-dependent and decreases sharply towards higher frequencies (> 10 kHz). For this reason, additional measures (shielding of the measurement site and spectral filters) must be taken to reduce high-frequency common-mode interference.

2.2.1.7 Frequency Response and Noise of Medical Measurement Amplifiers

The requirements for a measurement amplifier's amplitude- and phase-frequency response depend primarily on the desired measurement accuracy and the requirement for preserving the signal shape. In biosignals' acquisition, the measurement accuracy requirements are relatively low compared to technical measurement technology. Because the inter- and intra-individual fluctuations of the biosignals or their parameters can be 100% and the measurement conditions are rarely reproducible, measurement accuracy in the lower percentage range is usually sufficient. The amplitude-frequency response may show a ripple of up to 1 dB in the spectral working range. Furthermore, there is the question of the necessary spectral width of the operating range.

On the one hand, the spectral working range should not be much more comprehensive than necessary because the inherent noise (Eq. 2.18) increases directly proportional to the spectral width. On the other hand, the operating range must be broad enough so that the biosignals do not lose any relevant components. A typical working range is shown in Fig. 2.22.

In the low-frequency range, choosing the lower cutoff frequency is particularly difficult. One of the main circuit problems is related to the electrode voltage. Initially assumed to be static, the electrode voltage reaches up to several hundred millivolts, depending on the electrode material. It is, therefore, several powers of ten higher than the biosignals to be measured in the micro- and millivolt range. Even if identical electrodes are used in a derivation, the electrode voltage between

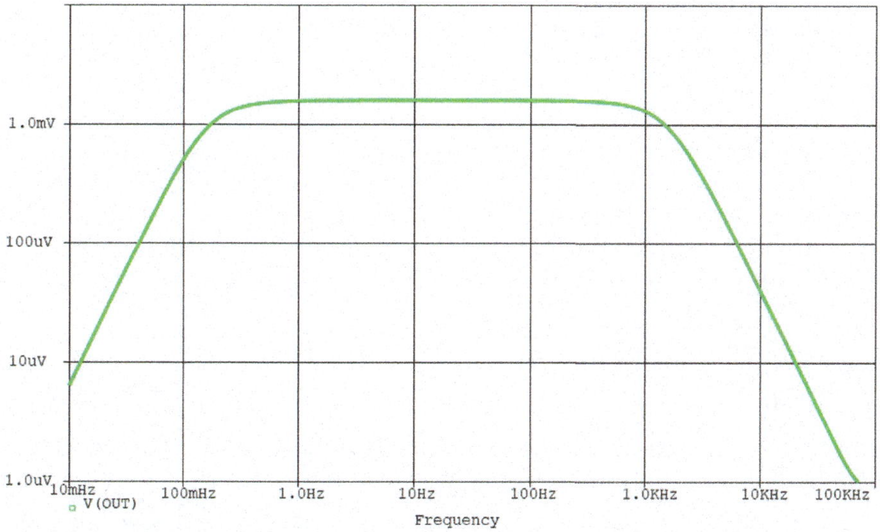

Fig. 2.22 Spectral operating range of a medical measurement amplifier. The lower cut-off frequency of 0.1 Hz is realized with a high-pass filter and the upper cut-off frequency of 2 kHz with a low-pass filter. Both are linear-phase second order (PSpice simulation BSV_2.17)

pairs of electrodes only partially cancels each other out. The contact points under the electrodes are different (moisture, salts, mechanical pressure, temperature). Even between identical electrodes, residual differences in electrode voltage occur that are significantly higher than the level of the biosignals. Even today, designers solve this problem by installing high-pass filters after the input stage, which has a relatively low amplification ($V_d = 10\ldots100$), which primarily suppresses the electrode voltage, see Figs. 2.23 and 2.24.

Some standards of medical measurement technology specify extremely low values as lower cut-off frequency, e.g., 0.05 Hz for the diagnostic ECG or 0.01 Hz(true DC) for the plethysmographic pulse curve. Such low cut-off frequencies are difficult to realize with analog filter technology, have poor dynamic behavior due to the high time constants, and the technological effort (extremely high capacities) is hardly justifiable. Therefore, the analog high-pass filter is increasingly dispensed entirely in such cases. For this, ADC with a very high bit width must

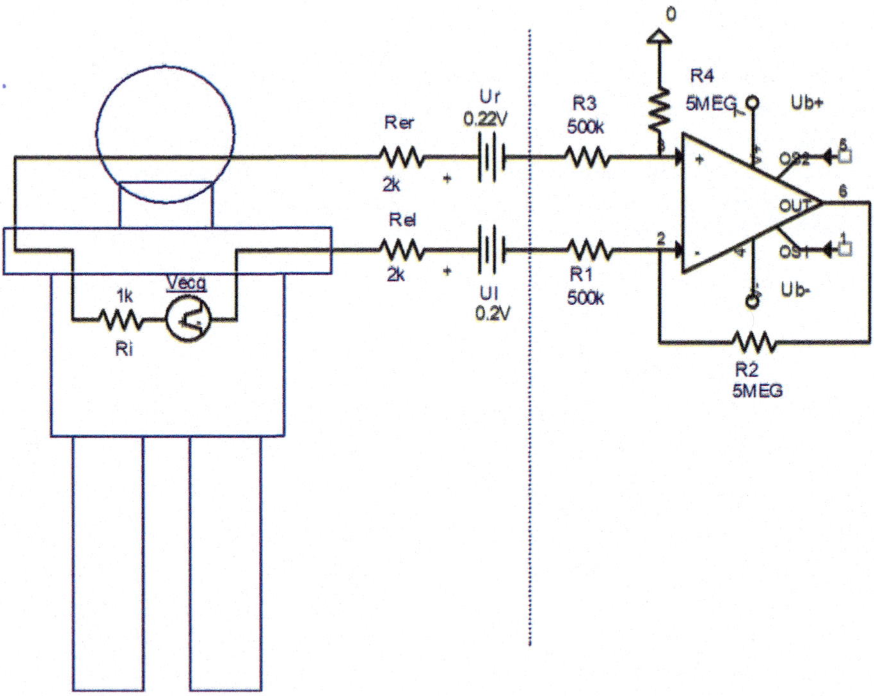

Fig. 2.23 Influence of the electrode voltage on the biosignal. Ur and Ul DC voltage sources simulate the electrode voltages at the contact points between electrodes and skin. Most electrode voltage cancels out since they are made of the same material. Nevertheless, the electrode voltages differ relatively strongly due to different conditions at the contact points. This voltage difference is additively superimposed on the biosignal and must (possibly) be suppressed with a high-pass filter after the first amplifier stage. (PSpice simulation BSV_2.43)

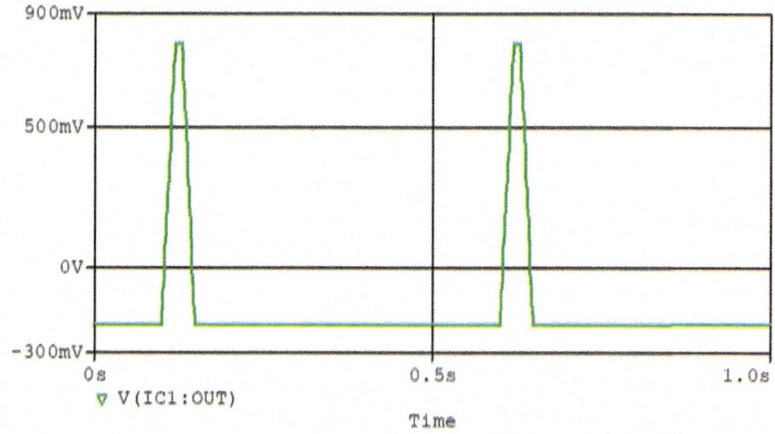

Fig. 2.24 Time course of the simulated ECG, considering the electrode voltages in the circuit according to Fig. 2.23. Due to the difference in the electrode voltages, the zero line is shifted downwards and could lead to limitation of the measuring amplifier if the difference is increased. Therefore, a high-pass filter must follow the first amplifier stage, or an extensive dynamic range must be ensured. With today's technology, such dynamic ranges are controllable, so problematic analog filters can be avoided

be used (20…24 bit) to control the electrode voltage and realize spectral filtering with the help of DSP (Digital Signal Processing).

2.2.1.8 Amplifier Noise

An important consideration when selecting the cutoff frequency of a high-pass filter is amplifier noise. Spectrally, noise comprises two major components—thermal and semiconductor noise. The thermal noise has a constant power spectral density (white noise). It depends on the bandwidth f, the absolute temperature *T*, and the ohmic resistance *R* according to Eq. 2.18 (k_B—Boltzmann constant). The bandwidth and the resistance can be influenced by design. The temperature is reduced only in exceptional cases and unique designs, e.g., helium-cooled magnetometers in biomagnetic measuring chambers. The noise characteristics result in circuit-technical consequences for the choice of the bandwidth and the real (ohmic) resistances, which should accordingly be as low as possible.

$$u_{eff} = \sqrt{4 \cdot k_B \cdot T \cdot R \cdot \Delta f} \tag{2.18}$$

In the low-frequency range, the so-called 1/f noise is added to the thermal noise. This noise is especially typical for MOSFET (CMOS), which are often used in amplifiers, at least in the input stages. The *1/f* noise has its power maximum at low frequencies down to 0.001 Hz, where its level is much higher than that of white noise and decreases with increasing frequency until it reaches the level of white noise in the decades between 1 and 100 Hz (Fig. 2.25). The inherent noise of an amplifier is one of the most critical operating parameters and, at the same time, a characteristic quality factor because the less noise in an amplifier, the higher the quality of the amplified signal. For this reason, people often try to set the lower cutoff frequency as far as possible towards higher frequencies so that the strong low-frequency inherent noise is less of an issue. However, this effort is contrary to the requirement to preserve as much as possible of the relevant properties of the biosignals. It has been known for a long time that some biosignals in the DC range (direct current) have essential features, e.g., the so-called SCP in the EEG (Slow Cortical Potentials). Therefore, attention to which amplifier has the required quality is generally necessary. The information about the noise voltage (given in V as an absolute measure of the peak or RMS value or in V^2/Hz as the power spectral density) alone is not sufficiently informative and meaningful. The measured and used spectral range should also be available as information. Today's leading-edge technology offers "full band" class measurement amplifiers that provide constant gain starting at DC and extending well into the kilohertz range. The choice of the upper cutoff frequency, on the other hand, is unproblematic. It depends on the desired spectrum of the biosignal.

2.2.1.9 Phase Frequency Response

Both cut-off frequencies of the operating range of the measuring amplifier are realized with analog filters, mostly with operational amplifiers or in SC technology (Switched Capacitors). In principle, there are many possibilities for analog spectral filtering. However, the question of which filters are suitable for biosignals, especially their phase frequency response, is of decisive importance. Biosignals' most diagnostically important property is their signal shape, which is both physiological and pathological. Therefore, the entire measurement path must not distort the signal shape. It can only be achieved if the group delay is constant, which requires a linear phase frequency response. Only a few filters fulfill this requirement. Analog filters are discussed in detail in Sect. 2.3, and digital filters in Sect. 5.2. The requirement for linear phase frequency response is explained in detail in Sect. 2.3.3.

In biosignal processing (BSP), there are often measurement tasks that require the use of nonlinear methods. For example, it is necessary to determine the instantaneous or spectral band power in EMG, EEG, and ECG. Since with/after nonlinear methods, the linear phase frequency response is no longer necessary, which is lost regardless of the nonlinearity, one must examine in the individual case up to which point of the processing chain linearity must prevail. A practically applicable criterion for using nonlinear methods and their position in the processing (sensor, amplification, filtering, processing, evaluation) results from the question of which point the superposition principle must function.

A typical analytical task, e.g., in studies at the neuronal level, is to determine how many different APs (action potentials) are measured in a culture and with what frequency (i.e., temporal frequency) they occur. The APs must be classified according to their signal form (Sect. 1.5.1). Therefore, the processing chain must have a linear phase frequency response up to this step. After the classification, the question of the temporal occurrence or frequency of the individual APs within the

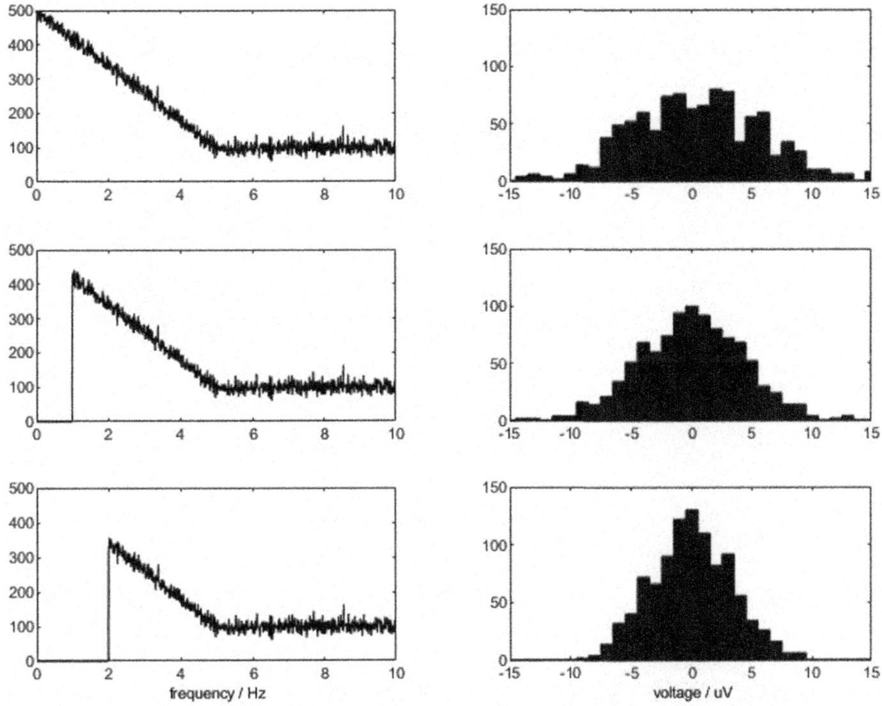

Fig. 2.25 Spectrum of amplifier noise (left column) and the corresponding empirical distribution density (right column). The top row shows a spectrum unbounded toward low frequencies, where a noise amplitude $U_{SS} = 34\ \mu V$ occurs (the subscript SS stands for peak-to-peak, a common noise parameter). High-pass filtering at 1 Hz (middle row) leads to the reduction of the noise amplitude to 27 μV. A limiting from 2 Hz (lower row) reduces the amplitude to 18 μV. Just by shifting the lower cutoff frequency from 0 to 2 Hz, the noise voltage could be almost halved but at the expense of the information content of the biosignal

classes becomes interesting. Here, the waveform is no longer critical, so in terms of evaluation, the sequence of APs of a course can be represented by a simple binary sequence. For this, a nonlinear binarization method is used since, from here on, only the time points of the individual APs are essential.

2.2.2 Differential Amplifier

As stated in the previous chapter, the task of a medical amplifier is to amplify the potential differences produced by biosignal sources in the body while common-mode signals are to be suppressed. The differential amplifier is particularly well suited for this task. The simplest version of a differential amplifier is shown in Fig. 2.26. It consists of two transistor stages coupled together at the emitter/source.

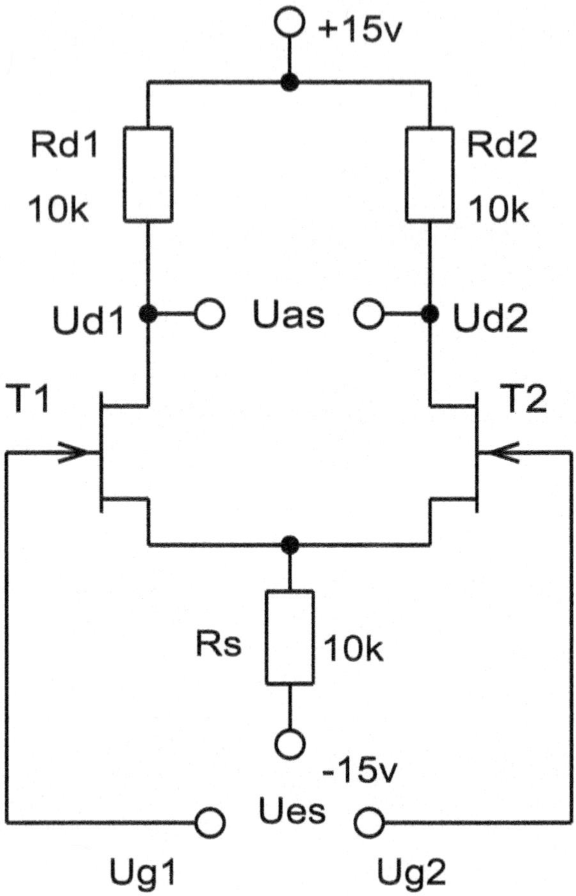

Fig. 2.26 Differential amplifier with SFET. $U_{g1,2}$ and $U_{d1,2}$ are ground-related, and U_{es} and U_{as} are ground-free as differential voltages

Compared to a simple transistor stage, the differential amplifier has several notable features, which will be discussed in detail here.

The common mode voltage is a voltage related to the ground, which is applied equally at two or more considered points of the circuit (with identical levels, in the case of AC voltages with identical time average, amplitude, frequency, and phase). Mathematically, the common mode, according to Eq. (2.19), is the instantaneous arithmetic mean of the considered voltages u_1 (t) and u_2 (t). It is valid for the input as well as for the output of a differential amplifier.

$$u_G(t) = \frac{u_1(t) + u_2(t)}{2} \tag{2.19}$$

A differential voltage is formed from the difference between two ground-related voltages or, after amplification, from another differential voltage. The mathematical formulation (Eq. 2.20) was adapted to the considered differential amplifier in

Fig. 2.26 (V_d is the differential gain).

$$u_{es}(t) = u_{g1}(t) - u_{g2}(t),\ u_{as}(t) = u_{d1}(t) - u_{d2}(t) = V_d \cdot u_{es}(t) \tag{2.20}$$

In the practical measuring technique, both voltage forms always occur mixed. Pure common mode and difference voltages are only used in the theoretical analysis. First, the differential amplifier is analyzed in the *pure differential drive* mode, as shown in Fig. 2.27. The starting point of the analysis is that the amplifier (identical transistors and drain resistors) and the differential signal source ud coupling are ideally balanced (resistors $R_{i1,2}$ are equal). The resistors $R_{i1,2}$ are necessary for the circuitry so that the gates of the transistors get a defined potential, most simply the ground potential. At the same time, they serve to arrange the signal source u_d symmetrically around the ground potential and, in this way, enforce the ground symmetry necessary for pure differential control. Furthermore, the following applies to the designation of voltages: one connection-related index (e.g., u_{g1}) designates a ground-related voltage, and two connection-related indices (e.g., u_{gs1}) designate the differential voltage between the respective connections.

For the derivation of the operating parameters, the differential amplifier is simulated with a linearized model (Fig. 2.29). The starting point of the observation is the signal current produced by the source u_d. One current path passes through resistors R_{i1} and R_{i2}, which, concerning this model, are intended to ensure that the intrinsically floating source voltage u_d is symmetrical about the ground (note: the electrical ground is not an ultimate signal sink here; from the perspective of the biosignal source u_d, it merely forms a connection between resistors R_{i1} and R_{i2}). The second signal path from and back to u_d passes through resistors R_{gs1} and R_{gs2}. Since they are equal in magnitude, the (ground-related) voltage u_s at the sources is zero. The resistor R_s plays no role in the differential control mode since the signal current does not flow through it. One can consider the node u_s as a virtual ground in the differential drive mode. The following, therefore, applies to the drain voltages

$$u_{d1} = -u_{gs1} \cdot S \cdot R_{d1}\ u_{d2} = -u_{gs2} \cdot S \cdot R_{d2}. \tag{2.21}$$

Equation 2.21 S is the slope (quantity initially from tube technology), and here it is the transmission of a FET (in mA/V). It indicates how strongly an input voltage influences the drain current.

Since the input voltages of the transistor stages are symmetrical about the ground potential, the following applies

$$u_{es} = u_d = u_{g1} - u_{g2} \Rightarrow u_{g1} = \frac{u_d}{2}, u_{g2} = -\frac{u_d}{2} \tag{2.22}$$

and with perfect symmetry for the output voltage

$$u_{as} = u_{d1} - u_{d2} = -S \cdot R_d \cdot u_{es}, \tag{2.23}$$

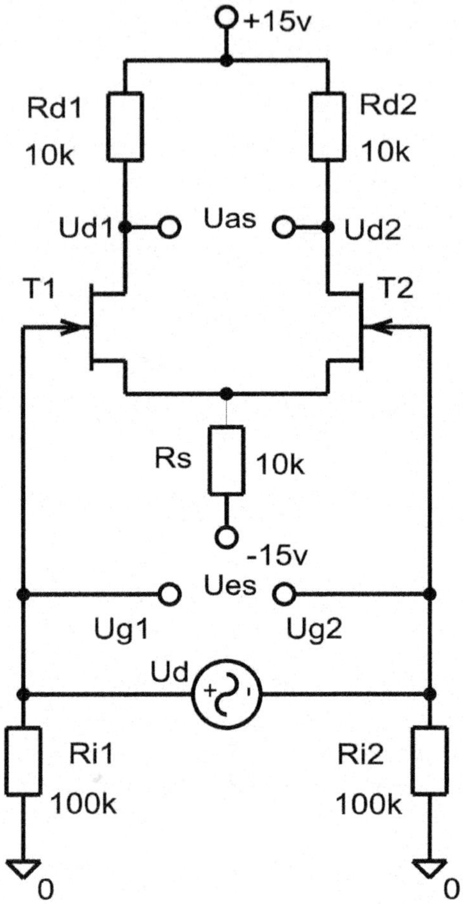

Fig. 2.27 A discrete differential amplifier built with transistors in the "pure differential control" operating mode. The resistors R_{i1} and R_{i2} are theoretically not necessary, but practically the gates of the transistors must have a defined DC voltage level via these resistors. The two resistors also force a ground-balanced input voltage for the true differential drive (PSpice simulation BSV_2.5)

from which the differential gain becomes

$$V_d = -S \cdot R_d \tag{2.24}$$

can be estimated ($R_d = R_{d1} = R_{d2}$). It should be noted that (2.21) only provides information about the AC voltage at the drains. The common mode voltage or DC component amount is not considered here (see Fig. 2.28.). Just as important as the intentional amplification of the differential voltage at the input of the differential amplifier is its ability to ignore the unwanted common mode voltage. The simulated circuit is shown in Fig. 2.30.

Due to the complete symmetry, the source resistor can be virtually divided into two equal, parallel, double-sized resistors, resulting in two factually independent, identical stages (Figs. 2.30 and 2.31). Therefore, to calculate the common mode gain, it is sufficient to consider only one of the two stages. The mesh equation of the input and the source resistor R_s shows that the source voltage is approximately

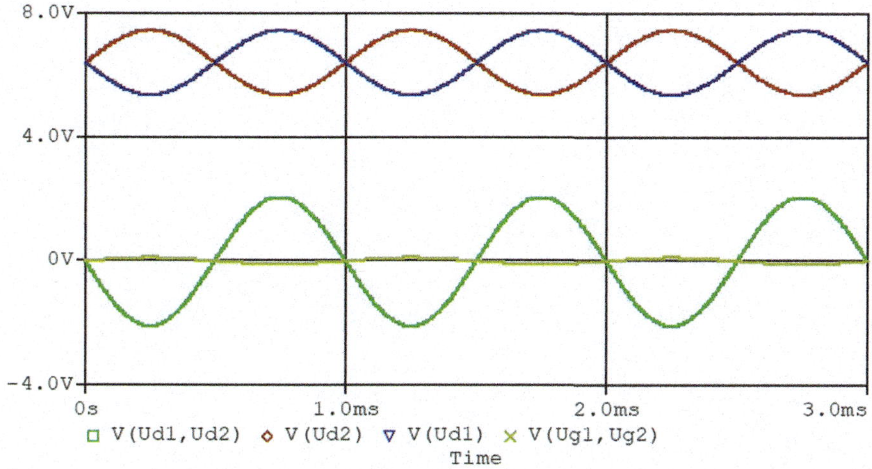

Fig. 2.28 The differential input voltage $u_d = u_{es}$ (yellow) applied to the input is ground balanced (forced by identical resistors R_{i1} and R_{i2}, Fig. 2.27). Particular attention should be paid to the location of the output voltages of the transistor stages: The ground-referenced drain voltages U_{d1} *and* U_{d2} (blue and red) have identical dc levels (time average), which are identical to the common mode ($U_G = 6.4$ V). The AC voltage components are of the same amplitude but in phase opposition, i.e., with the sign of the AC component reversed. The ground-free symmetrical output voltage U_{as} (green) results from the difference between the ground-related drain voltages U_{d1} and U_{d2} and is ground symmetrical

equal to the gate voltage (source follower, voltage follower principle), according to Eq. (2.25).

$$u_{g1} = u_{gs1} + u_s = u_{gs1}(1 + SR_{s1}) \approx u_{gs1} \cdot SR_{s1} \tag{2.25}$$

From the relationship for drain tension

$$u_{d1} = -Su_{gs1}R_{d1}, \tag{2.26}$$

results from (2.25) and (2.26) for the stress amplification at the drain

$$V_g \approx \frac{R_d}{2R_s}. \tag{2.27}$$

Both stages are identical in an ideally balanced amplifier, so the output differential voltage u_{as} would be zero independent of the input common-mode voltage. In practical measurement technology, however, both drive modes are always mixed. In addition, circuit technology cannot provide an ideally balanced differential amplifier. In the following, the differential amplifier in the mixed drive mode with the unbalanced coupling of signals and interference shall be considered realistically.

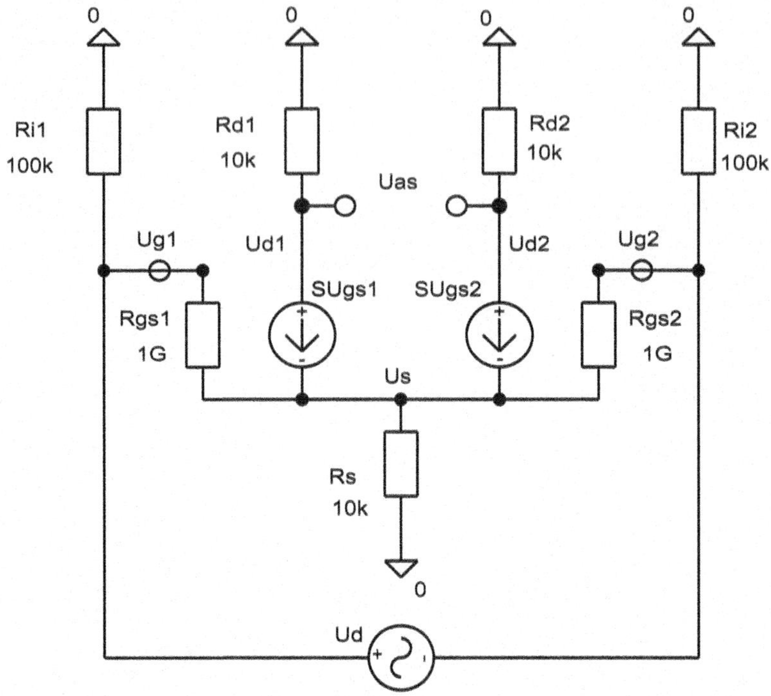

Fig. 2.29 .

A balanced differential amplifier with mixed drive and balanced common mode injection is shown in Fig. 2.32. Under these ideal conditions, the output differential voltage concerning the common mode is zero, as expected. The common-mode gain is approximately one, so the common-mode voltage from the input appears approximately the same as the ground-referenced drain voltage (Fig. 2.33). For the effective suppression of the common mode in the ground-referenced output, the source resistance would have to be increased significantly (dynamically, with a DC source) (Eq. 2.27).

For further considerations, an ideally symmetrical differential amplifier is assumed (Fig. 2.34). The effects of asymmetrically coupled common-mode voltages, always present under actual conditions, form the basis of the considerations. A common-mode voltage (e.g., the electric field of the mains voltage) acts on the measuring circuit from the outside—usually capacitively coupled in—according to its mechanism of origin. Resistors $R_{g1,\,2,3,4}$ simulate real impedance conditions when coupled with electrodes, and resistors $R_{i1,2}$ the internal resistance of the biosignal source. Note that resistor R_{g2} is 40% larger than resistor R_{g1}. This imbalance causes the common-mode voltage u_g to be coupled to the inputs asymmetrically, resulting in a differential voltage between the inputs in addition to u_d as if it were in series with the signal source u_d. This disturbing differential voltage can no longer be eliminated with circuit measures.

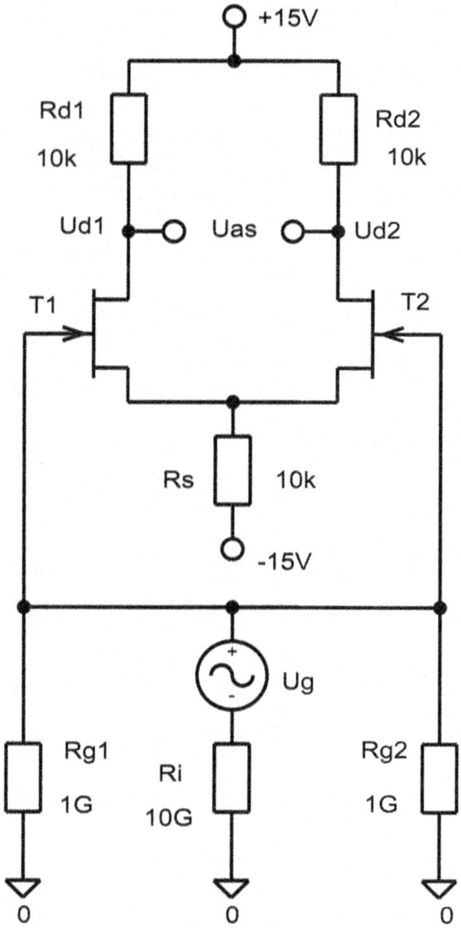

Fig. 2.30 Discrete differential amplifier in common mode operation. The resistors R_{G1} and R_{G2} are necessary to define the DC voltage for the gates of the FET. The resistor R_i of the common mode source U_g simulates natural conditions due to mains interference

The signal waveform of the output differential voltage u_{d1} -u_{d2} is shown in Fig. 2.35. In addition to the desired biosignal, the output voltage contains the amplified differential voltage of the disturbance that arose from the original common-mode voltage due to asymmetry at the input. The portion of the common-mode voltage that remains as a common mode is suppressed by the amplifier symmetry and eliminated in the output differential voltage. In practical measurement technology, one must also reckon that real differential amplifiers are never ideally balanced; even in integrated electronics, the ideal cannot be achieved. However, the circuit asymmetry causes less interference voltage than the asymmetry at the input when the common mode voltage is coupled (Exercise 2.7). This comparison leads to the realization that it is not practical to develop an ideal differential amplifier (operational amplifier) since the asymmetrical coupling of the common-mode interference at the input has a much more significant influence on the output voltage, but this cannot be reduced in terms of circuitry.

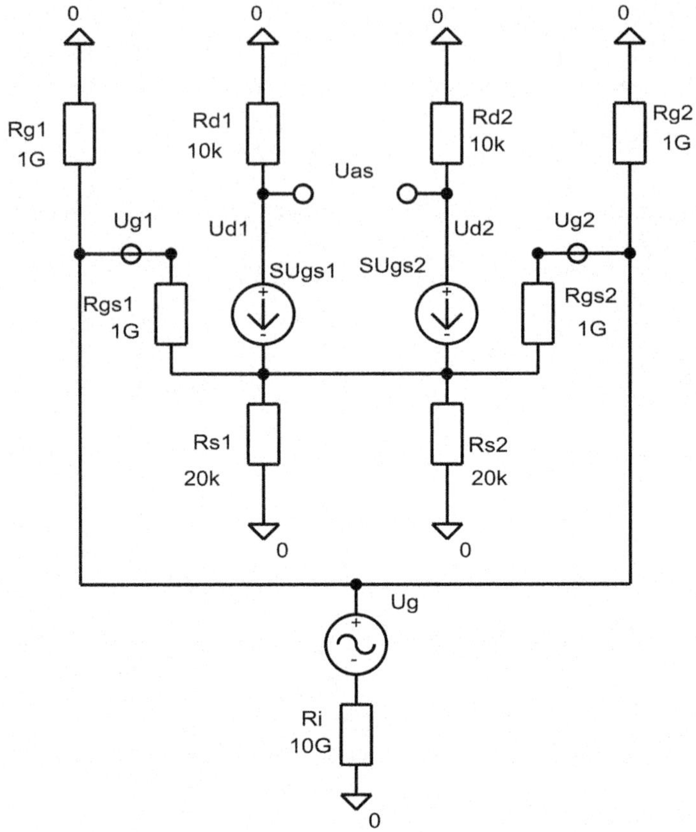

Fig. 2.31 Linearized equivalent circuit of the differential amplifier in common mode operation. Since the amplifier is symmetrical, the initial source resistor R_s can be divided into two double parallel resistors, R_{s1} and R_{s2}, without changing anything in the circuit

Besides the best possible amplifier symmetry, another circuit measure is necessary to minimize the influence of the common mode on the ground-related drain voltages: According to Eq. 2.27, it would be necessary to minimize the quotient. Theoretically, this would be possible by minimizing the drain resistance. Nevertheless, this directly influences the differential gain and is therefore out of the question as a parameter. The only alternative is to provide the source resistor with a substantial value, practically in the order of 1…10 MΩ. However, the sum of both the transistors' quiescent currents (direct current) also flows through the source resistor, which is normally in the 0.1…1 mA range. It would require a negative supply voltage in the range of Ub- $=$ RS*$(I_{S1} + I_{S2}) =$ 100 V to 10 kV to reach the necessary bias current. It is not feasible from a circuit point of view. On the other hand, a large real resistance would greatly increase the noise. The constant current source offers a solution: This has a very high (dynamic) internal resistance,

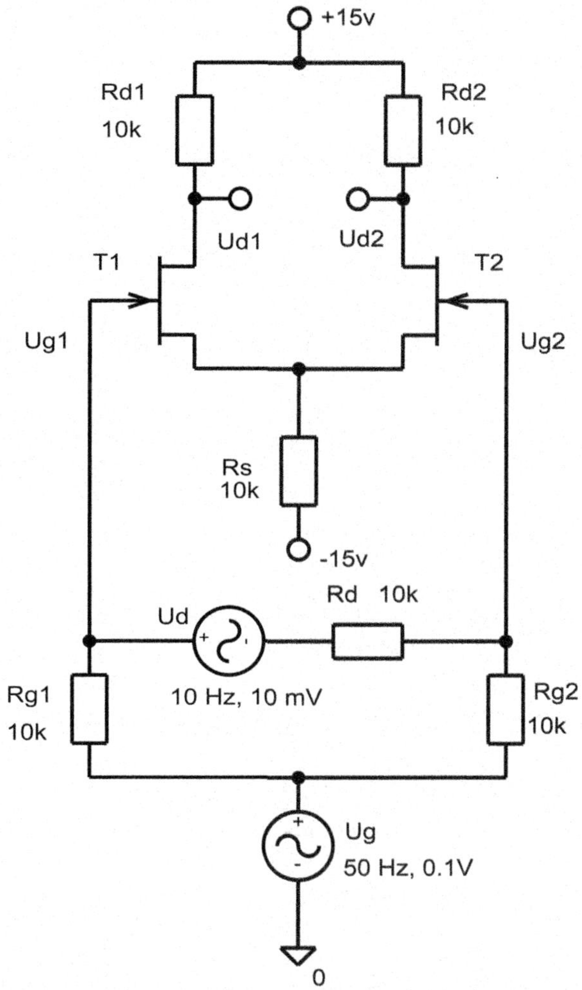

Fig. 2.32 Ideal differential amplifier with mixed (differential and common mode) control. The output differential voltage is zero, and the ground-related drain voltages show a strong common mode component (see Fig. 2.33) (PSpice BSV_ 2.20)

which is, in fact, equal to the transistor resistance R_{ds}, whereby only relatively low resistances are necessary for dimensioning (Fig. 2.36).

The differential amplifier is the basic structure used to build integrated operational amplifiers and other circuits (instrumentation amplifiers, isolation amplifiers). The integrated amplifiers still need other electronic functions, such as level shift, feedback, temperature compensation, current mirror, balanced output stages, and frequency response compensation. These stages are part of the fundamentals of analog circuit design, so the appropriate technical literature is recommended if needed. Furthermore, circuits are dealt with, which must fulfill specific medical measurement technology requirements.

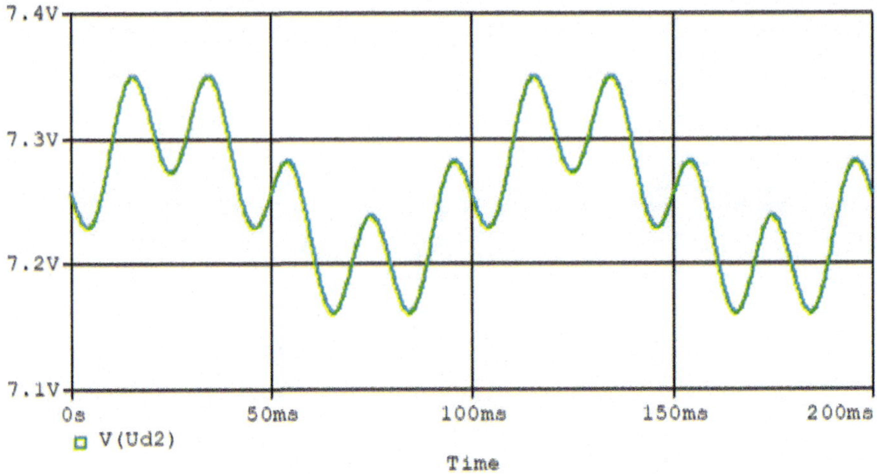

Fig. 2.33 Time characteristic of the voltage U_{d2} at the drain of T_2. The faster common-mode voltage (50 Hz) retains approximately its amplitude according to the common-mode gain given by Eq. 2.27. A factor of approximately 10 amplifies the slower differential voltage (10 Hz). This figure shows that even an ideally balanced differential amplifier cannot suppress the ground-referenced common-mode voltage from the input at its output

2.2.3 Operational Amplifier, Instrumentation Amplifier

Detecting biosignals in a disturbed environment. However, a discrete design—in addition to a multistage amplifier—is a constructional challenge, which today only makes sense for special constructions and is reserved for the research area. In conventional analog circuit technology, one makes use of a property of negative feedback systems that are well known in control engineering: If the feedforward gain of an open system (a system with no or interrupted feedback, "open loop") is very high (>100,000), the characteristics of the negative feedback system are defined by the feedback alone. It brings enormous advantages and simplification to dimensioning and realization compared to discrete circuits: One no longer must worry about the (nonlinear) parameters or selection of discrete active components. One can formulate the desired function mathematically precisely and implement it via feedback. Such a system, which on the one hand, amplifies only the desired potential difference and, on the other hand, has sufficiently high gain, is the operational amplifier (OV). Symbolically, the OV with a connected differential source is shown in Fig. 2.37. In further consideration, an ideal OV is first assumed with infinitely high amplification.

Suppose the OV would be used alone and without additional circuitry to amplify the desired differential voltage ud. In that case, it immediately reaches (in the real OV at a few microvolts, i.e., already by its noise) the limitation at the voltage supply $U_{b+,-}$. The aim is to amplify the input differential voltage with a defined and known factor. Therefore, negative feedback is introduced to achieve a defined

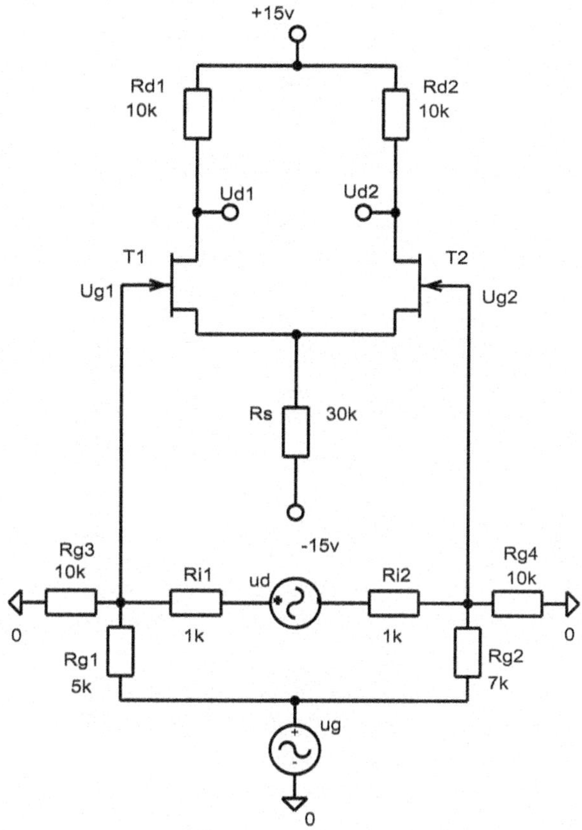

Fig. 2.34 Balanced differential amplifier with mixed control (differential and common mode) and asymmetrical coupling of the common mode (PSpice simulation BSV_2.21)

amplification level in the sense of the property, as mentioned above, of negative feedback. This approach is shown in Fig. 2.38.

When analyzing and dimensioning, it is helpful to build on the following principle: In a circuit with an ideal OV and negative feedback (connection from the output of the OV to the inverting input), the differential voltage between the two OV inputs is zero. From this, one could wrongly deduce that the output voltage of the OV should also be zero. Therefore one must proceed causally here: The input differential voltage of the OV (E_+ -$E-$) becomes zero only *after the* OV output is set to this condition. Based on this analysis principle, one can write for the circuit's output voltage according to Fig. 2.38.

$$u_d = u_+ - u_- = 0 \Rightarrow u_1 = u_2 + \frac{R_1}{R_1 + R_3}(u_a - u_2), \tag{2.28}$$

after resolution applies to the output voltage

$$u_a = \left(1 + \frac{R_3}{R_1}\right)u_1 - \frac{R_3}{R_1}u_2. \tag{2.29}$$

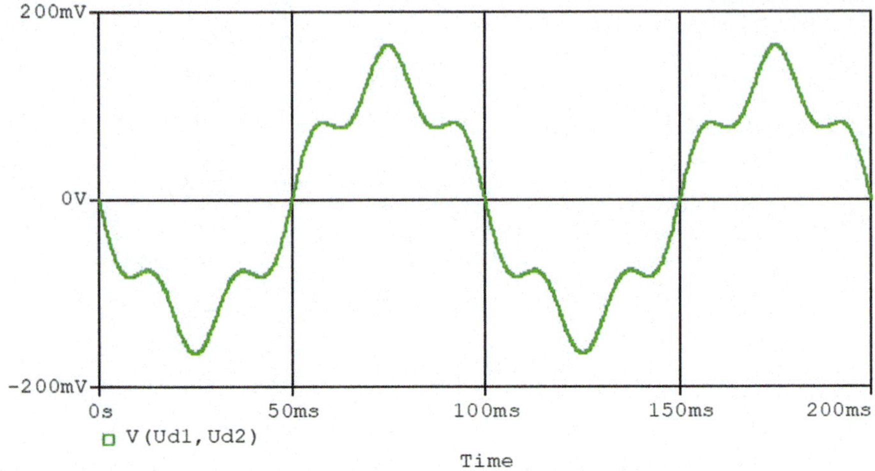

Fig. 2.35 Time characteristic of the output differential voltage of a differential amplifier with the asymmetrical coupling of a common-mode disturbance at the input. The disturbance in the output signal caused by the asymmetry is 20 mV, while the common mode voltage at the input is 100 mV strong. The symmetry eliminates the rest of the common mode voltage at the input

Now the circuit has a defined gain for both input voltages. However, the voltages do not form a real difference but have different weights. Thus, the problem of eliminating the common mode voltage at the input—remains unsolved. Therefore, the input voltage u_1 at the non-inverting input E_+ must also be given a degree of freedom, as shown in Fig. 2.39.

The following relationship applies to the output voltage according to Fig. 2.39:

$$u_+ = u_- \Rightarrow u_1 \frac{R_4}{R_2 + R_4} = u_2 + (u_a - u_2) \frac{R_1}{R_1 + R_3}. \tag{2.30}$$

After conversion from 2.30, the following is obtained for the output voltage

$$u_a = u_1 \frac{R_4(R_1 + R_3)}{R_1(R_2 + R_4)} - u_2 \frac{R_3}{R_1}. \tag{2.31}$$

According to Eq. 2.31, two degrees of freedom are related to the input voltages. These two degrees of freedom are necessary to ensure that the output voltage u_a depends only on the difference between the input voltages, not their common mode. From the condition for the common mode

$$u_1 = u_2 \Rightarrow u_a = 0 \tag{2.32}$$

substituting (2.32) into (2.31) yields

$$\frac{R_1}{R_3} = \frac{R_2}{R_4}. \tag{2.33}$$

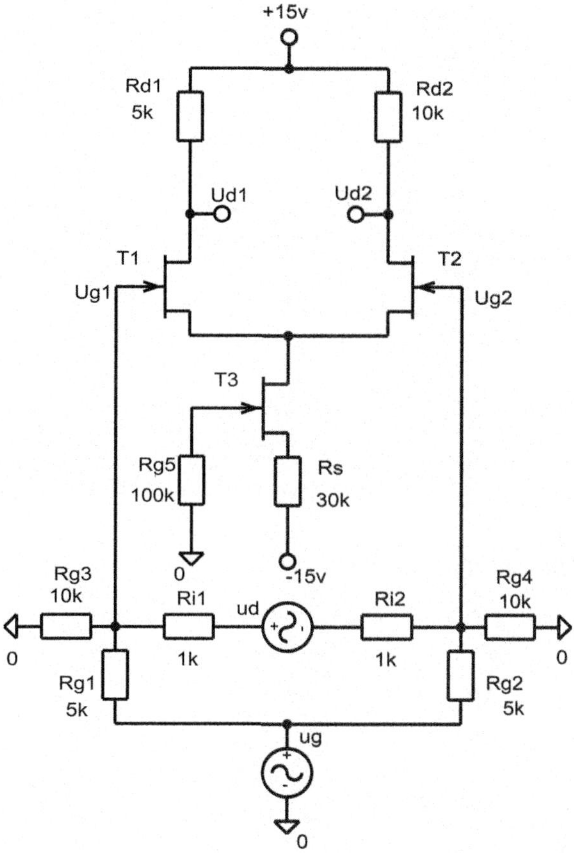

Fig. 2.36 Differential amplifier with a constant current source instead of the source resistor. The current source acts like a very high dynamic resistor on the operating parameters, which would not be realizable with a real resistor (PSpice simulation BSV_2.22)

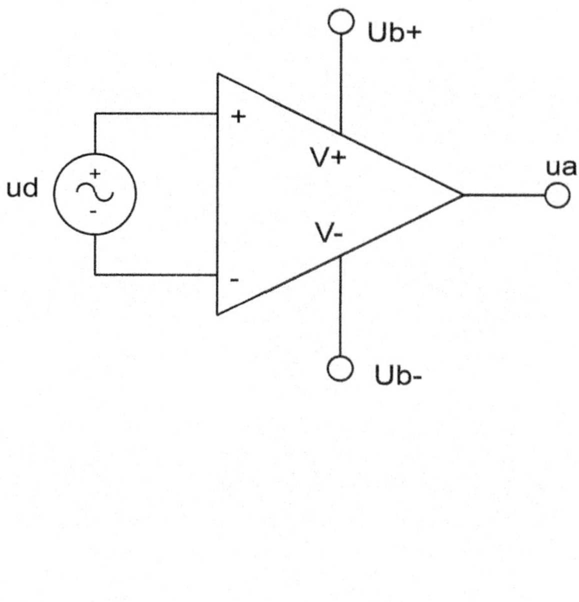

Fig. 2.37 Symbol of an OV and principle of operation: The output voltage is ground-related, $u_a = V_d.u_d$. Ideally, V_d is infinitely large. It is more significant than 10^5. The differential voltage u_d at the input of an amplifier usually results from the difference between two ground-related voltages

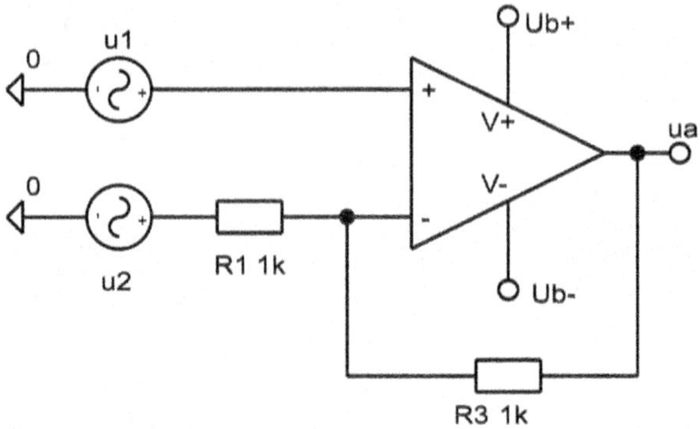

Fig. 2.38 OV with negative feedback for defined amplification of the difference between the ground-related voltages u_1 and u_2. The differential amplification of the system (OV) is theoretically infinite, so the realized amplification is given by the negative feedback alone to $(R_3 + R_1)/R_1$ (reciprocal value of the negative feedback factor). See PSpice simulation BSV_2.23

Fig. 2.39 Amplifier of the voltage difference between u_1 and u_2, where the weighting of both voltages is independently adjustable via the two voltage dividers $R_{2,4}$ and $R_{1,3}$ (PSpice simulation BSV_2.24)

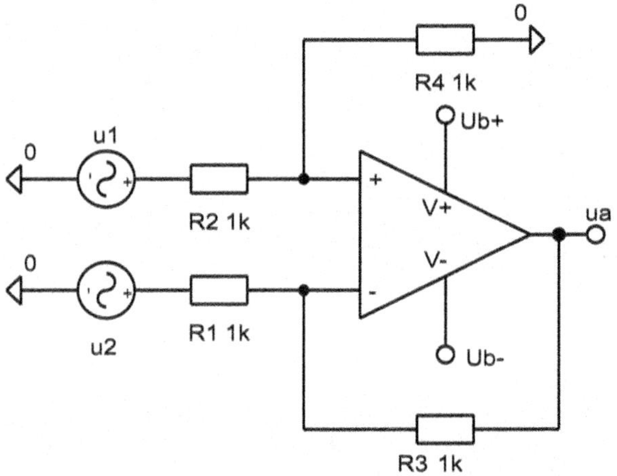

The resistors are selected for circuit reasons (offset compensation) so that $R_1 = R_2$ and $R_3 = R_4$. Thus, the following applies to the output voltage

$$u_a = (u_1 - u_2)\frac{R_3}{R_1} = (u_1 - u_2)\frac{R_4}{R_2}. \tag{2.34}$$

With (2.34) was achieved that the output voltage at the OV depends solely on the difference between the two ground-related input voltages, whereby this difference is amplified with a defined factor. The next step is to check whether the requirement for a sufficient input resistance is fulfilled (see Sect. 2.2.1.5 Input resistances of medical measuring amplifiers). Based on the circuit in Fig. 2.39, the input resistances can be estimated as follows: From the point of view of the differential voltage at the input, the differential input resistance under the assumption

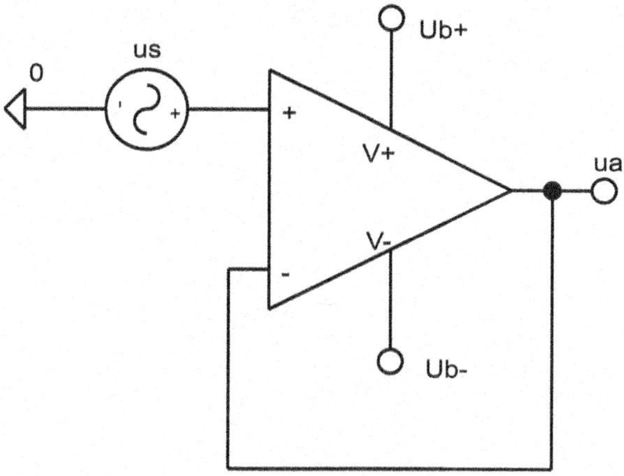

Fig. 2.40 Voltage follower with OV. Since the output sets the input differential voltage of the OV to zero, the output voltage u_a is equal to the input voltage u_s. The input resistance is theoretically infinite, practically equal to the OV's input resistance (catalog value). The output resistance is ideally zero, practically on the order of 100. This stage is also called an impedance converter (trans-impedance amplifier)

of an ideal OV is equal to the sum of the two resistors R_1 and R_2. It must always be assumed that the differential input voltage at the OV is zero, i.e., the two inputs are at the same potential as if they were virtually connected.

$$V_d \to \infty \Rightarrow u_{ed} = 0 \Rightarrow R_e = R_1 + R_2 \tag{2.35}$$

Since the real resistances are in the order of up to about 10 k, it is too little for medical measurement technology. Therefore, it is necessary to increase the input resistance significantly. Analog circuit technology offers an initially straightforward solution: the impedance converter with an OV, as shown in Fig. 2.40. The impedance converter can be connected in front of any input of the voltage differential amplifier to raise the input resistance to an acceptable level, as Fig. 2.41 shows.

In the following, the common mode rejection, *CMRR*, will be considered for the case of an ideal OV. The circuit in Fig. 2.41 is a multistage amplifier, with the two impedance converters forming the first (balanced) stage and the voltage differential amplifier forming the second stage. In a multistage amplifier, the first stage determines several essential characteristics (noise, drift) and common-mode rejection. Since the first stage in Fig. 2.41 contains impedance converters that function as voltage followers, both the common-mode and differential voltages are passed from the input to the ground-free differential output of the first stage. The first stage's common mode and differential gain are equal to one. Consequently, the *CMRR is* also one (Eq. 2.36). Concerning the requirements made in Sect. 2.2.1.6, this is entirely insufficient.

$$CMRR_1 = \frac{V_{d1}}{V_{g1}} = 1 \tag{2.36}$$

Therefore, realizing a sufficiently high differential gain already in the first stage is necessary because the common mode gain remains one. The differential gain can

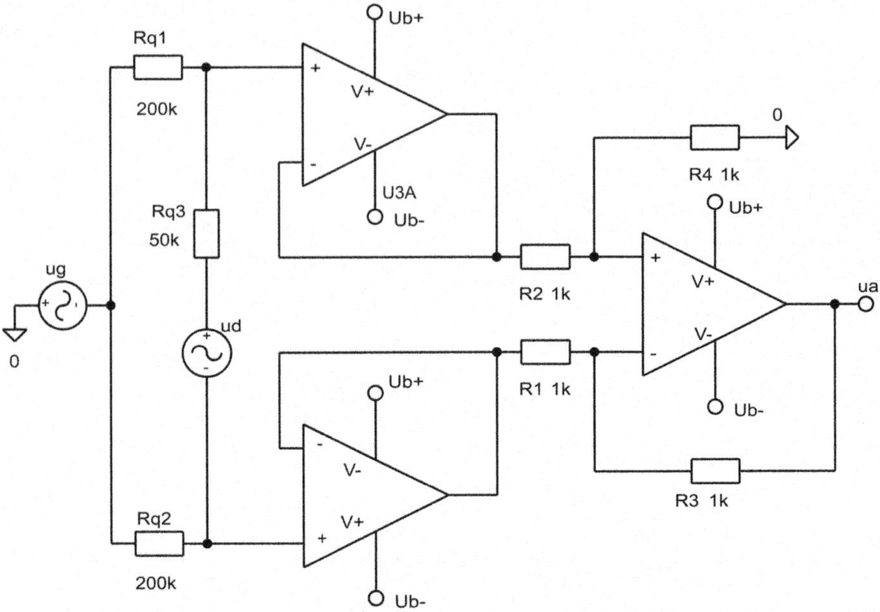

Fig. 2.41 Voltage differential amplifier with impedance transformers at the input to ensure a high input resistance. Common mode (u_g, ground referenced) and differential voltage source (u_d, floating) are simulated at the input. Resistors $R_{q1,2,3}$ simulate the internal resistances of the sources, where R_{q1} and R_{q2} are initially identical to ensure symmetrical coupling of the common-mode interference

be parameterized with a resistor network which forms a -connection between the two OV outputs and, simultaneously, is part of the negative feedback, as shown in Fig. 2.42. The differential gain can then be calculated from the cross current between the two OV outputs using Eqs. (2.37) and (2.38) to get (2.39).

$$i_Q = \frac{u_{R6}}{R_6} = \frac{u_{+IC1} - u_{+IC2}}{R_6} = \frac{u_d}{R_6} \tag{2.37}$$

$$u_{ad} = u_{aIC1} - u_{aIC2} = i_Q \cdot (R_5 + R_6 + R_7) = u_d \frac{R_5 + R_6 + R_7}{R_6} \tag{2.38}$$

$$V_d = 1 + \frac{R_5 + R_7}{R_6} \tag{2.39}$$

In Eqs. 2.37–2.39, u_{+ICi} is the voltage at the non-inverting input of OV *i*, u_{aICi} is the voltage at the output of OV *i*.

Since the first stage is still to amplify only the differential voltage, no component of the common-mode voltage caused by asymmetry may appear in the floating output voltage u_{ad}. This requirement leads to the resistor dimensioning $R_5 = R_7$.

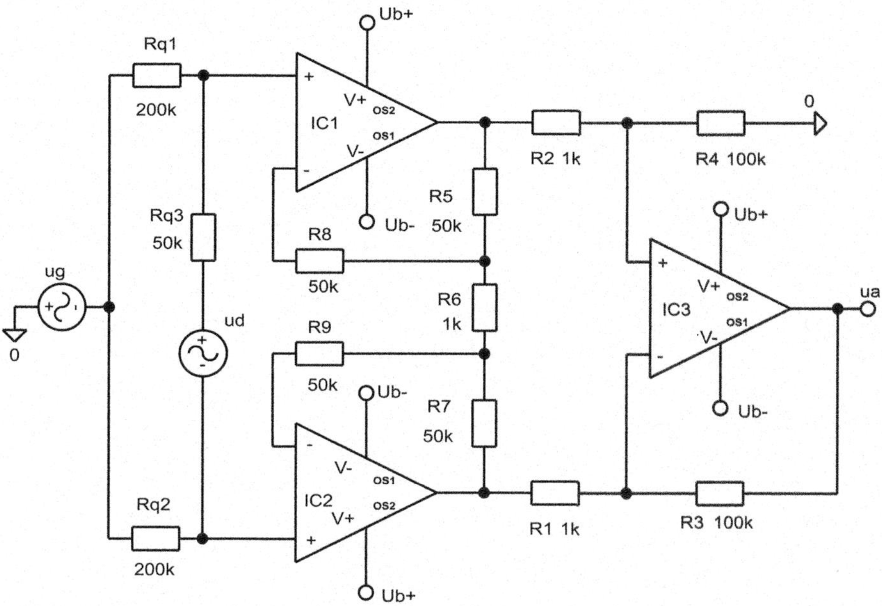

Fig. 2.42 In the first stage of the amplifier, in addition to its function as an impedance converter, defined differential gain was realized with the help of the resistors $R_{5,6,7}$. The resistors $R_{8,9}$ are functionally not necessary but valuable for offset reduction in bipolar technology. This circuit is called an instrumentation amplifier. (PSpice simulation BSV_2.27)

From Eq. (2.39) the value for the CMRR results directly of the first stage to (index 1 indicates the first stage)

$$CMRR_1 = \frac{V_{d1}}{V_{g1}} = 1 + 2\frac{R_5}{R_6}\bigg|_{R_5=R_7}, \tag{2.40}$$

which is only valid under the condition of an ideal OV. This value will be lower depending on the *CMRR* of the OVs used and the resistors' accuracy. The second stage of the instrumentation amplifier theoretically has an infinite *CMRR* since, here, the common mode gain is zero. Therefore, the stages are considered individually simulated (BSV_2.27). The desired differential voltage u_d has an amplitude of 0.8 mV (10 Hz) at the amplifier input (Fig. 2.42), and the interfering common-mode voltage u_g has an amplitude of 10 V (50 Hz), so the *SNR* (Eq. 2.17) at the input of the first stage is −82 dB. The common mode gain of the first stage is $V_{g1} = 1$. According to (Eq. 2.39), the differential gain is $V_{d1} = 101$, so mathematically, there is a $CMRR_1$ of 40 dB. With an input differential voltage of 0.8 mV (the differential voltage $u_d = 1$ mV is reduced at the voltage divider $R_{q1,2,3}$) and an output differential voltage of the first stage of 80 mV with a constant common-mode voltage of 10 V, an $SNR = -42$ dB is therefore achieved at the output of the first stage. At the output of the second stage (ground-referenced voltage u_a), a level of 9.1 V is measured at 10 Hz (desired signal), and a level of 32 mV is measured at 50 Hz (disturbance), which corresponds to an *SNR of* 49 dB. The difference in *SNR* between the output and the second stage's input is 91 dB. This difference is to be credited as the $CMRR_2$ of the second stage. It is not particularly good value; nevertheless, it is acceptable in typical cases. A total theoretical CMRR of 131 dB is achieved in the first stage. It is an appropriate value for medical measurement technology—of course, under the ideal conditions of a PSpice simulation. In practice, a maximum of 100 dB will be achieved with real components in this circuit.

The previous considerations assume ideal OVs and ideal simulation conditions. These assumptions are far from metrological reality. In the following, realistic operating parameters are assumed as a basis.

The differential gain comes very close to the ideal of infinite gain with its real value of about $V_d = 10^6$. However, this value is only sustainable in a limited spectral range. Already at the upper end of the biosignal spectrum (100–10 kHz), the gain of most OVs and instrumentation amplifiers drops dramatically (Fig. 2.43). The effect known from general electronics applies here: the bandwidth of a negative feedback amplifier increases with the strength of the negative feedback. If one assumes that amplifications of at least 1000 are required in medical measurement technology, the margin between amplification and bandwidth is small. According to the graph in Fig. 2.43, a differential gain of $V_d = 1000$ would result in a bandwidth of 0.001–1 kHz, meeting the minimum requirements for ECG and EMG. However, one must consider that here the aged OV of type 741 was analyzed to be able to use the student version of PSpice free of charge. Of course, noticeably more appropriate devices are available today (Analog Devices, National Instruments, Harris, Philips, Linear Technologies, Texas Instruments, Burr Brown). However, there will be no significant change in the bandwidth or the bandwidth amplification product.

The common mode rejection ratio CMRR varies between 60 and 120 dB, whereby components with the highest CMRR should be selected for medical instrumentation. Instrumentation amplifiers should achieve a *CMRR of* 120 dB. In design and construction, it should be noted that even with excellent OV, the *CMRR* decreases relatively rapidly with frequency because of the drop-off in differential gain (Fig. 2.44). It is mainly due to the unavoidable parasitic capacitances, which lower the large resistances required for high Vd and CMRR with increasing frequency due to their decreasing impedance. At relatively high frequencies of common-mode interference (e.g., 100 kHz for DC–DC converters), the *CMRR* already reaches shallow values (less than 40 dB). It can lead to the fact that in a poorly constructed measuring amplifier, or a measuring circuit with high-frequency disturbances from the environment, the higher frequencies become a greater problem than the otherwise usual mains disturbances. The variation of the common mode gain of the OV from Fig. 2.43 is shown as a function of frequency in Fig. 2.44. It can be seen that regardless of the differential gain, the drop starts at about 10 Hz but with a smaller increase than for the differential gain. When comparing the frequency-dependent curves in Figs. 2.43 and 2.44 can be seen that

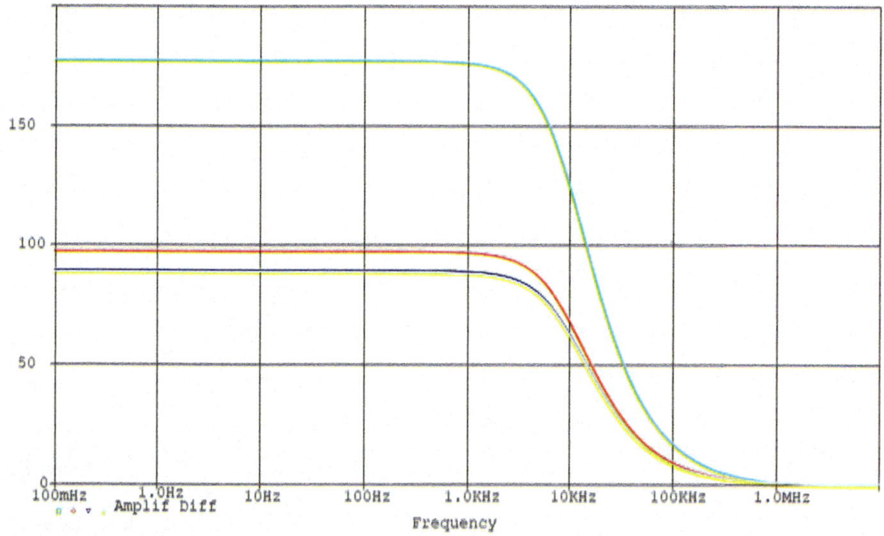

Fig. 2.43 Course of the differential gain of an OV as a function of the frequency and of the gain set by the negative feedback as a parameter: From top to bottom $V_d = 1000$, 101, 11, 2 (PSpice simulation BSV_2.28)

while the drop in the common-mode gain already starts at 1 Hz, the differential gain at $V_d = 1000$ only starts at about 1 kHz, but it is much steeper. The CMRR, as a quotient of the differential and common-mode gain, reaches its maximum in the range between about 1 and 1 kHz, then drops rapidly from the kink in the differential gain. This behavior is favorable for the expected common-mode interference frequency range (networks, tube 1 monitors, streetcars, airplanes) but insufficient for high-frequency interference above 10 kHz.

2.2.4 Isolation Amplifier

The isolation amplifier is a unique instrumentation amplifier whose input and output circuits and their power supplies are galvanically isolated, as shown in Fig. 2.45. The galvanic isolation in the isolation amplifier complicates the circuitry enormously since modulated AC voltages are required for both the power supply and the signal transmission. The circuit analysis in Fig. 2.45 explains why the additional transformer coupling for the galvanic isolation in the isolation amplifier is necessary since the amplifier is already galvanically isolated from the mains through the mains transformer. The measurement situation is shown in Fig. 2.46.

From a medical measurement technology perspective, the galvanic isolation in the isolation amplifier does not improve the measurement situation. On the contrary, the mains interference can increase due to the missing ground/earth connection. There is a relatively large stray capacitance C_{Trenn} (large common area

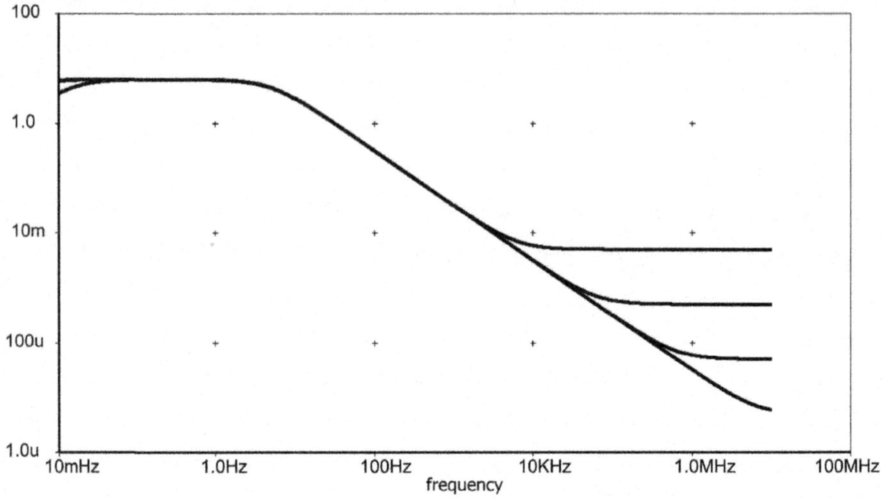

Fig. 2.44 Course of the common-mode gain V_g as a function of frequency and the differential gain V_d as a parameter of the same OV whose differential gain is shown in Fig. 2.42. From top to bottom, $V_d = 2, 11, 101, 1000$

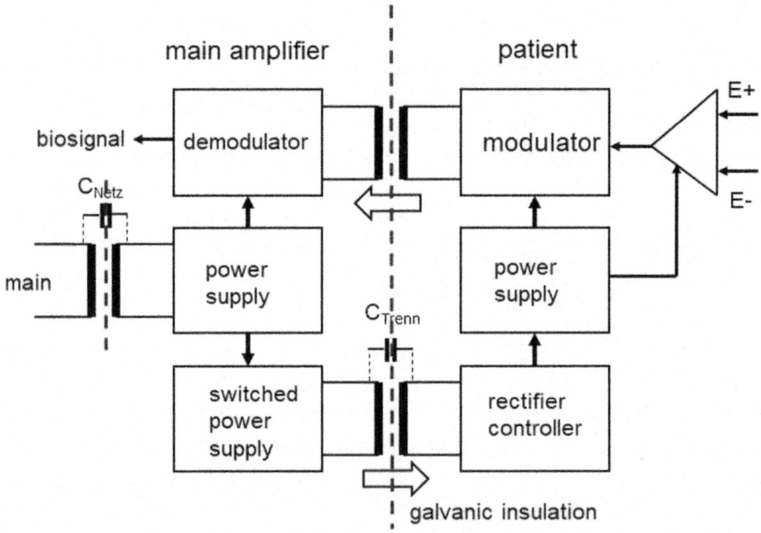

Fig. 2.45 Galvanic isolation in an isolation amplifier. Both the signal path and the power supply of the patient section are galvanically isolated from the mains. C_{Netz} represents the parasitic capacitance between the windings of the mains transformer C_{Trenn} and the transformer's windings in the switched-mode power supply

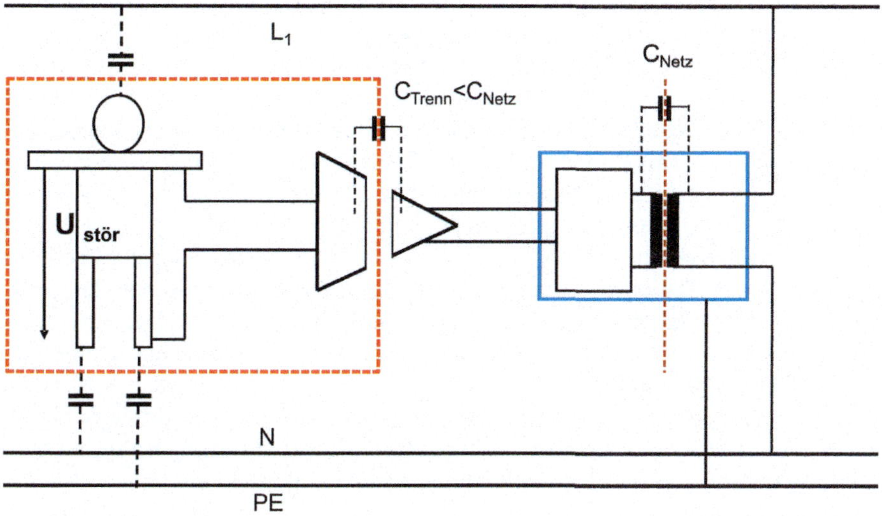

Fig. 2.46 Galvanic isolation and its effect on the measuring circuit: The patient is galvanically isolated from the mains and the earth. The capacitively coupled interference from the mains is still present. Due to the high-frequency energy transfer and smaller transformer dimensions, the AC connection to the mains via C_{Netz} has been reduced in the galvanic isolation of the isolation amplifier to C_{Trenn}

between the primary and secondary windings) between the primary and secondary windings of the mains transformer, which provides an AC connection between the mains and the measuring amplifier. The current from the mains can flow to earth through the stray capacitance of the windings and the patient and is potentially dangerous (in the cardio range, 10 A is already considered the upper limit). Since the switched-mode power supply operates around 100 kHz, its dimensions are smaller than the mains transformers.

Consequently, its parasitic capacitance is also significantly lower; finally, the patient is currently caused by it. For this reason, the galvanic isolation in the isolation amplifier is particularly relevant to safety; for measurement technology, it generally does not represent an improvement. Let us assume that the additional expenses for the isolation amplifier are considered in these methodological considerations. Notably, they are several-fold greater than those of conventional instrumentation amplifiers. A suitable isolation transformer instead of the usual mains transformer could be the more favorable alternative.

2.2.5 Guarding Technology

The interference acting on the measuring circuit from the environment can be reduced to some extent by shielding signal lines (shielded measuring cables). Especially low-frequency capacitively coupled and high-frequency electromagnetically coupled fields are effectively suppressed by shielding the signal lines. In this way, however, only the interference acting on the *signal line is* reduced, not the interference acting on the patient. Since the measurement cables are usually around one to two meters long and the area enclosed by the shielding above the signal line is relatively large, the shielding and the inner conductor form a capacitance of up to 100 pF. This capacitance is in parallel with the common mode input impedance, so it is reduced as the frequency increases (At a frequency of 100 Hz, the impedance of the capacitances $C_{1,2}$ in Fig. 2.47 is only 32 M). It deteriorates one of the most critical operating parameters of the measuring amplifier—the input resistance. Therefore, circuit measures are necessary to eliminate the reducing influence of the shielding on the input impedance. The cable capacitance is always physically present, and its reduction is impossible by design. However, it is possible to minimize the effect of capacitance. By flipping the grounded end of the capacitance to the common mode voltage u_g, only half the differential voltage remains across the capacitance. The capacitance would therefore appear as an infinitely high impedance for the common mode because no common mode current flows through it. The circuit realization of this so-called *guarding principle is* shown in Fig. 2.48.

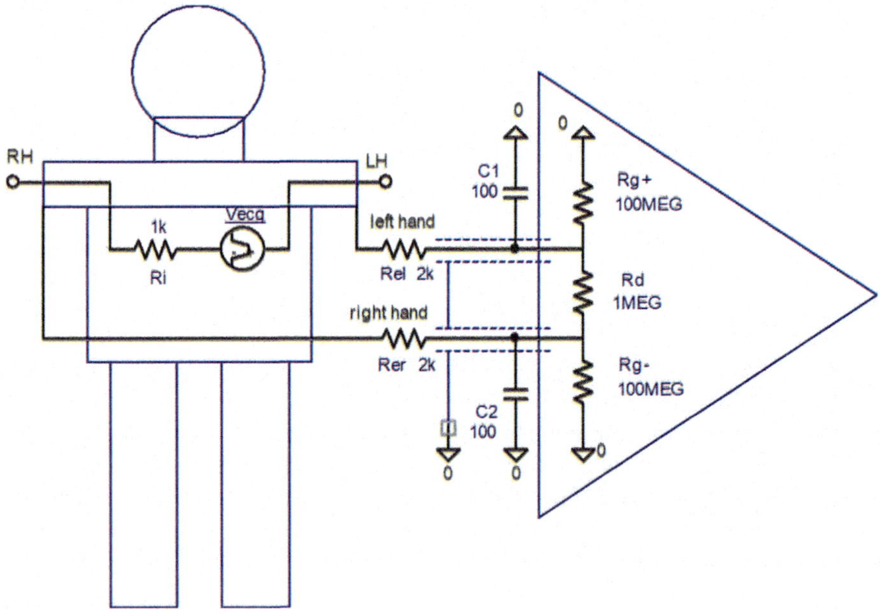

Fig. 2.47 Single-channel measurement arrangement for ECG acquisition. The connecting cables are shielded, and the shielding is grounded (shown dashed around the signal lines). The shielding with the signal lines forms the capacitances $C_{1,2}$, parallel to the input resistance $R_{g+,-}$ and reduces it (PSpice simulation BSV_2.31)

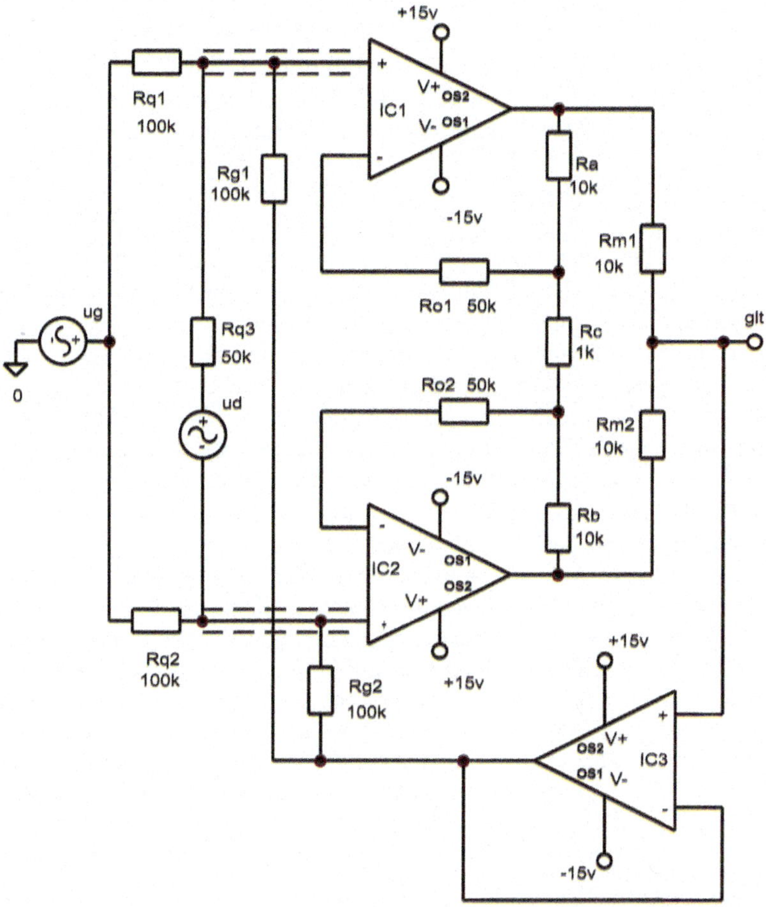

Fig. 2.48 Guarding technique with common mode signal. Resistors $R_{g1,2}$ simulate the impedance of the cable capacitance. The common-mode signal is obtained circuit-wise with $R_{m1,2}$ and fed to the cable shield via the impedance converter IC3. It ideally produces a zero voltage difference across the cable capacitance. The common mode voltage source does not "see" the capacitance. (PSpice simulation BSV_2.30)

Figure 2.48 shows the first stage of an instrumentation amplifier in which the ground-referenced common-mode signal is obtained with resistors R_{m1} and R_{m2}. The common-mode signal is fed to the cable shields (shown dashed around the signal lines) via the impedance converter IC3 for decoupling. The impedance converter must not have a gain higher than one because it forms *positive feedback,* and this would lead to instability or oscillations. Since the signal line and the cable shield carry the same common mode voltage, the impedance of the shield capacitance appears infinitely high from the input. In contrast, the capacitance is still physically present. This principle of dynamic boosting of impedance has been known for a long time and, in the early stages of analog circuitry development, was called *bootstrap.*

At this point, whether the shield also fulfills its initially intended protection function against interference coupling arises. In the conventional application, according to Fig. 2.47, the shield detects the interference and diverts it to the ground. In the guarding technique (Fig. 2.48), the shield is located at the output of an OV (IC3) and is therefore connected to the ground with low impedance via the output resistance of the OV (R_a by about 100). From the point of view of the interference coupling, the *guarding* has not changed anything essential. However, guarding preserved the common mode input resistance of the circuit, and the shield does not reduce it. Since the shield is not connected to the ground now, it could become a problem in terms of measurement and safety. Therefore, a second shield, grounded or connected to the ground, is placed around the cable and its guarding shield. Of course, a capacitance forms again between the two shields. However, this does not influence the signal line because the first shield protects it. In the circuit realization of the guarding technique, according to Fig. 2.48, it must be noted that the cable capacitance continues to act on the differential voltage. Since its impedance is frequency dependent, it will negatively affect the amplitude-frequency response of the measurement system. One accepts that the amplitude of the biosignal will decrease towards higher frequencies because the impedance decreasing with increasing frequency will reduce the differential input voltage. If high demands are placed on the amplitude-frequency response of the entire measurement chain, they must also be considered at this point. Here, the circuit variant shown in Fig. 2.49 would be preferable.

Each measurement channel can have its guard for particularly demanding measurement amplifier technology, as shown in Fig. 2.49. With this circuit alternative; the cable impedances are raised dynamically concerning the common mode and the differential voltage. However, the circuit complexity increases enormously in this variant for multi-channel measuring systems. Therefore, this effort is only justified in measurement systems that require high accuracy and constant amplitude frequency response over the entire measurement chain.

One option would be to attach the first amplifier stage directly to the sensor, thus eliminating the need for a guard. This variant will be discussed in the next chapter.

2.2.6 Active Electrodes

This circuit solution is based on a measuring amplifier having a low-impedance output. In contrast to high-impedance inputs, the output is insensitive to interference since this is conducted to ground via the low output resistance. In this case, the first amplifier stage is relocated directly to the sensor, so to speak, so that the signal connection is no longer disturbed to the same extent as was the case with the previous alternatives. Such an arrangement is shown schematically in Fig. 2.50.

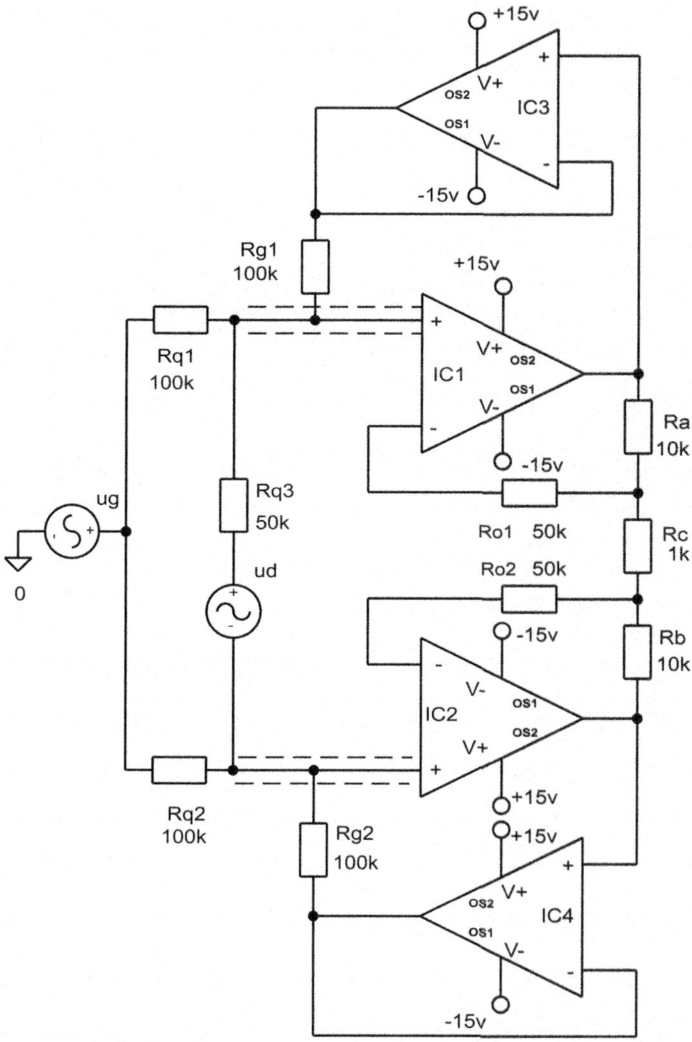

Fig. 2.49 Guarding technique protects each channel individually. The cable impedances are raised dynamically for both the common mode and the differential mode, so neither the common mode nor the differential voltage source "see" the cable impedance. This solution suits sensitive, accurate measurement amplifiers with high amplitude frequency response requirements. (PSpice simulation BSV_2.32)

Typically a common electrode forms the active electrode, and an OV or instrumentation amplifier is usually encapsulated in a compact housing to ensure biocompatibility. Although this solves the cable shielding problem, the main amplifier's input resistance requirements are no longer as high because of the upstream sensor amplifier. On the other hand, new methodological and circuit problems

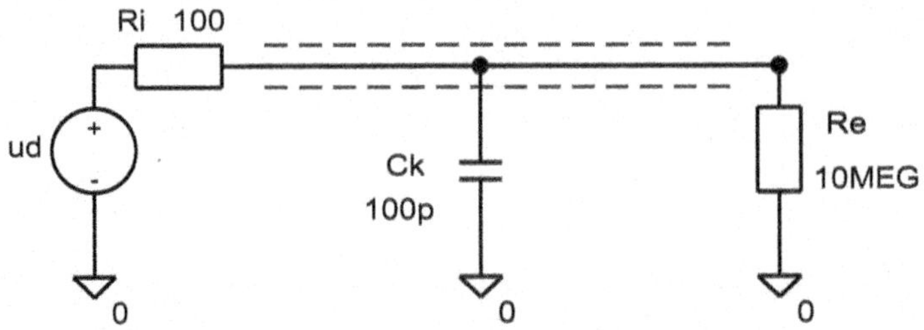

Fig. 2.50 Schematic representation of an active electrode: The signal source u_d represents the amplifier directly attached to the sensor. The cable capacitance C_k, the two resistors R_i (output resistor of the amplifier) and R_e (input resistor of the main amplifier) form a low-pass filter with a cut-off frequency of about 16 MHz, which is sufficient for biosignals. (PSpice simulation BSV_2.33)

arise: a passive electrode is electrically a single connection point on the body, and the potential difference between two or more electrodes is only formed in the measuring amplifier. An active electrode, however, requires either the electrical ground of the main amplifier (see Fig. 2.50) or, as a differential signal source, another connection on the body (see Fig. 2.47), including supply voltage. It means that in addition to the supply voltage, each active electrode must be supplied with a reference potential, whether that of the electrical ground or an adjacent electrode. It is a problem technologically and electronically. This problem can be simplified considerably if active electrodes are only designed as voltage followers (Fig. 2.51).

A circuit-technically and technologically demanding, but from the point of view of signal processing, a perfect and robust solution results from the well-known fact that digital or binary signals have the most significant resistance to interference. Therefore, it is practical to mount the amplifier and the AD converter on the electrode, which is necessary regardless. This solution already transmits digital data from the active electrode to the central unit.

2.3 Analog Filters

Even today, analog electronics and filtering are indispensable for signal quality as key components of the processing chain, despite the constantly evolving digitalization. The theory and practice of analog filtering are over a century old, and the theoretical foundations, mathematical apparatus, and design methods are well-worked and widely available in the literature. Significant contributions to analog filtering in recent years have come mainly from the technological field of analog integrated electronics and microtechnology, such as SC filters (Switched Capacitors), SAW filters (Surface Acoustic Wave Filters), or programmable filters. The following section will deal with the specific problems of biosignal processing in the medical field and the corresponding necessary theoretical foundations.

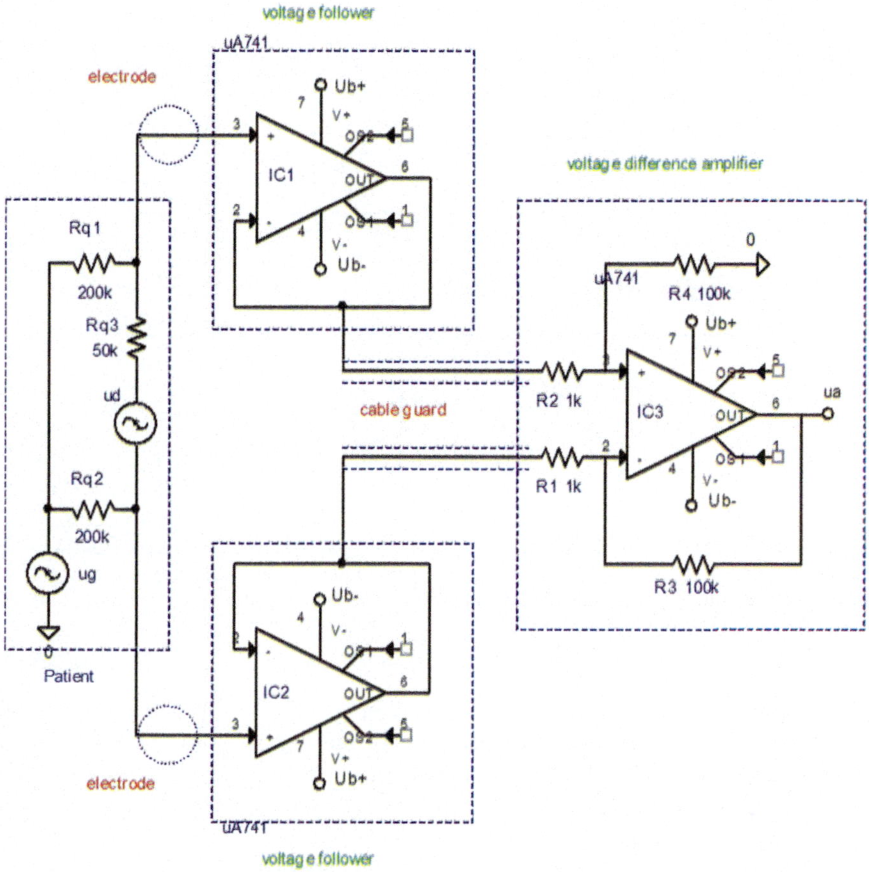

Fig. 2.51 Schematic diagram of a measurement amplifier with upstream input section as active electrodes (IC1 and IC2). When constructed as voltage followers, the active electrodes' necessary external circuitry and wiring include only an OV, two power supply lines, a ground line, and a signal line to the main amplifier. The inputs of the voltage followers are directly connected to the electrodes. (PSpice simulation BSV_2.34)

2.3.1 Basics

In analog electronics, the term filter refers to a frequency-selective circuit that attenuates unwanted frequency ranges and passes or emphasizes desired frequency ranges. Analog filters realize four essential functions of spectral filtering: low-pass, high-pass, band-pass, and band-cut (Fig. 2.52). Ideal filter characteristics are practically impossible to realize. Therefore, the first step is to define the limits (tolerance ranges) that a desired filter must satisfy (Fig. 2.54). Analyzing the filter characteristics in Fig. 2.52, we find that the bandpass and the bandstop do not necessarily have to be defined as separate filters. Both can be constructed by a

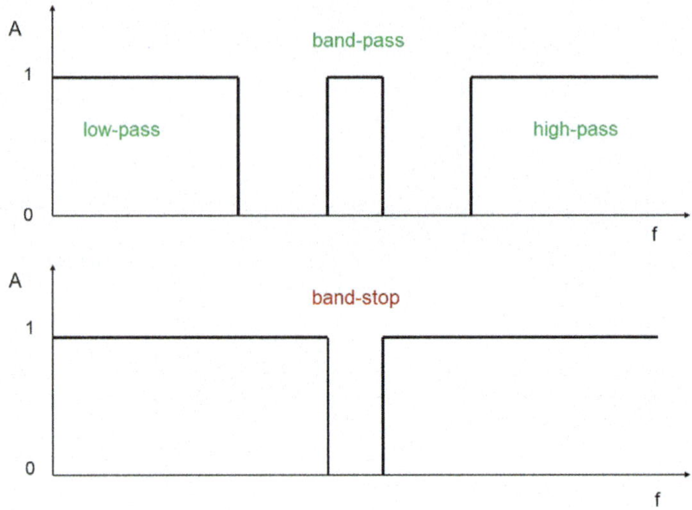

Fig. 2.52 Characteristics of ideal frequency filters: low pass, high pass, band pass, band stop. The transfer function A allows the desired frequency range to pass through unchanged (transfer factor equal to 1) and completely blocks the unwanted frequency range (transfer factor equal to 0). The phase frequency response is not changed

suitable combination of a lowpass and a highpass: the bandpass by overlapping the characteristics in a series connection and the bandstop by non-overlapping addition of the characteristics in a parallel connection (Fig. 2.53).

It is technically implemented in practice because construction is much easier this way. The ideal characteristics of analog filters shown in Fig. 2.52 are not realizable. Therefore, the limits (tolerance ranges) are first given, and feasible alternatives are investigated. A typical tolerance scheme for a lowpass filter is shown in Fig. 2.54. The lowpass may have a ripple in the passband of $A_{pass} = 3$ dB, and in the range between 250 and 350 Hz, its transmission should drop to at least $A_{stop} = -80$ dB. If no additional requirements were made, this lowpass could be realized with a 6th-order elliptic filter (filter order determines the steepness of filter edges), as shown (see Sect. 2.3.2).

Two filter classes are technologically distinguished in analog filtering: passive and active. The classification criterion is, therefore, whether the filter contains an active component (transistor, OV) or only passive elements. Passive filters are advantageous because they require nothing else except the components determining the transfer function. Since no active element exists, they are free from errors due to power supply or temperature fluctuations, semiconductor noise, and errors due to overdrive or limiting. In medical instrumentation, filters of at least 4th order are usually required (see Fig. 2.54). Since a passive network of 4th and higher order can become a very complicated circuit, its dimensioning and especially its realization are enormously costly. In addition, inductors, such as biosignals, are out of the question in a low-frequency range due to their constructive size. Passive

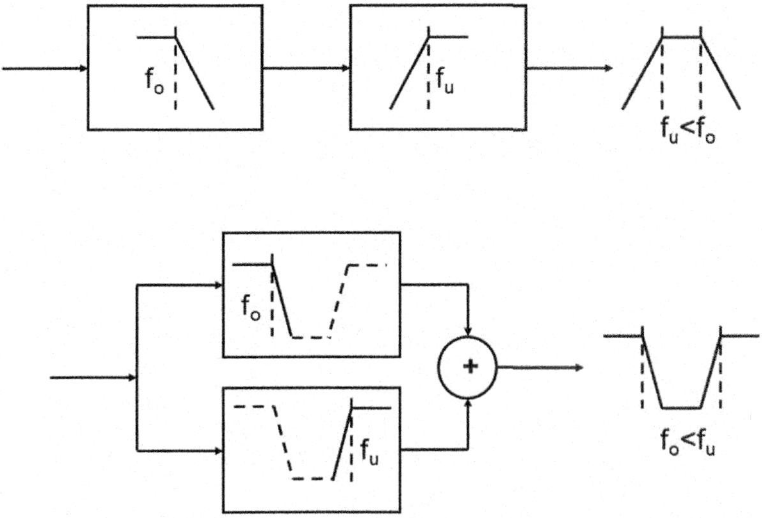

Fig. 2.53 Construction of bandpass (top) and bandstop (bottom) by combining a lowpass and a highpass. For the bandpass, the transfer functions are multiplied by the series connection. For the bandstop, the transfer functions are added by the parallel connection. From the point of view of the transfer function of the bandpass through the series connection, the order of the filters is unimportant. From a practical point of view, however, the outlined order—first low-pass, then high-pass—is more favorable. The interferences from the environment are broadband and far above the biosignal spectrum, so they should be suppressed as much as possible in the processing chain

biosignal filters can only be realized as RC networks (including new technologies), whereby the number of RC combinations doubles compared to theoretically necessary LC stages (one LC stage is a 2nd order system, and one RC stage is a 1st order system). However, this is not a technological problem because RC networks can be manufactured today with integrated technology to save space.

Figure 2.55 shows three variants of low-pass filters. Although the frequency-determining components are identical in each variant, the cutoff frequency depends on the specific circuit. It is partially because the cutoff frequency is defined as the frequency at which a 3 dB drop in the transmission is achieved (Fig. 2.56). The series connection alone reduces the cutoff frequency of the individual stages from 1600 to 1020 Hz (decoupled cascades). On the other hand, the mutual influence of the individual, non-decoupled stages comes into play. As a result, the additional time constant $R_1 C_2$ causes a further reduction of the cutoff frequency to 600 Hz.

Mathematically, the filter characteristics of the 2nd order low pass with non-decoupled stages can be formulated according to Eqs. (2.41) and (2.42), respectively ($T_1 = R_1 C_1$, $T_2 = R_2 C_2$, $T_{12} = R_1 C_2$).

$$G(\omega) = \frac{u_a}{u_s} = \frac{1}{1 + j\omega(T_1 + T_2 + T_{12}) + j^2\omega^2 T_1 T_2} \tag{2.41}$$

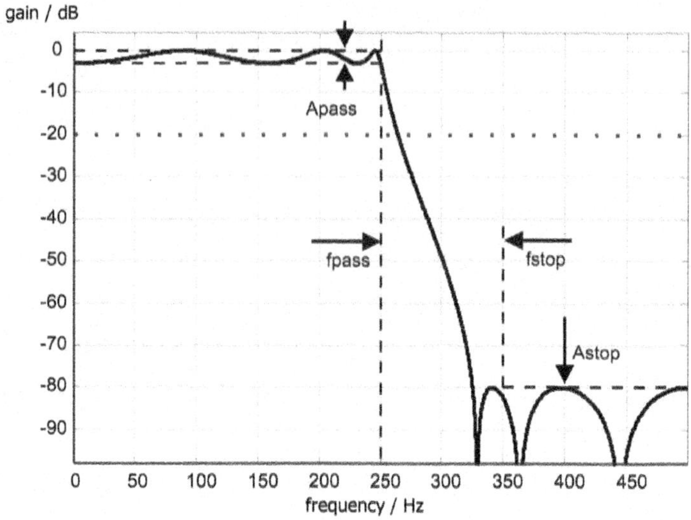

Fig. 2.54 Tolerance scheme of a low-pass filter. The corner frequencies f_{pass} and f_{stop} are decisive for the design, the permissible ripple in the passband Apass, and the required stopband attenuation A_{stop}. The filter edge is the transition from the passband to the stopband between the frequencies fpass and fstop. The steeper the filter slope, the closer the filter comes to the ideal, but the more difficult it is to realize. For example, the tolerance scheme was fulfilled with a 6th-order elliptical filter

$$|G(\omega)| = \frac{1}{\sqrt{\left(1 - \omega^2 T_1 T_2\right)^2 + \omega^2 (T_1 + T_2 + T_{12})^2}} \tag{2.42}$$

Equation 2.42 shows that the time constant $T_{12} = R_1 C_2$ additionally lowers the cutoff frequency due to the coupling of the two basic stages. However, this effect is not a problem since it can be considered in the dimensioning of the network. The passive network becomes a problem when the internal resistance R_i of the signal source and the load resistance R_a—as shown in Fig. 2.55—cannot be neglected and are unknown or variable. The two resistors directly influence the frequency-selective network (R_i in series with R_1, R_a in parallel with C_2). Consequently, there is no longer any control over the desired cutoff frequency. In practical measurement technology, this case can occur, for example, when low-pass filters are used at the input of a measurement amplifier to suppress high-frequency interference (Fig. 2.15). Since the impedances of the connection electrodes are *unknown* and can vary over two orders of magnitude (1–100 k), the cutoff frequency of the low-pass filters can also change in this dimension. For this reason, too, a low-pass filter at the amplifier input is not recommended.

In order to decouple cascaded (series-connected) stages from each other on the one hand and the filter from both the input and output resistance on the other, the use of active components is necessary. These are used as impedance converters or

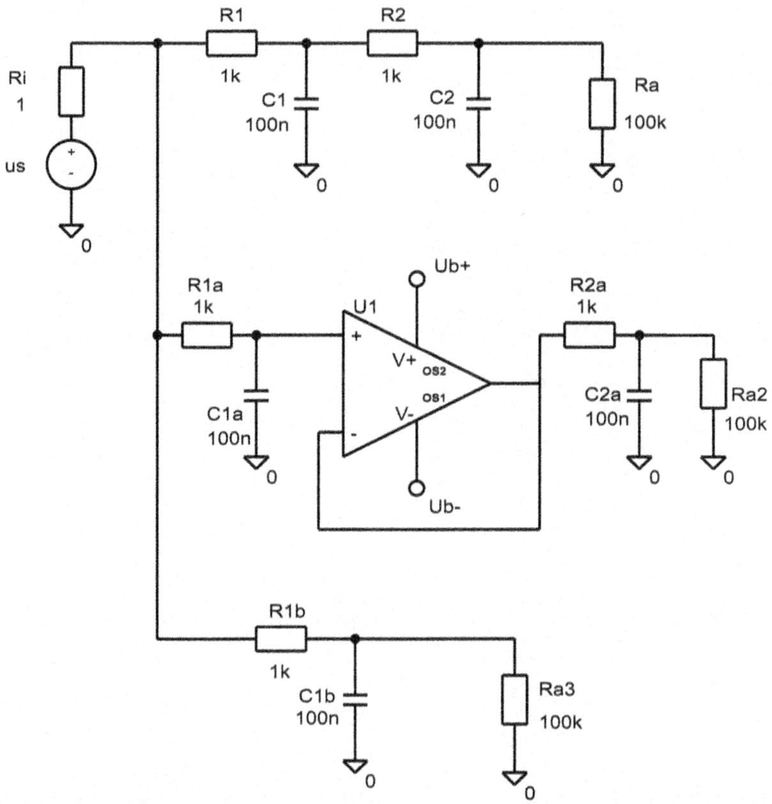

Fig. 2.55 Analog lowpass filters: The RC combination $R_{1b}\,C_{1b}$ of the 1st order lowpass results in a cutoff frequency of 1600 Hz. The decoupled 2nd-order series circuit with the same 1st-order lowpass filters ($R_{1a}\,C_{1a}$, $R_{2a}\,C_{2a}$) has a cutoff frequency of 1020 Hz. If the two stages are not decoupled ($R_1\,C_1$, $R_2\,C_2$), the mutual influence of the stages comes into play, so the cutoff frequency is only 600 Hz. The internal resistance of the signal source $R_i = 1\ \Omega$ and the load resistance $R_a = 100$ kΩ can be neglected compared to the frequency-determining elements. (PSpice simulation BSV_2.35a)

voltage amplifiers (high input resistance, low output resistance). Active filters are mainly designed with OV, which realizes the necessary decoupling to the outside and between the filter cascades. A typical example is shown in Fig. 2.57. The active filter consists of two filter cascades, a low pass and a high pass. The low pass limits the spectrum upwards, on the one hand, to avoid aliasing during the subsequent AD conversion, on the other hand, to minimize noise (see Sect. 2.2.1). The high pass suppresses the electrode offset voltage and slow artifacts. The stages are decoupled from each other and the output. To achieve decoupling from the biosignal source, whose internal resistance is fed directly into $R_{3,4}$, a voltage follower or amplifier would have to be connected in front of the filter. In the example in Fig. 2.57, the advantages of an active filter are visible: the individual cascades or filter functions

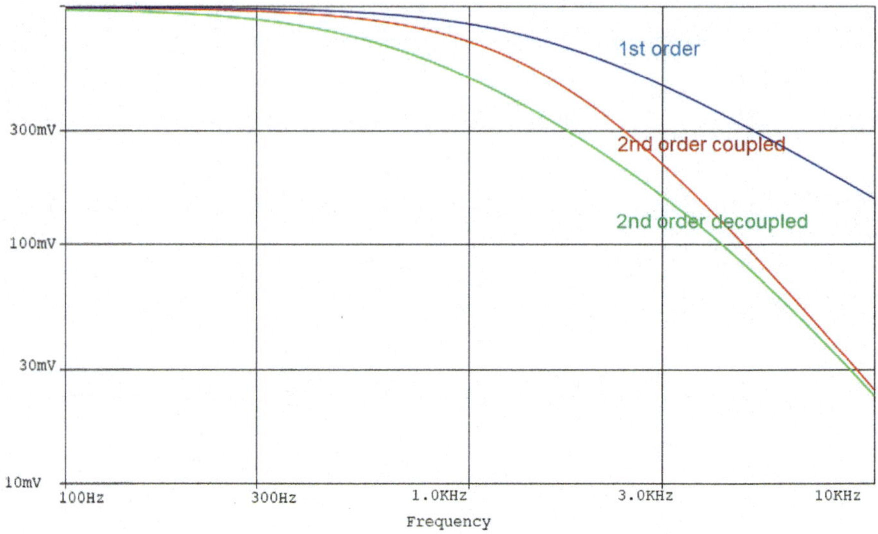

Fig. 2.56 Amplitude characteristics of low-pass filters from Fig. 2.55. 1st order low-pass filter with a cut-off frequency of 1600 Hz (blue), 2nd order low-pass filter with decoupled stages and the same time constants as for 1st order and a cut-off frequency of 1020 Hz (red), 2nd order low-pass filter with directly coupled 1st order stages and the same time constants with a cut-off frequency of 600 Hz (green)

can be dimensioned independently of each other, which is impossible with passive networks. Choose components using OV and suitable networks; in both 2nd-order cascades, the frequency-critical resistors and capacitors are identical. In addition, a gain can initially be set arbitrarily via negative feedback. The next chapter will show which influence the amplification has on the filter characteristic.

Of course, the active filters also have disadvantages. As shown in Sect. 2.2.3, the spectral operating range of OV is very limited. Above about 10 kHz, the filter function can no longer be guaranteed, which can be problematic, especially with high-frequency interference. In addition, OV produces low-frequency noise, which worsens the signal-to-noise ratio. Especially in medical measurement technology, the issue of resistance to strong interference (defibrillation peaks in ventricular fibrillation, massive high-frequency interference in RF surgery) is also highly relevant. Therefore, further measures are necessary to protect the semiconductors from excessive voltages compared to passive filters.

2.3.2 Active Filters with Operational Amplifiers

The most straightforward measure for decoupling frequency-selective networks from each other and the input or output is shown in Fig. 2.58. If the internal resistance of the signal source R_i can be neglected in the circuit, the following

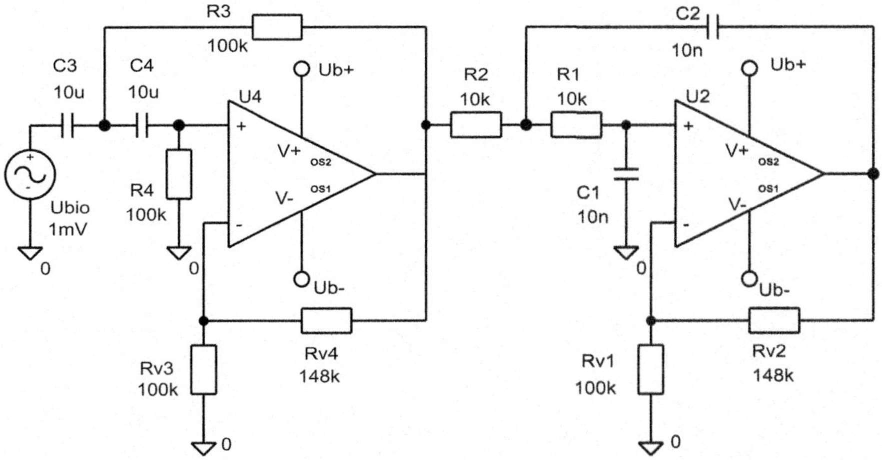

Fig. 2.57 Active analog filter with OV: The first cascade is a high pass, and the second a low pass. Altogether they result in a bandpass with cutoff frequencies $f_u = 0.2$ Hz and $f_o = 1.2$ kHz (PSpice simulation BSV_2.36). The high pass eliminates the high DC voltage of the electrode offset. The low pass limits the spectrum upwards for minimal noise and as an antialiasing filter before the possibly following AD converter

applies for the cut-off frequency of the low-pass filter

$$f_g = \frac{1}{2\pi R_1 C_1} \tag{2.43}$$

The 1st order stage shown in Fig. 2.58 can be cascaded to higher-order filters. However, it must be considered that each stage further reduces the attenuation by adding 3 dB at the cutoff frequency. For example, two cascaded stages, according to Fig. 2.59, have an attenuation of 6 dB at the same cutoff frequency of the individual cascades, so that the new cutoff frequency is lower with a 3 dB drop, in Fig. 2.59 at 1020 Hz.

Since filters with an order of four to eight are needed in analog biosignal processing, cascading of simple stages is not practical. Therefore, more efficient 2nd order stages follow (Fig. 2.60).

For the amplitude-frequency response of a 2nd order low pass filter (one stage in Fig. 2.60), the following applies

$$|G(\omega)| = 20\log_{10} k\sqrt{\frac{\omega^4}{\omega_g^4} + \left(d^2 - 2\right)\frac{\omega^2}{\omega_g^2} + 1}, \tag{2.44}$$

and for the phase frequency response

$$\varphi(\omega) = -\arctan\frac{d \cdot \left(\omega/\omega_g\right)}{1 - \left(\omega/\omega_g\right)^2}. \tag{2.45}$$

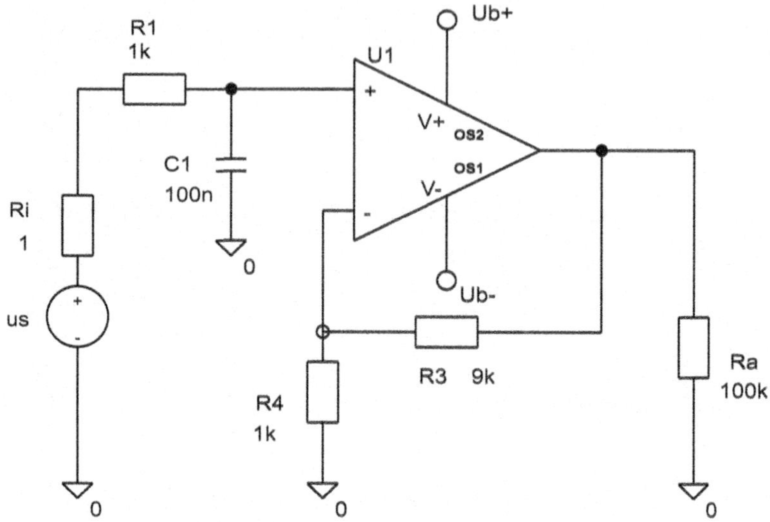

Fig. 2.58 The 1st order low pass filter R_1 C_1 is decoupled from the next stage with an OV. The gain is adjustable with R_3 and R_4 independently of the filter characteristic. The cutoff frequency is 1600 Hz

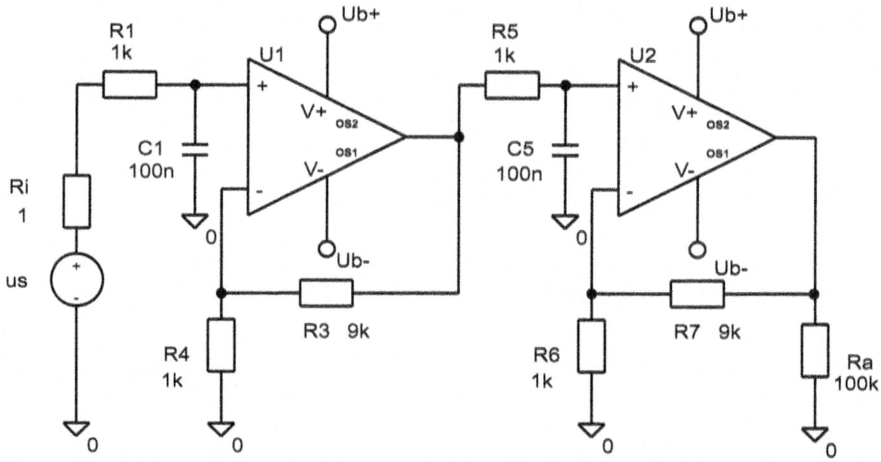

Fig. 2.59 Two cascades of 2nd-order low passes, 4th-order in total. Stages of 2nd order can be cascaded as often as desired, but it must be noted that the cut-off frequency of the whole cascade decreases with each additional stage. (see Fig. 2.61)

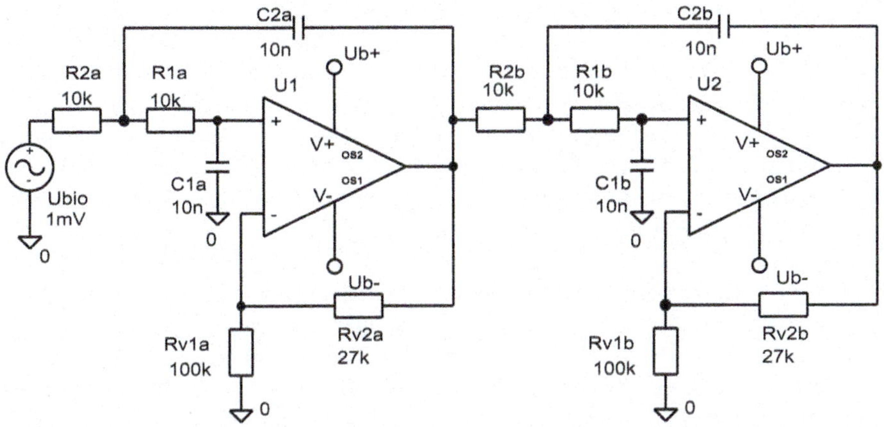

Fig. 2.60 Two cascades of 2nd-order low passes, 4th-order in total. Stages of 2nd order can be cascaded as often as desired, but it must be noted that the cutoff frequency decreases with each additional stage (see Fig. 2.61)

In Eqs. 2.59 and 2.60 $\omega = 2\pi f$, where $\omega_g = 1/(RC)$ is the low-pass cutoff frequency, d is the damping factor (ideal damping is achieved at $d = 1.41$), $R_{v2}/R_{v1} = 2\text{-d}$, gain $k = 3\text{-}d$, $R_1 = R_2$, $C_1 = C_2$.

If another cascade is added to the first stage (Fig. 2.60), the frequency responses multiply, and according to Eq. 2.59, the amplitude transfer in dB doubles. That is, the −3 dB drop at the cutoff frequency becomes a −6 dB drop after two stages, a −9 dB drop after three stages (6th order), and similar. The cutoff frequency—related to the −3 dB drop—moves further towards low frequencies. Thus, a cascaded

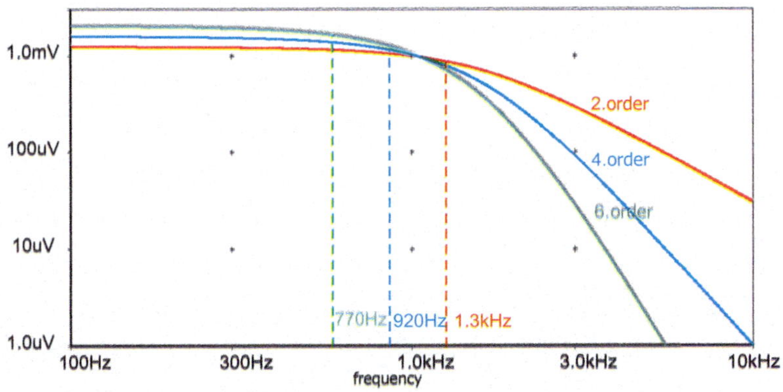

Fig. 2.61 Amplitude frequency responses of Bessel-lowpasses of 2nd order (after the first stage in Fig. 2.60), 4th order (after the second stage in Fig. 2.60), and 6th order. With identical components, the cutoff frequency is 1.3 kHz of the 2nd order, 920 Hz of the 4th order, and 770 Hz of the 6th order. The gains are 1.27, 1.61, and 2.05 in the same order

filter's cutoff frequency occurs where the cascade's drop is −3 dB/N, where *N is* the number of cascades. When dimensioning a cascaded filter, one must design the individual stages so that at the desired cutoff frequency, they have a drop of only −3 dB/N so that the filter comes to exactly −3 dB after addition. To do this, one can use Eq. 2.59 so that −3 dB/N is used for the amplitude, and the cutoff frequency of the individual stages is calculated from it. Now the question arises, which value should be set for the attenuation? The damping constant has the same meaning here as in oscillating circuits, i.e., in 2nd order networks. However, oscillatory circuits in the conventional sense are not used here, but the oscillatory capability is realized electronically with the help of the OV and positive feedback. Figure 2.62 shows the three typical amplitude characteristics of analog filters.

The circuit solution shown in Fig. 2.60 offers significant advantages, especially from a practical point of view: One can use identical components in all filter cascades for frequency selection, which is impossible in passive filters of conventional filter technology and other active filters. As can be seen from Eq. (2.44) and Fig. 2.62, with the aid of a single adjustable (programmable) resistor (R_{v1} or R_{v2}), one can use not only one of the filter types Chebyshev, Butterworth, or Bessel but also to tune continuously along the scale between the types as required. The following parameters can be used to dimension this circuit-based filter solution (Table 2.4).

For the resistors of the amplification chain $R_{v2} = (\varepsilon - 1)R_{v1}$, $u_a = \varepsilon \cdot u_e$, the cut-off frequency of a low-pass filter $f_g = \sqrt{\gamma}/2\pi RC$ for the cut-off frequency

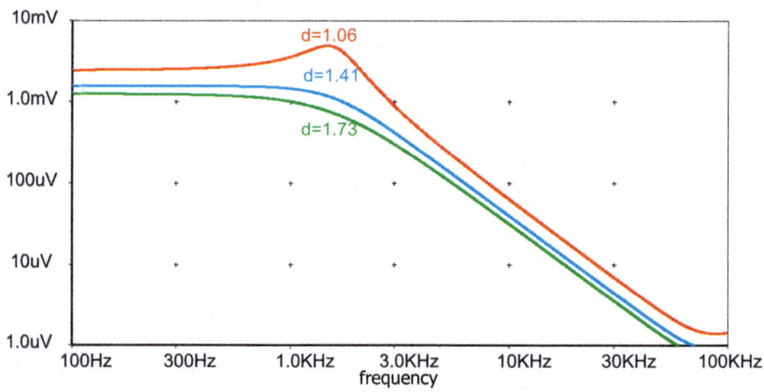

Fig. 2.62 Amplitude characteristics of a 2nd-order low-pass filter according to Fig. 2.60. The subcritical damping of $d = 1.06$ results in an overshoot at the cutoff frequency of 1 dB. It is the Chebyshev-type I filter (red). The damping of $d = 1.41$, called optimal, leads to the maximum flat amplitude (blue); it is typical for the Butterworth filter. The damping of $d = 1.73$ leads to the maximum flat or group delay, known as the Bessel filter (green). An attenuation of $d = 2$ would mean a separation of the two-time constants (decoupling of R_1 C_1 and R_2 C_2) and lead to the decoupled 2nd order filter (see Fig. 2.55 center)

Table 2.4 Operating parameters for dimensioning 2nd order filters

Bessel	Butterworth	Chebyshev (1 dB)
1.267	1.586	1.941
0.618	1.0	1.663

of a high-pass filter $f_g = 1/2\pi RC\sqrt{\gamma}$ is valid. A high pass can be easily realized utilizing the circuit considered so far: In the circuit shown in Fig. 2.60, the frequency-determining elements $R_{1,2}$ and $C_{1,2}$ are swapped for each other.

2.3.3 Phase Frequency Response

The shape of biosignals is of crucial importance in medical diagnostics. Physicians in cardiological, neurological, ophthalmological, audiological, somatosensory, motor, and related fields base their diagnosis primarily on qualitative characteristics of the recorded biosignals.

Therefore, one of the essential duties of measurement technology and biosignal processing is not to change the signal shape in any relevant way. Thus, the dependence on biosignal and phase frequency response arises. In order to demonstrate this relationship, the following is a consideration of the bandstop, also known as a "50 Hz filter", which has often been used in medical measurement technology. A circuit variant of such a bandstop is shown in Fig. 2.63. It is a 2nd order filter that can effectively suppress the line frequency. The frequency selective network consists of two T- links (Because of the shape of the circuit arrangement of C_2, C_3, and R_3 and R_1, R_2, and C_1, which form the letter T), a high pass, and a low pass with identical frequencies, which form a "notch filter". The amplification can be adjusted with R_{v1} and R_{v2}, influencing the filter type as with the low pass treated exemplarily.

The bandstop effectively suppresses the line frequency, and the attenuation reaches far more than 60 dB, sufficient for practical measurement (upper graph in Fig. 2.64). If an analog bandstop is to be used, then preferably at the beginning of the measurement chain so that an overload of the measurement amplifier is avoided early on (Fig. 2.1). The phase frequency response (Fig. 2.64, center) shows a linear course in sections, but also strong curvatures at 20 and 100 Hz. The shapes of the phase result in locally enormously increased group propagation times in the range of 10–300 Hz, i.e., in the middle of the spectrum of the biosignal (Fig. 2.64, bottom).

Starting with the observation of the time-varying spectrum of a typical ECG (Fig. 2.65), one notices: The R-wave or the QRS complex are pulse-like, short-time processes that are spectrally very broadband and reach energy components up to 200 Hz in the real signal (for the ECG in Fig. 2.65, a low-pass filter at 50 Hz was used). If such a signal is sent through the bandstop shown in Fig. 2.63, the low-frequency components of the ECG (0–10 Hz, see Fig. 2.64) are delayed by about 30 ms and the high-frequency components (10–40 Hz) are delayed by up

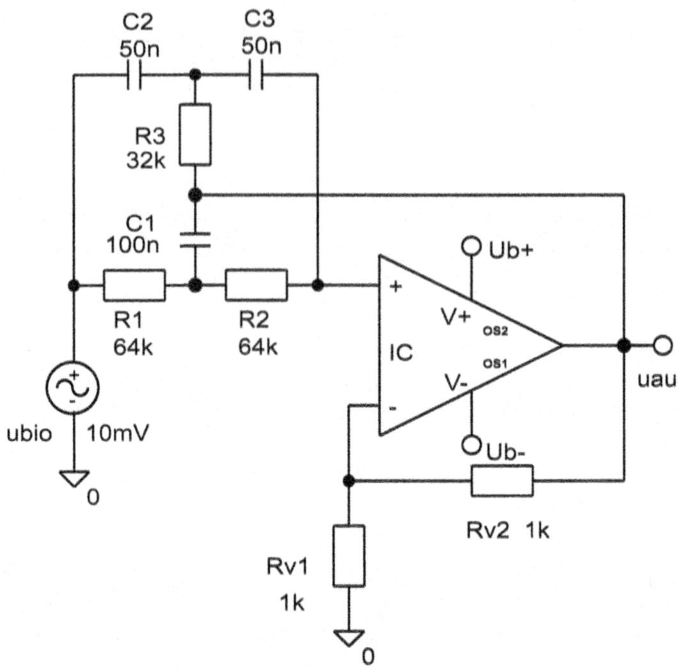

Fig. 2.63 A 2nd order bandstop for suppressing the mains interference at 50 Hz. For the cutoff frequency holds: $f_0 = 1/\pi RC$, where $R = R_1 = R_2 = 2R_3$, $C = C_1 = 2C_2 = 2C_3$ (PSpice simulation BSV_2.41)

to 70 ms. It means that the high-frequency components of the ECG are delayed twice as much as the low-frequency components.

Due to the more significant delay of the bandstop at higher frequencies, the high-frequency ECG components at the output of the filter appear later than the low-frequency components. As a result, the high-frequency components come to the fore only after the end of the QRS complex and are visible as a rapid after-oscillation (Fig. 2.66). The after-oscillation has nothing to do with the original ECG. The rapid after-oscillations ("ringing") simulate cardiac activity that does not exist in reality. It can have fatal consequences for the diagnosis and the therapy based on it.

From this example, it becomes clear how important the preservation of the signal shape is. Therefore, it must be ensured that all frequencies (groups) are delayed with as equal a delay time as possible, from which it follows; the phase frequency response must be linear in the spectral working range (Eq. 2.61).

$$\varphi(\omega) \approx k\omega \tag{2.46}$$

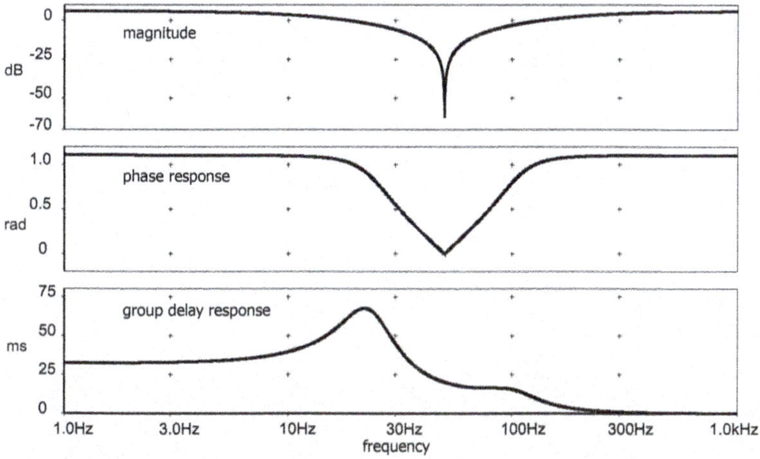

Fig. 2.64 The amplitude-frequency response of the bandstop from Fig. 2.63 (top) shows an acceptable attenuation of the mains interference of more than 60 dB. The phase frequency response (center) shows strong curvatures at 20 and 100 Hz, which can be attributed to an apparent nonlinearity of the phase response. The nonlinear phase response results in a strongly fluctuating group delay (bottom) in the middle of the biosignal region of the spectrum in the range between 10 and 300 Hz (the local maximum at 100 Hz turns out to be smaller than the maximum at 20 Hz because the linear representation, although they are both the same as local maxima). It leads to the shape change of the biosignal (PSpice BSV_2.41)

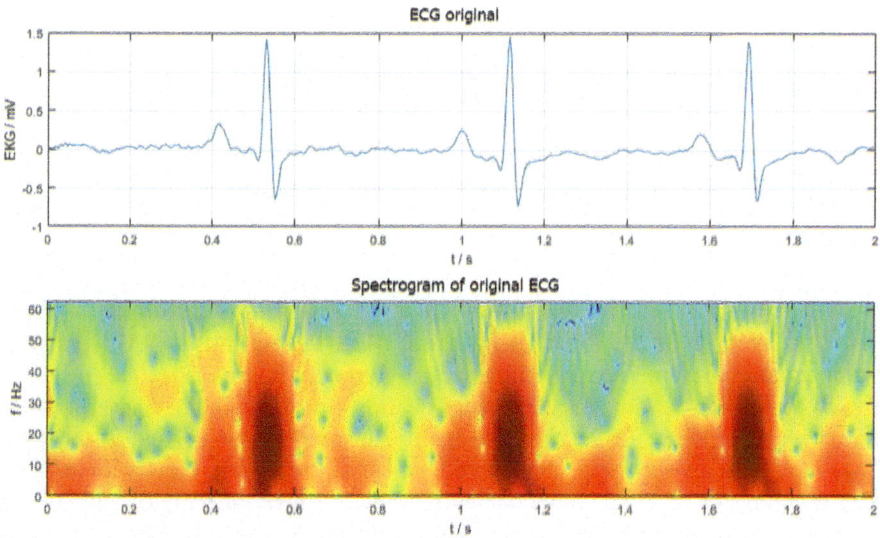

Fig. 2.65 ECG as time course (top) and in the spectrogram (bottom, time course of the spectrum). The colors represent the signal energy distributed in the time–frequency plane

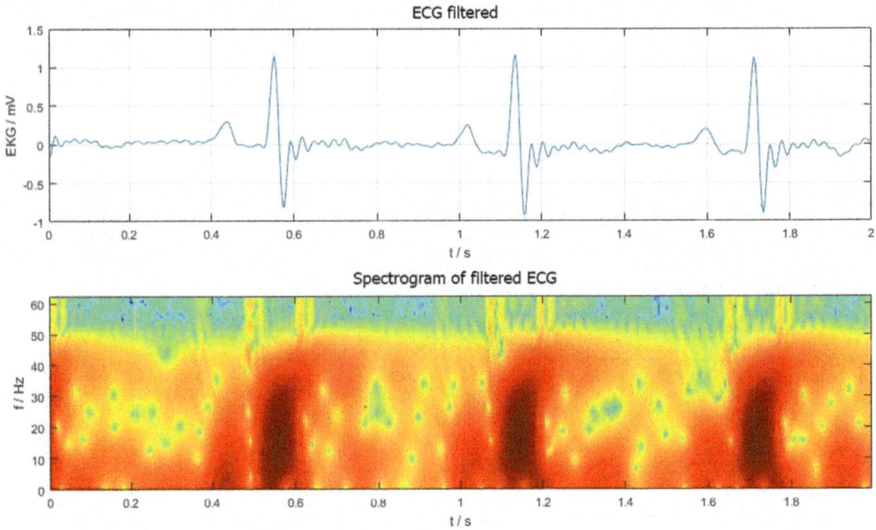

Fig. 2.66 Due to the different group delay times, the high-frequency parts of the ECG are delayed more than the low-frequency parts (spectrogram, bottom), compared with Fig. 2.65. This results in signal distortions, which appear as fast oscillations after the QRS complex (ST segment) in the ECG (top) as well as temporally shifted high-frequency energy (bottom)

These considerations conclude that for biosignals, which are used with their signal form for diagnostics, *exclusively linear-phase filters* (Bessel) must be used for signal processing. Since some manufacturers of medical measurement technology ignore this requirement, caution is also required in clinical use in this regard.

2.4 Exercises

2.4.1 Tasks

2.4.1.1 Analysis of the Influence of a Harmonic Disturbance on the Behavior of a Medical Measuring Amplifier

Let us assume a simple non-inverting OV amplifier. To do this, open the file BSV_2.1.sch in PSpice-Schematics. The amplification results mathematically from the ratio $V_d = (R_2 + R_1)/R_1 = 11$. The harmonic interference voltage at the input U_{netz} is 2 V, so a voltage of 22 V should appear at the output. However, since the supply voltage is 15 V, the amplifier output voltage enters the limit at about 14.6 V (Fig. 2.1). To display the voltage characteristics, we start the simulation with preset parameters. What is the maximum voltage allowed to appear at the input of the amplifier so that there is no distortion of the output signal? Examine the calculation with the help of the simulation by setting the input voltage U_{netz} in the voltage source to the calculated value. (Double-click on the voltage source symbol and change the parameter).

2.4.1.2 Balancing Operations

We can simulate a disturbing voltage jump with a switched voltage source and a 1st order high pass at the amplifier input, as shown in simulation BSV_2.42. The simulated signal consists of a square wave with a rising edge at 100 ms and a falling edge at 600 ms with 1 V amplitude. This voltage waveform is applied to a high-pass filter with a time constant of 100 ms. The equalization process corresponds to an e-function in the first power, from which the process is completed after a duration of about five-time constants. Start the preset simulation BSV_2.42. How does the time course change with a time constant ten times larger? What does this imply for the transmission of short pulses in the amplifier?

2.4.1.3 Galvanic Interference Coupling, Insulation Resistance

Taking into account the superposition principle, we will exemplarily investigate the effect of the mains voltage via the insulation resistance on a signal line of the measuring circuit (Fig. 2.7). In the typical case (dry air, dry device), we can assume an insulation resistance of at least 1 Γ (BSV_2.6). Simulate the coupling of the mains disturbance and measure the part of the disturbance in the output signal of the amplifier. Next, simulate the reduction of the insulation resistance due to high humidity or condensation after setting up a cold device after transport in a warm room. To do this, set the insulation resistance R_{is} to one-tenth of its original value. How does the reduced insulation affect the output signal? Place another insulation resistance of 0.1 GW between the mains and the other signal line of the measuring circuit. What effect will the extended simulation have on the output signal? What is the signal-to-noise ratio of the output signals in each simulation? Note that the electrical operating ground is isolated from the ground fault (R_{is3}) as safety regulations require (medical device must be completely galvanically isolated from the mains).

2.4.1.4 Galvanic Interference Coupling, Ground Loop

The second possible path of galvanic interference coupling is via the ground line or the earth line (Fig. 2.9). For the sake of simplicity, the ground symbol stands for both—for the electrical ground or the earth. From the point of view of possible network installations, investigate when a ground loop can occur and when it is relevant regarding measurement and safety (PSpice simulation BSV_2.7, a). This analysis requires basic knowledge of the types of network installations, which can be obtained from the relevant literature if required.

2.4.1.5 High-Frequency Interference

High-frequency interference is particularly problematic in medical measurement technology because it cannot be shielded in practice. The measures usually used in technology, such as low-pass filters at the input, cannot be implemented even in the measurement amplifier. Therefore we will analyze how these interferences can affect the measurement electronics. The AM-demodulator (Fig. 2.13) was replaced in the circuit BSV_2.10b with a simple transistor stage to simulate reality as well as possible. In PSpice, investigate how the coupled interference affects the spectral range of the biosignals. In particular, analyze the influence of the amplitudes from the LF and HF components of the AM signal. For this purpose, first, set an amplitude of the AM signal at the output of the OV to about 2 mV with the resistor R_2 = 100 ι νορδερ to control the following transistor stage as linearly as possible in the small signal range. After the simulation, plot the time course in PSpice, then plot the spectrum using the FFT. Zoom the range between 0 and 100 Hz to examine the disturbance (amplitude logarithmic). Both signals are pure harmonics, so they are visible in the spectrum as needles, if they are present at all. Otherwise, only random signals or noise from the components will be visible. Perform this analysis repeatedly for a strong disturbance, setting R_2 to about 1 kΩ. Further, try to make the transistor stage less sensitive to these disturbances by circuitry measures. As already discussed, AM demodulation occurs at nonlinear characteristics (PN junction), so the goal is to linearize the transistor stage. In circuit terms, the most effective means of linearization is negative feedback.

2.4.1.6 Signal-To-Noise Ratio, SNR

In measurement technology, several demands are made on the signal and interference characteristics to obtain reliable measured values. Since it must be assumed that the level to be measured is not known in advance and can also vary over several orders of magnitude, no level limits can be set concerning the interference. A useful measure for assessing the quality of the measured value is the signal-to-noise ratio (SNR), which indicates how large the ratio is between the signal to be measured and the interference. The required SNR depends on the specific measurement task; in medicine, an SNR of at least 40 dB is expected at the output. Calculate what gain (common mode and differential mode gain) and what *CMRR* a measurement amplifier must have to meet this requirement for the ECG (1 mV) and an assumed common-mode interference of 1 V at the input. What is the permissible common-mode interference voltage at the input if a measurement amplifier with a *CMRR of* only 60 dB is present?

2.4.1.7 Discrete Differential Amplifier

The differential amplifier is particularly well suited for amplifying differential signals and suppressing common-mode signals. The most critical parameters of the differential amplifier and the characteristics achievable with real measurement setup are investigated in the following: With the help of the simulations BSV_2.21 and BSV_2.22, determine the operating parameters V_d, V_g, *CMRR* of an ideally balanced discrete differential amplifier. Examine the floating (between drains) and floating (d1 or d2) output voltages separately. Start with an initially ideally balanced signal injection (let all resistors in the input network be balanced). In the next step, investigate how input asymmetry affects the output voltages. To create the input unbalance, change one of the input resistors influencing the common mode (e.g., set Rg1 to 7 kΩ). How does replacing resistor R_S with a constant current source affect the *CMRR*? In reality, one must expect and estimate in advance asymmetries in both the coupling of signals and noise and in the circuit itself. Compare how external and internal asymmetries affect amplifier characteristics.

2.4.1.8 Instrumentation Amplifier

For common-mode rejection, the symmetry of the multistage differential amplifier and the CMRRis of crucial importance. Investigate the influence of the differential gain of the first stage of an instrumentation amplifier on the CMRR in the two treated impedance converter and amplifier variants. What influence do common resistance tolerances in the instrumentation amplifier (5%, 1%, and 0.1%) and asymmetries in coupling the common mode noise (variability of impedances up to 100%) have on the common mode response? Use simulations BSV_2.26 and BSV_2.27. Compare the consequences of internal (in the amplifier) and external asymmetries (coupling conditions of the disturbances).

2.4.1.9 DRL Technology

The DRL (Driven Right Leg) technique reduces power disturbances due to common mode by active current injection into the right foot (for ECG leads; for EEG at the head). Apart from the fact that this technique is safety-relevant and therefore controversial (active current injection, neither diagnostically nor therapeutically justified), the aim is to investigate whether and how effective it is. For several years, there have even been integrated circuits on the market that offer this option. Further, intensive work is being done to establish it in EEG and other electrical biosignals. However, over 90% of the solution approaches fail due to a wrong or insufficient electronic/signal analytical model. Based on the preconceived electronic/signal analytic model, examine whether and in what domain the DRL technique is effective. Further, investigate how much the DRL technique contributes to reducing network disturbance in the effective range. To perform the simulations, LTSpice is required, which is (currently) available for free download.

2.4.2 Solutions

Descriptions and saved simulation results of the solutions are contained in.doc and.pdf files in the directory/exercises of this chapter and can be used supplementary for control and better understanding. Their numbering is identical to the numbering of the exercises.

2.4.2.1 Analysis of the Influence of a Harmonic Disturbance on the Behavior of a Medical Measuring Amplifier

From the simulation, the upper limit of the output voltage at which limitation occurs is $U_{aus} = 14.6$ V. For safety from nonlinearities near the limit, the maximum output voltage is set to 14 V. Since the gain is known, the maximum input voltage can be calculated to $U_{eimax} = Uausmax/V_d = 14\ V/11 = 1.27\ V$. We enter this value as a parameter in the voltage source U_{netz} and simulate for control.

2.4.2.2 Balancing Process

With a time constant of the high-pass filter of 100 ms and the pulse width of 500 ms, the equalization process almost completely decays at the falling edge. Suppose the time constant becomes more significant (in reality, up to the order of 10 s) or the pulse width narrower. In that case, the equalization process becomes less and less active, so the pulse is transmitted more and more unaltered. Qualitatively, it follows that with the exclusive high-pass behavior of the measuring amplifier, pulses whose width is smaller than the time constant of the high-pass are transmitted almost unaltered.

2.4.2.3 Galvanic Interference Coupling, Insulation Resistance

In practice, a device is often set up at room temperature after being transported in a cold environment. The humidity in the device increases considerably, and even a layer of condensation can form. It dramatically reduces the insulation resistance, not infrequently below one-tenth of its original value. In this simulation, the value should be reduced from 1 to 0.1 GΩ. This results in an increase of the disturbance in the output signal from 0.7 to 7 mV. According to Eq. 2.6, the SNR deteriorates from 39.5 to 19.5 dB, a value no longer acceptable in metrology. Starting from the original simulation with $R_{is} = 1$ GΩ, an insulation resistance of 0.09 GΩ is simulated on the other signal line. According to the natural conditions, the insulation resistances are never identical, so a difference of 0.01 GΩ results here. As seen from the circuit (see Sect. 2.2.3), a voltage difference is formed between the two inputs of the measuring amplifier. In the case of ideally identical insulation resistances, their effect would cancel each other out. If the insulation resistances are unequal, their difference also affects, in this case, 0.01 G. Therefore, the noise component in the output signal is 118 mV, and the SNR drops further to −5 dB, an unacceptable value for medical measurement technology.

2.4.2.4 Galvanic Interference Coupling, Ground Loop

For simulation, we start the circuit BSV_2.7. First, we will check if there is a ground loop. It is often very difficult or impossible in practice. It only works if a reference potential free of interference is available with certainty, e.g., a direct connection to earth or the power supply star point. Furthermore, it must be possible to start up devices supplied via the ground loop in a controlled and targeted manner because otherwise, no measurable voltage drop would occur.

In this simulation, the interference-free reference potential is the analog ground marked with a triangle symbol, and the amplifier's operating ground U_0 is located in the middle of the voltage supply between U_{b+} and U_{b-}. Suppose we place a voltage marker at any node in the simulated circuit. In that case, the mains interference is visible everywhere, most strongly at the amplifier's operating ground with an amplitude of 3.2 mV. The potential measurement of the operating ground against undisturbed reference is the surest indicator of whether the measurement circuit is disturbed via a ground loop. However, the resistance of the operating ground to the star point or earth must be measured simultaneously to ensure that there is a low-resistance connection R_m and that the operating ground is not disturbed by stray capacitances alone.

Now the question arises of how to eliminate the effect of the ground loop. The answer is simple: The operating ground of the amplifier must not have a conductive connection to the power supply on the side of the mains. It would supposedly be taken care of with the mains transformer. However, Class I equipment must be grounded through the protective earth conductor. However, in TNC-type networks (common protective and neutral conductor PEN), the protective conductor carries the operating current of the mains supply, so in TNC networks, the ground loop is always and unavoidably present. The only way to avoid the disturbance resulting from ground loops is to galvanically isolate the patient part from all conductors (i.e., also from the protective conductor) of the power supply (see Sect. 2.2.4 Isolation amplifiers). It is not decisive from a medical measurement technology's point of view. However, it is still relevant from the point of view of safety whether the galvanic isolation of the patient part also guarantees safety. The answer must be more satisfactory: one must expect the patient—intentionally or unintentionally—to be grounded. Therefore, the voltage generated by the ground loop can be directly carried over to him. For this reason, the TNC network is prohibited in rooms used for medical purposes to supply power to medical equipment. This prohibition does not apply in the domestic area, and therefore this is to be classified as a safety-relevant defect.

2.4.2.5 High-Frequency Interference

When $R_2 = 100\Omega$, a voltage of 3 mV is present at the input of the transistor stage and 300 mV at its output, so the gain is $V = 100$. The stage operates in the range of the so-called small signal modulation. This range is characterized by the input and output signals being much smaller than the transistor characteristics. Therefore, the transistor is assumed to be linear around its operating point in this range. This assumption is necessary to analyze and dimension the stages with linear mathematical methods. However, as our simulation shows, one often forgets that the nonlinearities are still present—i.e., also in the small signal range. After the simulation, one can initially see an almost ideal AM signal in the time course of the PSpice. If we perform a spectral analysis with the help of the FFT, we can see at least six harmonics of the AF signal in the range of 0–100 Hz. It is the spectral evidence of the nonlinearities already effective in generating the AM signal (multiplier, OV), especially in the transistor stage. If we set an input voltage at the transistor of 30 mV with $R_2 = 1$ k, the transistor should deliver an AC voltage of 3 V at the output corresponding to its gain of $V = 100$. It will do this, but we can already see that the signal is no longer symmetrical, i.e., distorted non-linearly. The assumption of small signal modulation is no longer valid here. Spectral analysis using the FFT shows levels between 4 and 10 mV for harmonics of the AF signal, representing a massive disturbance even for the strongest biosignals and making the biosignal unusable. Subsequent filtering will not improve because, in typical cases, the LF component is unknown and certainly not harmonic, as in this simulation.

Since effective noise suppression outside the amplifier is hardly possible, we can try to reduce it with circuitry measures. Without providing proof at this point, we assume that negative feedback generally leads to an improvement in linearity. In the simulated circuit, we can introduce negative feedback in the transistor stage simply by canceling the AC short circuit of resistor R_{e4}. For this purpose, we delete the capacitance C_4 or set it to a negligibly small value, e.g., 1 pF. Due to the now effective negative feedback, the gain naturally decreases, in this example, to about $V = R_{c4}/R_{e4} = 1$. It reduces the AF component to about 0.1 mV, which is still not a very good value, but two decades better (originally 10 mV) than in the original circuit.

From the circuitry point of view, it is possible to reduce the influence of unwanted AM demodulation by linearizing the measurement electronics with the help of negative feedback, even if this should be at the expense of additional gain stages.

2.4.2.6 Signal-To-Noise Ratio, SNR

The ECG amplitude is usually understood to be the peak value of the R-wave. It is, on average, 1 mV, so for the usual analog level for further processing with 1 V, a differential amplification of $V_d = 1000$ will be necessary. At the output, the *SNR* should be 40 dB or 100 so that the permissible interference at the output may reach an amplitude of 1 V/100 = 10 mV. Therefore, the common mode gain must not exceed $V_g = 10$ mV/1 V = 0.01. It follows for the CMRR = Vd/Vg = 1000/ 0.01 = 10^5, corresponding to 100 dB.

The permissible interference must be reduced significantly if a measurement amplifier with a CMRR of only 60 dB is available. Here can be used the advantage of the logarithmic representation—the multiplications and divisions of the transmissions are reduced to the simple addition: SNR_E/dB + *CMRR/dB* = SNR_A/ dB, or detailed as follows

$$SNR_A = \frac{U_{Aus}^{Diff}}{U_{Aus}^{Gleich}} = \frac{V_d}{V_g} \cdot \frac{U_{Ein}^{Diff}}{U_{Ein}^{Gleich}} = CMRR \cdot SNR_E.$$

This specific example applies:

$$SNR_E/\text{dB} = SNR_A/\text{dB} - CMRR/dB = 40\text{ dB} - 60\text{ dB} = -20\text{ dB}.$$

It follows that at a level of the ECG of 1 mV, the disturbance may be at most 1 mV/10 = 0.1 mV, which is hardly feasible in practice. With the measurement of the EEG we would get with this weak *CMRR*, we would get into the microvolt range, where the amplifier noise already lies so that the requirement for 40 dB would not be technologically feasible from the interference point of view. The requirement for a *CMRR of* at least 100 dB is sufficiently proven for practice.

2.4.2.7 Discrete Differential Amplifier

In this simulation, we are dealing with a discrete differential amplifier where it is not yet clear how the output voltage is decoupled, and we consider both alternatives—floating (balanced, index S) U_{AS} and floating output voltage U_{AM} (index M). According to the circuit, $U_{AS} = U_{d2} - U_{d1}$, $U_{AM} = U_{d2}$ (or $U_{AM} = U_{d1}$). The common-mode output voltage is given index G. The differential-mode output voltage is given index D. After the simulation of BSV_2.21, we can read from the time history for the amplitude of the output voltage $U_{ASD} = 130$ mV and $U_{ASG} = 0$ V. Reading the amplitudes of the ground-referenced output voltage in the time domain is difficult because of the additive composition of the DC component, the slow differential voltage (10 Hz) and the fast common-mode voltage (50 Hz). Therefore we transfer the representation with the FFT into the frequency domain, where each component is represented individually with its corresponding amplitude. In the spectrum, we can then read for the amplitudes: $U_{AMG} = 22$ mV (34 mV) and $U_{AMD} = 66$ mV. The following applies concerning the voltage sources for the massless amplifications: $V_{DS} = U_{ASD}/U_D = 13$, $V_{GS} = U_{ASG}/U_G = 0$. Therefore, the $CMRR_S$ is theoretically infinite. For the ground-related gains, $V_{DM} = U_{AMD}/U_D = 6.6$ and $V_{GM} = U_{AMG}/U_G = 0.22$. This results in $CMRR_M = V_{DM}/V_{GM} = 30$ and 30 dB, respectively, for the ground-related common-mode rejection. It is a value that would be unacceptable in medical metrology, but here we have only a single-stage differential amplifier for now. Let us compare the results with those of the differential amplifier with a constant current source. For this purpose, we start the simulation BSV_2.22. Nothing changes for the floating output voltages and the associated parameters. $U_{AMD} = 66$ mV and $U_{AMG} = 12.4$ mV for the ground-referenced voltages, so a clear improvement of the common-mode behavior can be seen here. As expected, the differential voltage has not changed because it does not depend on the source resistor. The common mode voltage was reduced to about half by dynamically increasing the source resistor. The gains are $V_{DM} = U_{AMD}/U_D = 6.6$ and $V_{GM} = U_{AMG}/U_G = 0.12$, resulting in values of 55 and 35 dB, respectively, for the $CMRR_M$. It follows that the constant current source could improve the *CMRR* to almost double.

Now we look at the influence of the asymmetry on the operating parameters. The asymmetry in the coupling of the common mode interference is simulated with the reduction of the resistance R_{g2} (or R_{g1}) to half. We can read the output voltages' amplitudes in the spectrum: $U_{ASD} = 124$ mV, $U_{ASG} = 62$ mV, $U_{AMD} = 62$ mV, and $U_{AMG} = 45$ mV. It follows that even with an ideal differential amplifier, the asymmetry at the input in this specific case can cause the ratio of the differential and common-mode signals to drop from theoretical infinity to $U_{ASD}/U_{ASG} = 2$ and 6 dB, respectively, in the floating output, and from 55 to $U_{AMD}/U_{AMG} = 1.4$ and 2.8 dB, respectively, in the ground-referenced output. At this point, however, we must be aware that these relatively poor values are not due to the circuitry but to the real unbalanced coupling. Now we would like to contrast the asymmetry of the coupling, which cannot be influenced by us, with the electronically given asymmetry in the amplifier: To do this, we simulate the circuit (symmetry restored at the input) in such a way that the drain resistor R_{d1}, which is deliberately exaggerated, is set to half. After simulation, we obtain the amplitudes at UASD = 100 mV, UASG = 6 mV, UAMD = 67 mV, and UAMG = 12 mV from the spectrum. The gains are $V_{DS} = U_{ASD}/U_D = 10$, $V_{GS} = U_{ASG}/U_G = 0.06$, $V_{DM} = U_{AMD}/U_D = 6.7$, and $V_{GM} = U_{AMG}/U_G = 0.12$, giving values of 166 and 44 dB for $CMRR_S$ and 56 and 35 dB for $CMRR_M$, respectively. Suppose we compare these values with those when the coupling is asymmetric. In that case, we find that the (unrealistically strong) circuit asymmetry has an almost negligible effect on the operating parameters compared with the asymmetry when the disturbance is coupled at the input. Although the circuit asymmetry significantly reduces the ground-free gain, the ground-related gain remains unchanged.

This analysis concludes that extreme circuit optimization of the electronics concerning symmetry will only bring a significant benefit if one cannot simultaneously make the measurement circuit before the amplifier sufficiently symmetrical—and in practice, one usually cannot.

2.4.2.8 Instrumentation Amplifier

Simulation BSV_2.26 shows that even with ideal simulation conditions (symmetrical noise coupling, identical resistors in the amplifier), ideal common-mode rejection is impossible in PSpice. It is because the used OV of type 741 also offers only a limited CMRR, which becomes visible at the ground-related output. The SNR at the input is

$$SNR_E = \frac{u_{DE}}{u_{GE}} = \frac{1\,\text{mV}}{10\,\text{V}} = 10^{-4} \equiv -80\,\text{dB},$$

SNR measured at the output (in the frequency domain after the FFT) is

$$SNR_A = \frac{u_{DA}}{u_{GA}} = \frac{90\,\text{mV}}{32\,\text{mV}} = 2,8 \equiv 9\,\text{dB}.$$

It follows that the CMRR of the instrumentation amplifier in the impedance converter variant has a value of

$$CMRR = \frac{V_D}{V_G} = \frac{u_{DA}u_{GE}}{u_{DE}u_{GA}} = \frac{SNR_A}{SNR_E} = 2,8 \cdot 10^4 \equiv 80\,\text{dB} + 9\,\text{dB} = 89\,\text{dB}.$$

From this calculation, it follows that the CMRR of the instrumentation amplifier, according to BSV_2.26, is about 90 dB, which is an unsatisfactory value given today's requirements. Note that if the SNR at the input and output of the amplifier is known, the CMRR can be found by simply subtracting the values given in dB.

From the simulation according to BSV_2.27, in which the input stage (first amplifier stage with IC1 and IC2) has a higher differential gain ($V_d = 100$) than in the simulation according to BSV_2.26 ($V_d = 1$), the $SNR_A = 8.47\ \text{V}|_{@10\ \text{Hz}}/268\ \text{mV}|_{@50\ \text{Hz}} = 30$ dB, thus a $CMRR = 110$ dB. It is an acceptable value for the practical measurement technique, which essentially results from the good differential gain of the first stage.

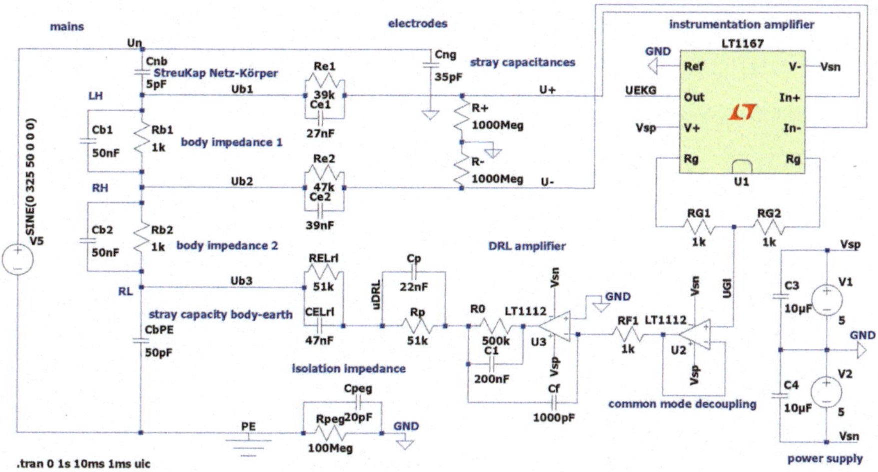

Fig. 2.67 Simulated DRL circuit in LTSpice. By varying the real unknown values of the electrode impedances and the parasitic impedances between the circuit and the mains, a range can be empirically determined in which the DRL technique works. However, it also becomes clear that it only marginally improves the mains disturbance in the output of the amplifier and even worsens it in specific configurations

2.4.2.9 DRL Technology

The electronic model was created and tested with LTSpice (currently freely available). In particular, it should be noted that this model also considers the patient's torso (impedance between left hand and right foot; Cb1, Rb1), which is methodologically inadmissibly largely ignored in the literature. Suppose one varies the model in the simulation according to reality (Cb1, Rb1, Cb2, Rb2, Cnb, Cng, CbPE, Cpeg, Rpeg, RELrl, CELrl, as well as the electrode impedances). In that case, a range can be determined experimentally in which the DRL circuit works (Fig. 2.67).

The level of the mains disturbance at the output of the instrumentation amplifier U_{EKG} is used as a quality criterion of DRL effectiveness. This model shows that the mains disturbance at the output U_{EKG} does not decrease even when the DRL circuit is proven to work, and under certain conditions, it even increases. This study shows that even if the DRL technology works, it does nothing to reduce the mains disturbance; at worst, it increases it. A detailed circuit analysis inevitably concludes that it is not the common mode that is the problem, but—as is always the case—the interference component above the patient that has become the difference.

Acquisition, Sampling, and Digitization of Biosignals

3

3.1 Biosignal Acquisition

This chapter deals with acquiring biosignals from the point of view of technically oriented electrophysiology and transferring the biosignals into a computer-compatible form. From the signal analysis point of view, the human being, as a living object, is a spatial entity with numerous and always active, spatially distributed signal sources. Their electrical activity is to be obtained as non-invasively as possible from the body surface, the magnetic activity contactless at a certain distance from the body. In acquiring electrical activity, positioning the electrical ground of the measurement, operational amplifier, and the reference of the derivation systems for medical interpretation plays a central role. The questions of grounding, the ground connection, and referencing, which are equally crucial for technicians and physicians, are dealt with here.

3.1.1 Derivation Technology

From an electrical point of view, the human body is a closed, electrolytic volume conductor in which local (nerve bundles) and extended current sources (heart) are always active at numerous spatial positions. These sources generate current loops in the entire volume conductor, which create potential differences inside and on the body surface (Fig. 3.1). These potential differences are amplified with differential, isolation, or instrumentation amplifiers (see Chap. 2). In addition to the ground-free differential inputs, such an amplifier also has an electrical operating

---

Supplementary Information The online version contains supplementary material available at https://doi.org/10.1007/978-3-662-67998-2_3.

P. Husar and G. Gašpar, *Electrical Biosignals in Biomedical Engineering*,
https://doi.org/10.1007/978-3-662-67998-2_3

ground that must be galvanically connected to the patient (not necessarily directly but conductively, e.g., via a resistor network). Without this connection, the differential inputs would be at an undefined potential to the ground, driving the amplifier into limiting. A technically simple possibility (Fig. 2.22) is to connect the ground directly via a ground electrode. The ground electrode should not be connected too far from the measuring circuit to avoid unnecessary common mode interference. For ECG, it is the right foot by default; for EEG a point on the head surface or near the head (neck, forehead, nose, ear). It is important to note—especially for the following considerations on the reference problem—that the physical connections shown in Fig. 3.1 for a channel must always be present. However, the ground connection can also be high-impedance and indirect since only a potential reference should exist, and no measuring current must flow. A minimum technical solution for ECG acquisition is shown in Fig. 3.2. The extremity derivative, named after Einthoven, is created by the difference formation between the hands and the left foot.

In contrast, the right foot is connected to the electrical ground of the measuring amplifier and possibly grounded. Since the potential differences are formed between, in each case, two electrically active and anatomically equal-ranking points, one speaks of a *bipolar* derivative. In the search for possibilities to represent the electrical activity in the body unipolar, one tries to produce an *indifferent* electrode. The most common approach is to form the arithmetic mean of two or more channels using a voltage divider, assuming that this is electrically inactive

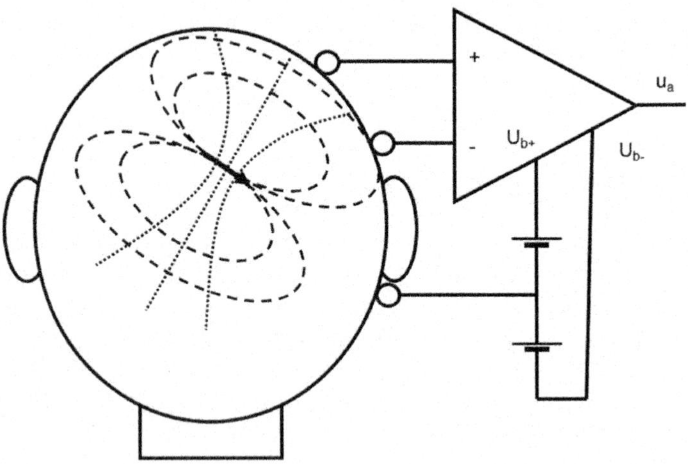

Fig. 3.1 The brain's current source (marked by a directional arrow) generates spatial currents, represented by streamlines (dashed). The current flow generates potentials in the volume, and the equipotential lines are marked with dots at the surface. Potential differences can be measured between sensors at the surface. The electrical ground of the amplifier (midpoint of the supply voltages U_{b+} and U_{b-}) must also be connected to the patient

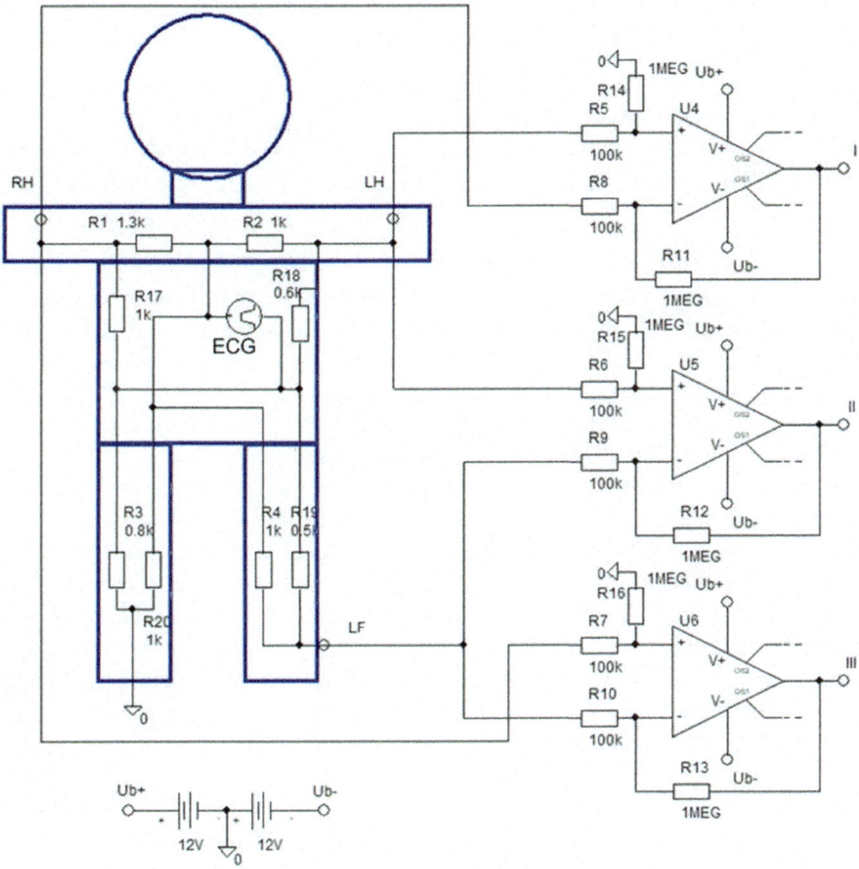

Fig. 3.2 ECG derivation according to Einthoven. The limb leads are recorded as potential differences between both hands LH and RH, and the left foot LF. The right foot is (usually) connected to the electrical ground of the measurement amplifier. The ECG signal source and the resistor network simulate the patient's internal currents and potential differences. The Einthoven derivatives I (LH–RH), II (LF–RH), and III (LF–LH) are created by differencing and amplification using operational amplifiers (PSpice simulation BSV_3.1). The model does not include preamplifiers of the differential amplifiers with high input resistance because of the node limitation of the student version of PSpice. Therefore, during the simulation, it is necessary to make sure that the differential amplifiers have a low input resistance (order of magnitude 100 kΩ)

or at least minimally active (for theory, see Sect. 3.1.2). Under certain conditions, this approach is theoretically correct.

A derivative using electrodes is always *bipolar* in the metrological sense since it is formed as the difference of at least two potentials tapped from the electrically active surface of the human body or the interior of the body.

Accordingly, an activity measurement of the functionally interesting neuronal structures (sensory and motor centers) on the body surface would be unipolar. The measurement task of an actual potential measure (unipolar activity) is familiar to many fields. Medical measurement technology seeks information on the investigated neuronal structure independent of diverse, functionally uninteresting reference areas. An example of the circuit realization of this approach is shown in Fig. 3.3 (known in electroencephalography as the reference "connected ears" and in cardiology as the indifferent electrode of the Goldberger or Wilson lead). With the help of identical resistors, one forms a virtual electrode from two or more derived, mass-related potentials. Under ideal conditions of the medium (homogeneity, isotropy) and the electrode arrangement (symmetrical and sufficiently densely distributed over the entire surface), the virtual electrode would lie in the geometric center between the discharge points electrically inactive. However, these assumptions and metrological practice must still be fulfilled in reality, so virtual (average reference) electrodes cannot be considered electrically inactive.

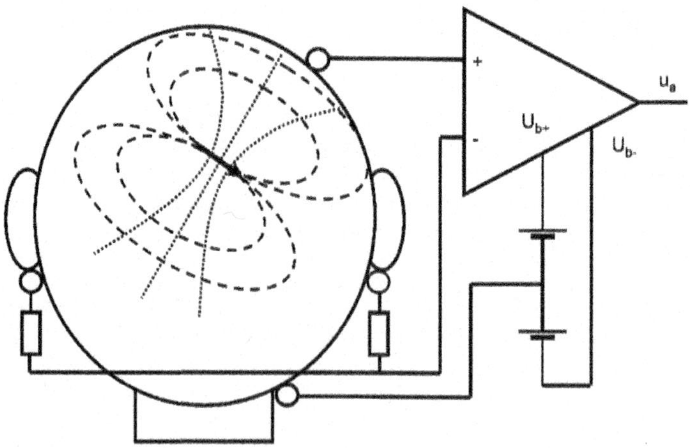

Fig. 3.3 Formation of a virtual reference electrode: a reference is formed from two or more channels (connected ears) via identical resistances, which are assumed to be *indifferent*. Under this assumption, such derivatives are called *unipolar*. However, this assumption needs to be revised and can lead to misinterpretations

The problem of reference, ground connections, and grounding are complex and of fundamental importance not only from the point of view of biosignal acquisition but also for the electrical safety of patients. Before further explaining the rejection technique, here is an overview of the function and feasibility of the ground, electrical ground, and reference connections.

3.1.1.1 Grounding

Grounding is safety-relevant and necessary for devices of protection class I. The patient is generally not grounded (galvanic isolation from the supply system). However, it must always be assumed that he may be grounded for safety and metrological analyses, even unintentionally. Grounding is optional for signal acquisition and is also not helpful since it can reduce the signal quality (see Sect. 2.1.3 Coupling of interference into the measuring arrangement). An example of the effect of grounding is shown in Fig. 2.10, where the ground connection at the patient's foot is to be interpreted as a high-impedance ground. Therefore, grounding should be avoided in the rejection technique (protection class II, battery operation, mobile technique) if possible. According to current standards, any galvanic connection of the patient circuit to the power supply network must be avoided for safety reasons. However, the measurement technology must always reckon with a high-impedance (insulation resistance), a capacitive (stray and parasitic capacitances), or also an inductive (supply cable vs. measurement cable) coupling.

3.1.1.2 Electrical Ground

A galvanic connection between the patient and the electrical ground of the measuring amplifier is functionally necessary. Without a ground connection, the amplifier inputs would float in space, especially in the case of MOSFET input stages, and the amplifier could behave uncontrolled. This potential danger stems from the characteristics of a field-effect transistor: the gate is an insulated conductive surface that forms a capacitor with the substrate. If the gate does not have a defined potential, the capacitor will charge randomly, controlling the transistor current in an uncontrolled manner. A galvanic connection to the ground is necessary for the gate to have a defined potential. This connection does not necessarily have to be made directly via a cable; it can also be realized indirectly and with high impedance via the actual circuit. The ground connection of the measuring amplifier can therefore be made either directly (Figs. 3.1, 3.2 and 3.3) or circuit-wise via an *internal* ground as well as indirectly via the circuit (Fig. 3.4).

3.1.1.3 Reference

A reference is not necessary for signal acquisition. The technically simplest solution is a measurement set-up in which differences are formed from all derivation points on the patient against the electrical ground. Since, from the point of view of the lead technology, every lead is bipolar; a reference initially has only a purely declarative character. For unipolar derivations (in the medical sense), however, it is necessary to define one of the derivation points as a reference and to relate

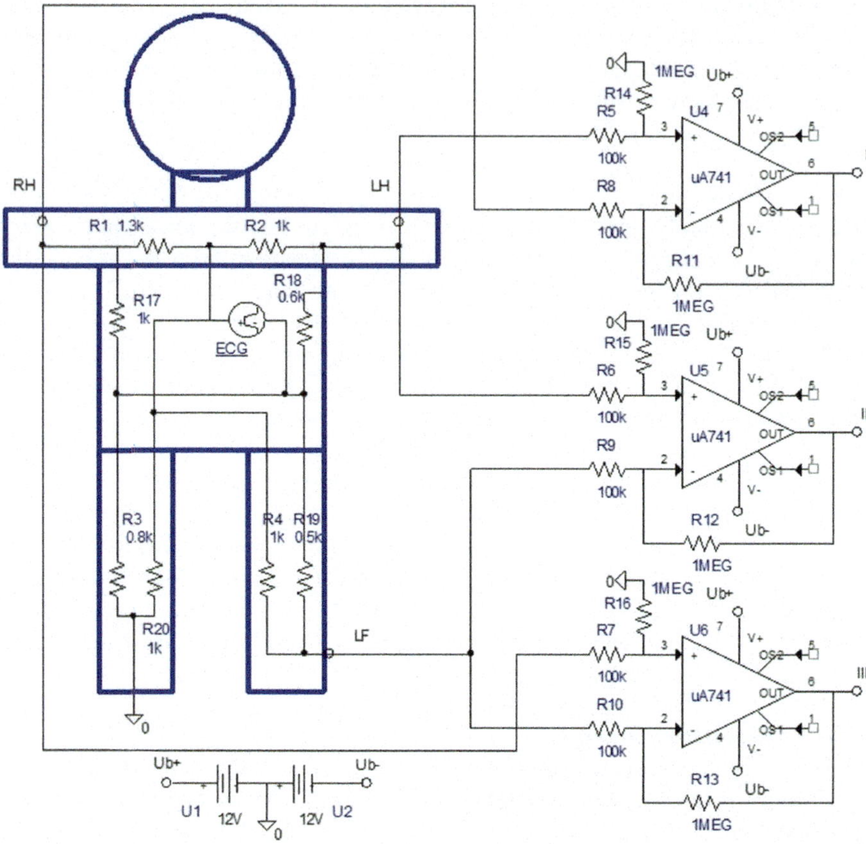

Fig. 3.4 Internal electrical ground is formed by the simultaneous arithmetic mean of all limb potentials using identical resistors (R_{21}, R_{22}, R_{23}). Apart from the limb electrodes, no other connections to the patient are necessary. The same solution can be used to form a reference electrode for unipolar leads (Wilson). In this case, the star point of the resistors would not be grounded but would create the standard input potential for the measurement amplifiers. If the star point is used as a reference electrode, the galvanic connection of the patient via the input circuit of the differential amplifiers is still ensured via $R_{5,6,7}$ or $R_{14,15,16}$

all other potential differences to this point. A connection can also be realized by circuitry (Fig. 3.3) or by calculation (para. 3.1.2) as a virtual reference point. Solutions create an internal electrical ground (without a patient electrode) or a virtual reference circuitry (Fig. 3.4).

Although the basic lead systems for recording ECG, EEG, EMG, or EOG are largely standardized concerning electrode positioning, there is disagreement and uncertainty in questions of referencing and grounding as well as electrical ground in the medical and technical field. To simplify the handling of the measurement technology and minimize erroneous discharges, it makes sense to solve these problems in advance regarding circuitry. A plausible approach is shown in Fig. 3.4. The need for a patient-side ground connection is eliminated by having the connections form a star point via equal resistors connected to the amplifier's electrical ground. The same principle could be applied to creating a virtual reference (Fig. 3.3), in which the star point is not connected to the ground but is a common reference for all other derivative points. Such a reference is called CAR (Common Average Reference) and is already irreversibly built into some commercial instruments. However, CAR can lead to erroneous derivations and misinterpretations for some questions (anatomy-related, spatially resolved EEG) (see Sect. 3.1.2), so it should generally be rejected as a permanently installed option of a derivation system. It is advisable to thoroughly check a recording system intended for use in this respect and, if necessary, to exclude it from a diagnostically relevant application.

3.1.2 References in Biosignal Acquisition

As explained in the previous chapter, from the point of view of measurement technology, every derivative is bipolar. In electrophysiology, however, bipolar and unipolar differentiate between bipolar and unipolar leads. A unipolar lead means potential differences between active lead points and a desired inactive reference, while a bipolar lead is a potential difference between two active lead points. Electrophysiologically, both types of derivatives are of interest. Unipolar leads represent more global activity, projecting between the lead points and the usually relatively distant reference. Bipolar derivatives represent more local activity, naturally representing spatially shorter projection distances. The assumption of an inactive reference point is not necessary for virtual references (except for the Hjorth-derivation, which is a computed derivation without a reference, see Reference-free derivations) and is hardly fulfilled for physically existing reference points not at all. It is also why physicians, even in the case of unipolar derivations, always want to know which derivation point was used exactly as a reference during the measurement. Implicitly, they treat unipolar derivatives as bipolar ones with the above projection difference.

Regarding referencing, electrophysiology lead systems can be classified into three groups: Lead systems with real and virtual references add reference-free leads.

3.1.2.1 Derivation Systems with Real References

The reference is *declared a* physically present recording point on the body surface. Commonly used for EEG recordings are earlobes (A_1, A_2) for unipolar interpretation or positions of the 10–20 standard system Cz or Fz (Sect. 3.1.4) and selected parts of the electrode matrix for EMG recordings (Fig. 3.5).

3.1.2.2 Derivation Systems with Virtual References

The reference potential is formed by circuitry (Fig. 3.3) or by calculation (Eq. 3.1) and used as a virtual reference electrode. By default, electronically created

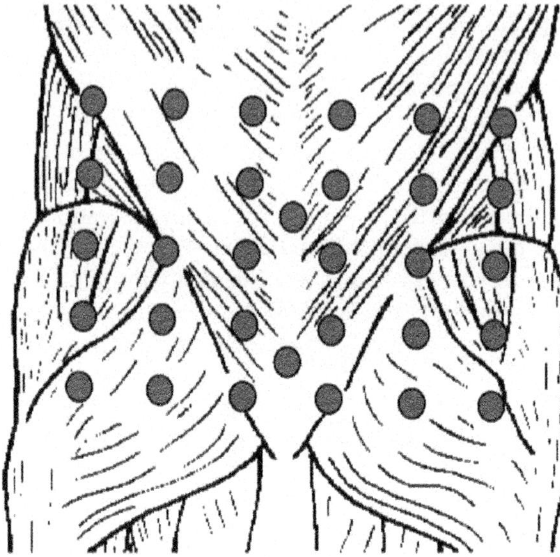

Fig. 3.5 Electrode matrix for EMG potential mapping over the lower part of the back. In principle, depending on the question, any discharge point can be selected as a reference. As a general rule, prominent points are suitable as references: edge or center of the electrode matrix

references are used for ECG leads; according to Wilson and Goldberger, an electronic or computational common reference is used as CAR for EEG and EMG derivations.

$$car(k) = \frac{1}{M}\sum_{i=1}^{M} x_i(k) \tag{3.1}$$

In Eq. 3.1, k is the time index, and i is the index of the derivative points.

$$x_i^c(k) = x_i(k) - car(k) \tag{3.2}$$

In Eq. 3.2, k is the time index, i index of derivative points, $x_i(k)$ unipolar derivatives from which the *car(k) is* formed, *car(k)* the reference according to Eq. 3.1, $x^c{}_i$ (k) derivatives related to the common reference *car*(k).

References of individual derivatives according to Eq. 3.2 has some positive effects on signal acquisition. The CAR, according to Eq. 3.1 corresponds to all measuring channels' common mode signal of coupled disturbances. If it is electronically possible to form the CAR as close as possible to the input of the measurement amplifier and to subtract it from the channels according to Eq. 3.2 (see Guarding-technique), this would correspond to a very effective common mode suppression. It is essential concerning coupled interference from the environment (mains, tube monitors). According to the same principle, endogenous but interfering biosignals distant from the recording site are attenuated in the measured biosignal, e.g., artifacts due to eye and eyelid movements in occipital EEG. However, CAR can lead to electrophysiologically problematic analysis effects that will be addressed later. Therefore, it should *not be irreversibly* built-in hardware-wise but should only be available as a *computational option for* signal analysis.

3.1.2.3 Reference-Free Derivations

All technical and signal-analytical referencing problems would be solved if the activity of the biosignals could be recorded without reference. However, it has already been shown that a reference potential is always necessary for metrological recording. However, one could apply an operation to the measured data in which the reference potential is eliminated. In principle, differentiation is suitable for this purpose, where the reference to absolute values is lost.

$$\varphi_{ir} = \varphi_i - \varphi_r \tag{3.3}$$

φ_i and φ_r are potentials at locations i and r, φ_{ir} are the technically measurable unipolar potential differences.

First of all, one can start from a simple but realistic model of a three-channel unipolar derivative (Fig. 3.6). Unipolar potential differences between the discharge points and the reference electrode are measured (Eq. 3.3). In the following, one assumes the simplification that the measuring points lie on a line and the reference is orthogonal to this line at infinity. One can form the first spatial derivative, making the unipolar derivatives become bipolar derivatives (Eq. 3.4).

$$\varphi_{ix} = \frac{\mathrm{d}}{\mathrm{d}x}\varphi_{ir} \approx \left\{\varphi_{0r} - \varphi_{1r};\ \varphi_{0r} - \varphi_{2r}\right\} \tag{3.4}$$

The reference to reference r is lost with the differentiation, but the electrical activity is still formed as a potential difference at two derivative points. Further differentiation is necessary to obtain a reference-free derivative that relates

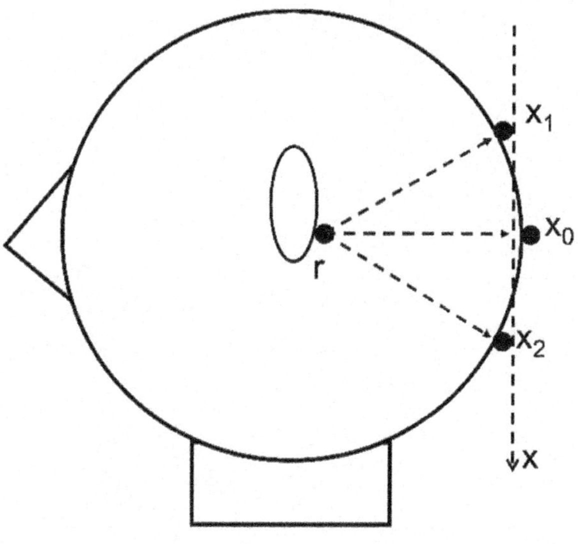

Fig. 3.6 Three-channel unipolar derivative with reference at the ear. For simplicity, the derivative points are assumed to lie on the x-axis line, and x_0 is the array's center. In unipolar metrological acquisition, one measures the potential differences $\varphi_{1r} = \varphi_1 - \varphi_r$, $\varphi_{2r} = \varphi_2 - \varphi_r$, $\varphi_{0r} = \varphi_0 - \varphi_r$

the electrical activity to a single derivative point x_0 (Eq. 3.5). The expression, according to Eq. 3.5 is the one-dimensional variant of the Laplace operator.

$$\varphi_{Q_x} = -\frac{d^2}{dx^2}\varphi_{ir} \approx (\varphi_{0r} - \varphi_{1r}) + (\varphi_{0r} - \varphi_{2r}) \tag{3.5}$$

After the second differentiation, a potential is obtained, which measures the electrical activity at the derivative point x alone$_0$ and is therefore reference-free. This quantity is called the *source potential* because it measures the current source density (CSD, Current Source Density). Figure 3.7 shows the determination of the CSD in one dimension: There are electrically active current sources and current sinks in the body, which generate corresponding potential curves at the surface. The second spatial differentiation of the potential curves reaches its maxima over the current sources and its minima over the current sinks. The potential curves are measured concerning a reference r at the derivative points x_0, x_1, and x_2, corresponding to spatial sampling. The calculation of the source potential according to Eq. 3.5 can be extended to the two-dimensional case (Fig. 3.8, Eq. 3.6):

$$\begin{aligned}\varphi_Q &= -\left(\frac{\partial^2}{\partial x^2} + \frac{\partial^2}{\partial y^2}\right)\varphi_{xy} \approx (\varphi_{0r} - \varphi_{1r}) + (\varphi_{0r} - \varphi_{2r}) \\ &\quad + (\varphi_{0r} - \varphi_{3r}) + (\varphi_{0r} - \varphi_{4r}) \\ &= 4\varphi_{0r} - \varphi_{1r} - \varphi_{2r} - \varphi_{3r} - \varphi_{4r}\end{aligned} \tag{3.6}$$

According to G. 3.6, the relation is known as the *Hjorth operator*. It is valid for sensor arrangements with equal distances to the considered derivation point and approximately planar projection of the electrical activity. It can be extended to any number of sensors with different distances by weighting the individual potentials according to their distance from the considered discharge point. With the Hjorth derivative (Fig. 3.8), a reference-free product (calculated or realized by circuitry) is thus available, fulfilling the theoretical assumptions well. However,

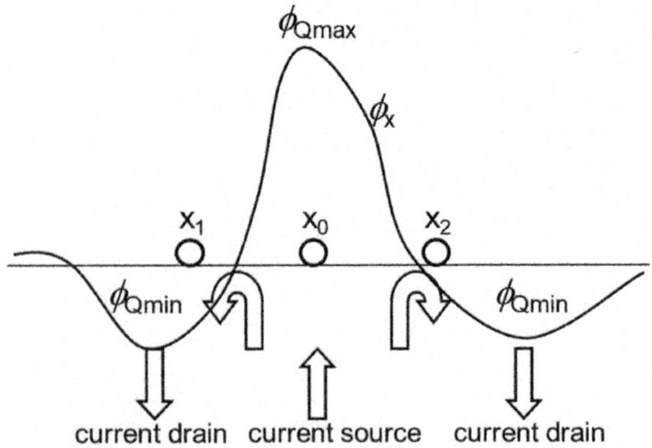

Fig. 3.7 One-dimensional model for the source potential. Electrically active current sources and sinks generate a potential curve on the surface. The second difference of the potential curve (curvature) reaches local maxima over the current sources and local minima over the current sinks. The sensors realize spatial scanning

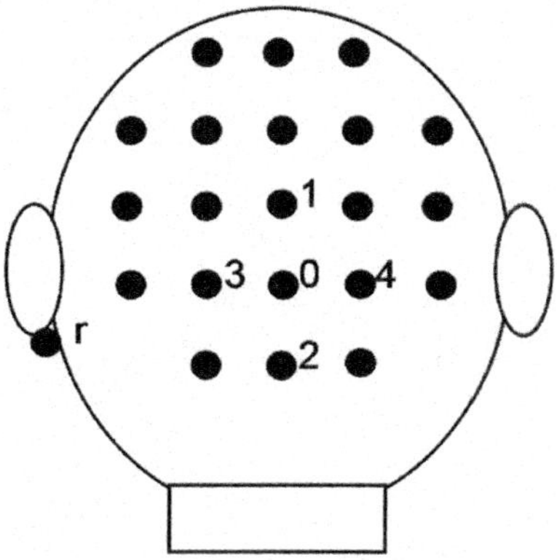

Fig. 3.8 Two-dimensional model for the source potential. The potentials at the leakage points are measured unipolar against the reference *r*. The distances between the discharge points are constant. For each electrode considered, the source potential is given by $4\,\phi_0 - \phi_1 - \phi_2 - \phi_3 - \phi_4$. This way, a reference-free image of the electrical activity at the surface is obtained. The edge effects reduce the number of reference-free points from 21 to 9

some significant disadvantages arise. In Fig. 3.7, according to the spatial sampling theorem, the density of the derivative points would have to be higher than that of the current sources. It is not realizable in practice; one deliberately accepts the sampling theorem's violation.

Furthermore, it is clear from the model that only radial current sources or only radial parts of current sources can be detected. This disadvantageous effect on investigations in which predominantly tangential current references occur. The sensor arrangement, according to Fig. 3.8, reduces the total area of the reference-free derivative by one sensor row at the edge of the matrix. It leads to a loss of information. According to Eq. 3.6, nine source potentials result from the original 21 potential derivatives.

3.2 Biosignal Sampling

3.2.1 Spectral Characteristics of the Scan

The sampling of signals is a mathematically and technologically well-elaborated topic covered by numerous publications. Therefore, only issues directly related to biosignals are selectively discussed here.

Mathematically, sampling corresponds to multiplying the input signal by a Dirac pulse sequence (Fig. 3.9):

$$y(t) = x(t) \cdot \sum_{n=-\infty}^{\infty} \delta(t - nT_A) \qquad (3.7)$$

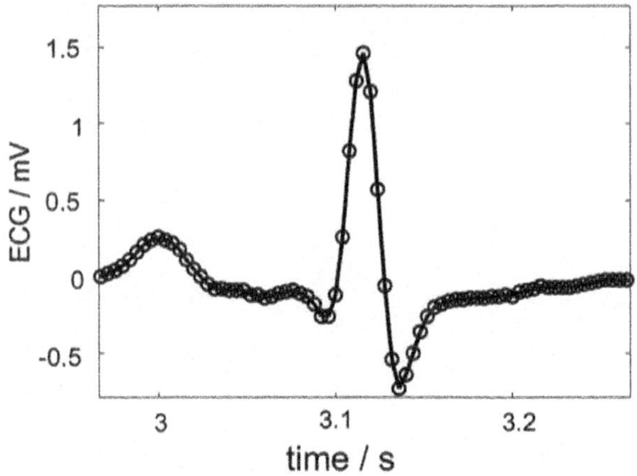

Fig. 3.9 A real ECG multiplied by Dirac pulses (marked by circles). The sampling period is $T_A =$ 8 ms

In Eq. 3.7, x(t) is the input signal, $y(t)$ is the sampling signal, (t) is a Dirac pulse, and T_A is the sampling period. The multiplication of signals in the time domain corresponds to the convolution of their spectra in the frequency domain:

$$Y(f) = X(f) * \frac{1}{T_A} \sum_{n=-\infty}^{\infty} \delta\left(f - n\frac{1}{T_A}\right) = \frac{1}{T_A} \sum_{n=-\infty}^{\infty} X\left(f - n\frac{1}{T_A}\right) \tag{3.8}$$

The spectrum of the Dirac pulse sequence also consists of Dirac pulses whose spacing is $1/T_A$. The original spectrum of the original signal after sampling is periodic with the harmonics of the sampling rate (Fig. 3.10). This can lead to overlapping (aliasing) of the spectra can occur if the sampling rate is lower than twice the maximum frequency of the original signal. Therefore the following must apply

$$T_A \leq \frac{1}{2f_{\max}}. \tag{3.9}$$

The condition according to Eq. 3.9 is the *sampling theorem.* If not met, the spectra overlap after sampling, leading to misinterpretations in the *baseband,* which cannot be corrected. A typical example is shown in Fig. 3.11.

The ECG is often influenced by respiration in the heart rate (respiratory arrhythmia) and the amplitude of the QRS spikes and waves. The latter can be attributed to amplitude modulation in terms of signal theory (see Sect. 2.1.3). In a correctly sampled ECG (Fig. 3.11), the amplitude change due to respiration is visible as an envelope. If the sampling condition is violated, the overlapping parts of the spectrum distort the signal in the baseband uncontrollably (e.g., moiré in images with fine structures, tight shirt pattern in film/television) and thus simulate courses that do not exist in reality. After violating the sampling condition, the same sampled ECG is shown in the lower part of Fig. 3.11. The cardiac actions (QRS complexes)

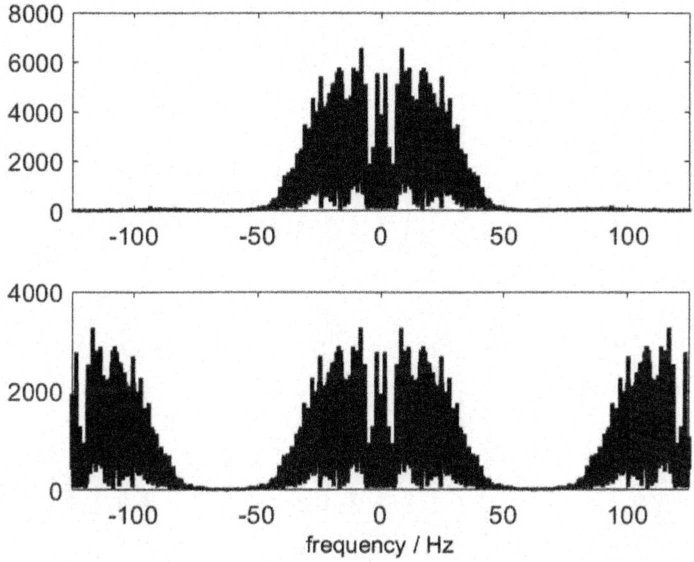

Fig. 3.10 Spectrum of the real (top) and the sampled ECG from Fig. 3.9 (bottom). The sampling period was $T_A = 8$ ms. The ECG was filtered with a 2nd order lowpass filter with a cutoff frequency of 20 Hz before sampling

are preserved, but their amplitudes are stochastic. The envelope generated by respiration and, thus, an essential diagnostic feature was wholly lost. Added to this is the stochastic character of the amplitudes, which can lead to false diagnostic conclusions. For the above reasons, selecting a sampling rate as high as possible is necessary or limiting the signal spectrum to be sampled with an antialiasing filter (lowpass filter) so that the sampling condition (Eq. 3.9) is met. A lowpass filter with a cutoff frequency as low as possible is always helpful since it also suppresses interference lying above the biosignal spectrum. However, it must meet the requirements for the phase frequency response (see Sect. 2.3.4).

One often tries to relativize the demand for linear phase frequency response by working with a much higher sampling rate than necessary. Then a very steep antialiasing lowpass is used, whose phase frequency response is nonlinear but far outside the biosignal spectrum. This situation is shown in Fig. 3.12. Suppose the antialiasing lowpass L(f) is to pass only the biosignal spectrum B(f). In that case, it must also have a linear phase at the expense of the relatively low edge steepness (Bessel filter). The steep lowpass $N(f)$ (mostly Butterworth, rarely Chebyshev) limits the biosignal spectrum effectively. However, due to its nonlinear phase, it must be spectrally significantly higher than Bessel $L(f)$. Because the bandwidth is much larger, the noise power also increases. The conclusion is that a high slope of the antialiasing filter is bought by higher noise. Which of the two effects is better in a specific design must be decided on a case-by-case basis. As a general rule, noise should be minimized.

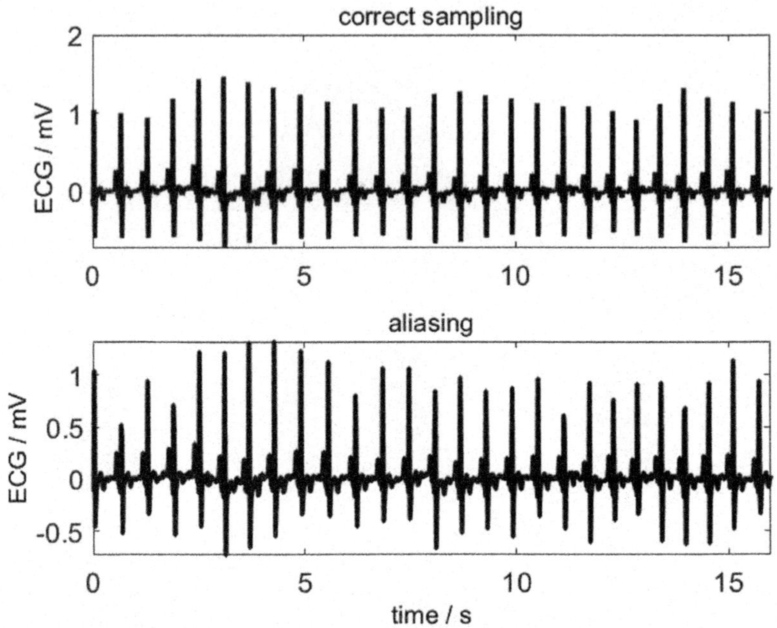

Fig. 3.11 Correctly sampled ECG (top) and undersampled ECG (aliasing, bottom). While in the upper image, the envelope (connecting curve of the R-spikes in the positive area and the S-spikes in the negative area) is pronounced because they originate from respiration (amplitude modulation, see Sect. 2.1.3), in the lower image, the spikes show a stochastic course. The uncontrolled change of the waveform at the bottom is due to violating the sampling condition

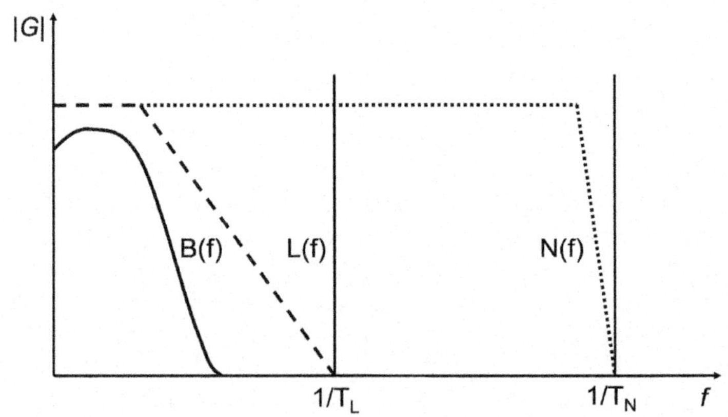

Fig. 3.12 The signal $B(f)$ to be sampled can be filtered with an antialiasing lowpass $L(f)$ with linear phase and flat filter edge and sampled with period T_L. It can be filtered with a steeper filter $N(f)$ but a nonlinear phase at a higher frequency and sampled with a shorter period T_N. The nonlinearity of the phase of $N(f)$ need not affect $B(f)$ if T_N is small enough (at least one decade smaller than T_L). However, any interference spectrally between $L(f)$ and $N(f)$ may affect the measurement

Nevertheless, if one wishes for a steep filter slope $N(f)$, which is inevitably accompanied by a nonlinear phase (Butterworth, Chebyshev, Cauer), the cutoff frequency must be chosen high enough. Then the nonlinearity of the stage does not affect the biosignal spectrum $B(f)$. After analog filtering with $N(f)$, the signal can be digitally filtered (see section Digital Filtering) and subsampled (the sampling rate is reduced). Methodically, both possibilities—filtering with $L(f)$ or $N(f)$ and undersampling—are not equivalent. Since $N(f)$ has a nonlinear phase, it remains nonlinear even after undersampling. Therefore, a Bessel-lowpass with a minimum necessary cutoff frequency is preferred in principle.

From the point of view of interference resistance, the alternative $L(f)$ is to be preferred: Once an interference between $L(f)$ and $N(f)$ reaches the amplifier, it can override it and make the measurement worthless. Another major disadvantage of the $N(f)$ variant is that, due to the dependence of the noise on the bandwidth (Eq. 2.18, see Sect. 2.2.1), the noise is significantly higher than with the $L(f)$ filter. The conclusion of these analyses is:

> The sampling rate should not be higher than necessary, and the antialiasing lowpass should have a linear phase frequency response (Bessel).

Some commercial solutions try to solve the problem of noise in the $N(f)$ variant by digitally filtering the sampled signal after sampling with T_N and then subsampling it (see section Digital Filtering) (see section Digital Filtering). The noise is reduced, but the phase becomes nonlinear in the signal $B(f)$ range. As a result of the undersampling, the signal spectrum $B(f)$ *is* spread and finally lies in the nonlinear transmission range of the filter $N(f)$. This effect can be seen in Fig. 3.12: If we first assume both alternatives, we have, on the one hand, a signal filtered with $L(f)$ and sampled with T_L, and on the other hand, a signal filtered with $N(f)$ *and* sampled with T_N (here with $T_L = 2T_N$). After subsampling to half the sampling rate, the characteristic $N(f)$ compresses to half in the direction of the frequency axis and lies precisely above the characteristic $B(f)$ with the consequence of the nonlinear phase, which was far to the right before subsampling (Fig. 3.13). It is also clear from Fig. 3.13 that a digital filter for spectrum limiting before undersampling cannot do anything in terms of nonlinear phase except prevent possible violations of the sampling condition. These considerations lead to the following consequences for development and practical analysis: One should always use linear-phase filters (Bessel) should be used, even at the expense of the edge steepness. The lowest possible sampling rate should always be used to minimize noise, including the filter edge, which satisfies the sampling condition, assuming the linear phase of all filters.

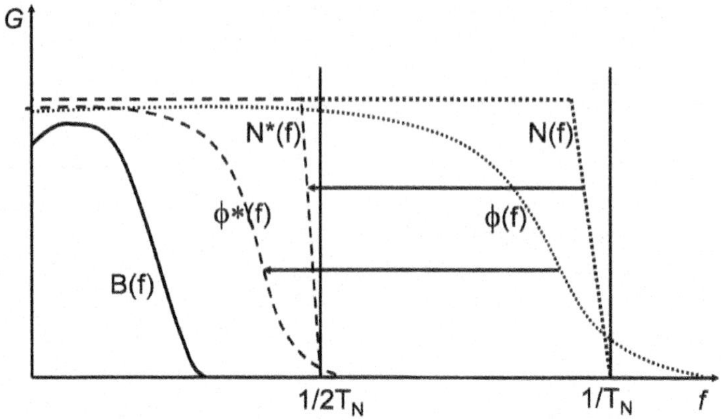

Fig. 3.13 Effect of undersampling on the half sampling rate: The amplitude-frequency response of the lowpass compresses from $N(f)$ along the frequency axis to half $N^*(f)$, and the filter edge becomes steeper. The phase frequency response also compresses from (f) *to half* (f) to lie in the middle of the signal spectrum as an amplified nonlinear phase

3.2.2 A Sampling of Bandlimited Signals

Biosignals are generally multichannel analog modulated and transmitted signals, especially in wireless telemetry systems. A typical scenario arises in the project to monitor three ECG channels wirelessly in the central office of a rehabilitation clinic while the cardiological patients move freely outside. A simplified block diagram of such a system is shown in Fig. 3.14. A three-channel ECG lead (Einthoven) is amplified, and each amplitude is modulated onto a low-frequency subcarrier (kilohertz range). The AM-modulated ECG channels are adjacent in the spectrum, called frequency division multiplexing. They are summed up and frequency modulated onto a primary high-frequency carrier (megahertz range). Then the frequency mixture is amplified to a sufficient power and broadcast via the antenna. On the receiving side, these steps must be completed in reverse order. Due to high-speed digitization, a signal processor can process all operations after the FM demodulator (frequency demultiplexing, AM demodulation). The signal coming from the FM demodulator must still be sampled first. The spectral position of an AM-modulated ECG channel is shown in Fig. 3.15. To sample according to the sampling theorem (Eq. 3.9), the sampling period for the AM signal would have to satisfy the condition (the bandwidth of the ECG spectrum here is 120 Hz)

$$T_A < \frac{1}{2f_{\max}} = \frac{1}{2(500\ \text{Hz} + 120\ \text{Hz})} = 0.8\ \text{ms}.$$

From this relation, a minimum basic frequency of the sampling rate of 1240 Hz follows. It is technically not a problem to realize that sampling. It can become a problem with considerably more channels and higher frequencies. Therefore, the question must be answered whether the requirement according to Eq. 3.9 is a

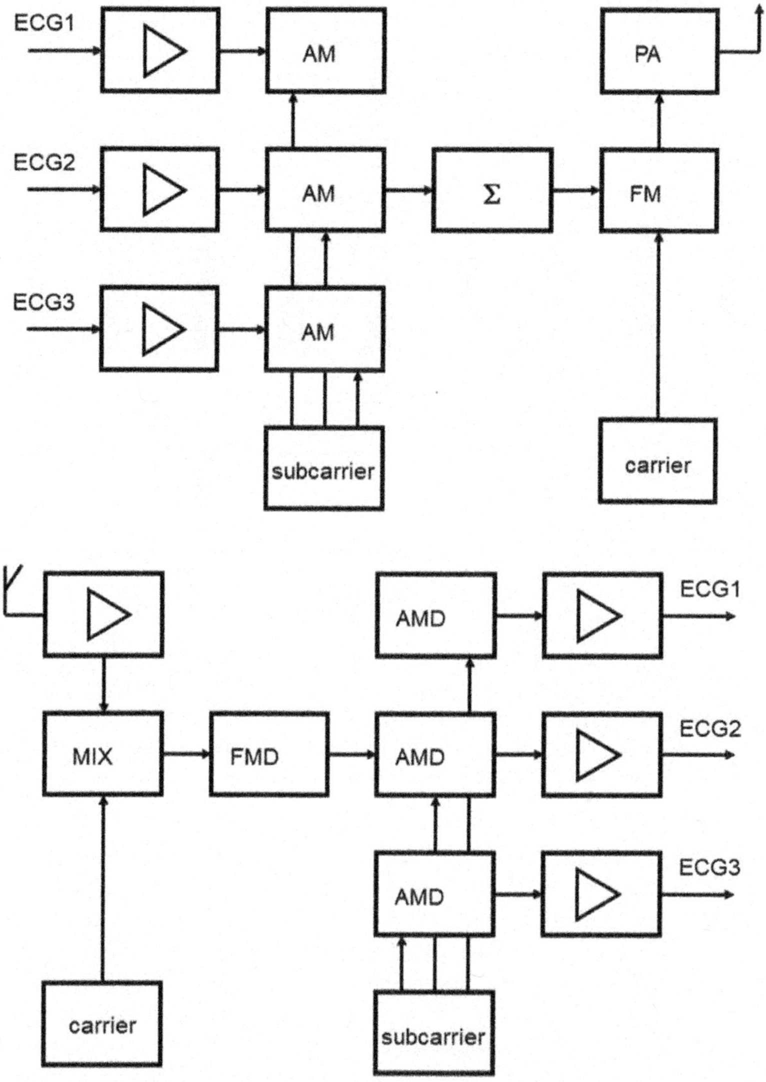

Fig. 3.14 Block diagram of the transmitter part (top) and the receiver part (bottom) of a system for biotelemetry. AM are amplitude modulators, FM frequency modulators, and PA power amplifier (top). FMD and AMD are frequency and amplitude demodulators. Signal processing after FMD can be done digitally, e.g., in a DSP or FPGA

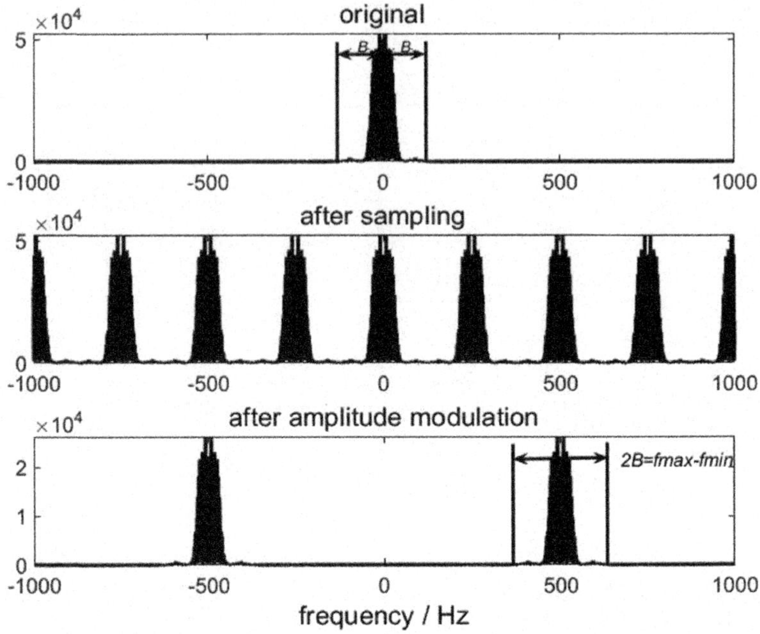

Fig. 3.15 Spectral position of the original signal in the baseband (top) and the signal sampled with $T_A = 4$ ms (middle). The amplitude-modulated signal on a subcarrier of 500 Hz is shown below. Note that the original spectrum is mirrored around the subcarrier (bottom), as it is during sampling (middle). However, the spectra have no periodicity (below) because the subcarrier is a harmonic signal and thus produces only one needle in the spectrum

necessary and *sufficient* condition. From what has been said, this condition is sufficient: if it is satisfied, the aliasing does not occur. The comparison of the spectra in Fig. 3.15 center and the bottom shows that the original spectrum is mirrored around the harmonics of the sampling rate and the carrier signal, respectively, both after sampling and after amplitude modulation (lower and upper sideband). However, there is no periodicity in the AM signal as in the sampling signal. The spectrum of the AM-modulated ECG coincides with one of the spectra of the sampled signal when the frequency of the AM carrier is equal to a harmonic of the sampling rate. Therefore, for the analysis, it does not matter at which sampling rate the AM-modulated ECG is sampled if it is equal to the integer multiple of the minimum sampling rate in the baseband. Therefore, the AM signal on a carrier of 500 Hz can also be sampled at the baseband sampling rate of 240 sps (sps—samples-per-second, see Fig. 3.10) because the sampled signal is processed in the baseband. Thus, for the sampling condition, it is not the highest frequency present in the signal that is important. Still, the actual bandwidth B of the signal to be sampled (Eq. 3.10). Similar to the antialiasing lowpass, the spectrum must be limited with a bandpass before sampling.

$$T_A < \frac{1}{2B} = \frac{1}{f_{\max} - f_{\min}} \tag{3.10}$$

Applied to this concrete example, this means that even the AM signal, whose highest frequency is 620 Hz, can be sampled at the original sampling rate of 125 sps since its width does not exceed 62.5 Hz. The sampling condition is given in Eq. 3.9 is *not a necessary* condition for bandlimited signals. For bandlimited signals, the sampling condition, according to Eq. 3.10, applies.

3.2.3 Scanning in Multichannel Systems

Sampling systems are technically composed of sample and hold (S&H) elements. In addition to the primary function of sampling, holding is also necessary (analog storage of the sampled value) since the subsequent AD converter requires a specific time for digitization (Fig. 3.16). In multichannel systems, the technical effort of digitization can be reduced enormously by using only one AD converter for all channels. The individual channel signals appear sequentially in time at the input of the AD converter, and they are converted in time division multiplex. The technical effort can be further reduced by applying time division multiplexing before sampling (Figs. 3.17 and 3.18). The function not necessarily to be executed simultaneously is shifted from simultaneous processing to sequential processing on the time axis. Thus the function must be present only once instead of several times.

Sequential sampling can minimize the effort, especially for multichannel systems (256-channel EEG). However, this hardware structure can lead to signal analysis problems. Between adjacent channels, the time offset is

$$T_V \le \frac{1}{N} T_A, \tag{3.11}$$

between the first and the last channel $(N - 1)T_V$, T_A is the sampling period, and N *is* the number of channels. However, simultaneous values are required in many topographic and spatial signal processing methods. In the example of CAR, the effects of the time multiplex are shown. In the calculation rule (Eq. 3.1) for the CAR time-shifted sample values, enter:

$$c(k) = \frac{1}{M} \sum_{i=1}^{M} x_i(kT_A - (i - 1)T_V). \tag{3.12}$$

In Eq. 3.12, k is the time index, i *is the* channel index, T_A is the sampling period, T_V is the time offset between two adjacent channels, x_i signal in channel i, M number of channels, $c(k)$ is the CAR.

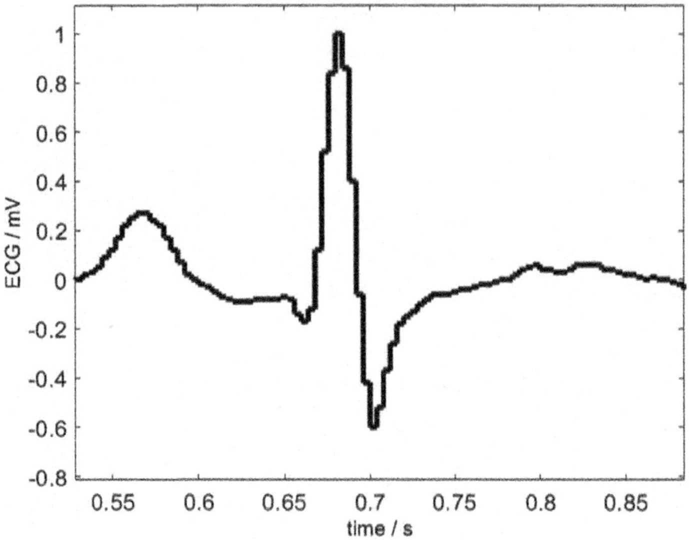

Fig. 3.16 Sampled biosignal (ECG) after S&H with analog voltage levels

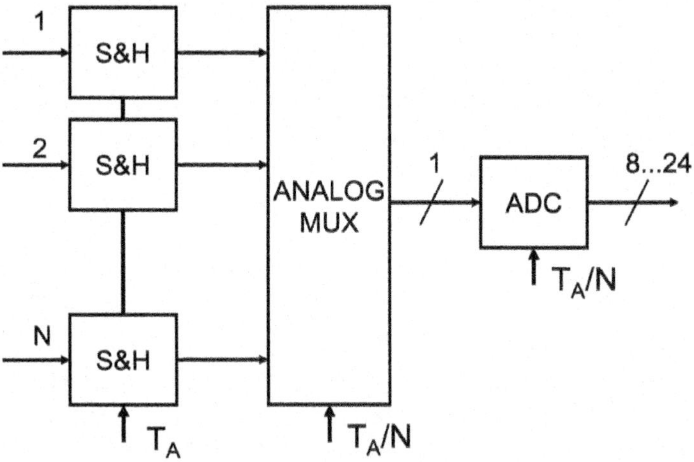

Fig. 3.17 Multichannel system with simultaneous sampling. All channels are sampled simultaneously and fed sequentially to the AD converter in a time division multiplex

After the Fourier transform into the frequency domain, the following applies

$$C(f) = \frac{1}{M} \sum_{i=1}^{M} X_i(f) e^{-j2\pi f(i-1)T_V}. \tag{3.13}$$

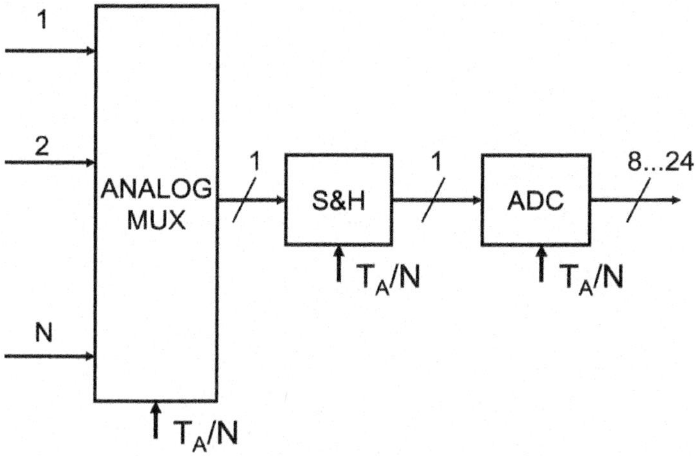

Fig. 3.18 Multichannel system with sequential sampling. The channel signals are sequentially sampled, buffered, and converted via the analog time multiplexer

Under the simplifying assumption that all channel signals are identical in shape, the phase part of the relation, according to Eq. 3.13, can be interpreted as a transfer function:

$$G(f) = \frac{1}{M}\sum_{i=1}^{M} e^{-j2\pi f(i-1)T_V}. \tag{3.14}$$

According to Eq. 3.14, the transfer function's magnitude corresponds to the amplitude-frequency response of a lowpass filter. The lowpass filter reaches the first zero in the amplitude-frequency response at $= 1/MT_V$. Since Eq. 3.11 applies to the offset, the first zero of the amplitude-frequency response of $G(f)$ lies above the first harmonic of the sampling rate and reaches at the Nyquist frequency a drop of 3 dB. This value is not critical because the biosignal at the Nyquist frequency should be suppressed by at least 60 dB regardless to avoid aliasing. Nevertheless, sequential sampling should be well-checked and used with caution. In some special applications, especially in ECG acquisition, time shifts of a few fractions of the sampling period play an important role. For example, a standard 12-channel ECG is sampled at 1000 sps per channel ($T_V = 1$ ms). This results in a sampling period T_A of at least 12 ms or a sampling rate of 83.3 sps. If a sampling period of 12.5 ms is selected, the first harmonic of the sampling rate is 80 Hz, and the Nyquist frequency is 40 Hz.

Due to the sequencing, the time offset between the first and the last channel can reach up to 11 ms. This latency would be too large an error for specific applications (pacemakers). The time offset and, thus, also the phase shift in the frequency domain are usually known. Therefore, it can be compensated by a phase correction,

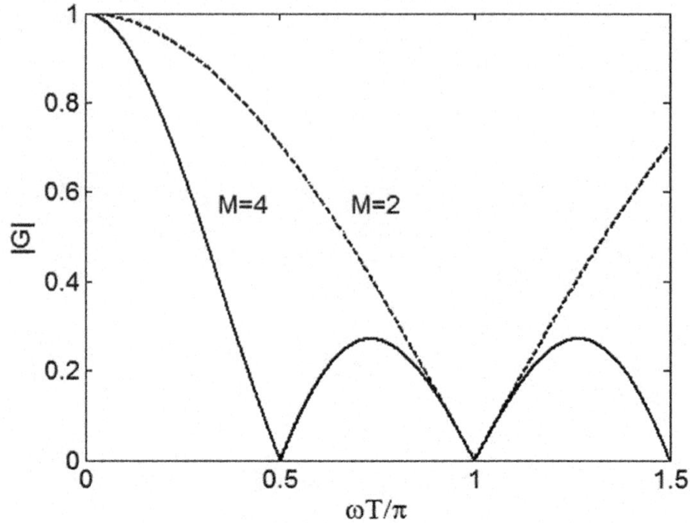

Fig. 3.19 Amplitude frequency response of the transfer function of the CAR after sequential sampling. The lowpass reaches the first zero at a time offset of T_V /2M

according to Eq. 3.15, even without knowledge of the biosignal.

$$X_i^*(f) = X_i(f)e^{j2\pi f(i-1)T_V} \tag{3.15}$$

The phase correction is correct from the point of view of signal theory and compensates in real signals for the shifting T_V caused by the sequencing. However, this correction could be more capable of real-time since the FFT usually requires a sufficient block length. In addition, problems in calculating the Fourier transform are caused by the transform itself (edge effects, leakage) and by the signal character (transient, ECG, and EMG impulsive). These considerations conclude that simultaneous sampling should be considered for a few channels for biosignal acquisition (M < 30), and sequential sampling should only be considered for a huge number of channels (M > 100), but with acceptable consequences for signal analysis (Fig. 3.19).

3.3 Digitization of Biosignals

In this chapter, selected digitization methods relevant to biosignal processing are discussed. Biosignals are low-frequency processes (upper cutoff frequency up to about 10 kHz) with significant dynamics (up to 120 dB). Accordingly, AD converters with low sampling rates and large bit widths are required.

3.3.1 Integrating Transducers

The basic principle of integrating converters is that the input voltage is integrated until a threshold is reached, while the integration time can be measured digitally. According to Fig. 3.20, OVA integrates the input voltage U_e during the time T_I (Eq. 3.16), and OVB compares it with the reference voltage U_{ref}.

$$U_i = -\frac{1}{R_i C_i}\int\limits_{T_I} U_e(t)\mathrm{d}t \tag{3.16}$$

The integration time T_I is measured digitally by counting clock pulses between the start of integration and the time the integrated input voltage U_i reaches the reference voltage U_{ref} (Fig. 3.21). The number of pulses counted is a digital measure of the input voltage. Integrating methods have two significant advantages from the point of view of signal analysis. First, noise and most disturbances are stochastic or periodic processes with zero time average. Therefore, temporal integration leads to the suppression of unwanted components. On the other hand, integration over a specific time T_I has a comb-filter character from a spectral point of view (see section Digital Filters). Thus, the integration time T_I can be set so that, for example, the harmonics of the primary frequency are suppressed (Fig. 3.22).

However, the actual integration time T_{AD} of the conversion does not correspond to the fixed time T_I; it is always shorter. The length of T_{AD} is determined by the time when the comparator stops the counter, so it depends on the actual input voltage (Eq. 3.17).

$$T_{AD} = \frac{U_{ref}}{\overline{U}_e} R_i C_i \leq T_I \tag{3.17}$$

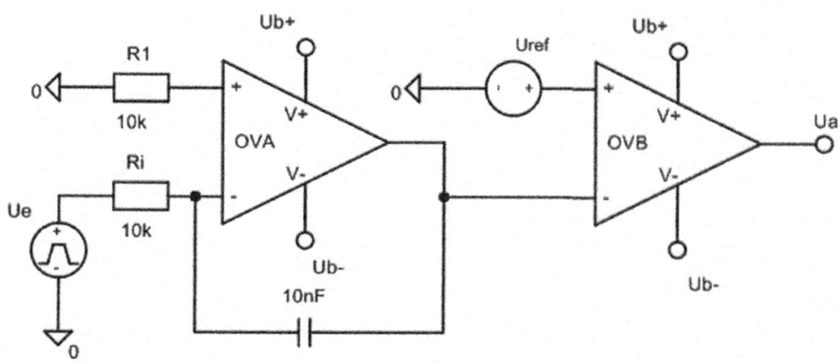

Fig. 3.20 Analog part of an integrating converter. The input voltage U_e is integrated with the time constant $R_i C_i$ starting at zero and compared with the reference voltage Uref in the comparator. The time between the start of integration and reaching the reference voltage is measured digitally by counting clock pulses

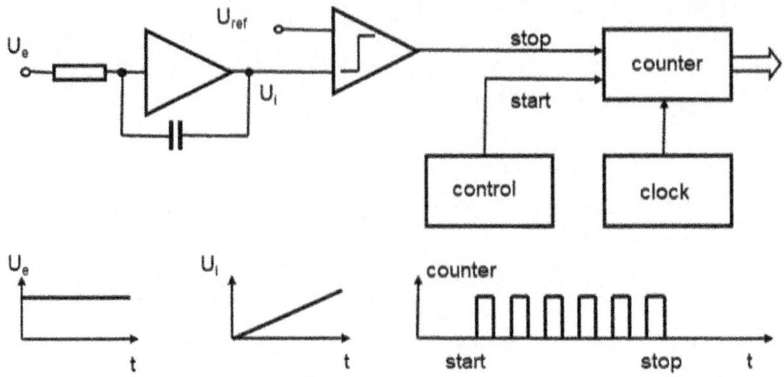

Fig. 3.21 Block diagram of an integrating A/D converter using the single-ramp method. The controller starts the integration and the digital counter. If the integrated voltage U_i reaches the reference voltage U_{ref}, counting is stopped and restarted

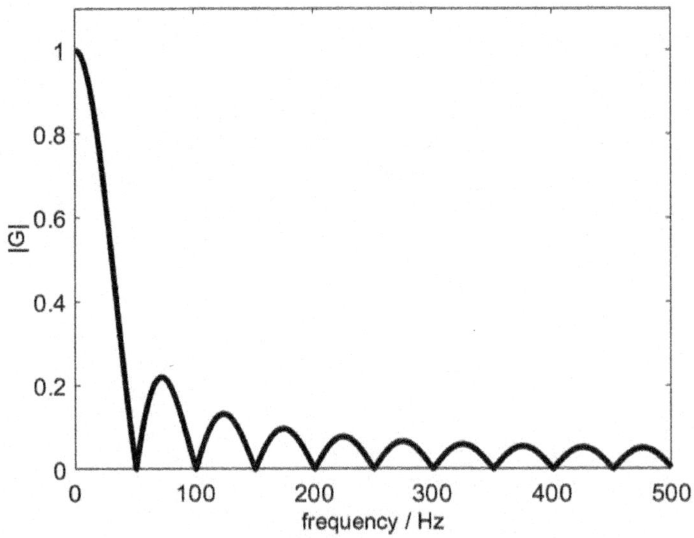

Fig. 3.22 The temporal integration over a fixed time has a comb filter character in the spectrum. With an integration time of $T_{AD} = 20$ ms, all harmonics of the mains frequency are eliminated

Despite the inequality in Eq. 3.17, suppression of unwanted frequencies can be achieved by choosing T_I as a multiple of the frequency period to be suppressed. Concerning the suppression of interference, the two-ramp method offers a better solution. From the relation Eq. 3.17 seen that the measured time T_{AD} depends on the time constant RiCi. Thus, the devices' tolerances and variations determine the conversion's accuracy. Even with precise and stable components, with the single-ramp method and integration times of up to one second, an effective resolution of at most ten bits can be achieved. The dual-slope conversion method offers a solution to the problem of fluctuations (drift) and the non-constant integration time. Figure 3.23 shows the block diagram of the dual ramp method. The first phase T_{i}, in which the input voltage U_{e} is integrated, is constant and can be adjusted according to the previous considerations to suppress unwanted interference frequencies. For example, if Ti is 100 ms, all harmonics of the fundamental wave of 10 Hz are eliminated.

$$U_i(T_i) = \frac{1}{RC}\int_{T_i} U_e(t)\mathrm{d}t = \frac{T_i}{RC}\overline{U}_e \tag{3.18}$$

$$T_{AD} = RC\frac{U_i(T_i)}{U_{ref}} + T_i = T_i\left(\frac{\overline{U}_e}{U_{ref}} + 1\right) \tag{3.19}$$

In the second phase, the internal, undisturbed reference voltage U_{ref} is integrated, and the time T_{AD} reaches zero is measured digitally. Since T_i integration time is constant, the integrated voltage U_{i} depends solely on the input voltage U_{e} and on time constant RC (Eq. 3.18). The subsequent opposite integration with the constant reference voltage U_{ref} cancels the influence of the time constant RC according to Eq. 3.19. Thus the parameter fluctuations of the components could be

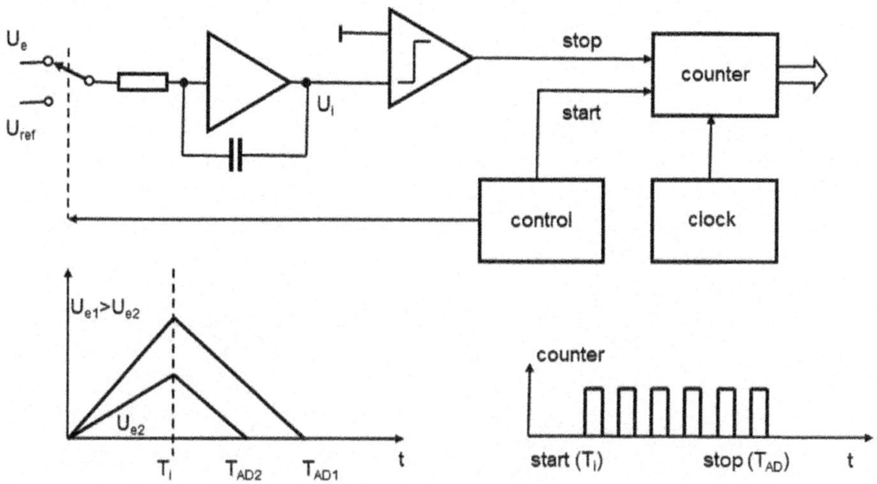

Fig. 3.23 Integrating AD converter using the two-ramp method. The integration time T_{i} is constant. The voltage U_{i} corresponds to the time integral of the input voltage U_{e} after T_{i}. In the second integration phase, the input voltage equals U_{ref}, so the time TAD reaches zero is directly proportional to the input voltage U_{e}. The clock counting starts at T_{i} and ends, controlled by the comparator, when the zero line is reached at T_{AD}

more effective, and an effective resolution of up to 17 bits is achieved. Integrative converters, however, could be faster, achieving sampling rates of less than 100 sps. Their effective resolution is sufficiently high for most requirements, making them suitable for slow biosignals and parameters with a dynamic range of up to 80 dB. A disadvantage of integrative conversion methods is that the conversion time is variable and depends on the specific input voltage. Therefore, one must always expect the longest integration time or the highest input voltage, although this is never utilized.

Typical applications are the measurement of body temperature, weight, respiration, or oxygen saturation. Integrating transducers are technologically less demanding and are often manufactured directly with drivers for displays, making them particularly suitable for portable devices.

3.3.2 Successive Approximation

The conversion principle is that a digitally weighted voltage U_{DAC} (Figs. 3.24 and 3.25) successively (stepwise) approximates the input voltage U_e. In principle, different approximation methods are possible. Based on the objective that the absolute digitization error must be at most $U_{LSB}/2$ (quantization noise) and the conversion time should be minimal, the converter approximates starting with the MSB successively to the LSB. For each bit, the current-voltage U_{DAC} is compared with the input voltage U_e and the current bit is set or reset to always be below the input voltage. It ensures that the voltage U_e is approached with a monotonic progression from U_{DAC}, and the best possible approximation is obtained after passing through all bits.

Since a specific time is required for the approximation, it must be ensured that the voltage at the comparator input U_{AH} does not change during this time. For this purpose, the input voltage U_e must be sampled and held (S/H module in Fig. 3.24). It is a disadvantage compared to integrating methods since noise and interference cannot be reduced here. However, the SAR (successive approximation register) method has significant advantages over integrating methods. The analog functional parts S/H and comparator are mainly independent of fluctuations and external influences, and the DA converter is also very stable concerning its analog part. The conversion time is short and constant due to the primarily digital and digitally controlled circuitry. The SAR converters achieve sampling rates from 10^4 sps to 10^6 sps and bit widths up to about 16 bits. Thus, due to their conversion speed, they have sufficiently good resolution and dynamic range for most biosignals and are also suitable for multichannel time-division sampling systems.

3.3.3 Delta-Sigma Conversion

A dynamic range of up to 120 dB is required for some biosignals. For example, one would like to record evoked potentials with amplitudes in the microvolt range

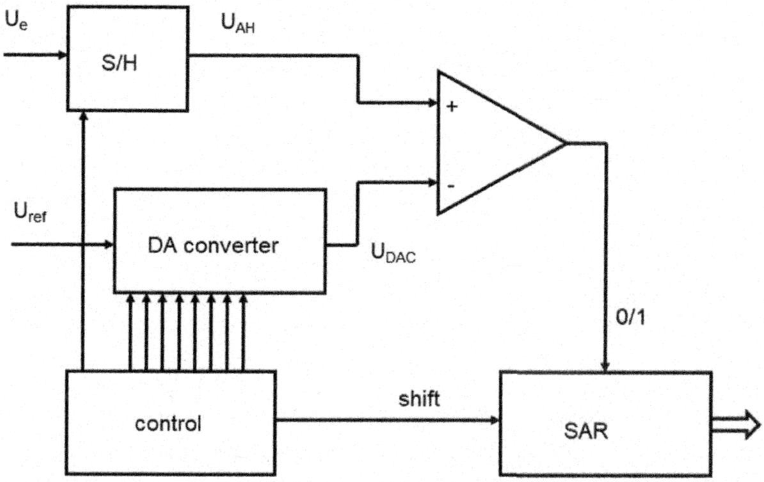

Fig. 3.24 AD conversion by successive approximation. Starting with the voltage U_{DAC} for the MSB, the input voltage U_e is compared, and the corresponding bit is set. The shift register SAR (Successive Approximation Register) contains the digital equivalent of U_e after all bits have been compared

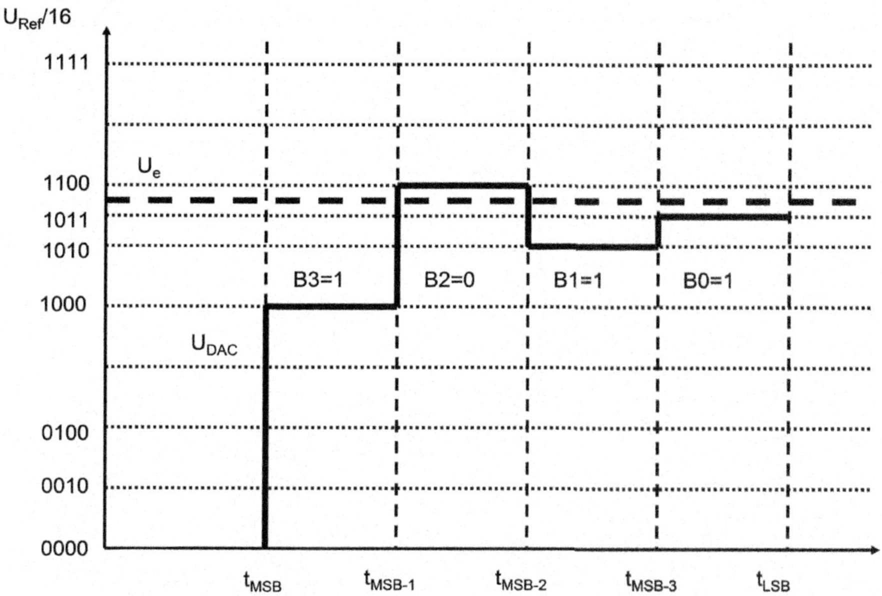

Fig. 3.25 Successive approximation in the timing diagram. Starting with the most significant bit MSB, the comparator compares the input voltage U_e with the output voltage U_{DAC} of the DA converter (DAC) at times $t_{MSB\text{-}k}$. If U_e is above U_{ADC}, the bit is set; otherwise, not

with a true DC EEG system. Since the electrode voltage (difference between electrodes due to drift) can be several hundred millivolts, the required dynamic range reaches $1{:}10^6$ or 120 dB. The integrating and SAR converters cannot cover this range for technological reasons (accuracy and stability of the conversion-critical components). The two methods discussed so far are based on simultaneously executed functions in the time domain (sampling or time division multiplexing of channel signals). The resolution, quantization noise, SNR, and accuracy depend directly on the bit width. Since the analog functional parts reach their limits with, at best, 0.01% accuracy and stability, only a maximum of 16-bit conversion width can be achieved. However, it is possible to *generate* digital information serially or sequentially. It has to be distinguished from AD converters with serial data transmission.

As already shown in Sect. 3.2.3, the basic approach is to sequence an operation that is not necessarily to be executed in parallel. One shifts the hardware-wise complex problem of N-fold parallel processing to the technically considerably simplified problem of serial processing in time sequence (time multiplex). The consistent implementation of this idea results in a one-bit converter whose primary function is known in communications engineering as a delta modulator (δ) (Figs. 3.26 and 3.27). From the input signal $x(t)$ and its integrated binary variant $x_i(t)$, the difference is formed, which is evaluated for its sign with a comparator (1-bit quantizer). The sign appears clocked at the output of a D flip-flop (1-bit memory) as a binary sequence. This sequence qualitatively corresponds to a signal after pulse width modulation; its time average value yields the input signal again. When reconstructing (demodulating) a delta-modulated signal (Fig. 3.28 center), the operations performed during modulation (Fig. 3.28 top) must be reversed.

First, delta modulation integrates into negative feedback, which affects the overall transfer function, like differentiation. Therefore in the demodulator, integration is necessary. Secondly, the comparator binarizes the difference (sign) in the modulator and passes it on the clock. Therefore, high-frequency clock components

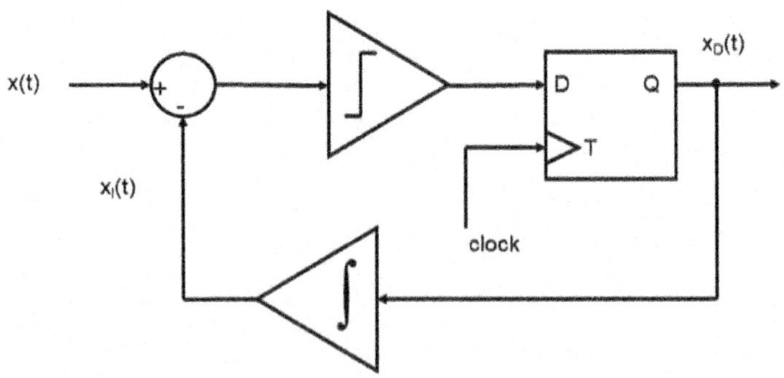

Fig. 3.26 Delta modulator, a block diagram

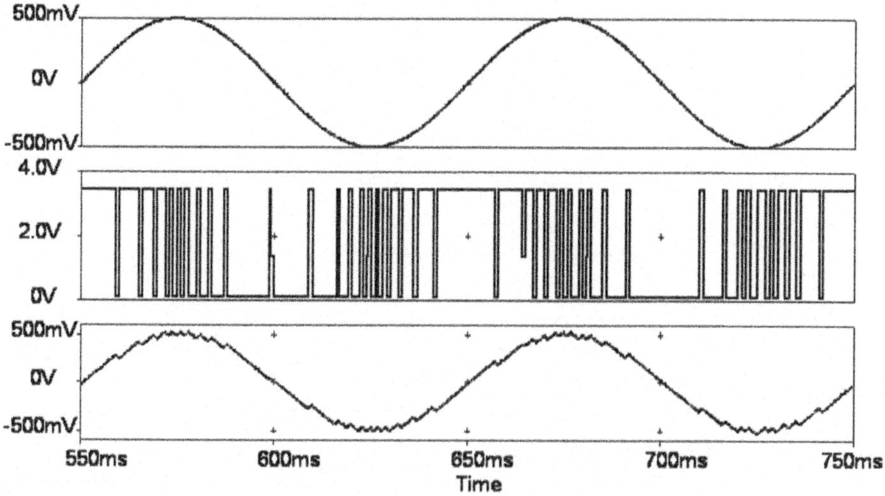

Fig. 3.27 Time courses of the signals in the delta modulator: the input signal $x(t)$ (sine wave) forms a difference with the integrated output signal $x_D(t)$. It is compared in binary form; a one-bit quantizer is present. The signal information is coded in the time interval of the output pulses

remain in the demodulator after integration, which must be eliminated by a low-pass filter. Thus, the integration from the demodulator can be shifted in front of differentiation in the modulator. It will not change anything in signal processing (linearity, superposition principle). Finally, both integrations—the one in the negative feedback and the one before the difference formation—can be combined and placed after the difference formation. Thus the total function becomes more straightforward, and in the demodulator, one needs only a low pass (Simulink simulation BSV_3_1.mdl). This function is called a *delta-sigma modulator* because of the different formations and subsequent integration.

$$Y(j\omega) = (X(j\omega) - Y(j\omega))\frac{1}{j\omega T} \Rightarrow \frac{Y(j\omega)}{X(j\omega)} = \frac{1}{1 + j\omega T} \tag{3.20}$$

$$Y(j\omega) = -Y(j\omega)\frac{1}{j\omega T} + N(j\omega) \Rightarrow \frac{Y(j\omega)}{N(j\omega)} = \frac{j\omega T}{1 + j\omega T} \tag{3.21}$$

The integration in the modulator causes two favorable effects regarding the spectral properties. On the input signal, it acts like a low pass and can already be used as a simple antialiasing filter (Fig. 3.29, Eq. 3.20). On the quantization noise; it acts like a high-pass filter, which can be dimensioned to minimize the noise in the used frequency band (Fig. 3.29, Eq. 3.21). Noise minimization is also urgently required with the required dynamic range of 120 dB.

By connecting several delta-sigma stages in series, the filtering effect (Eq. 3.21) on the quantization noise can be increased considerably, thereby increasing the

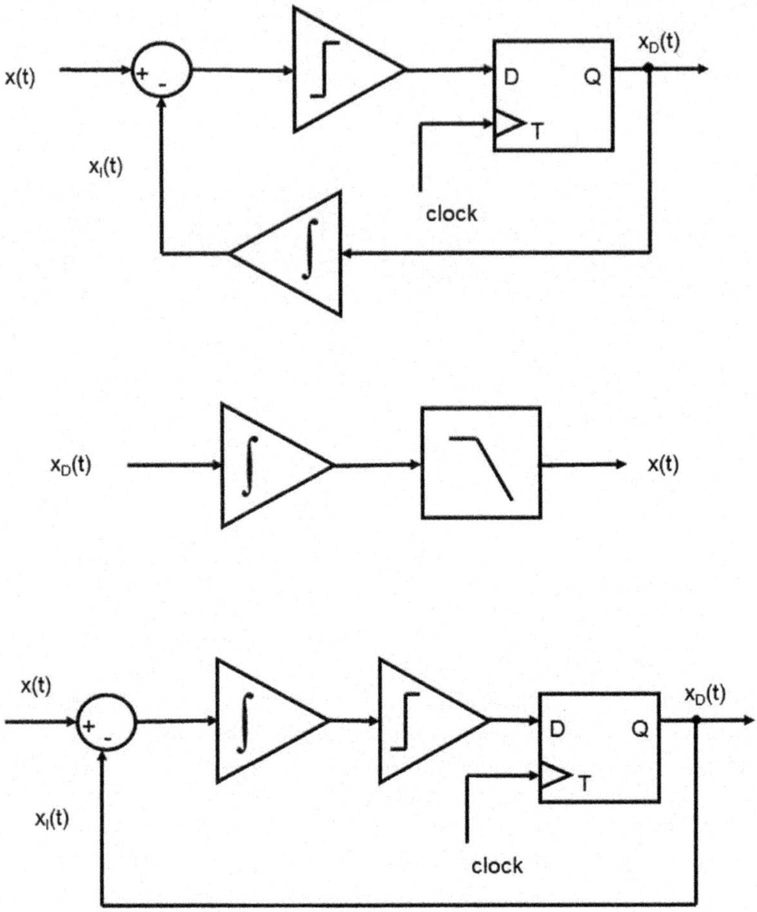

Fig. 3.28 Delta modulation and delta-sigma modulator: the input signal $x(t)$ is converted into the binary sequence $x_d(t)$ by a delta modulator (top). For reconstruction (demodulation), it must be integrated and filtered with a lowpass filter (middle). The linear integration of the demodulator can be placed in front of the differentiation of the modulator. Thus the demodulation is reduced to lowpass filtering. Then one can combine the integrator at the input of the modulator with the integrator in its feedback and apply it to the differential signal (below). That is a delta-sigma modulator

SNR and the dynamic range (Fig. 3.30). In medical practice, delta-sigma converters of up to fourth order are sufficient. The limiting parameter for further SNR improvement is the inherent noise of the measurement amplifier and the electrodes used, which generate a total noise voltage of up to 10 microvolts. Therefore, further reduction of the quantization noise below this level is needed to improve the signal quality further.

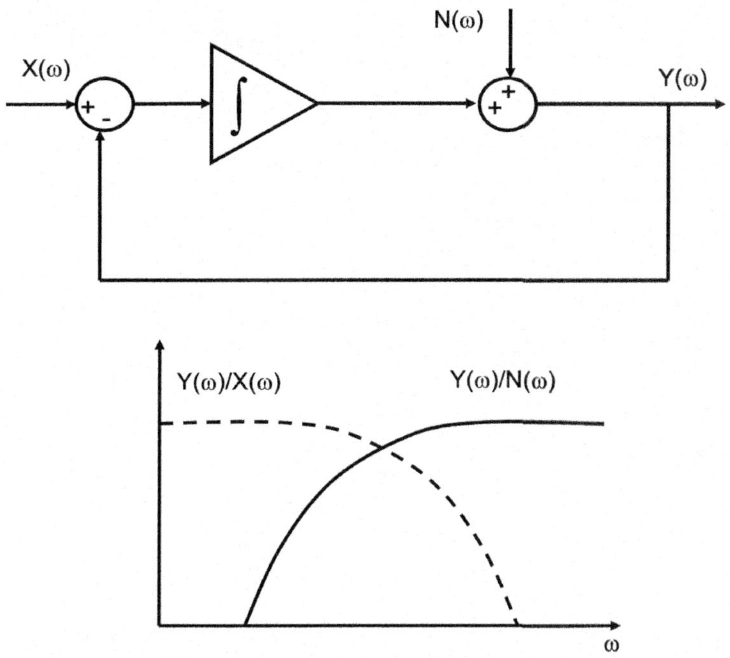

Fig. 3.29 Filter effect of the integration in the delta-sigma modulator: the input signal acts like a lowpass filter and on the quantization noise, like a high-pass filter. The comparator was replaced by summation with additive quantization noise for linear signal analysis

The input signal $x(t)$ of a $\Delta - \sum$ modulator is encoded in the binary output sequence $x_D(t)$ in such a way (Fig. 3.28) that the time average of this sequence corresponds to the input signal. Therefore, for signal reconstruction (demodulation), a simple interpolation filter with a lowpass filter would suffice. It is necessary to convert the bit sequence into parallel binary information to be used as a converter. For this purpose, the time average is first formed—as in the reconstruction—but here digitally. One counts the number of logical ones in a time window of the length of *N-bit* places. It reduces the sampling rate by a factor of *N*, and one obtains a binary word with $\log_2 N$ bits. A typical example is shown in Fig. 3.31.

The $\Delta - \sum$-Modulator, for example, works with a sampling rate of 256 ksps. It has the spectral characteristic of a lowpass filter (Eq. 3.20). The average value over 64 binary values is digitally determined from the bit sequence and filtered. It corresponds to the spectral characteristic of a comb filter with 16-bit width and a 64-fold subsampling. Further, it is corrected with the subsequent FIR filter (Fig. 3.31 right) to keep the characteristic filter flat in the baseband, and the data is subsampled by a further factor of 4. In total, a width of 24 bits and a sampling rate of 1 ksps are achieved for biosignals.

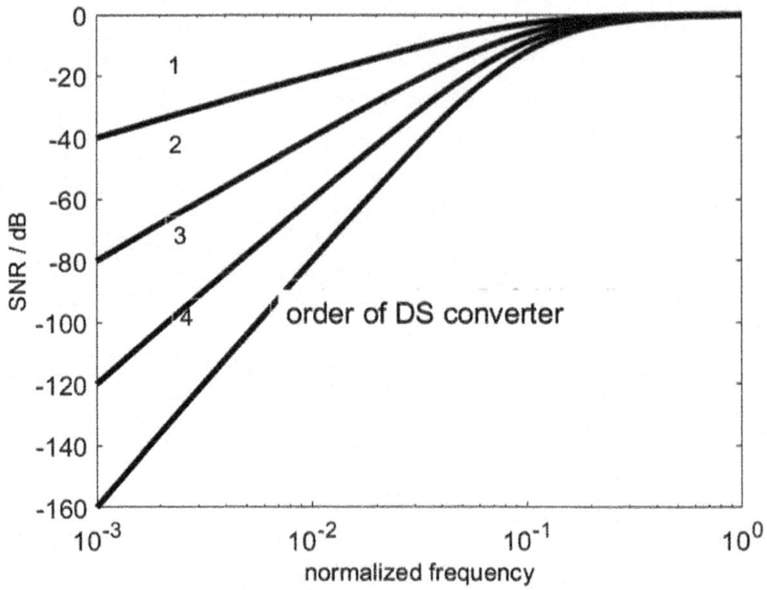

Fig. 3.30 High-pass effect of a $\Delta - \sum$-converteron the quantization noise as a function of the order of the converter (number of $\delta\sigma$) with identical integration time constant

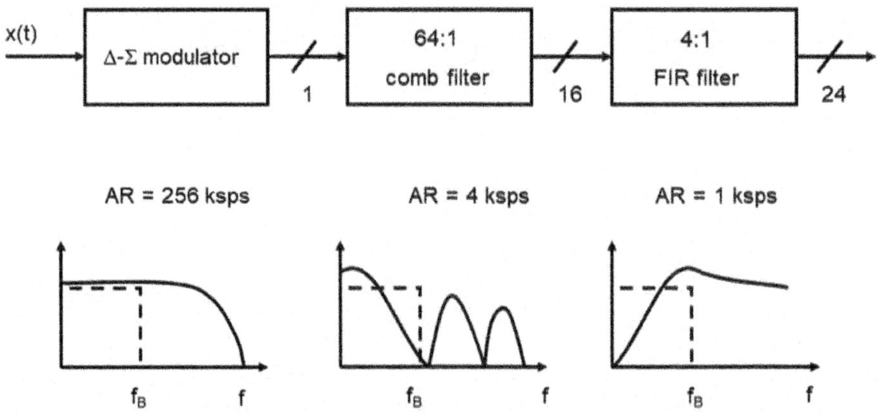

Fig. 3.31 Block diagram of a $\Delta-\sum$-converter as an example for the conversion of biosignals. The dashed area represents the baseband f_B of a biosignal. The lowpass effect of the $\Delta-\sum$-modulator must not affect the baseband of the biosignal, so a sufficiently high sampling rate must be selected. The subsequent digital filtering over 64 binary values has the spectral effect of a comb filter (center), affecting the biosignal spectrum. Therefore, this must be corrected in the subsequent FIR filter (right). At the same time, the final width of 24 bits is achieved with a further subsampling by a factor of 4

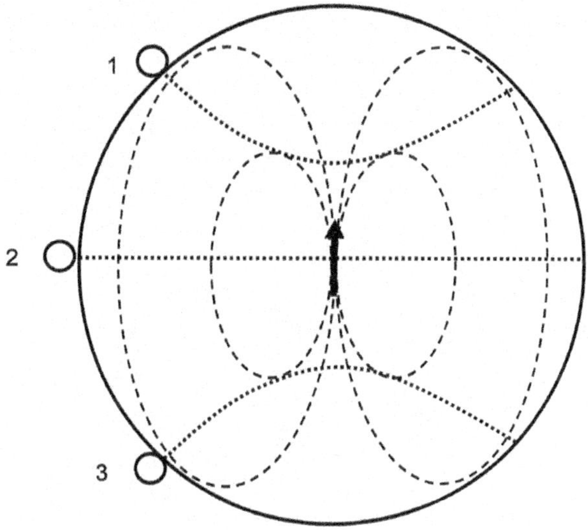

Fig. 3.32 Model of a homogeneous, isotropic, conductive sphere with an electric current source. The contemporary lines are dashed, and the equipotential lines are dotted. The potential differences ϕ_{12} and ϕ_{23} form two bipolar derivatives, which are equal in magnitude in this model

3.4 Exercises

3.4.1 Tasks

3.4.1.1 Potential Distribution on the Surface of a Spatial Conductor

Assuming a closed, homogeneous, isotropic, spherical, spatial conductor, in which there is an almost point-shaped electric current source and from which no current can flow to the outside, the value of the sum of all surface derivatives is to be tested. It is assumed that a sufficient number of sensors (electrodes) with identical distances to each other are arranged above the sphere's surface so that the sampling theorem is fulfilled. It is recommended to start from a straightforward model and generalize the findings afterward. The simplest model of the potential distribution is shown in Fig. 3.32.

What signed value does the sum of the potential differences between the measuring points 1 and 2 or 2 and 3 reach? Can this finding be generalized to arbitrary source positions and a sensor network above the surface?

3.4.1.2 The Electrical Ground of a Measuring Amplifier and Body Connection

Biosignals are derived from the body's surface without grounding as potential differences. Since each differential amplifier has an electrical operating ground, this must also have a galvanic connection to the patient's body. In principle, this is shown in Fig. 3.1. In the case of an ECG lead, as shown in Fig. 3.2, is a ground connection and ground always required on the patient's right foot in terms of circuitry? What is the minimum condition that must be met? Is an "internal ground," as shown in Fig. 3.4, necessary from a circuitry point of view?

3.4.1.3 Influence of the Electronically Formed CAR on Signal and Disturbance

Investigate the influence of CAR in an extremity lead of the ECG. How do network disturbances affect the monopolar and bipolar leads? Use the model shown in Fig. 3.33. What would be the effect of connecting the common reference CAR to the operating ground of the measurement amplifier?

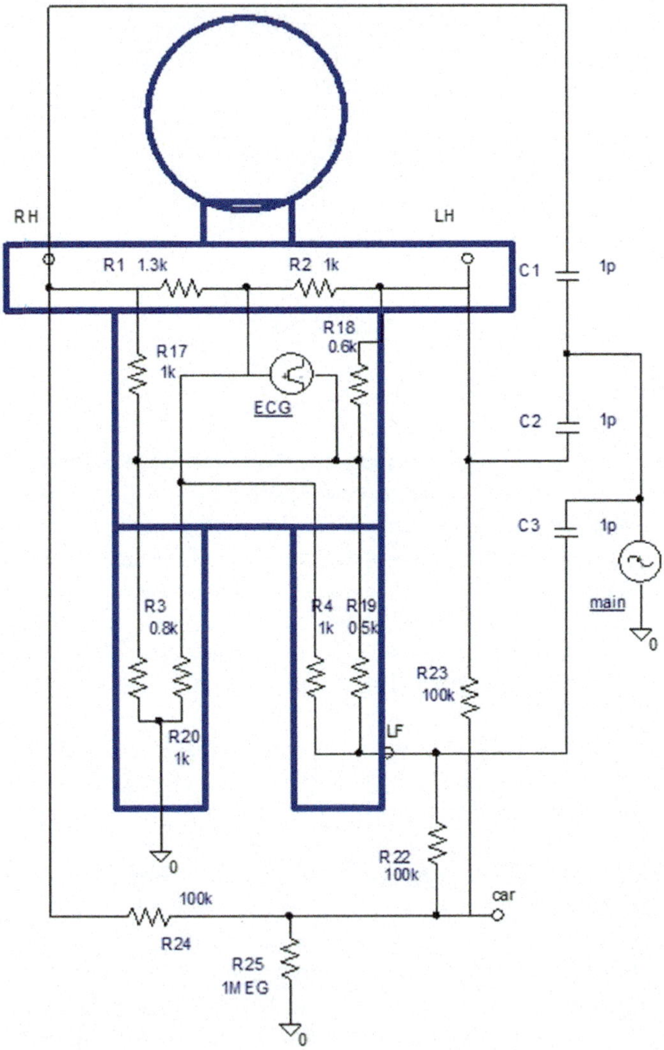

Fig. 3.33 Circuit model of an extremity lead of the ECG. The capacitances $C_{1,2,3}$ represent the stray capacitances of the extremities concerning the installation of the mains. The resistors $R_{22,23,24,25}$ form the simultaneous common reference (PSpice simulation BSV_3.3)

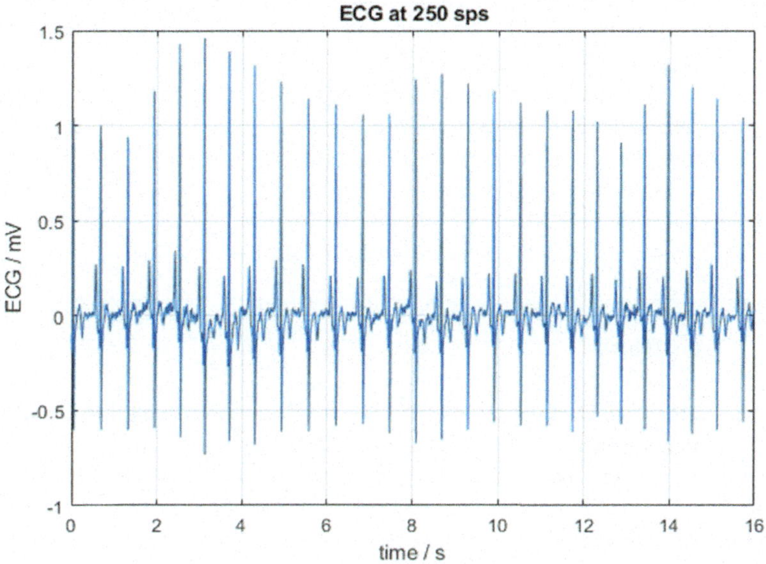

Fig. 3.34 Original recording of an ECG with a sampling rate of 250 sps

3.4.1.4 Influence of the Computationally Formed CAR on Signal and Disturbance

Investigate the influence of CAR on a simulated multichannel EEG recording. For this purpose, 16 electrodes are arranged in a square matrix. All electrodes receive typically distributed white noise with similar levels. Four electrodes contain additively to the noise a simulated transient EP (evoked potential), as shown in Fig. 3.34. Calculate the CAR and relate all electrode signals to the reference to compute a unipolar derivative. It is possible generating the simulated data using the Matlab function BSV_3_4.m or load already-developed data from the Matlab file BSV_3_4.mat. Explain the influence of CAR as the basis of a unipolar derivative on disturbances and desired signals.

3.4.1.5 Influence of the Sampling Rate on the Signal Shape

According to the sampling theorem (Eq. 3.9), the sampling period must be shorter than the reciprocal value of twice the signal frequency. It applies to all unmodulated biosignals since their spectrum starts at zero. Since the data is already sampled and digitized, different sampling rates can be simulated by simple reduction. For details on sampling rate reduction, see the chapter Digital Filtering. For this task, it is sufficient to realize the reduction of the sampling rate by omitting the corresponding number of samples. For example, a reduction by a factor of 3 always omits two samples and includes only every third. Investigate the influence of the reduction factor on the waveform of an ECG. Use the ECG from the Matlab file e23.mat (Fig. 3.34) for analysis.

3.4.1.6 Influence of the Antialiasing Filter on the Waveform

Before signal sampling or sub-sampling, the signal must be limited in bandwidth with an antialiasing filter to comply with the sampling theorem. The antialiasing filter must have a linear phase frequency response to avoid signal shape distortion. Investigate the influence of a Butterworth-lowpass filter on the waveform of an ECG concerning the cutoff frequency. For the analysis, use the Simulink function analogfilter_sampler.mdl to simulate analog filtering and sampling and Matlab to analyze digitized data (BSV_3_6_solve.mdl). In this task, the sampling will be done with 256 sps and the digitization with 12 bits.

3.4.2 Solutions

3.4.2.1 Potential Distribution on the Surface of a Spatial Conductor

The potential distribution on the surface of a spatial conductor with an internal current source can be described using appropriate methods and the mathematical apparatus of field theory. The following considerations can also determine the value of the potential sum: the electric current source generates symmetrical streamlines, assuming first the simple surface model shown in Fig. 3.32. These, in turn, produce voltage drops in the homogeneous and isotropic conductor, which are described by equipotential lines. Due to the ideal symmetry, the horizontal equipotential line (to sensor 2) marks the zero potential. Consequently, the potential differences ϕ_{12} and ϕ_{23} are magnified but with opposite polarity. It is possible to move the derivative points 1 and 3 symmetrically over the edge of the surface conductor. Thus, the vectorial sum remains zero (Fig. 3.35). This property will not change if the third dimension extends the area conductor to a volume conductor, according to Fig. 3.32. A general electric current source will not be in the center. It is evident from the model in Fig. 3.07 that there must always be a zero potential line on the body surface. Assuming no current leaks out, the sum of the positive potentials above the zero likely line must be identical to that of the negative potentials in the opposite body region below the zero potential line (Fig. 3.33). This fact can now be extended to any source positions and current directions in the volume conductor. In conclusion, the potential sum over the surface of a spatial conductor is equal to zero under the mentioned conditions (Fig. 3.36).

From these considerations, any current source in the closed-volume conductor generates a potential distribution whose areal sum is zero-valued on its surface. According to the superposition principle, this can be extended to several sources. This approach applies to discrete spatial sensing (surface sensors) if the spatial sensing theorem is satisfied.

3.4.2.2 The Electrical Ground of a Measuring Amplifier and Body Connection

The circuit-based requirement regarding the electrical ground of the measurement amplifier is that it has a galvanic connection to the patient's body. It is necessary, especially for amplifiers with FET inputs; otherwise, the inputs would float

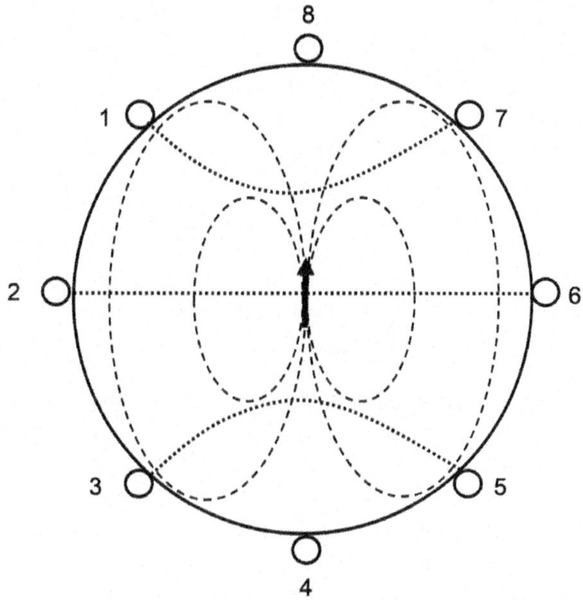

Fig. 3.35 Extension of the surface model to a symmetrical sensor network. The equipotential lines are symmetric in magnitude about the horizontal axis (zero potential line). In the upper half, the potentials are positive, and in the lower half negative; therefore, their sum cancels on the surface

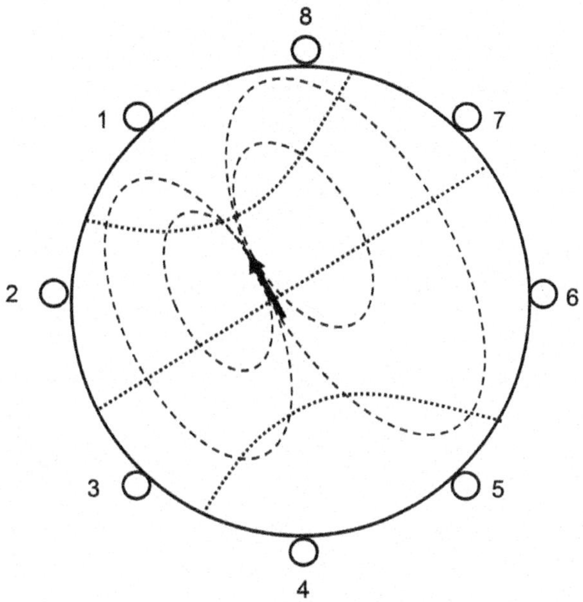

Fig. 3.36 Model of a general current source in a volume conductor. The current loops (dashed) and, therefore also, the equipotential lines (dotted) are distributed asymmetrically in the body. Since the zero potential line (dotted across the current source) separates the two polarity regions from each other, the (discrete) potential sum of the positive region must be identical in magnitude to the potential sum of the negative region

potential-free in the air. If an amplifier, as shown in Fig. 3.2, is used to derive biosignals, a ground connection at the right foot is unnecessary. Via the resistors $R_{14,15,16}$, and $R_{5,6,7}$, the amplifier has galvanic contact with the patient's body, no matter how large the resistors are. Typically, these differential amplifiers are preceded by another input stage (see Chap. 2).

Nevertheless, the principle remains the same. Should groundless differential amplifiers with FET input be used, the requirement for a ground connection would not be fulfilled. Then either the amplifier ground must be connected to the right foot (Fig. 3.2) or an "internal ground" must be formed (Fig. 3.4). For detailed circuit analysis, simulate the circuits einthoven.sch, einthoven_1.sch, einthoven_fet.sch and einthoven_1_fet.sch.

3.4.2.3 Influence of the Electronically Formed CAR on Signal and Disturbance

The PSpice simulation for the circuit shown in Fig. 3.33 shows the following signal characteristics of the unipolar limb derivatives (vs. the CAR) and the common reference CAR (Fig. 3.37). The CAR reaches amplitudes of about 0.17 mV and contains, for the most part, network disturbance. If one forms "unipolar" leads, i.e., potential differences between extremities and the CAR, using this virtual reference, they show the expected simulated ECG course for the right hand (center) and the left hand (bottom). However, the leads are still partially disturbed by the network, but much less so than without the CAR. In the signal form of the CAR, one can recognize ECG parts. Therefore, the difference in the extremity signals with the CAR also attenuates the desired signal. If the CAR were additionally connected to the electrical ground of the measuring amplifier, i.e., if an internal ground were generated, this measure would, in principle, have positive effects on the common mode behavior of the amplifier. Since the electrical ground of the amplifier would resonate with the interfering mains component, the actual level of the common interfering mode would be lower overall. Of course, this approach only shows the effect if the amplifier ground is not connected to the mains.

3.4.2.4 Influence of the Computationally Formed CAR on Signal and Disturbance

The CAR is formed as a simultaneous arithmetic average (Eq. 3.1, ensemble average), to which all channels are referenced (Eq. 3.2). This leads to the fact that a common mode disturbance (capacitively coupled network disturbance) is partially eliminated if the disturbance levels in the channels are not too different. This behavior, which is entirely intentional in the case of interference, can lead to unwanted changes when analyzing desired signals. With the CAR, one must always consider that each channel delivers the Nth part (N—number of sensors) to the CAR. By reference, this part appears with reversed sign in all other channels, although it was not contained there before the referencing. It is exemplarily shown in the following simulation. Synthetic data of length 750 contain in 16 channels uncorrelated noise (BSV_3_4.m, .mat) and in the last four channels, a transient EP with equivalent level and phase (Figs. 3.34 and 3.38). If the CAR is calculated

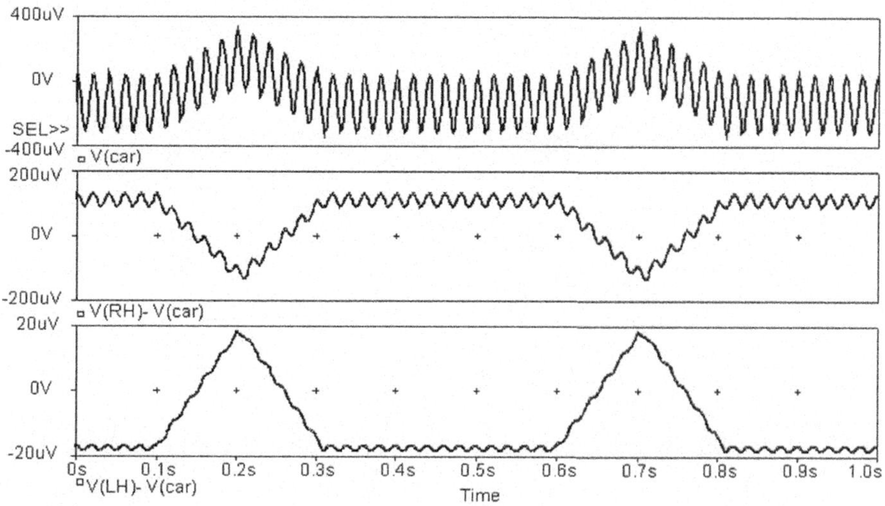

Fig. 3.37 Signal curves of the common reference CAR (top), the unipolar limb derivatives of the right (middle), and the left (bottom) hand

according to Eq. 3.1, and this is then subtracted from the channels according to Eq. 3.2 is subtracted from the channels, and the waveforms change significantly (Fig. 3.39). The signal amplitude in channels 13–16 becomes smaller (to 75%), but parts of the transient EP (25%) appear with reversed signs in channels 1–12, where they were initially absent. That can result in fatal misinterpretations for further analysis. For example, during source localization, a bipolar source can be detected this way, although, in reality, none is present. In extreme cases—rare in biosignal analysis—a strong signal under a single sensor can affect all signals of the sensor matrix in this way.

Also, the comparison of the mean signal power with the help of the variance shows a clear shift within the sensor matrix (solution with the Matlab function BSV_3_4_solve.m):

Channels 1 to 8
var(data) 0.9523 1.0458 1.0043 1.0048 0.9668 1.0705 0.9707 0.9616
var(dcar) 1.5527 1.4587 1.5868 1.5042 1.4859 1.4367 1.4400 1.4930

Channels 9 to 16
var(data) 1.0287 0.9613 1.1333 1.0618 9.4514 9.4470 9.3730 9.1526
var(dcar) 1.3794 1.5102 1.3966 1.4026 5.3405 5.8320 5.5260 6.0022

The change in the waveform in the channels and the average signal power can be interpreted as the desired signal being artificially spread across the CAR being

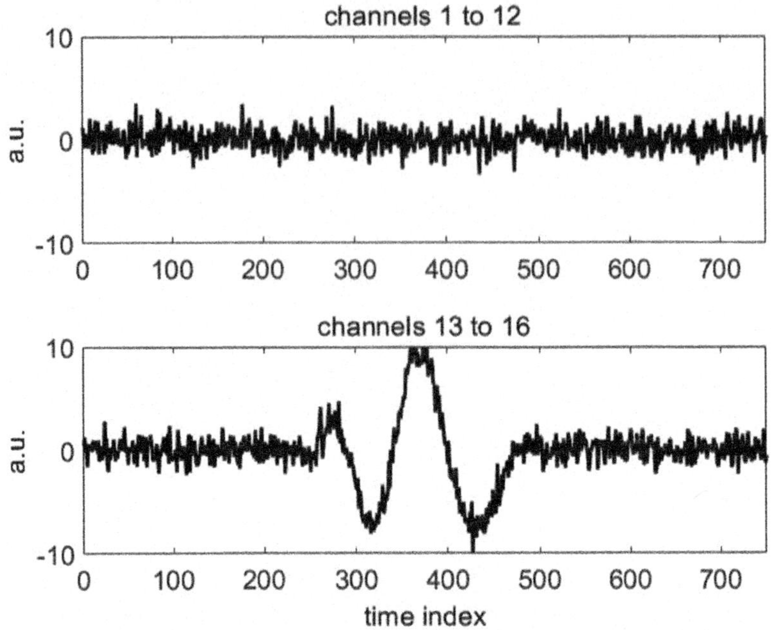

Fig. 3.38 Time histories at the electrodes of a square matrix with pure noise in channels 1–12 and an additive mixture of noise and transient EP in channels 13–16. Data were simulated, therefore, do not contain physical units. Channels 1–12 contain signal components of the transient EP (reversed sign) which were not included before referencing

artificially spread to all sensors. In contrast, the average signal power is leveled (Fig. 3.40).

3.4.2.5 Influence of the Sampling Rate on the Signal Shape

Different sampling rates can be realized on already sampled signals by reducing (increasing) the initial rate by a factor. An exemplary ECG, which was sampled with 250 sps, is available in the file e23.mat. Predefined functions can be found in the Matlab function BSV_3_5_solve.m. The original ECG is shown in Fig. 3.41, and its spectrum is in Fig. 3.41. In the time course of the ECG, respiration presents itself as the envelope of the R and S spikes and has a period of about 4 s. The harmonics of the cardiac action as a multiple of about 1.7 Hz and the fundamental of the respiration at about 0.25 Hz can be seen in the spectrum (Fig. 3.42). Lowpass filtering below the Nyquist frequency would be necessary for correct undersampling in signal analysis before rate reduction. Here, however, the violation of the sampling theorem is to be simulated, so the lowpass filter was deliberately omitted. The spectrum of the ECG (Fig. 3.42) shows that a sampling rate reduction already by a factor of 2 will violate the sampling theorem. How it affects the waveform is shown in the following figures.

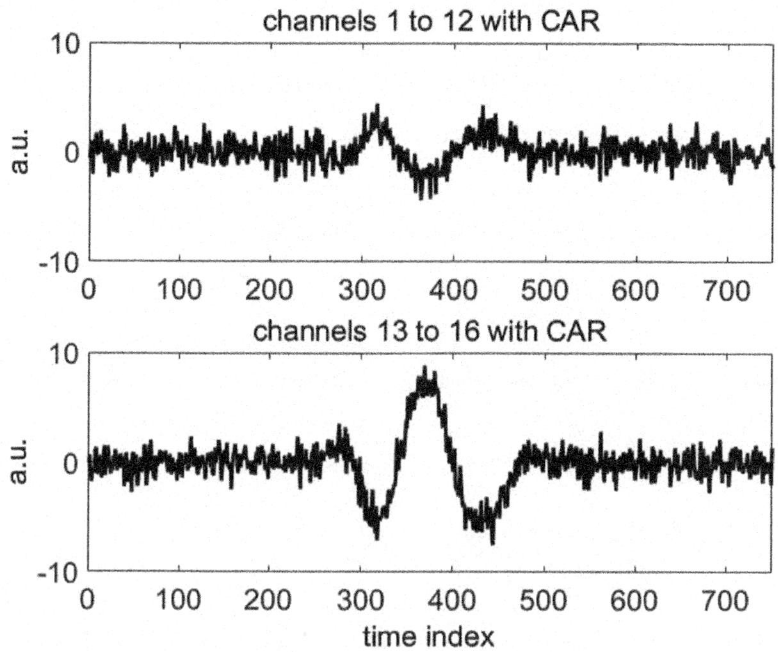

Fig. 3.39 Time courses at the electrodes of a square matrix after referencing with the common reference CAR. Channels 1–12 contain signal components of the transient EP (reversed sign) which were not included before referencing

This study concludes that even a slight violation of the sampling theorem can lead to impairments of the waveform. Since the waveform is diagnostically fundamental, a violation of the sampling theorem cannot be accepted in practice.

3.4.2.6 Influence of the Antialiasing Filter on the Waveform

To analyze the influence of the antialiasing filter on the waveform, the same ECG is used for good comparability as in task 5. In the simulator (Fig. 3.45), the actual ECG is read in from the file ekg.mat and filtered with the Butterworth lowpass filter to be varied. The sampling rate should be 256 sps (block sampling TP ECG), and the conversion width (block AD converter) 12 bit. The spectrum of the original ECG is shown in Fig. 3.42. It is already filtered, where one can estimate the highest relevant frequency as about fmax = 40 Hz. Since, in this case, the Nyquist frequency is $f_N = 128$ Hz, one puts the cutoff frequency of the Butterworth-lowpass filter as close to it as possible. It is known that Butterworth filters have a nonlinear phase, especially near their cutoff frequency, so the cutoff frequency should be as far away from the signal as possible but must be below the Nyquist frequency. The ECG for this alternative for a cutoff frequency 100 Hz is shown in Fig. 3.46. The waveform is correctly reproduced despite the nonlinear phase of the Butterworth lowpass filter. It is because the cutoff frequency of 100 Hz and the

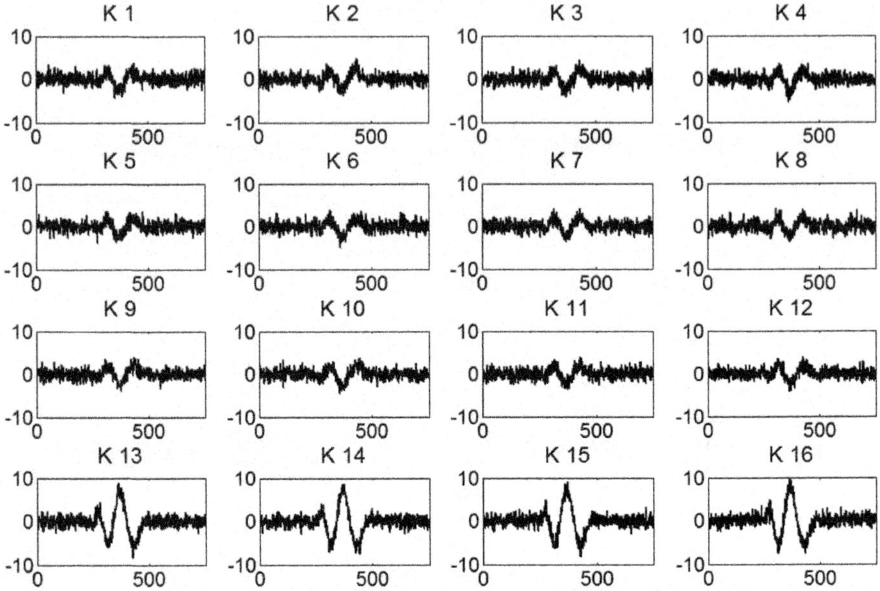

Fig. 3.40 Effect of referencing with the CAR on the waveforms of the channels. A comparison of the waveforms with the original waveforms in Fig. 3.34 shows that while the EP is attenuated in channels 13–16, it appears with the opposite sign in channels 1–12, where it was initially absent

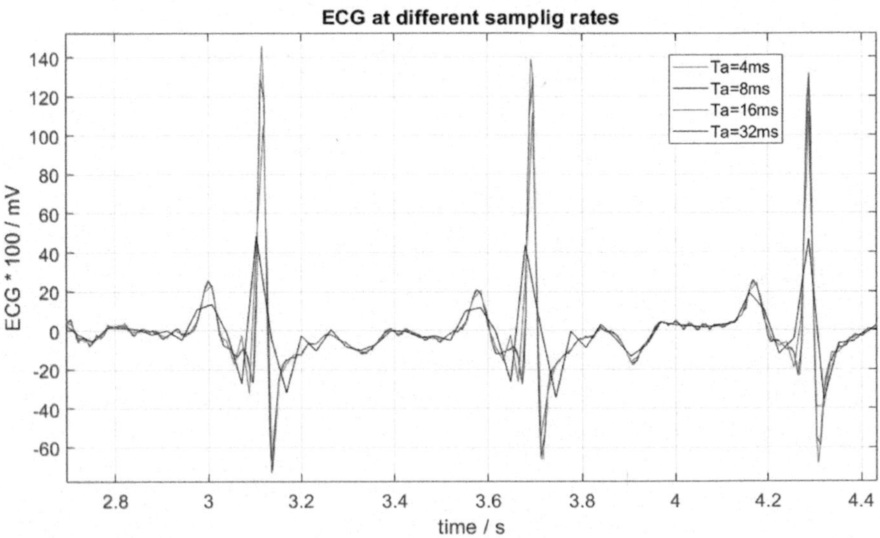

Fig. 3.41 Short-term recording of an ECG. The heart rate was about 1.7 s^{-1} or 101 min^{-1}. Significant changes in the waveform can already be seen when the sampling rate is halved

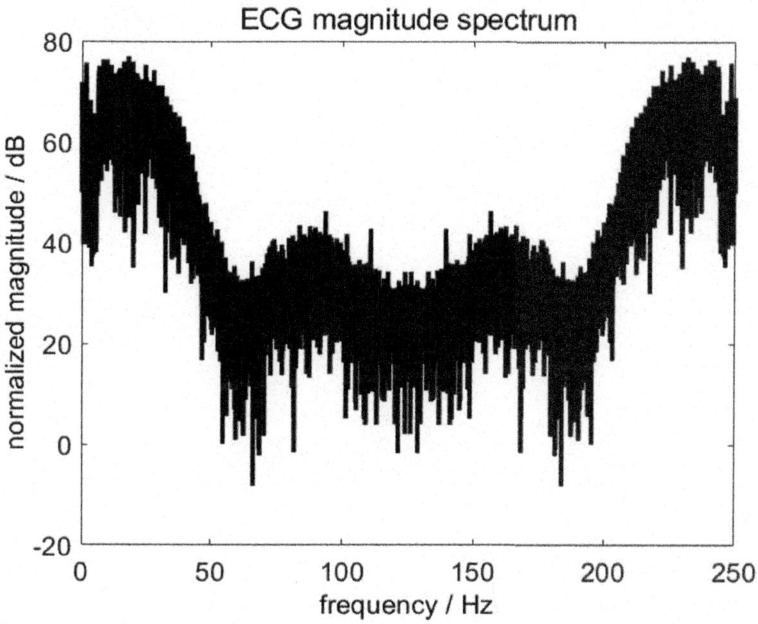

Fig. 3.42 Original spectrum of an ECG (e23.mat) sampled at 250 sps; the Nyquist frequency is 125 Hz

nonlinear part of the phase are still far from the ECG. The ECG is noisy and visible between the T and P waves at times of inactive ECG. This noise predominantly produces an unnecessary frequency band between 40 and 100 Hz. The waveform is distorted after filtering at 30 Hz compared to the ECG shown in Fig. 3.46. The S-wave has a higher amplitude after the S-wave high-frequency reverberations occur, which are not present in the original signal. They are the result of the nonlinear phase of the filter in the spectrum of the ECG, which entails a longer group delay at high frequencies. The high-energy, high-frequency components of the QRS complex appear later and cause reverberation (also known as ringing). The noise is not significantly weaker despite lowering the cutoff frequency from 100 to 30 Hz, only lower frequency. It is, therefore, attributable primarily to the quantization noise of the AD converter (Figs. 3.43, 3.44 and 3.47).

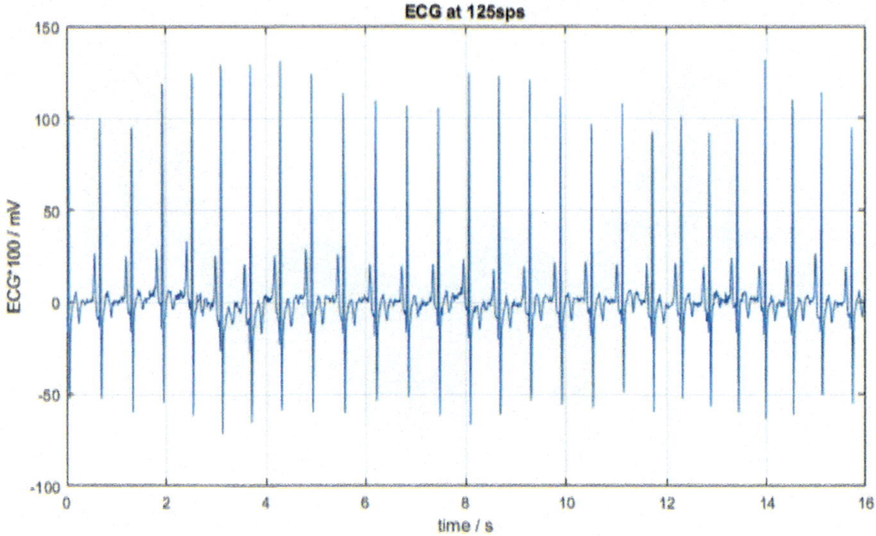

Fig. 3.43 The original ECG from Fig. 3.41 was undersampled by a factor of 2 without an antialiasing filter. Changes can be seen in the envelope of the R-spikes

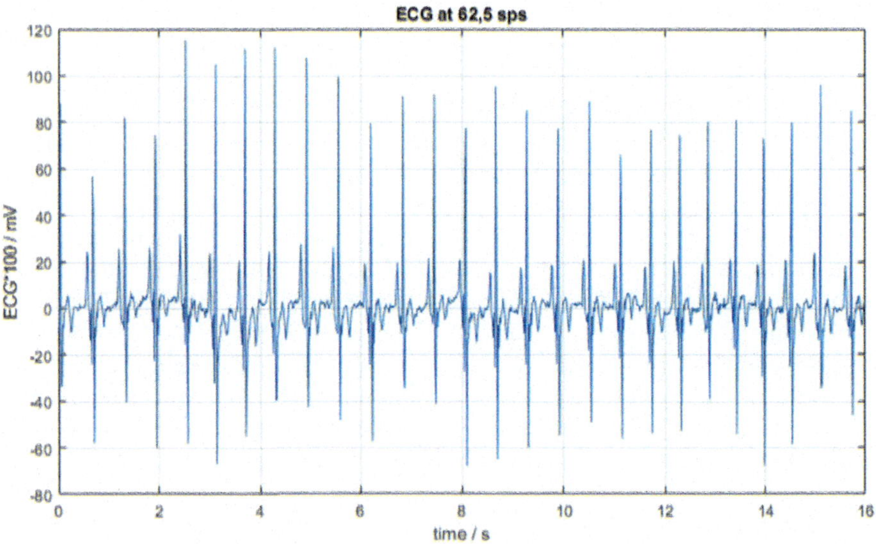

Fig. 3.44 The original ECG from Fig. 3.41 was undersampled by a factor of 4 without an antialiasing filter. The envelope of the R-prongs is now just a chaotic process. The undersampling has so disfigured the R-peaks that they can no longer be evaluated

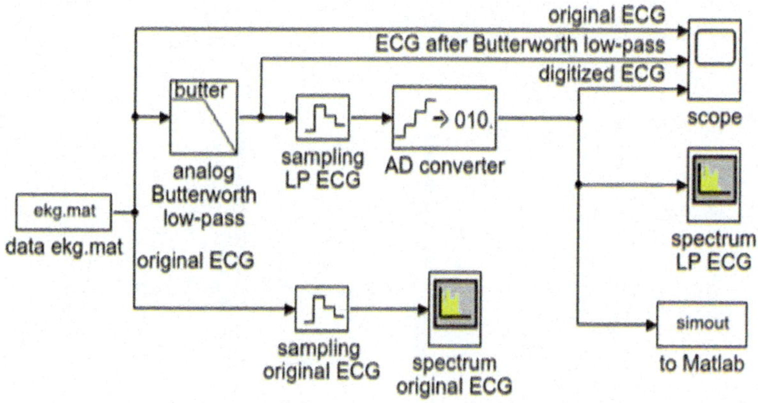

Fig. 3.45 Simulation of analog lowpass filtering (antialiasing), sampling, and digitization in Simulink. The time courses represent "Oszi," the spectra of the sampled original signal "Spectrum Original ECG," and the filtered, sampled, and digitized signal "Spectrum TP ECG"

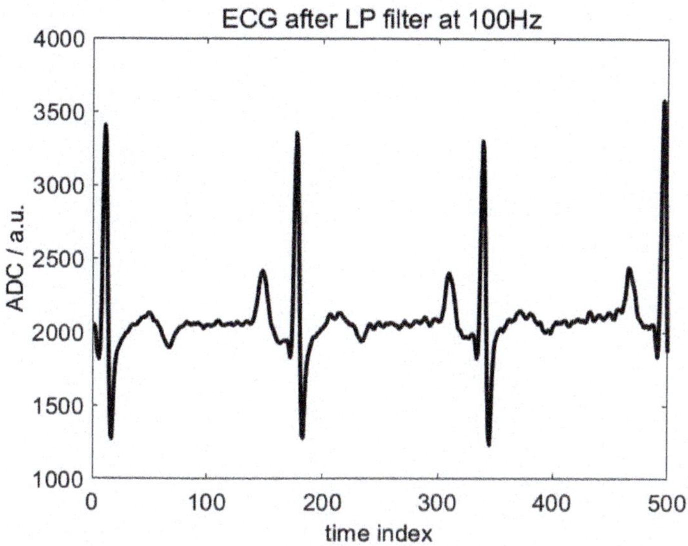

Fig. 3.46 ECG sampled at 256 sps, filtered with Butterworth lowpass at 100 Hz

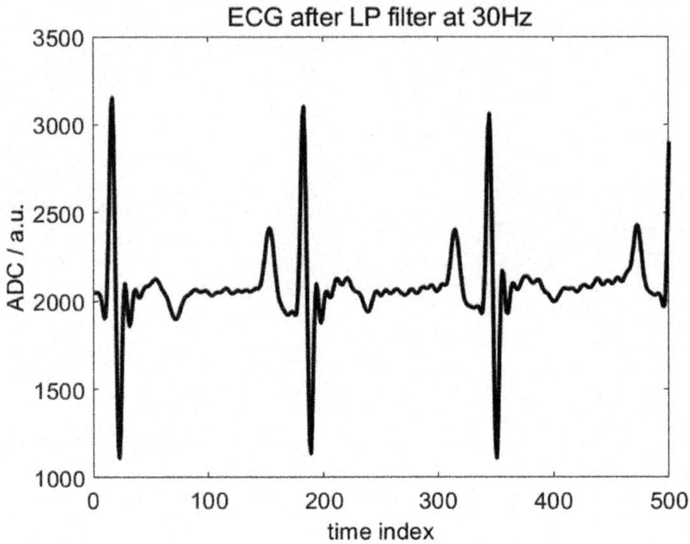

Fig. 3.47 ECG sampled at 256 sps, filtered with Butterworth low pass at 30 Hz

Reference

Husar, P. (2010). *Biosignal processing*. Springer.

Part II

Time and Frequency Analysis, Digital Filtering

4 Time, Frequency, and Compound Domain

Traditionally, medical diagnostics focus on the time courses of the body's signals. It is also justified and makes sense since the knowledge accumulated over decades is stored like a database with typical patterns in physiological cases and pathological changes. Nevertheless, for some biosignals, at least a rough spectral classification is made from the beginning (spectral bands of the EEG), but the frequency is determined in the time domain. Empirical methods of time–frequency analysis are also used implicitly, albeit on a subjective basis (inspection of heart sounds and breath sounds by the doctor). In addition to time series analysis methods, spectral and time–frequency analysis are also finding increasing application and acceptance, especially in computer-assisted analysis. For technical reasons, sensor arrays are also being used increasingly (2D and 3D signals), so methods of spatiotemporal analysis are also being used, especially in imaging. Finally, attempts are being made in the research field to develop suitable composite representations for biosignals, which allow a composite representation in space–time-frequency (spatiotemporal spectral analysis) or space–time phase (correlation analysis). The four groups of methods mentioned are dealt with in this chapter.

Supplementary Information The online version contains supplementary material available at https://doi.org/10.1007/978-3-662-67998-2_4.

P. Husar and G. Gašpar, *Electrical Biosignals in Biomedical Engineering*,
https://doi.org/10.1007/978-3-662-67998-2_4

4.1 Signal Analysis in the Time Domain

The typical courses of the physiological and pathological ECG, EEG, and EMG form the diagnostic basis of cardiologists and neurologists. The aim of a digital biosignal analysis should therefore be, on the one hand, to preserve the accumulated knowledge and use it for method development and, on the other hand, to automate the routine work of the medical practitioners and, in this way, relieve them.

4.1.1 Feature Identification

In biosignal-based diagnostics, the task for the medical practitioner is to identify known physiological and pathological patterns of time courses in a recording and to evaluate them diagnostically. In signal analysis, the task of the first phase is, therefore, feature identification as part of pattern recognition. For the solution approach, they can be divided into two groups:

- *deterministic*, where the course sought is known qualitatively and quantitatively, e.g., ECG or EMG in physiological and known pathological cases,
- *fuzzy ones*, where the course sought is at best known qualitatively, e.g., the EEG in stages of wakefulness and sleep phases.

In practice, biosignals are always subject to the influence of technical disturbances (see Chap. 2) and subjectively caused artifacts. Therefore, pattern recognition methods must identify the desired parts of the biosignals and artifacts to correct them or at least exclude them from the evaluation.

4.1.1.1 Artifacts: Creation, Detection, and Elimination

Artifacts are products of biological and technical processes that do not originate from the physiological or pathological signal source under investigation and the ambient noise.

The signal obtained from a medical measuring system can first be described in a simplified way with an additive signal model (Eq. 4.1).

$$x(t) = s(t) + n(t) + a(t) \tag{4.1}$$

In Eq. 4.1, $x(t)$ is the measured signal, $s(t)$ is the biosignal of interest, $n(t)$ is noise and technical interference, and $a(t)$ is artifacts.

Noise and technical interference are dealt with in Chap. 2, Amplification and analog filtering in medical measurement technology. By nature, there cannot be a sharp distinction between the purely technically generated disturbances and disturbances in the patient's immediate environment. Therefore, artifacts can initially be divided into two classes:

Artifacts of biological origin

- Cardiac activity in the EEG: $s(t)$ = EEG; a(t) = ECG
- Muscle activity in the ECG: $s(t)$ = ECG; a(t) = EMG
- Eye movements in the EEG: $s(t)$ = EEG; a(t) = EOG

Artifacts of technical origin

- Electrode potential and its drift
- Movements of the patient and the measurement set-up
- Energy application on the patient (HF surgery, diathermy)

Class	Example
Extreme values and outliers	110% SpO_2 in pulse oximeter, 10 mV in ECG
Slow fluctuations of the baseline	Head rotation during EEG measurement
Rapid fluctuations of the baseline	Muscle activity in the ECG
Non-linearities in the biosignal	ECG amplitude change due to breathing

The classes are arranged from top to bottom according to the difficulty of detection and elimination.

This list shows that, especially for artifacts of biological origin, it depends on the specific measurement task in which biosignal is desired and which is considered an artifact. For example, the ECG is once the desired signal disturbed by the EMG, but in an EEG measurement, it is an artifact. Therefore, in signal theory, a distinction is made between desired and undesired signals between signal and noise. A typical measurement situation of an EEG derivation with optical stimulation is shown in Fig. 4.1. While the technical disturbances are controllable due to their predominantly periodical character, the detection and especially the elimination of artifacts are problematic. The biological and technological artifacts are unique, transient in the metrological sense, and not reproducible. To obtain a solution approach for signal analysis, it (always) makes sense to translate the language of the medical professional into an algorithm in a suitable way. Building on this pragmatic approach, the following classes result in the detection of artifacts (see above classes):

Extreme Values and Outliers

In biostatistics, we must distinguish between extreme values and outliers: Extreme values are unusually high or low values of a biosignal with a low probability of occurrence. Thus, they are physiologically and pathologically not impossible. An example of this is a systolic blood pressure value of 180 mmHg. Although this

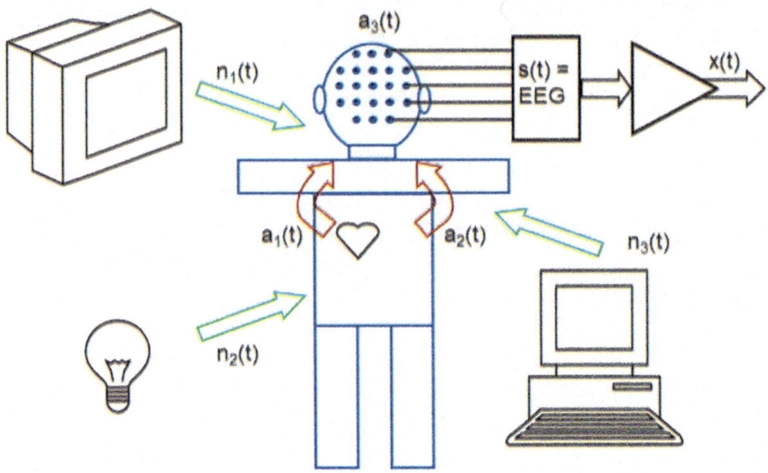

Fig. 4.1 Disturbances and artifacts in an EEG recording during optical stimulation with a monitor. According to the model in Eq. 4.1, the patient's EEG is affected by disturbances from the stimulation monitor $n_1(t)$ (image frequency), from the power supply $n_2(t)$ (mains frequency), from the control computer $n_3(t)$ (data transport), as well as artifacts of a cardiac activity a_1(t) (ECG), respiration and muscle activity $a_2(t)$ (EMG) and electrode drift $a_3(t)$

is unusually high, it can certainly occur. Outliers are impossible readings with an artifact as their cause, such as a blood pressure value of 500 mmHg. In the simplest case, a peak detector or an amplitude discriminator detects both values. Analytically, an artifact in the sense of an extreme value or an outlier is considered to be detected when the biosignal exceeds a predefined threshold (Eq. 4.2).

$$\begin{aligned} x(t) < \tau,\ H_0 : a(t) = 0 \\ x(t) < \tau,\ H_1 : a(t) > 0 \end{aligned} \tag{4.2}$$

In Eq. 4.2, $x(t)$ and $a(t)$ correspond to the measured biosignal. According to the signal model in Eq. 4.1, the artifact is a discriminant threshold, and H_0 and H_1 are hypotheses of a statistical test for detection decisions with a given statistical uncertainty.

The mathematically formulated detection of an artifact in Eq. 4.2 is challenging to implement in practical analysis. Specifying exact range limits for extreme values and outliers for any biosignal is impossible. It is also why, for example, the alarm limits for all monitored parameters can be set in wide ranges on patient monitors in intensive care units. The question of which value of blood pressure, heart rate, oxygen saturation, and respiration is considered extreme. Therefore, the physician must decide on an alarm limit based on each concrete case. In the sense of the detector, according to Eq. 4.2, the physician determines the decision threshold based on his knowledge and experience (frequency distributions of $x(t)$ under the hypotheses H0 and H1). The bottom line is that no generally valid limits can be

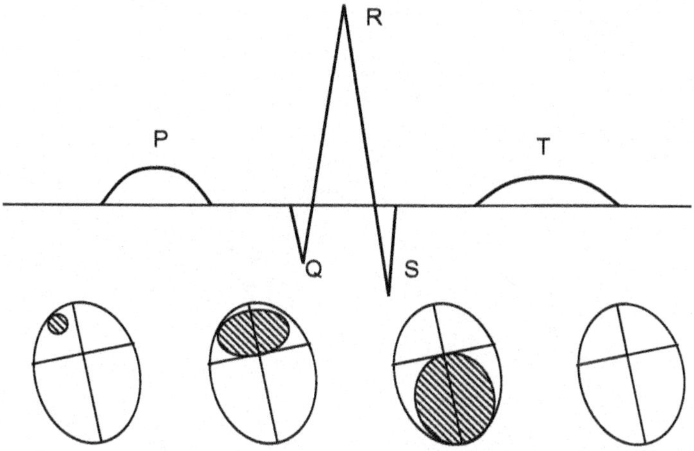

Fig. 4.2 Relation of the single-channel ECG leads to excitation phases in the heart: during the P wave, excitation spreads from the sinus node through the atria. The atria are fully excited in the interval between the end of the P wave and the beginning of the Q wave; the potential difference is at zero level. The same effects appear in the ventricles between the end of the S wave and the beginning of the T wave. Electrically excited structures are shown hatched

given for parameters. They are individually different and must therefore be treated as a variable by the signal analysis.

Slow Fluctuations of the Baseline

The term baseline (currently baseline) has been used in medical diagnostics for a long time but is not clearly defined from a technical point of view. Signal analysis understands the baseline as a parameter progression along level zero, whether it is a measured variable itself or a variable derived from it. In medical terms, the baseline is understood as the level of a variable that occurs at intervals of inactivity of the examined organ or as its temporal mean value. This more open interpretation leads to misunderstandings. The baseline problem will be explained using an ECG section as an example (Fig. 4.2). If one assumes a single-channel lead, it can be regarded as the first spatial difference. Concerning the excitable structures, this means that the difference is almost zero both in the resting state and in the state of maximum excitation. In addition, in medical measuring amplifiers, a high-pass filter is used at the input to suppress the electrode voltage. Such a high-pass filter has a differentiating behavior for low-frequency biosignals. Thus, spatially and temporally differentiating components have an effect simultaneously, which finally leads the complicated three-dimensional electrical field of the heart to a zero difference at times of maximum excitation. It relativizes the concept of the baseline in biosignals. Qualitatively similar behavior of the baseline can be observed in EEG derivations. The basic line arises solely from a high-pass filter at the amplifier input. It has been known for several years that the EEG also has diagnostically relevant DC components lost through the high pass.

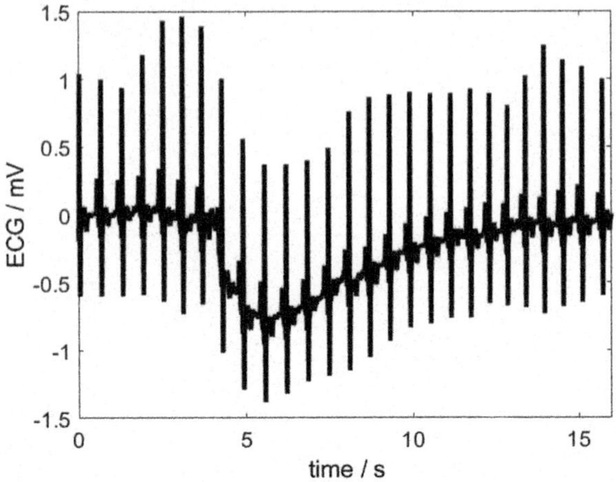

Fig. 4.3 Slow fluctuations of the basic line due to motion artifacts. The compensation process had a time constant of 1 s in the rise, 3 s in the fall. Spectrally, the artifact is below 1 Hz

This way, a zero-value low-frequency activity is simulated, although DC-EEG and SCP are present. Slow baseline fluctuations are artifacts whose duration or period is approximately seconds. The most frequent cause of the fluctuations is movement artifacts. Every patient's movement directly affects the transition impedances of the applied electrodes and the electrode voltage, which shift the biosignal from its resting position. In the case of susceptible measurement technology (EEG monitoring technology), even movements of the personnel or changes in the measurement set-up (interventions in the equipment technology) lead to movement artifacts. Identifying this class of artifacts with statistical detectors is hardly possible due to their random transient character. A typical example of slow fluctuations of the baseline due to motion artifacts is shown in Fig. 4.3.

Slow fluctuations can be eliminated without detecting them during the measurement or later in the analysis. If one assumes they are spectrally below 1 Hz, they can be significantly reduced by an appropriate high-pass filter setting at the amplifier input. If such an artifact does not overdrive the amplifier, the high-pass filtering can be done digitally and offline. However, realizing that such high-pass filtering can lead to severe losses of potentially critical diagnostic features of the biosignal is essential. Guidelines from several medical societies (cardiology, electrophysiology) even reject such a measure. In individual cases, therefore, it must be examined whether eliminating slow fluctuations through a high-pass filter is diagnostically justifiable (Figs. 4.4, 4.5 and 4.6).

The high-pass suppresses significant parts of biosignals in addition to the unwanted artifact. Therefore, the filtering question must be oriented to the specific task. For example, high-pass filtering before R-wave detection would be helpful and practical for determining the heart rate. In contrast, it should not be applied as a preliminary stage for ECG measurement.

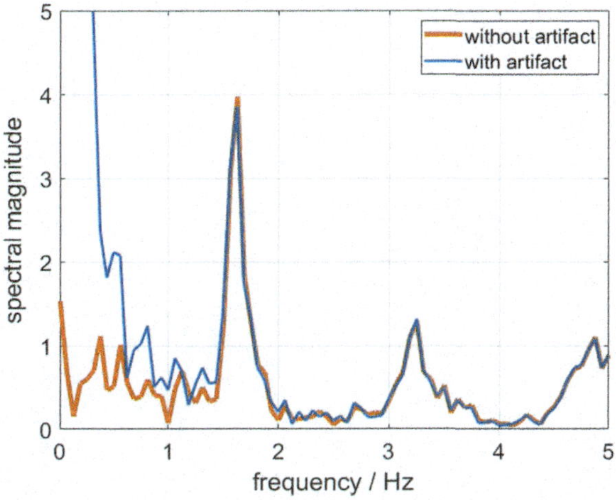

Fig. 4.4 Amplitude spectrum of an undisturbed ECG (red) and the same ECG (blue) disturbed by a motion artifact (Fig. 4.3). The harmonics of the heart rate are 1.6, 3.2 and 4.8 Hz, the harmonics of the respiration are 0.25, 0.5, 0.75 Hz. Due to the movement artifact, the low-frequency component increases strongly below 2 Hz

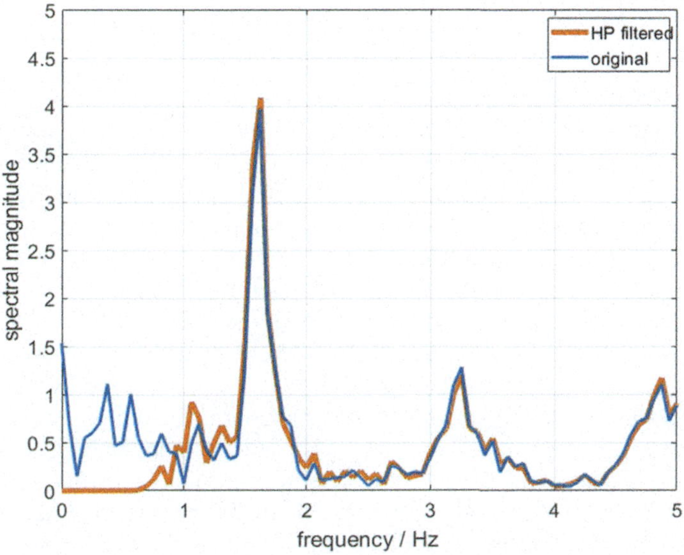

Fig. 4.5 The low-frequency artifact was reduced with a digital high-pass filter (red). As a result, the low-frequency parts of the original spectrum (blue) are also lost. The effect is shown in Fig. 4.6

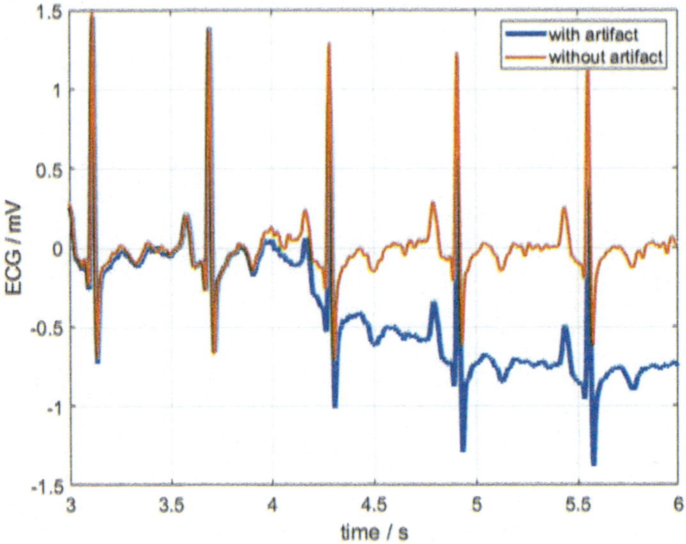

Fig. 4.6 The motion artifact starts at about time $t = 4$ s (blue). The artifact is primarily eliminated by high-pass filtering (red). It relatively increases the high-frequency component in the ECG, which can be observed as rapid fluctuations in the rest phases between the end T and beginning P wave (*uebung_4_1_m*)

Rapid Fluctuations of the Baseline

Rapid fluctuations of the baseline are periodic and stochastic courses that do not originate from the examined biosignal because they are of high frequency or occur in inactive periods of the biosignal. A typical example is shown in Fig. 4.7. ECG lead I, according to Einthoven, contains a high-frequency stochastic signal, which is particularly visible between the spikes of the ECG. These muscle artifacts are spectrally very broadly distributed (up to 10 kHz) and naturally have a stochastic character. The usual electrophysiological signals EEG, ECG, EOG, and EP (evoked potentials) range spectrally from 0 Hz to a maximum of about 100 Hz, with the signal energy reaching its maximum at the lower edge of the spectrum. The interfering muscle artifacts have their maximum (depending on the mechanical tension) between about 100 Hz and 1 kHz and reach up to about 10 kHz; most of the energy of the artifacts can be reduced with a suitable low-pass filter.

Figure 4.8 shows the ECG from Fig. 4.7 after low-pass filtering at a cut-off frequency of 30 Hz. The muscle artifacts are almost eliminated, while the shape of the ECG is preserved. Low-pass filtering of an EEG disturbed by muscle artifacts shows the same effect.

Non-linearities in the Biosignal

According to Eq. 4.1, this category includes artifacts that do not correspond to the signal model but disturb the investigated biosignal. A typical example is shown in

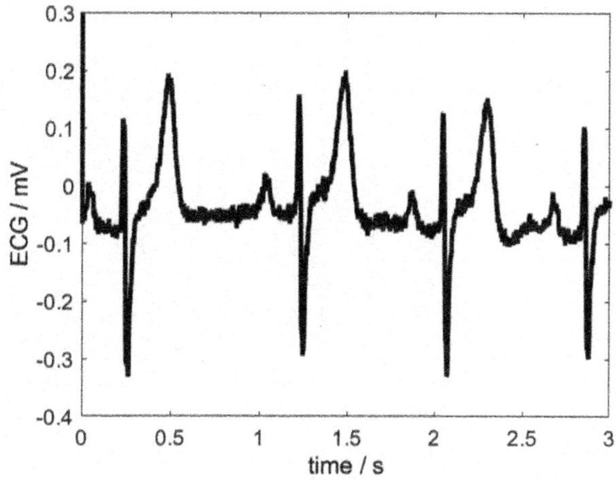

Fig. 4.7 ECG lead I according to Einthoven with muscle artifacts. The spectrum of muscle potentials extends to the Nyquist frequency of 250 Hz

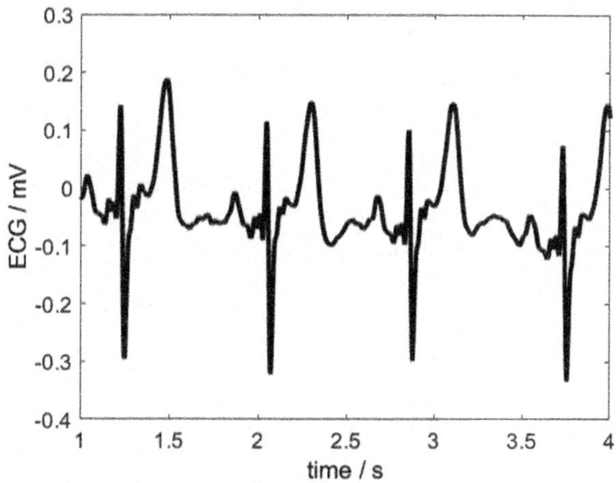

Fig. 4.8 ECG lead I, according to Einthoven (from Fig. 4.7), after a low-pass filtering with the cut-off frequency of 30 Hz (the recording starts after one second because the digital filter causes edge effects)

Fig. 4.9. The amplitudes of the R and S spikes form an envelope whose period is about 5 s. This effect is obviously due to the influence of breathing on the ECG. Quantitative measurement of the ECG becomes problematic due to this disturbance. Since the respiration period is known, it could be eliminated with a high-pass filter. Figure 4.10 shows that the fundamental wave of the respiration is at 0.2 Hz, so it would have to be separated from the fundamental frequency of the ECG at 1.1 Hz with a suitable high pass. A high-pass filter (cut-off frequency at 0.8 Hz) reduces respiration by at least 40 dB. Thus, its influence on the ECG is reduced (Fig. 4.10). However, the time course of the filtered ECG shows almost the same course as in Fig. 4.9. The amplitude modulation explains the ineffectiveness of the high-pass filtering concerning the envelope (see Chap. 2).

According to the signal model in Eq. 4.1, respiration is not additively superimposed on the ECG but multiplicatively. The ECG is amplitude modulated by

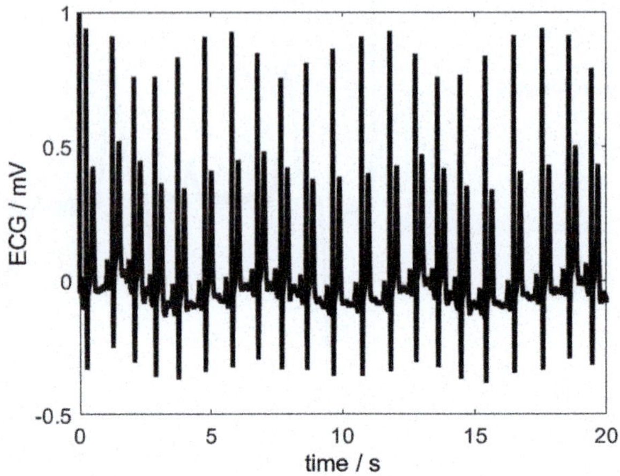

Fig. 4.9 ECG lead V6. The envelope of the R and S spikes shows a period of about 5 s and originates from respiration. It nevertheless remains after high-pass filtering. The breathing signal is amplitude modulated on the ECG. It is close to the heart rate's harmonics and above the high-pass filter's cut-off frequency (see Fig. 4.10)

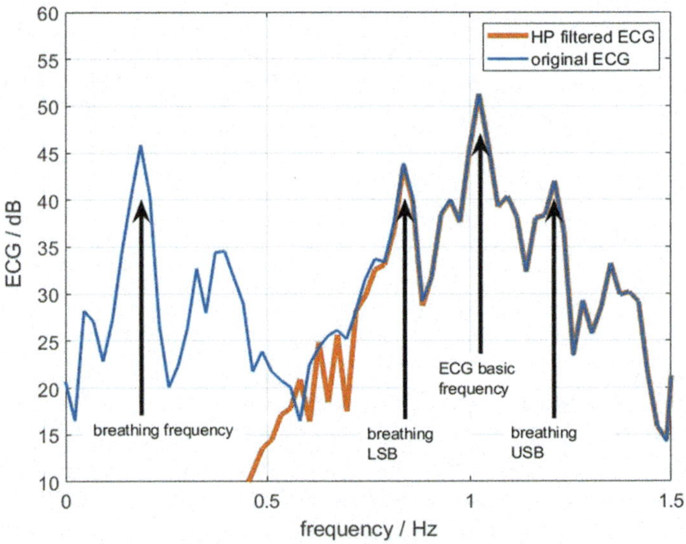

Fig. 4.10 Amplitude spectrum of an ECG V6 in dB. The fundamental wave of the respiration is at 0.2 Hz, that of the ECG at 1.02 Hz. Note that there are two symmetrical local maxima around the fundamental frequency of the ECG at a distance of 0.2 Hz (respiration in LSB and USB). After high-pass filtering with a cut-off frequency of 0.8 Hz, the spectral part of the fundamental wave of respiration disappears. Still, the sidebands of the respiration around the ECG fundamental frequency remain (*uebung_4_2.m*)

respiration. It can also be seen from the fact that the respiration harmonics are symmetrically arranged around the fundamental wave of the ECG at 1.1 Hz (Fig. 4.10). This property of amplitude modulation is also known in communications technology as a sideband. For this influence of the artifacts on the investigated biosignal, the signal model from Eq. 4.1 must therefore be extended:

$$x(t) = A \cdot (s(t) + a(t)) + M \cdot s(t)a(t) + n(t) \tag{4.3}$$

In Eq. 4.3, A is the additive mixing of the biosignal and the artifacts, M is the factor for the multiplicative mixing, and $n(t)$ is the additive noise.

In the case according to Figs. 4.9 and 4.10, one sets the additive part $A = 0$ and the multiplicative part $M = 1$ in the signal model according to Eq. 4.3. This approach follows from the realization that the time course of the signal filtered by high-pass filter has hardly changed. The additive part is, therefore, negligible. From this, it can be concluded that the influence of breathing can only be removed by multiple bandpasses (comb filter) for the harmonics of the heart rate. From the communication engineering point of view, one would filter out the carrier without sidebands. However, this would only result in one (or more) harmonic oscillations that hardly change their amplitude. For details, see exercises.

4.1.1.2 Deterministic Feature Identification

Several biosignals have a deterministic time course whose parameters show natural fluctuations, but the signal shape remains essentially constant (ECG, pulse wave/plethysmogram). In pathological cases, the signal shape changes. It can often be described deterministically (extrasystoles, pathological pulse wave). This class includes, in particular, biosignals directly related to the cardiovascular system: ECG, pulse wave, blood flow, blood pressure, or ICG (impedance cardiogram). The deterministic approach will be explained using the ECG as an example (Fig. 4.11). In the ordinary physiological course, the waves P and T are single-phase, and the QRS complex is three-phase. The waves and the intervals between the waves lie in specific time windows. To recognize a regular cardiac action, it is, therefore, necessary to determine the waves' start and end times and monitor the course of the waves. Determining a wave's start and end times or edge is a complex task for real biosignals with no clear baseline, and they are noisy or disturbed. The wave boundaries are not clearly defined and disappear in the noise. For estimation of the start and end times, the following possibilities can be used first:

Fixed Level Threshold

A fixed threshold is set for the beginning and end of a wave. In Fig. 4.11, it could be a level of $x_{abs} = 0.02$ mV for the ECG. This criterion is straightforward and can also be realized in real-time, which is advantageous, for example, for devices in small spaces and fast analysis (pacemakers). However, the times vary considerably because this threshold is frequently exceeded by interference, noise, and artifacts (Fig. 4.11).

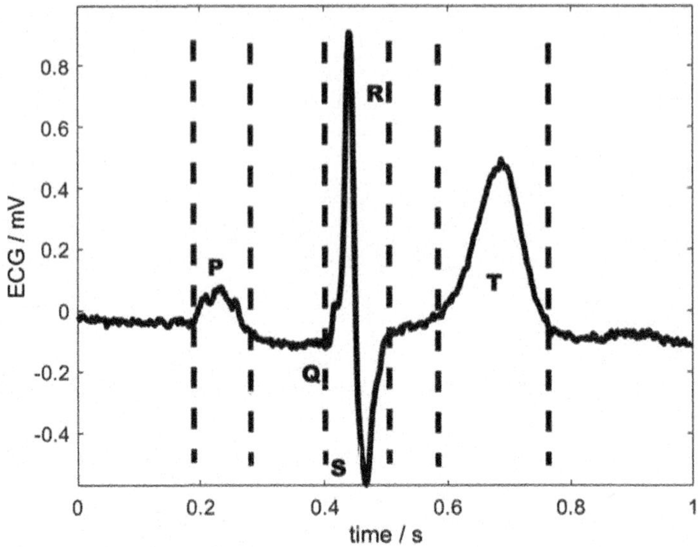

Fig. 4.11 Sections of a regular time course of the ECG lead aVF. Criteria for pattern recognition are the time intervals between the dashed lines in which the waves occur

Relative Level Threshold

The principle known from the technical field (oscilloscope technology) for determining the start/end time at 10% or 90% of the maximum level is used here. This measuring principle considers the well-known effect that especially the lower and the upper 10% of an edge run relatively flat and are superimposed by noise and therefore cause fluctuations in the time measurement. In technical areas (data transmission, digital circuitry), the maximum level is known so that the threshold of 10%/90% can be set in advance. With biosignals, the amplitude is only known in exceptional cases (action potentials), so the wave detection can only take place after the recording and is, therefore, only conditionally suitable for real-time processing. Before threshold determination, the amplitude must be detected. It is possible with several approaches; see Sect. 4.1.2 Parameter determination.

Tangent-Based Threshold Determination

The fixed and relative thresholds are sensitive to noise and disturbances and can therefore be subject to strong temporal fluctuations. However, the tangent-based approach to determining the start/end times of waves (Fig. 4.12) is electrophysiologically based: The flank of a wave represents the highest projected speed of excitation propagation, in electro-technical terms, the peak of a charge shift. Since excitation propagates along anatomically stable pathways, the temporal occurrence of flanks is directly linked to the neuronal activity under investigation. Therefore, wave edges are much less sensitive to interference than level-based wave detection. This property is also known in communications engineering and can be observed in

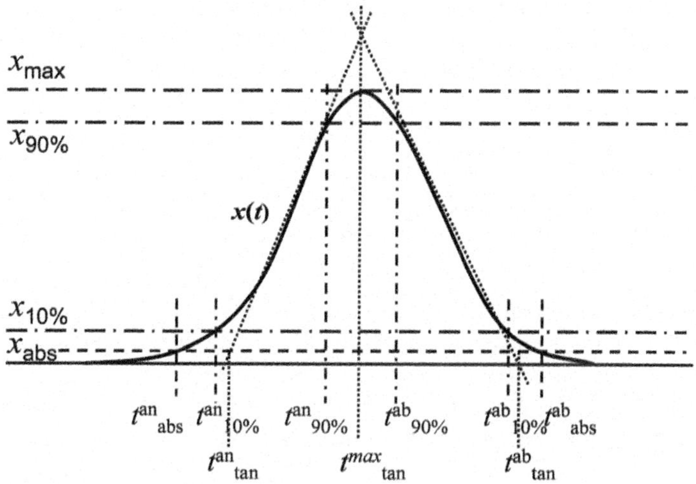

Fig. 4.12 Determination of a wave's starting/ending times and the time of the maximum: x_{abs} is the absolute threshold, x_{max} is the maximum level of the wave, $x_{10\%}$, and $x_{90\%}$ the respective relative levels of x_{max}. The upper indices at t^{an}, t^{max}, and t^{ab} indicate the times of the rising edge, the maximum, and the falling edge; the lower indices t_{abs}, $t_{10\%}$, $t_{90\%}$, and t_{tan} the respective method by which they are determined

comparing frequency and amplitude modulation. Suppose one compares the time points from the beginning/end of the wave in Fig. 4.12, which are determined by the respective methods. In that case, the following can be observed: The time points are naturally different and strongly dependent on noise and interference for the two level-based methods (fixed and relative threshold). The safest method is the tangent-based determination of the time points, which works with much higher levels. The disturbance would have to reach the size of the wave for a false measurement to occur. However, this method requires much more computational effort than level-based methods, so it can only be used to a limited extent for real-time processing. The determination of the tangent is described in Sect. 4.1.2.

4.1.2 Determination of Curve Parameters

4.1.2.1 Amplitude Detection

The detection of amplitudes aims to determine the maximum local magnitude of a level and the time of its occurrence. The task can be mathematically formulated with the help of the known relations for the extrema search (Eq. 4.4).

$$\frac{dx(t)}{dt} = 0 \rightarrow \begin{cases} \frac{d^2x(t)}{dt^2} < 0, \; x_{\max} = x(t) \\ \frac{d^2x(t)}{dt^2} > 0, \; x_{\min} = x(t) \end{cases} \tag{4.4}$$

The relationship, according to Eq. 4.4 is suitable for theoretical analyses but can also be implemented in analog electronics (analog signal processors) in some exceptional cases. For computer-aided analysis, the transition to the discrete-time range (sampling period normalized to 1) is necessary (Eq. 4.5):

$$\frac{\Delta x(n)}{\Delta n} = 0 \rightarrow \begin{cases} \frac{\Delta^2 x(n)}{\Delta n^2} < 0, \; x_{\max} = x(n) \\ \\ \frac{\Delta^2 x(n)}{\Delta n^2} > 0, \; x_{\min} = x(n) \end{cases}. \tag{4.5}$$

To determine the derivatives of $x(n)$ precisely at the point n according to Eq. 4.5 is not possible at all because they can only be determined approximately by differences. For the first derivative, there are already two possible differences to choose from (Eq. 4.6), which are approximately proportional to the theoretical gradient after the normalization of the sampling period:

$$\frac{dx(t)}{dt} \approx \frac{\Delta x(n)}{\Delta n} \approx \begin{cases} \Delta_{-1}(n) = x(n) - x(n-1) \\ \Delta_{+1}(n) = x(n+1) - x(n) \end{cases} \tag{4.6}$$

To estimate the first derivative, a combination of the two discrete differences from Eq. 4.6 will therefore be suitable, most simply their arithmetic mean (Fig. 4.13), also known as the central difference quotient:

$$\frac{\Delta x(n)}{\Delta n} \approx \frac{\Delta_{-1}(n) + \Delta_{+1}(n)}{2} = \frac{1}{2}(x(n+1) - x(n-1)). \tag{4.7}$$

Note that in Eq. 4.7 the value $x(n)$ no longer appears at all due to the formation of differences in the equation. Spectrally, this corresponds to low-pass filtering (undersampling) with the first zero at $f = 1/2\mathrm{T_A}$, where $\mathrm{T_A}$ is the sampling period. Especially with noisy biosignals, a more robust smoothing may be necessary, i.e., a lower cut-off frequency of the lowpass filter. For this, the estimation of the first derivative must be extended to several samples:

$$\frac{\Delta x(n)}{\Delta n} \approx \frac{\sum_{k=-L+1}^{0} \Delta_{-1}(n-k) + \sum_{k=0}^{L-1} \Delta_{+1}(n+k)}{2L} = \frac{1}{2L}(x(n+L) - x(n-L)). \tag{4.8}$$

The cut-off frequency of the smoothing, according to Eq. 4.8 is $f = 1/2\mathrm{LT_A}$. However, differentiation also has a high-pass effect. Figure 4.13 shows exemplary spectral properties of differentiators according to Eq. 4.8. From order 3, a comb filter forms spectrally. From the point of view of sampling (see Sect. 3.2, Sampling of biosignals), the formula according to Eq. 4.8 corresponds to undersampling by a factor of 2L. Therefore, an antialiasing (low-pass filter) with a cut-off frequency of f = 1/2LTA must be placed before the differentiation. The low-pass filter and the differentiator together form a bandpass. For a practical realization, it would be more effective to use a bandpass regardless so that the formula, according to

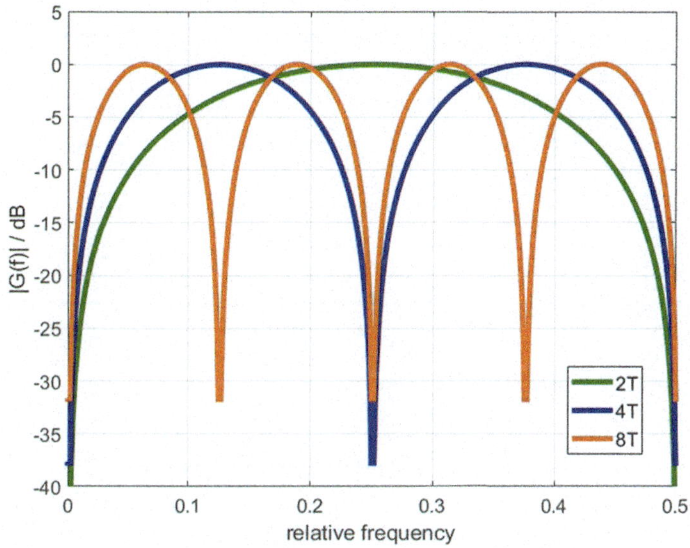

Fig. 4.13 Spectral transfer function of differentiators with orders L = 2T, 4T, and 8T

Eq. 4.8 would be indirectly included. However, the high-pass part must correspond to the course of differentiation (see Chap. 5 Digital Filtering).

The calculation of the first derivative does not need the arithmetic mean, as indicated in Eq. 4.8. If the discrete-time differences $\Delta_{+,-1}(n)$ is considered as a sample, other statistical measures, such as the median or the mode, can also be used to estimate the derivative. For more details, see the chapter Analysis of stochastic processes.

Regardless of the methods to determine the first derivative, it can only improbably become zero-valued in the discrete-time and digital domain. Concerning temporal discretization, one must always assume that noise is present, which is not reduced by the sampling. Although the noise is mean-free, individual samples are real numbers, so their probability of occurrence is zero-valued. Regarding digitization, the theoretical occurrence probability of an exact zero decreases sharply with the bit width of the AD conversion. With today's conversion widths of 24 bits, it is theoretically 6×10^{-8}. Suppose the theoretical condition of a zero-valued first derivative for extrema search is applied to real biosignals (Eq. 4.4) and a large conversion width. In that case, there is hardly any chance of identifying the extrema. In the opposite extreme case of a very low sampling rate (below 100 sps) and a low conversion width (below 10 bit), the first difference often reaches zero value, but also when there is no extreme value. It follows that one should not look for the identity of the first difference with zero but for a zero crossing. For the practical analysis, the task is to determine as precisely as possible from discrete-time values of the first difference (Eq. 4.5) the time a sign change occurred (Fig. 4.14). Real biosignals are noisy and disturbed, whereby the differentiation increases the

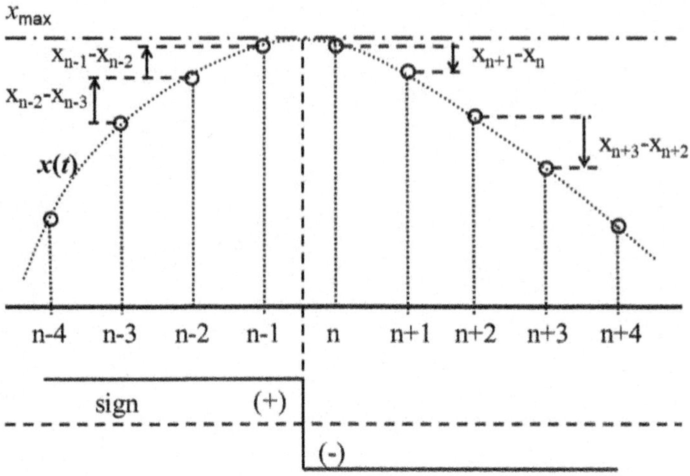

Fig. 4.14 Determining the first differences from discrete-time signal values. The sign of the first derivative changes at the time of maximum, and the sign of the first difference changes at the discrete-time n − 1 or n

noise component even more. Therefore, although the first difference follows the qualitative course of the signal (Fig. 4.15 center), due to the high noise component.

The sign changes in a stochastic process (Fig. 4.15). Rules must be established for the sign change so that the biosignal is recognized and does not react to every zero noise crossing. These rules are not universally valid; they must consider typical courses. For detection of the amplitudes of R and S spikes in the ECG, the following shape-dependent rule for sign changes ($T_A = 2$ ms) can be established (Eq. 4.9).

$$\begin{aligned}
&\sum_{k=0}^{4} \operatorname{sgn}(\Delta_{+1}(n-k)) > 4 \cap \sum_{k=1}^{5} \operatorname{sgn}(\Delta_{+1}(n+k)) < -4 \Rightarrow x_{max} = x(n) \\
&\sum_{k=0}^{4} \operatorname{sgn}(\Delta_{+1}(n-k)) < -4 \cap \sum_{k=1}^{5} \operatorname{sgn}(\Delta_{+1}(n+k)) > 4 \Rightarrow x_{min} = x(n)
\end{aligned} \tag{4.9}$$

The shape-dependent evaluation of the sign change according to Eqs. 4.9 and 4.10 is particularly necessary if a biosignal has rest phases in which noise and disturbances are present, for example, in the ECG, needle EMG or EOG (Fig. 4.16).

In particular, for the automatic detection of R-waves in the ECG, there are several shape-dependent detectors, some of which are also real-time capable and, therefore, suitable for pacemakers. Pan and Tompkins have designed an efficient method. Pan and Tompkins (1985) which is shown schematically in Fig. 4.17. Since its first publication, this method has been frequently modified and adapted to unique problems. The bandpass is narrow and low-frequency, with 5–11 Hz at

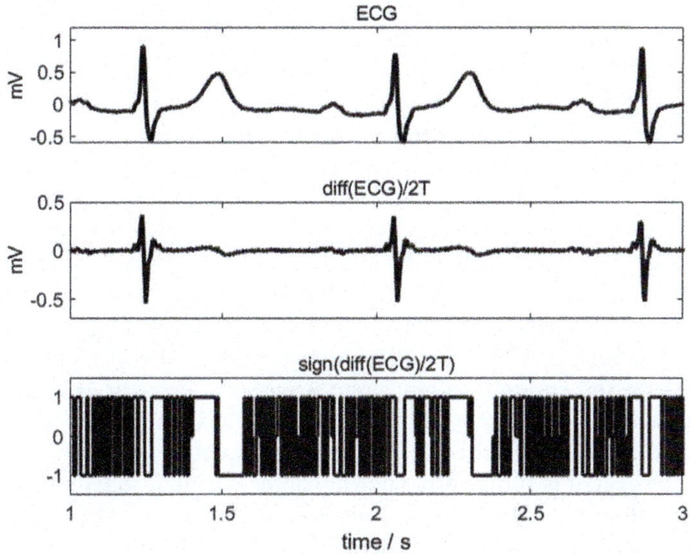

Fig. 4.15 The ECG (top), its first difference over two sampling periods (middle), and the sign of the first difference (bottom). Due to the noise, the sign shows more of a stochastic course except for the QRS complex and the T-wave ($T_A = 2$ ms)

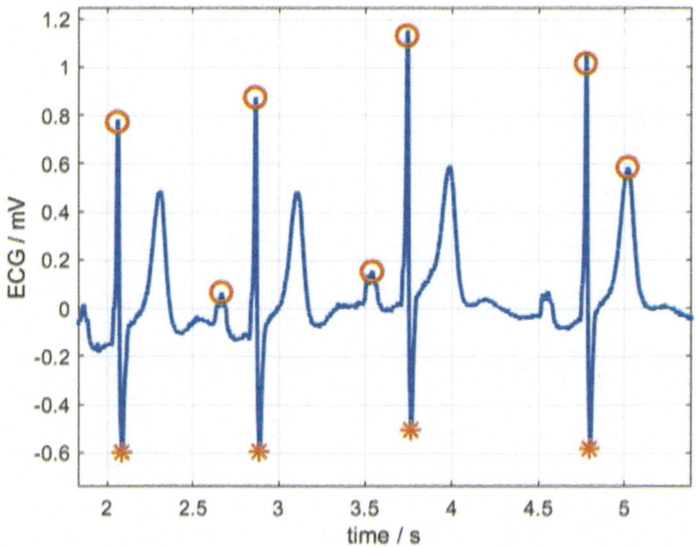

Fig. 4.16 Amplitude detection of local maximum (o) and minimum (*) in an ECG (aVF). Sign criterion 5−/5+. R and S amplitudes are reliable, that of P and T waves only sporadically

the usual sampling rates (500…1000 sps), so it is not straightforward to realize. A possible variant consists of a series connection of a low-pass filter with a cut-off frequency of 11 Hz and a high-pass filter with 5 Hz. According to Pan and Tompkins, the normal ECG has the most significant energy in this frequency band. The differentiator and the squarer provide the instantaneous power of the first derivative so that it is an energetic measure of the edges of the QRS complex. The differentiator is unique because it only differentiates in the range up to about 30 Hz but then drops steeply downwards (Fig. 4.18). This preprocessing ensures that only the spectral content of the QRS complex appears in the first derivative. The spectral range of the bandpass and the differentiator specified by Pan-Tompkins, according to Eq. 4.10, define a fixed value of 200 sps results for the sampling rate. The differentiator (Fig. 4.18, Eq. 4.10), the bandpass, or the MA lowpass must be recalculated for the current sampling rate to use this detector at other sampling rates. Alternatively, (practically more clear) the signal sampled with the current sampling rate must be converted to 200 sps (resampling).

Today, however, it is sufficiently known that the energetic maximum of the ECG lies between 10 and 30 Hz. While all blocks (as functions) can be preserved, the bandpass is adjusted according to this knowledge (Fig. 4.19).

$$d(n) = \frac{-x(n-2) - 2 \cdot x(n-1) + 2 \cdot x(n+1) + x(n+2)}{8T_A} \tag{4.10}$$

In Eq. 4.10, $x(n)$ is the measured and bandpass filtered ECG, T_{A} is the sampling period, and $d(n)$ is the detector signal of a detected R-wave.

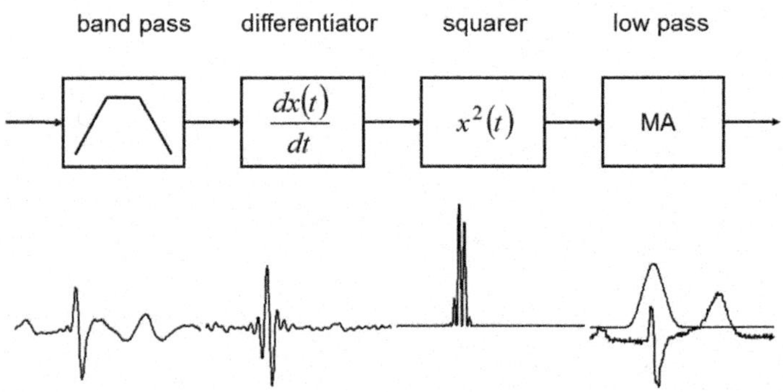

Fig. 4.17 ECG recognition according to Pan and Tompkins (Pan & Tompkins, 1985). Moving Average (MA) states that the output signal can be discriminated with a detection threshold. The bandpass filters the band between 5 and 11 Hz in which the most significant energy portion of the QRS complex is located (according to Pan & Tompkins. Today, it is known that the energetic maximum is between 10 and 30 Hz). The differentiator is described by Eq. 4.10. The moving average (MA) averages the signal power over 80 ms. All time curves represent the respective output signals of the function blocks. The graph at the MA compares the original ECG and the detector output

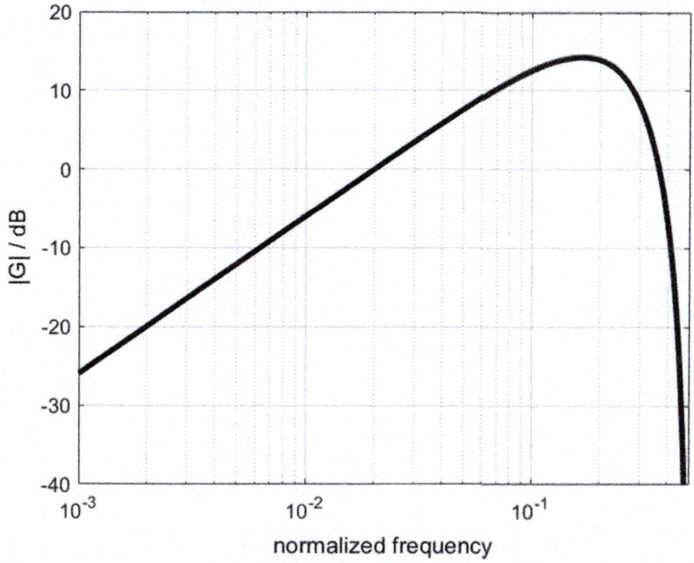

Fig. 4.18 Transfer function of the differentiator according to Eq. 4.10. In the range up to about 0.15 of the relative frequency, it works like an ideal differentiator, after that, like a steep low-pass

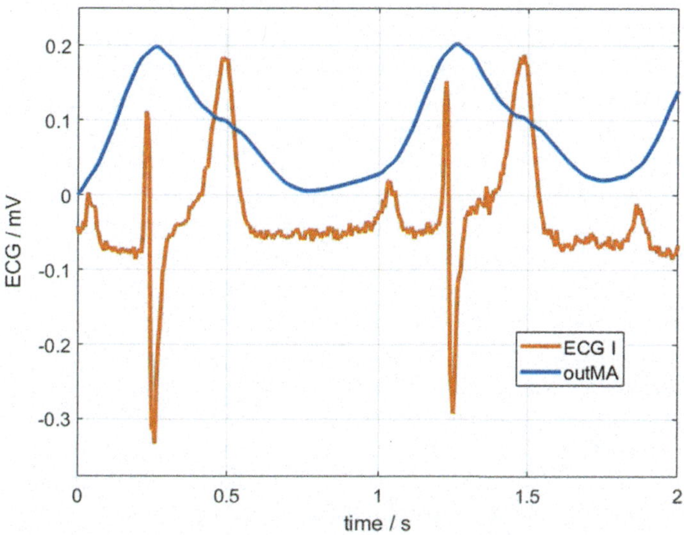

Fig. 4.19 The original ECG I compared to the detector signal at the moving average output (Fig. 4.17). The detector signal can be discriminated with a fixed or adaptive threshold. The maximum energy at the MA output coincides in time with the QRS complex, while the relatively strong T-wave is not detected

4.1.2.2 Edge Detection

As explained in Sect. 4.1.1.2, tangent-based threshold determination methods are more reliable than level-based methods. Theoretically, a tangent can be set at any point in the waveform, whereby its rise corresponds to the first derivative (Eq. 4.11). The quotient in the discrete-time case is proportional to the actual increase after normalization to sampling period 1.

$$\tan(\alpha)|_t = \frac{dx(t)}{dt} \approx \frac{\Delta x(n)}{\Delta n} \tag{4.11}$$

Each wave edge with a continuous course also has a turning point, which can be determined mathematically with the help of the second derivative (Eq. 4.12).

$$\frac{d^2x(t)}{dt^2} = 0 \approx \frac{\Delta^2 x(n)}{\Delta n^2} \Rightarrow w = n \tag{4.12}$$

In Eq. 4.12, $x(t)$ is the signal under investigation, n is the time index, and w is the time index at the inflection point of the slope.

With real signals, however, one cannot assume that the turning point is in the middle of the flank as desired. On the contrary, it can be located along the path between the local extremes. Since the waveform is not detailed in advance, even with deterministic biosignals, the tangent to the flank can be laid using the least squares method (linear regression). Beforehand, the range for the straight line must be determined automatically (see amplitude detection, threshold detection) or interactively. A typical task is measuring the ST line (Fig. 4.20). The evaluation of the first difference shows that the S-wave reaches its maximum at index 136 and its end at index 149. The polynomial for the regression line thus contains the coefficients p = (0.0416, −6.2474).

This way, all three subsections necessary for collecting the ST path are calculated. The elevation of the ST segment, which generally must not be greater than 0.2 mV, can be checked very precisely and reliably. The check is, therefore, almost independent of different interpretations of the baseline (baseline). According to medical interpretation, the baseline begins by definition at the beginning of the ST segment, i.e., at the end of the S-wave. This baseline does not generally coincide with the baseline of a high-pass filtered ECG, as shown in Fig. 4.20. Therefore, it is always necessary to adjust the interpretation of the medical interpretation in addition to the signal analytically correct detections and linear approximations. The measurement of biosignals is even more complex, which naturally does not show inactivities in time, for example, the EEG or the EMG.

Even after high-pass filtering, one cannot assume that the analytical zero level corresponds to the baseline. A typical task in the analysis of EP consists in measuring the waves concerning their amplitudes and latencies. Since the flanks of the waves are much less sensitive to disturbances than to the relatively flat local extrema, the tangent method is the most robust. The results are robust because the biosignals are superimposed with noise, so level-based extrema search has high

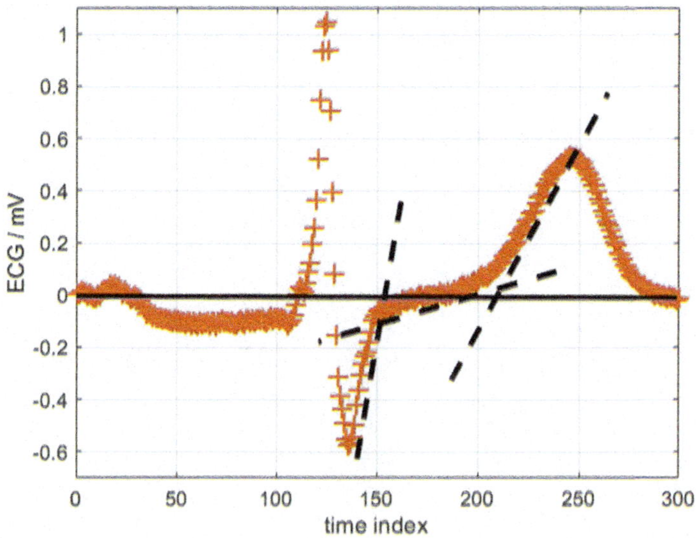

Fig. 4.20 Tangent method for determining the start/end times of the S wave and the T wave in the ECG, as well as the level of the S-T wave. The tangents are offset parallel to the curve sections for better visibility. No high-pass filter may be used beforehand to correctly measure the ST segment, as this signal section is spectrally very low-frequency (< 0.05 Hz)

statistical uncertainty. The times of the edges are anatomical/physiologically predetermined, and the noise hardly comes into play here. Therefore, the fluctuations of the edge determination are about one order of magnitude (one-tenth) below the changes in extreme values. From a computational point of view, edge determination is the most complex method. However, the tangent method does not entirely solve the baseline problem. For independence from the uncertain baseline, the amplitudes are estimated as potential differences of adjacent waves, e.g., N90P120 (see Fig. 4.21), P120N170, or according to their order in EP with N1P1, P1N2. Combining the tangent method with the measurement of amplitude differences of adjacent waves is the safest method concerning noise and interference as well as baseline fluctuations.

4.1.2.3 Determination of Frequency and Rate

For the spectral analysis of signals, one usually uses suitable transformations and methods for estimating the power spectral density. In the case of biosignals, the measurement task often consists of determining a "dominant" frequency (also incorrectly referred to as instantaneous). That is present in only one or a few periods. For example, in the waking state EEG, short-wave trains occur (a few periods) that are diagnostically interesting but are lost in the noise in a spectral analysis due to integration over a long time.

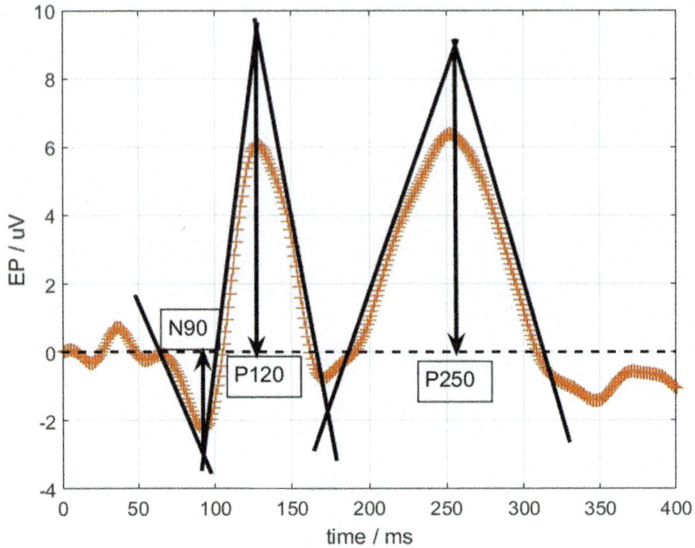

Fig. 4.21 VEP (visual evoked potential) with characteristic waves. A possible wave designation is based on the amplitude's polarity and latency (runtime after stimulus). The tangent method is much more robust and reliable than the search for a zero-valued first difference (extreme value detection)

Only individual heart periods with their reciprocal value are included in their calculation, so spectral analytical methods are useless here. The task is, therefore, to determine the frequency or rate directly from the time course of a biosignal. Figure 4.22 shows the first six seconds of an ECG recording (lead III) after the R-wave amplitudes have been algorithmically detected and the heart periods measured. The period duration changes here to about 20%, typical for a healthy organism.

The original recording was 43 s long, so one can already give statistics about the heart rate in the form of a histogram (Fig. 4.23). Since the current heart rate is calculated as the reciprocal of the heart period, it is only known over time at the end of the respective heart period. It is, therefore, not equidistant on the time axis. For spectral analysis of HRV, the values must be available during sampling. Furthermore, it is necessary that the area between the rate data, which in the form according to Fig. 4.24 still represents a non-equidistant pulse sequence, is also filled with adequate values. One of the simplest methods, also widely used, is a linear interpolation between the rate values (Fig. 4.25). Linear interpolation is relatively simple to perform, but it has signal-analytical problems. The most critical shortcoming is that at times of the determined heart rate, linear interpolation can cause (Berger et al., 1986). Kinks in the course occur at times of the determined heart rate due to the linear interpolation (right- and left-sided derivatives are not the same), which leads to impairment of the Fourier transforms.

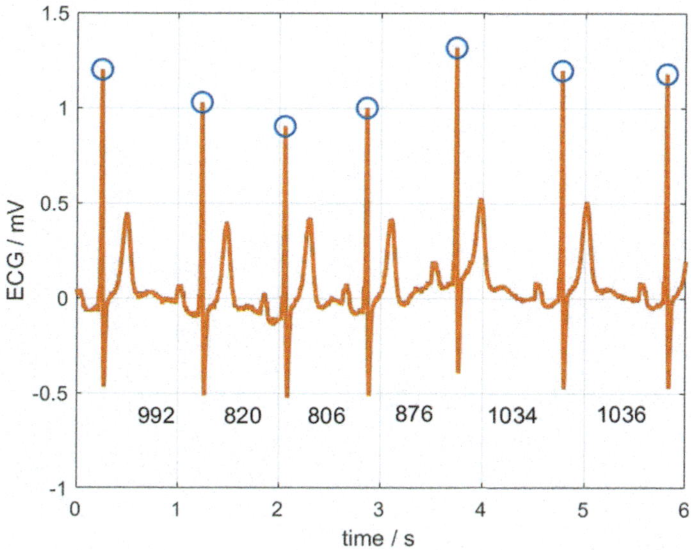

Fig. 4.22 ECG (lead III) after amplitude detection of R-waves and measurement of the R-R distances in milliseconds (cardiac periods). The reciprocal value of each period gives the instantaneous heart rate

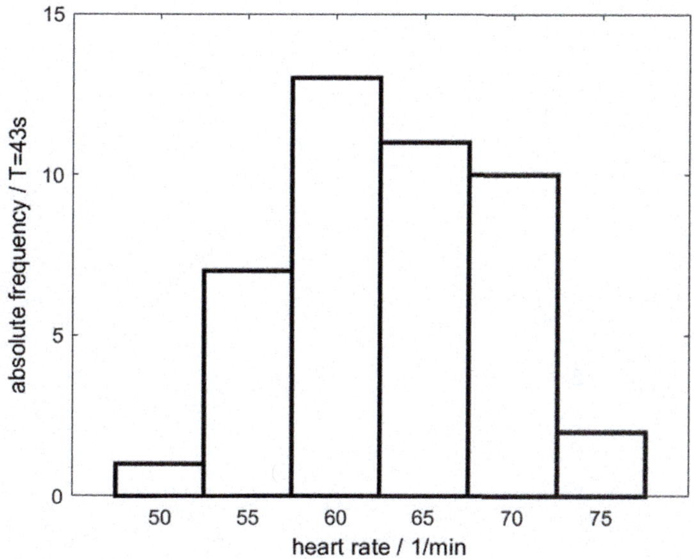

Fig. 4.23 Histogram of the heart rate of an ECG lead over a recording duration of 43 s

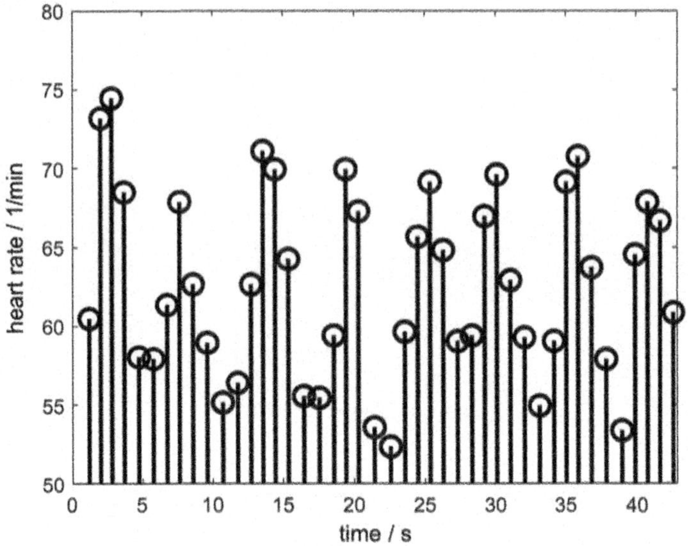

Fig. 4.24 Time course of the heart rate of the ECG derivation according to Figs. 4.22 and 4.23. The period of the heart rate change is about 5 s and can therefore be attributed to respiration. However, a trend and fluctuations are also visible over the measurement period, so other regulatory mechanisms of the organism are also effective. The time intervals of the discrete rate values are not equidistant

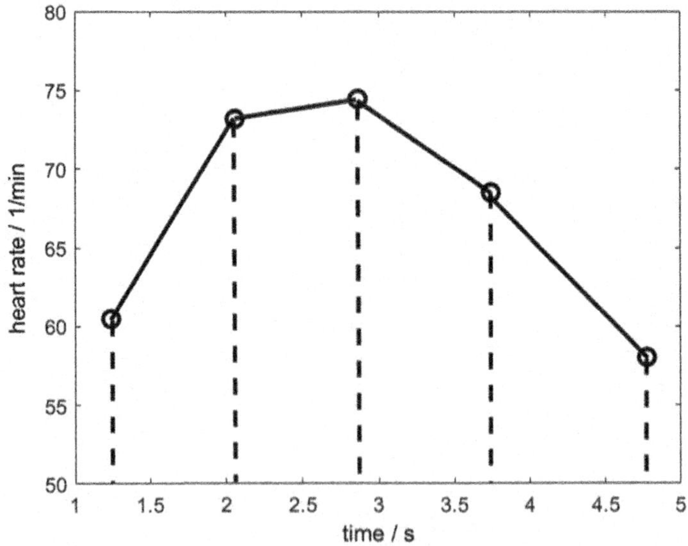

Fig. 4.25 Linear interpolation of the current heart rate. The interpolated course is formally sampled at the desired rate. The already existing sampling points are filled with the interpolated data

Therefore, methods should produce a smooth (differentiable) course for calculating intermediate heart rate values. Another approach to calculating HRV is the combined AM/FM signal model of the fundamental wave of the heart rate and the modulation signals (Eq. 4.13).

$$ekg(t) = a(t) \cdot \sin(\omega_0 t + hrv(t)). \tag{4.13}$$

In Eq. 4.13, $ekg(t)$ is the modeled fundamental wave of an ECG derivative, $a(t)$ is the AM modulation signal, ω_0 is the fundamental frequency of the ECG, and $hrv(t)$ is the HRV of interest.

Based on the signal model according to Eq. 4.13, the HRV is of interest, while the influence of the AM must be eliminated. As in communications engineering, this is achieved by keeping the amplitude of the fundamental wave constant. Practically, one replaces the original ECG with a cosine wave with the period of the current distance R-R, as shown in Fig. 4.26. The cosine interpolation in the R-R periods ensures the model signal is continuous and differentiable. The spectral analysis leads directly to information about the heart rate and its variability (Fig. 4.27). Following the signal theory, the spectrum of HR/HRV contains many sidebands. However, in the modeled signal, according to Eq. 4.13, their level is less than – 20 dB concerning the spectral bands of HRV, so it is not relevant for determining HRV.

With some biosignals (EEG, pulse wave), the fundamental frequency can be read directly from time by counting periods. It is common to measure the time of several consecutive periods or to count the number of periods in an appropriate

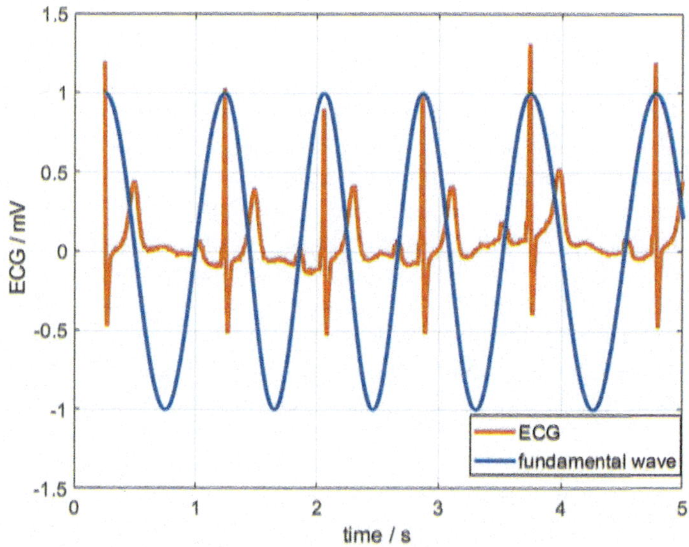

Fig. 4.26 The fundamental wave of the ECG (red) models the harmonic carrier signal (blue)

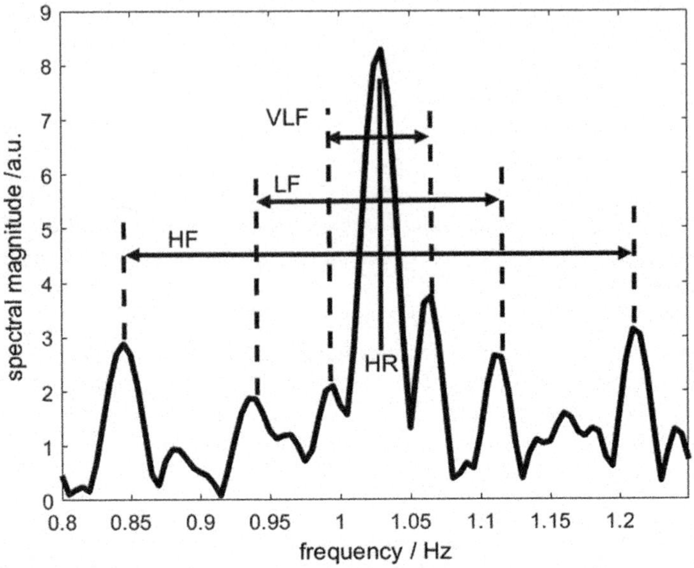

Fig. 4.27 Heart rate HR and heart rate variability HRV. Symmetrically around the heart rate are the bands of HRV: VLF (very low frequency, <0.04 Hz), LF (low frequency, 0.04 Hz…0.15 Hz), and HF (high frequency, 0.15 Hz…0.4 Hz). The classification of the spectral bands corresponds to the AHA (American Heart Association) recommendations

time interval. An averaged fundamental frequency is obtained (Eq. 4.14).

$$\overline{f}_0 = \frac{1}{\frac{1}{N}\sum_{i=1}^{N}\left(t\{x_m(i+1)\} - t\{x_m(i)\}\right)} \tag{4.14}$$

In Eq. 4.14, $x_m(i)$ is a curve feature of the biosignal (local maximum, local minimum, zero crossing, inflection point), i is the time index of the feature, $t\{x_m(i)\}$ is the time of occurrence of a feature $x(i)_m$.

According to Eq. 4.14, averaging reduces the random error by a factor of $\sqrt{N}$. In practical analysis, N *is* in the range of $10 < N < 100$, although higher values of N naturally give higher statistical confidence. For lower values of $N < 20$, using the median instead of the arithmetic means in Eq. 4.14 is recommended, as this is particularly robust with few measured values. The choice of a suitable curve feature x_m is decisive for the estimation quality. Except for the ECG's R-wave, local biosignal extremes show a relatively flat course. Hence, the determination of the associated time has a significant statistical error due to noise. As explained in the previous chapter, edges in the signal course are statistically more reliable regarding time measurement than all other curve features. Therefore, for determining the fundamental frequency according to Eq. 4.14 under the condition of a zero mean signal, the zero crossings should be selected if possible. A typical measurement situation is shown in Fig. 4.28.

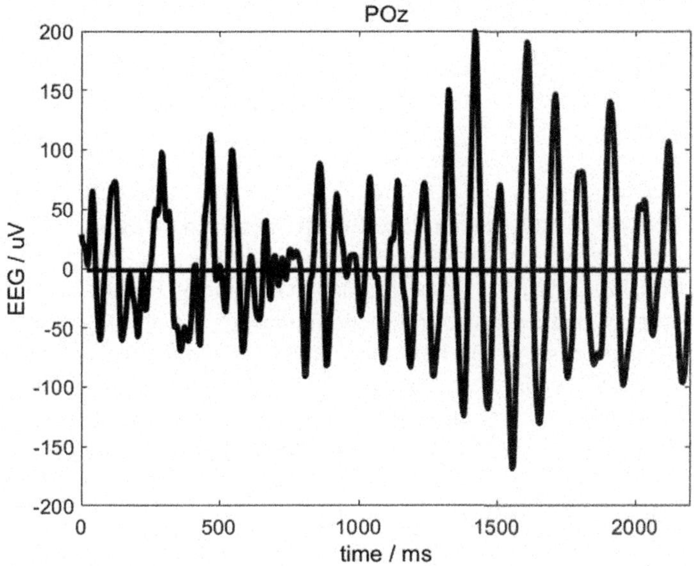

Fig. 4.28 Section of a unipolar EEG recording at position POz. A typical wave train in the- band runs in the right part (t = 1000 ms…2200 ms). Frequent determination by counting the zero crossings is possible for this highly reliable wave train. According to Eq. 4.14, the average fundamental frequency is $f_0 = 11$ Hz. From a signal analysis point of view, counting is only possible and helpful for almost harmonic oscillations or only for narrowband signals

A wave train runs in the right part of the EEG section in the α-band. Since the EEG is usually recorded with a high-pass filter at the amplifier input, it is a zero-mean signal. One chooses a time window where this wave train is present, here with $t =$ 1000 ms…2000 ms, and detects the zero crossings of the rising or falling edges. The normalization factor N in Eq. 4.14 must be doubled to evaluate zero crossings. In the left part of the image in Fig. 4.28, the EEG runs in the-band. This band is relatively high-frequency and is characterized by a partially stochastic character. Therefore, the method described here is unsuitable for frequency determination in the β-band.

Note: The terminology regarding biosignal frequencies, especially in EEG, is inconsistent and leads to controversial discussions and interpretations. In the sense of an accurate engineering description, the terms fundamental wave, fundamental frequency, harmonic, and harmonics introduced in the signal analysis are used here. Suppose one applies this view to the upper example of determining the fundamental frequency based on counting zero crossings. In that case, the count results in the fundamental frequency (fundamental wave) by applying Eq. 4.14. However, the wave train in the band is not a pure harmonic but a signal with a (narrow) continuous spectrum. A continuous spectrum contains infinite frequencies, so it would be incorrect to use the term instantaneous frequency and similar terms at this point. Strictly speaking, the instantaneous frequency can only exist with harmonic mono-components, such as frequency-modulated carriers, that eventually change their frequency. In a sense, instantaneous frequency is used to characterize the component in biosignals that is of interest and has the highest power or the best SNR.

4.2 Signal Analysis in the Frequency Domain

The transition from the time to the frequency domain is beneficial when the theoretical system, signal analysis, and spectral analysis are necessary or methods of practical signal processing can be implemented more effectively in the frequency domain. Therefore, the Fourier transform in its theoretical continuous and usable discrete form is dealt with in the following.

4.2.1 Fourier Transform

The Fourier transform (FT) is generally known and sufficiently investigated so that only properties and characteristics relevant to biosignal processing are dealt with here. The basis is the continuous FT and its inverse variant (IFT) according to Eq. 4.15 (the convolution is marked with an *).

$$X(f) = \int_{-\infty}^{\infty} x(t) \cdot e^{-j2\pi ft} dt, \quad x(t) = \int_{-\infty}^{\infty} X(f) \cdot e^{j2\pi ft} df \tag{4.15}$$

4.2.1.1 Properties of the FT and IFT

Multiplication of Two-Time Functions

$$z(t) = x(t) \cdot y(t) \underset{IFT}{\overset{FT}{\rightleftarrows}} Z(f) = X(f) * Y(f) \tag{4.16}$$

According to the signal model, according to Eq. 4.3, strong disturbances have a multiplicative effect on the measurement signal. If the interference is harmonic or at least periodic, it can be eliminated from the measurement signal with complex methods. Such a measurement situation is shown in Fig. 4.29. If one of the signals, according to Eq. 4.16, is harmonic, the spectral convolution causes a shift and mirroring of the other spectrum. If both spectra are continuous and overlap—always in the case of biosignals—their separation is no longer possible (see Fig. 4.27).

Convolution of Two-Time Functions

$$z(t) = x(t) * y(t) \underset{IFT}{\overset{FT}{\rightleftarrows}} Z(f) = X(f) \cdot Y(f) \tag{4.17}$$

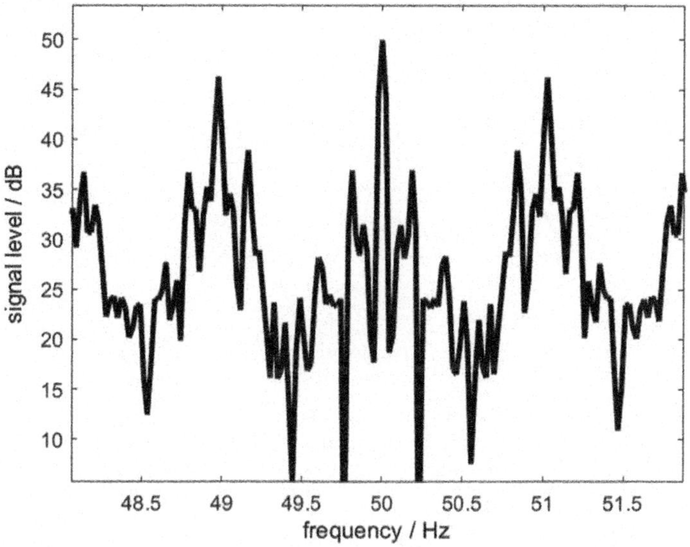

Fig. 4.29 Amplitude spectrum of an ECG recording (Einthoven III), which was strongly disturbed by the network and had a multiplicative effect on the ECG. Since the disturbance is harmonic, the convolution, according to Eq. 4.16, causes a shift and mirroring of the ECG spectrum symmetrically around 50 Hz

In the continuous-time domain, signals and systems can be described with the help of their Fourier transforms. A typical section of analog biosignal processing is shown in Fig. 4.30. A measurement signal passes through an analog high-pass to separate the electrode voltage and a low-pass to limit the band. Since the transfer functions in the frequency domain are known, one can easily describe each system output. A description in the time-continuous range with the help of convolution is not usual; it is mainly reserved for theoretical analyses.

Time Shift of a Time Function

$$x(t - t_0) \underset{\overleftarrow{IFT}}{\overset{\overrightarrow{FT}}{}} e^{-j2\pi f t_0} X(f) \tag{4.18}$$

The time shift of a biosignal is significant when time-critical processes are processed (on-demand pacemaker) or coincidence is a necessary condition (ensemble averaging in sensor arrays). If the time shift is known, it can be corrected a posteriori using the relationship Eq. 4.18 (see Sect. 3.2.3 Sampling in multichannel systems). Timeshift does not influence the amplitude spectrum of a signal (translation invariant), so it has no significance in spectral analyses (EEG band power, HRV). Figure 4.31 shows how a time shift affects the phase frequency response

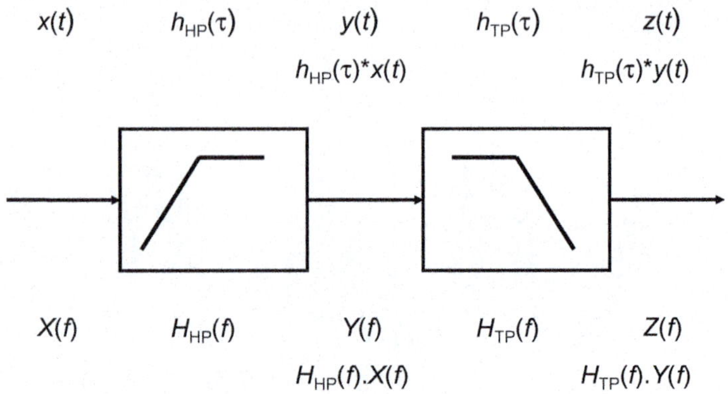

Fig. 4.30 A signal processing chain consists of a series of signals ($x(t)$, $y(t)$, $z(t)$) and systems (h_{HP} (τ), $h_{TP}/(\tau)$). According to the relationship Eq. 4.17, in the frequency domain, one multiplies the transfer functions; in the time domain, one convolves over an independent auxiliary variable of time

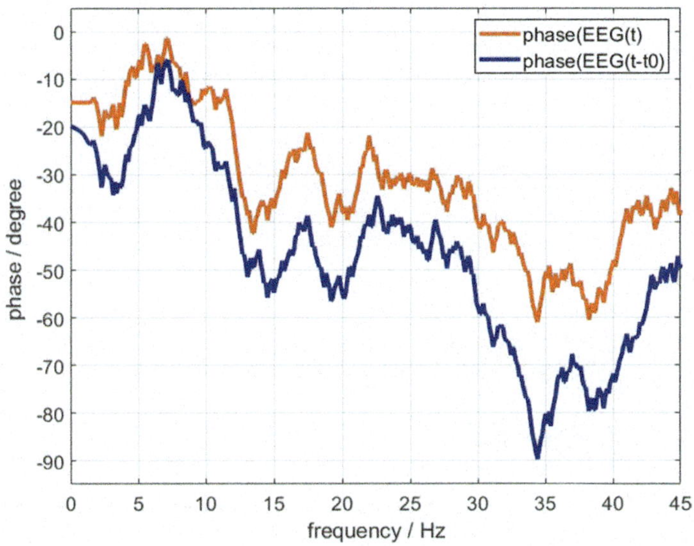

Fig. 4.31 Phase frequency response of an EEG section. The original measurement signal (red) was delayed by 400 ms due to various system run times (blue). The amplitude spectrum remains unchanged, and the phase response shifts according to Eq. 4.18

of a biosignal. It should be noted that the phase frequency response essentially does not change qualitatively but decreases linearly with increasing frequency, according to Eq. 4.18.

The Frequency Shift of a Time Function

$$x(t) \cdot e^{j2\pi f_0 t} \underset{\overleftarrow{IFT}}{\overrightarrow{FT}} X(f - f_0) \tag{4.19}$$

The frequency shift of a time function corresponds in message terms to the amplitude modulation of the harmonic carrier with the frequency f_0 with a modulation signal $x(t)$. It can be interpreted as a particular case of the relationship according to Eq. 4.16. An undesired frequency shift due to a harmonic disturbance is shown in Fig. 4.29, an intended frequency shift dealt with in Sect. 3.2.2, sampling of band-limited signals (Fig. 3.15).

Scaling of a Time Function

$$x(at) \underset{\overleftarrow{IFT}}{\overrightarrow{FT}} \frac{1}{|a|} X\left(\frac{f}{a}\right) \tag{4.20}$$

In its practical implementation, scaling is known as zoom. While it can be realized continuously in many areas (optics, physics), it plays a purely theoretical role for biosignals in its continuous version. In biosignal processing, it is of fundamental importance in its discrete-time variant (sampling, undersampling, and oversampling).

Time Mirror of a Time Function

$$x(-t) \underset{\overleftarrow{IFT}}{\overrightarrow{FT}} X(-f) \tag{4.21}$$

The mirroring of a time function is, of course, not realizable in actual biosignal processing. Still, it forms the theoretical basis for several signal analytical approaches (see Sect. 4.2.1.3 Convolution in the time domain and correlation function). The effect of time mirroring is shown in Fig. 4.32. While the amplitude-frequency response remains constant, the phase changes its sign. Therefore, one can use time mirroring, especially in the correlation analysis, since the correlation function and the convolution differ formulaically only in the sign of the shift (Eq. 4.25). Since this is known, the sign reversal can be corrected at any point in the (linear) signal processing.

4.2.1.2 Windowing

Even in theoretical signal analysis, it cannot be assumed that the integration limits of the FT are realizable. Therefore, one must consider a finite signal from the outset. Mathematically, a signal section corresponds to the implicit windowing of a signal with a rectangular window, according to Eq. 4.22 (Fig. 4.33).

$$x_w(t) = rect(t) \cdot x(t),\ rect(t) = \sigma(t - t_m + T/2) - \sigma(t - t_m - T/2) \tag{4.22}$$

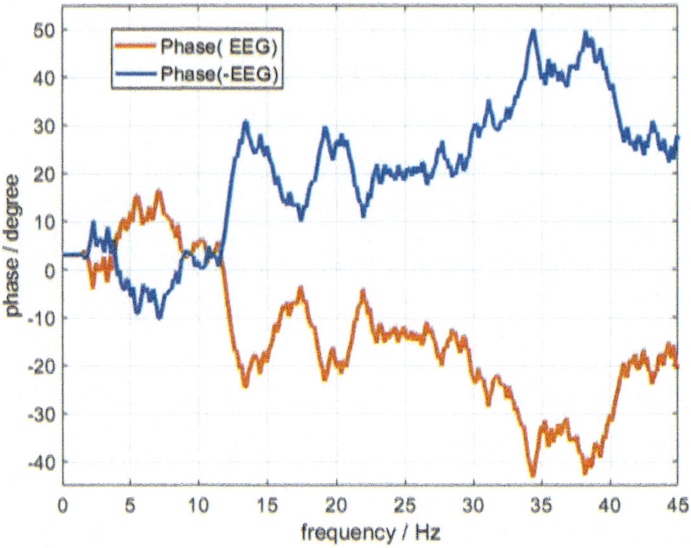

Fig. 4.32 Phase frequency response of an EEG section with strong- and -waves (red) and its temporally mirrored variant (blue). While the amplitude-frequency response remains constant, the phase changes its sign. A subsequent correction is, therefore, feasible at any time by reversing the sign

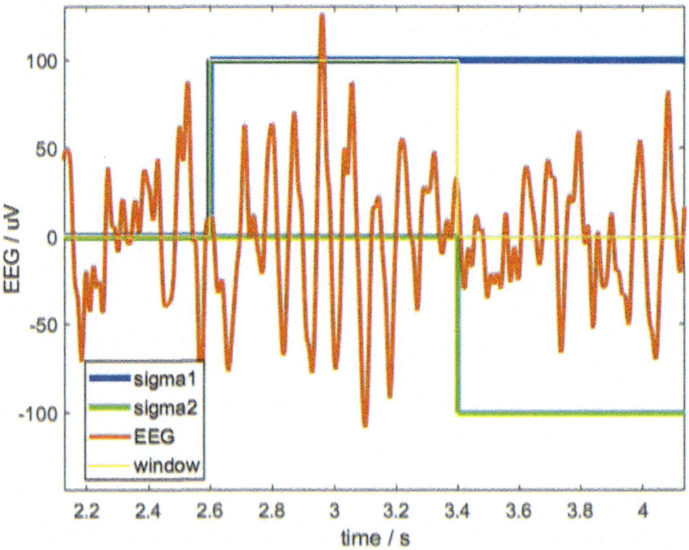

Fig. 4.33 Realisation of a signal section in the EEG between the integration limits $t_1 = 2.6$ s and $t_2 = 3.4$ s with the help of the step function $(t - t_{1,2})$ according to Eq. 4.22 ($t_m = (t_1 + t_2)/2$, $T = t_2 - t_1$)

Table 4.1 Selected window types and formulas for the continuous time function

Window type	Formula for $w(t)$
Blackman	$0.42 - 0.5\cos(2\pi t/T) + 0.08\cos(4\pi t/T)$
Bartlett	$1 - 2\lvert t\rvert/T$
Hamming	$0.54 - 0.46\cos(2\pi t/T)$
Hanning	$0.5(1 - \cos(2\pi t/T))$

In Eq. 4.22 σ is a unit step, t_m is the temporal center of the window, and T is the window length.

The FT of the window function according to Eq. 4.22 and Fig. 4.33 gives the sinc function

$$rect(t) \xrightarrow{FT} \frac{\sin(\pi fT)}{\pi fT} e^{-j2\pi t_m f} = \sin c(fT) e^{-j2\pi t_m f}, \tag{4.23}$$

where T *is* the window length, and t_m is the window center. For the general case of t_2 and t_1, consider Eq. 4.18. According to Eq. 4.23, the transfer function is shown in Fig. 4.34. The transmission reaches zero for integer multiples of the reciprocal window length. This effect can be explicitly used for specific signal processing problems, such as suppressing disturbing periodic interferences (mains frequency, the frame rate of monitors, see Sect. 3.3.1, integrating AD-converters). The zero crossings are always present regardless of the window shape, and their position depends solely on the window length. Since the transfer function of the window is convolved with the signal spectrum according to Eq. 4.16, it should ideally have the shape of a Dirac pulse to avoid changing the spectrum. In practice, this would mean that the window length should be as large as possible (theoretically infinitely long), which is not feasible for reasons of practicality. Therefore, one tries to design the window shape with signal-specific spectral properties (Table 4.1).

For discrete-time (sampled biosignals), replace t with n and T_A with N (Fig. 4.35).

The best-known window types are shown in Figs. 4.35 and 4.36. As shown in Fig. 4.34, the transfer function contains a main maximum and several secondary maxima. The goal of a window function is to keep the main maximum as narrow as possible and to suppress the secondary maxima as much as possible. Since both are not feasible simultaneously, a suitable compromise must be sought for each signal type. Comparing the presented window types (Fig. 4.36), the following can be observed: Blackman suppresses the side maxima in the spectrum best but has the broadest main maximum, and Bartlett has the narrowest main maximum but the worst suppression of the side maxima. Hanning shows a slightly wider main maximum than Bartlett but the second-best suppression of the secondary maxima. For most cases, therefore, Hanning is the best alternative. Hamming has the second-best width of the main maximum but an almost constant suppression of the secondary maxima, whereby the first two are better suppressed than Hanning. Thus, Hamming is more suitable for good spectral resolution. The rectangular window—which is implicitly always present if one does not want to use any of

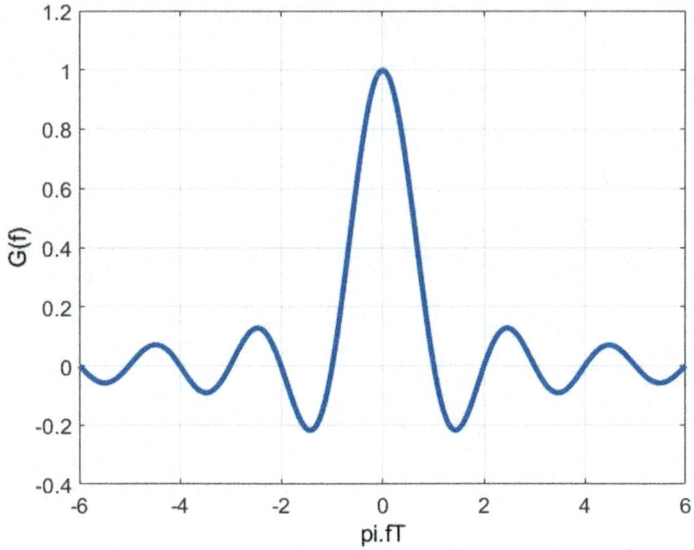

Fig. 4.34 Transfer function of a spectral rectangular window in the time domain. For integer multiples of the reciprocal window length, the transmission reaches zero

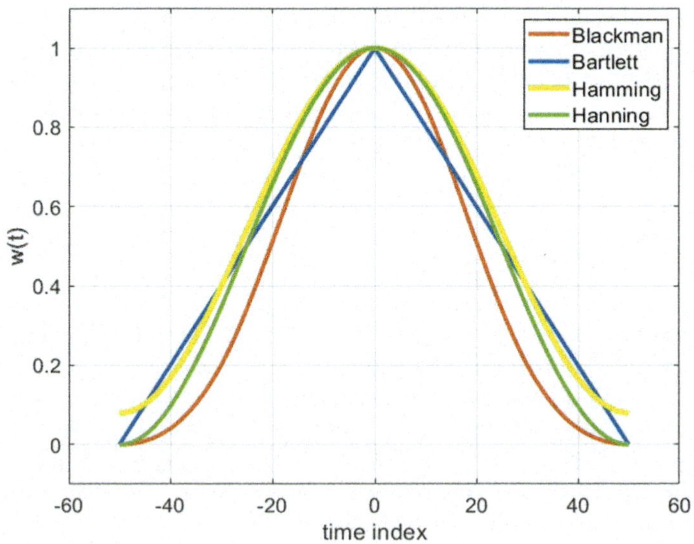

Fig. 4.35 Blackman, Bartlett, Hamming, and Hanning window types with window length 101

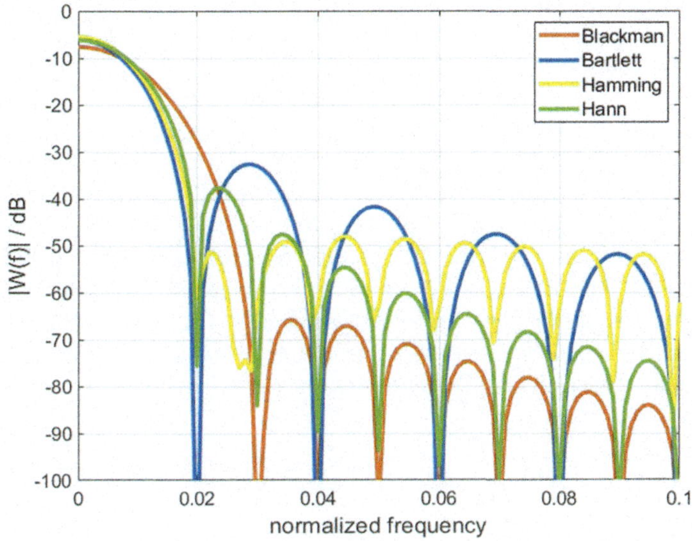

Fig. 4.36 Spectral characteristics of the window functions according to Blackman, Bartlett, Hamming, and Hanning

the windows mentioned—has the narrowest "main lobe" (main maximum) of all window types but the most significant variance of the spectral estimate.

4.2.1.3 Convolution in the Time Domain and Correlation Function

Two-time functions' cross-correlation function (CCF) is calculated as follows (Eq. 4.24).

$$\rho_{xy}(\tau) = \int_{-\infty}^{\infty} x(t)y(t+\tau)dt \tag{4.24}$$

If one of the time functions in Eq. 4.24 is mirrored on the time axis, one obtains its convolution integral (Eq. 4.25).

$$\rho_{xy}(\tau) = \rho_{yx}(-\tau) = \int_{-\infty}^{\infty} x(t)y(-t-\tau)dt = x(t) * y(-t) \tag{4.25}$$

Transferring Eqs. 4.24 and 4.25 into the frequency domain, we get

$$\psi(f) = X(f) \cdot Y(-f). \tag{4.26}$$

With the help of the IFT, the CCF can therefore be calculated according to Eq. 4.27.

$$\rho(\tau) = FT^{-1}\{X(f) \cdot Y(-f)\} \tag{4.27}$$

According to Eq. 4.25, the CCF is a convolution of two-time functions, one mirrored on the time axis. Therefore, according to Eq. 4.27, the CCF can also be calculated indirectly from the FT of the particular time functions. This theoretical derivation will later be used for the efficient calculation of the PPF with the aid of the FFT. The PPF (f), according to Eq. 4.26, is also called *cross-power density* in the spectral analysis of stochastic signals. (Eq. 4.28). If one applies the relations according to Eqs. 4.24–4.27 to a single signal $x(t)$, one speaks of the autocorrelation function (AKF) and *auto power density* (Eq. 4.29).

$$\rho_{xy}(\tau) \underset{IFT}{\overset{FT}{\rightleftarrows}} X(f) \cdot Y^{*}(f) = |X(f)| \cdot |Y(f)| \tag{4.28}$$

$$\rho_{xx}(\tau) \underset{IFT}{\overset{FT}{\rightleftarrows}} X(f) \cdot X^{*}(f) = |X(f)|^{2} \tag{4.29}$$

In Eqs. 4.28 and 4.29, $\rho_{x,y}$ is the root mean square in the sense of statistics, equivalently the KKF/AKF in system analysis of deterministic signals. A * for complex quantities denotes the conjugate complexes. Y(−f) = Y* (f) applies to real signals.

According to Eq. 4.26 and Eqs. 4.28 and 4.29, the relations are also known as the *Wiener-Khinchine-theorem.* The theorem states that both ways of calculating the power density spectrum are equivalent: First, via the FT to calculate the complex spectrum (periodogram) and then form the magnitude square (direct method), or secondly, to first calculate the PFT and then transfer this to the frequency domain via the FT (indirect method). In practical analysis, however, these methods are not equivalent. The connection between the correlation function (Eq. 4.24) and the temporal convolution (Eq. 4.25) also forms the theoretical basis for the interpretation of the FT as a correlation between a harmonic oscillation (analysis function in the complex pointer, Eq. 4.15) and the examined signal.

The duality of the FT exists between the power density and the correlation function but is no longer with the examined signal as a time course. Due to the magnitude squares in the direct method and the calculation of the correlation function (second statistical moment) in the indirect method, the phase information is lost so that the examined signal can no longer be reconstructed. An illustrative example is provided by the elementary analytical signals of Dirac's pulse and white noise (Fig. 4.37). This example shows that based on the spectral power density alone, a conclusion about the time signal is no longer unambiguous. At this point, it becomes clear how essential the phase information is for signal analysis and synthesis: it determines the signal shape and amplitudes.

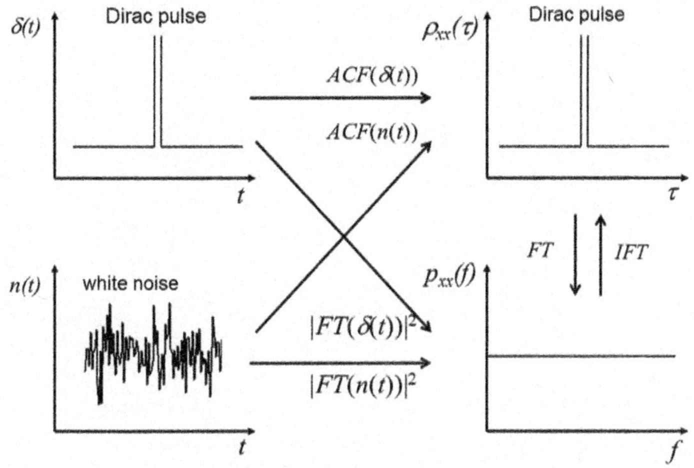

Fig. 4.37 Calculation of the power spectral density of the Dirac pulse and white noise via the ACF (indirect method) and the magnitude square of the FT (direct method). The duality of the FT exists between the AKF and the power density. A reconstruction of the time signal is no longer possible because the phase information is lost

4.2.2 Discrete Fourier Transform

4.2.2.1 Continuous and Discrete Time

Biosignals are sampled (not yet digitized) after the acquisition, analog filtering, and amplification. Therefore, it is necessary to set up a corresponding set of instruments of discrete mathematics for discrete-time biosignals. If one assumes the time-continuous FT (Eq. 4.15) and inserts a sampled signal, the formulation according to Eq. 4.30 results. A continuous spectrum is obtained after multiplying the time signal x(t) with the Dirac pulse train.

$$\begin{aligned} X(f) &= \int_{-\infty}^{\infty} x(t) \sum_{n=-\infty}^{\infty} \delta(t-nT) e^{-j2\pi f t} dt \\ &= \int_{-\infty}^{\infty} \sum_{n=-\infty}^{\infty} x(nT)\delta(t-nT) e^{-j2\pi f t} dt \\ &= \mathbf{F}F\left\{ \sum_{n=-\infty}^{\infty} x(nT)\delta(t-nT) \right\} \end{aligned} \tag{4.30}$$

In Eq. 4.30, $X(f)$ is the Fourier transform of $x(nT)$, T is the sampling period, nT is the sampling times, $x(nT)$ is the values of the signal $x(t)$ at sampling times nT, f is the continuous frequency, (t) is Dirac momentum at time t.

The sampling approximates the infinitely narrow time interval *dt* from Eq. 4.15 into the constant sampling period *T*. Thus, the continuous integration is replaced with a sum over the sampling times (Eq. 4.30). At the same time, for reasons of practicability, one first applies a rectangular window (Eq. 4.31). The reciprocal of the sampling rate replaces the sampling period T in the exponent.

$$X(f) = \sum_{n=0}^{N-1} x(nT)e^{-j2\pi n\frac{f}{f_A}} \tag{4.31}$$

In Eq. 4.31, n is the time index; this starts at zero to ensure equivalence with the time axis; f_A is the fundamental frequency of the sampling rate, and N is the number of samples.

The FT of $x(nT)$ can initially be calculated for infinite frequencies f despite the sampling. The more frequencies are calculated, i.e., the closer these frequencies are to each other, the less adjacent Fourier coefficients differ. Thus they are very similar or strongly correlated in the sense of second-order statistics and, therefore, redundant. Therefore, a correlation-free or geometrically expressed orthogonal transformation is required to calculate the FT and a minimum necessary amount of data. If their scalar product is zero-valued, two-time series are uncorrelated—or orthogonal as vectors (Eq. 4.32).

$$\rho_{xy}(0) = \sum_{n=0}^{N-1} x[n] \cdot y[n] = \vec{x} \cdot \vec{y} = 0 \tag{4.32}$$

Since the window length is necessarily an integer multiple of the sampling period, Eq. 4.32 can only be satisfied if the orthogonal decomposition vectors are also an integer multiple of the sampling period and identical to the window length (see Integrals of Trigonometric Functions). Equation 4.31 can be transformed in terms of the transition from a continuous to a discrete frequency as follows (Eq. 4.33):

$$X\left(k\frac{f_A}{N}\right) = \sum_{n=0}^{N-1} x(nT)e^{-j2\pi n\frac{k\frac{f_A}{N}}{f_A}} \tag{4.33}$$

It follows from Eq. 4.33 that the fundamental frequency of the sampling rate f_A is no longer critical but integer discrete parts f_A/N. One can therefore normalize Eq. 4.33 to the frequency resolution f_A/N and obtain a formulation independent of the concrete sampling rate (Eq. 4.34). It is the *Discrete Fourier Transform* (DFT).

$$X[k] = \sum_{n=0}^{N-1} x[n]e^{-jkn\frac{2\pi}{N}} \tag{4.34}$$

In Eq. 4.34, k is the normalized discrete sampling rate index, n *is* the time index of the sampling points with distance T_A, and square brackets are used for frequency and time discrete quantities.

According to the duality of the FT (Eq. 4.15), the inverse DFT is as follows (Eq. 4.35):

$$x[n] = \frac{1}{N}\sum_{k=0}^{N-1} X[k]e^{jkn\frac{2\pi}{N}} \tag{4.35}$$

The relations are given in Eqs. 4.34 and 4.35 are the dual discrete formulations of the FT. The transformation factor 1/N in Eq. 4.35 can also be used in advance in Eq. 4.34, e.g., to indicate the actual amplitude in the spectrum. Mathematically, it can be used in either of the relations 4.34 and 4.35 and possibly symmetrically via $\sqrt{N}$; only the duality must be preserved.

In Eqs. 4.34 and 4.35, the reference to the sampling period T_A was lost. On the one hand, this facilitates the algorithms through discrete numerics since, for DSP (digital signal processing), only vectors or matrices—number sequences and number fields—are available for processing. In many cases, however, the result of an algorithm must be transferred back into the analog world. For this purpose, the fundamental frequency of the sampling rate must be restored (e.g., music CD, video DVD, radio and television, and similar), or the number sequences must be brought to the correct time axis (signal reconstruction).

If one introduces a rotation factor (Eq. 4.36), the DFT can be expressed as follows (Eq. 4.37):

$$W_N = e^{-j\frac{2\pi}{N}} \tag{4.36}$$

$$X[k] = \sum_{n=0}^{N-1} x[n] \cdot W_N^{kn},\ x[n] = \frac{1}{N}\sum_{k=0}^{N-1} X[k] \cdot W_N^{-kn}, \tag{4.37}$$

and in effective matrix notation

$$\mathbf{X}_N = \mathbf{W}_N\mathbf{x}_N,\ \mathbf{x}_N = \mathbf{W}_N^{-1}\mathbf{X}_N. \tag{4.38}$$

In Eq. 4.38, it should be noted that the IDFT (Inverse Discrete FT) requires the inverse matrix WN, which would otherwise be costly. Based on Eq. 4.36, only its conjugate complex version is needed for the inverse matrix and a constant factor of 1/N.

4.2.2.2 Discrete Fourier Transform as Cross-Correlation

The formula Eq. 4.34 can be expressed using trigonometric functions as follows (Eq. 4.39):

$$X(k) = \sum_{n=0}^{N-1} x[n] \cdot \cos\left(kn\frac{2\pi}{N}\right) - j\sum_{n=0}^{N-1} x[n] \cdot \sin\left(kn\frac{2\pi}{N}\right)$$

$$= \mathrm{Re}\{X[k]\} + j\mathrm{Im}\{X[k]\} \tag{4.39}$$

In Eq. 4.39, both terms (real part, imaginary part) are element-wise products of two-time series. That is also the way to calculate the cross-correlation. If one replaces the trigonometric functions in Eq. 4.39 with a generalized time series $y[n]$, one can formulate (Eq. 4.40):

$$r_{xy}[0] = \sum_{n=0}^{N-1} x[n] \cdot y[n]. \tag{4.40}$$

From a statistical point of view, the Fourier coefficients according to Eq. 4.41 for $k = 0$, the first coefficient corresponds to the arithmetic mean of the signal. All others ($k > 0$) represent the cross-correlation between the signal and the harmonic at point k for a zero-value shift.

$$\begin{aligned} k = 0 &\Rightarrow X[0] = \mu_x \\ k > 0 &\Rightarrow X[k] = r_{\mathbf{W}_N^k \mathbf{x}}[0] \end{aligned} \tag{4.41}$$

According to Eq. 4.40, the relationship describes the cross-correlation between two processes. If one of the two processes is a zero mean signal, the cross-correlation is identical to the cross-covariance. Since the trigonometric functions are zero-symmetric, the Fourier coefficients can be interpreted as cross-covariances (Eq. 4.41). This interpretation is essential for many statistical analyses, as demonstrated in the following example.

Discrete Fourier Transform of an ECG Recording

The absolute frequency distribution of digitized values of an ECG recording with 21,500 readings (43 s) is shown in Fig. 4.38. The distribution shows a typical pattern for the ECG: The absolute maximum (mode, value of the most significant probability of occurrence) lies near the baseline and represents the idle times in the ECG. The R and S spikes and the T wave have relatively high amplitudes, but their values are represented less frequently. That leads to local maxima (modes) in the histogram with low frequencies. Such a distribution is highly unsuitable for statistical analysis. According to the DFT, the Fourier coefficients are approximately normally distributed (Gaussian distribution, Fig. 4.39). This positive effect occurs with a signal length from about N = 100 and is due to the Central Limit Theorem. Since the DFT is a linear transformation, many statistical analyses can be carried out on normally distributed Fourier coefficients instead of unfavorably distributed time values (this applies to all biosignals).

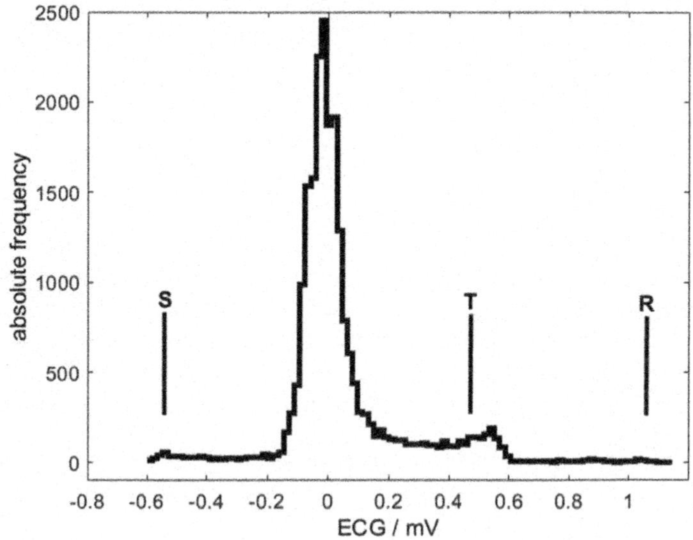

Fig. 4.38 Absolute frequency distribution (histogram) of digitized ECG readings of a recording with 21,500 sampling points. Such a distribution is highly unsuitable for statistical analyses. Local maxima (modes) over the spikes R and S and the wave T are hardly identifiable using the histogram

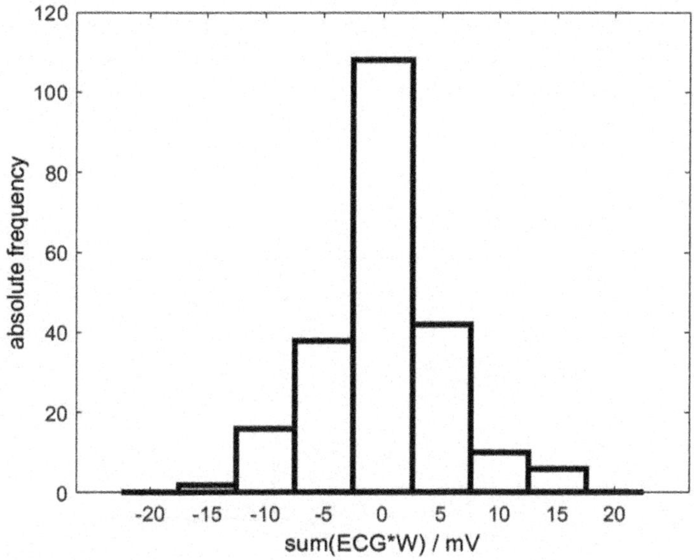

Fig. 4.39 Absolute frequency distribution (histogram) of 215 Fourier coefficients calculated from a sum of 100 element-wise products each. The normal distribution is approximately achieved

4.2.2.3 Discrete Fourier Transform as a Filter Bank

The term $y[n]$ in Eq. 4.40 is a harmonic signal in the DFT, so the following applies accordingly

$$z[m] = \sum_{n=0}^{N-1} x[n]y[n+m] \Rightarrow z[n] = \sum_{m=-M}^{M} x[n]y[n-m]. \tag{4.42}$$

With Eq. 4.42, the equivalence of the cross-correlation and the convolution with a time-gated signal becomes apparent. Since $y[m]$ can be interpreted as the impulse response of a transversal digital filter, which is identical to the harmonic analysis function, the DFT of length 2M + 1 for a signal of length N is, in fact, a filter bank. Accordingly, the DFT represents 2M + 1 parallel filters, each passing the current frequency k (Fig. 4.40). These filters are not ideal; they have a finite bandwidth and side maxima.

Considering the DFT as a parallel configuration of spectral filters according to Fig. 4.41, one can observe and conclude the following:

1. Except at k = 0, the maxima (main lobes) are – 6 dB, i.e., a reduction of the amplitude to half. That is because the DFT (its magnitude) is an even function. The amplitudes are therefore divided in half between the positive and negative frequency ranges.
2. The transfer functions (filter characteristics) spectrally overlap with their neighbors by 50% each. It means that the uniqueness of the spectral decomposition is not always ensured. Frequency components that do not fall precisely on the supporting points of the DFT (in Fig. 4.41, the integer multiples of 1/N =

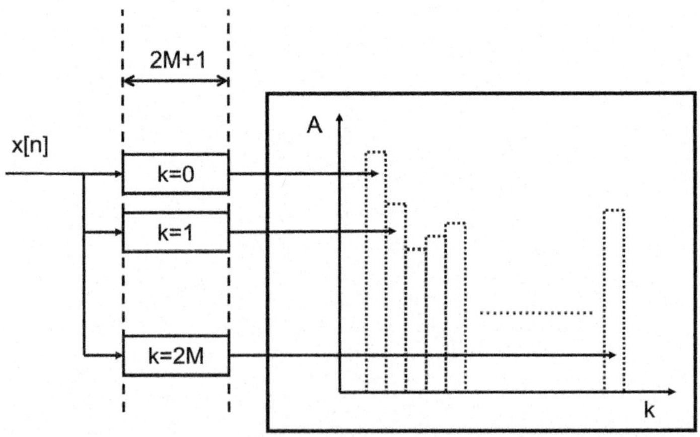

Fig. 4.40 Interpretation of the DFT as a filter bank. Each harmonic oscillation of the discrete frequency k represents the impulse response of a transversal filter. Thus, at each frequency k, the corresponding signal component is selectively filtered out and is applied in parallel at the output of the filter bank

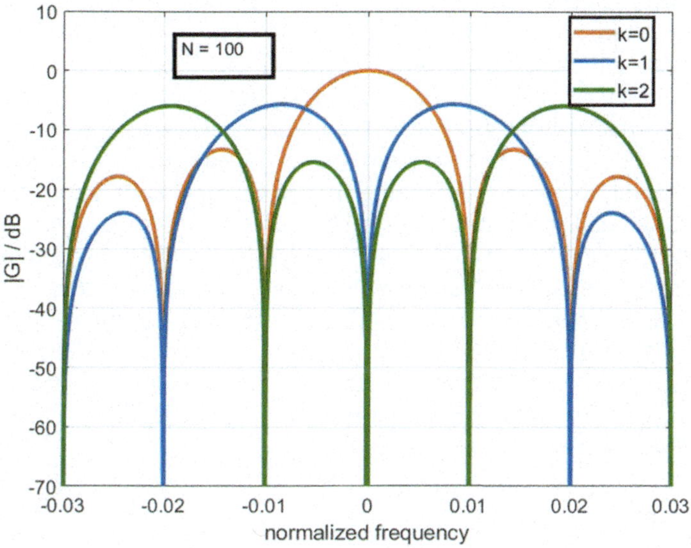

Fig. 4.41 Transfer functions of the DFT coefficients for N = 100. DFT interpolation points lie on the integer multiples of the relative frequency of 1/N = 0.01. The passbands of the filters spectrally overlap 50% with their neighbors

0.01) capture neighboring passband curves. A harmonic oscillation with a relative frequency of 0.015 (15 Hz at 1000 sps) is credited with three coefficients: The DC component ($k = 0$) and the first two harmonics $k = 1$ and $k = 2$ (and others). It is evident with the equal component: its first side lobe in the positive range is above the relative frequency of 0.015. That is not a methodological error; the harmonic oscillation at 0.015 is missing half a period to fill the window, or there is half a period too much, so it has an equal component. That is, therefore, rightly credited to the DC component but damped by about 12 dB.

3. The DFT is a linear transformation, so it must map all frequency components, even those that do not fall precisely on the supporting points of the DFT. If the sum of the DFT coefficients must result in the original signal (IDFT), the spectrum must also have a continuous course in the sum. This property is ensured by overlapping the filter functions in Fig. 4.41.

4.2.2.4 Calculation of the DFT via the FFT

FFT (Fast Fourier Transform) refers to algorithms that calculate the DFT more effectively and faster than is possible according to the formulae Eqs. 4.34 and 4.35.

Note: Contrary to its name, an FFT is not a transformation. It is merely a fast algorithm for calculating the DFT. In other words: An arbitrary FFT is a fast algorithm that achieves the same result as a conventional DFT by reducing the computational redundancy.

To increase the effectiveness of a calculation, one must first examine whether there are redundancies in the conventional formulae because only then can the calculation be carried out more efficiently. One can interpret the rotation factor according to Eq. 4.36 as follows (Fig. 4.42): The full rotation of the unit circle is divided among N angles. Assuming that N is a power of 2, due to the even function cosine and the odd function sine, the circle consists of octants, so that quantitatively at most N/4 (two values for each point in an octant), different values are necessary to describe it. Thus the redundancy can be reduced mathematically to 1/8 for the time being. From the relationship, according to Eq. 4.37, it is further evident that the exponent—$k.n$ always projects onto only one of the positions of the unit circle (Fig. 4.42). It shows further potential for reducing redundancy. Most algorithms of the FFT are based on a different approach (Eq. 4.43), with which the redundancies mentioned are also reduced.

$$\begin{aligned} X[k] &= \sum_{n=0}^{N/2-1} x[2n] W_N^{2kn} + \sum_{n=0}^{N/2-1} x[2n+1] W_N^{(2n+1)k} \\ &= \sum_{n=0}^{N/2-1} x_1[n] W_N^{2kn} + W_N^k \sum_{n=0}^{N/2-1} x_2[n] W_N^{2nk} \\ &= \sum_{n=0}^{N/2-1} x_1[n] W_{N/2}^{nk} + W_N^k \sum_{n=0}^{N/2-1} x_2[n] W_{N/2}^{nk} = X_1[k] + W_N^k X_2[k] \end{aligned} \tag{4.43}$$

The principle of the procedure (named after Cooley and Tukey, the mathematical basis of Gauss), according to Eq. 4.43, is that the input sequence is halved to even and odd indices of the time series. For one stage of this division at N = 4,

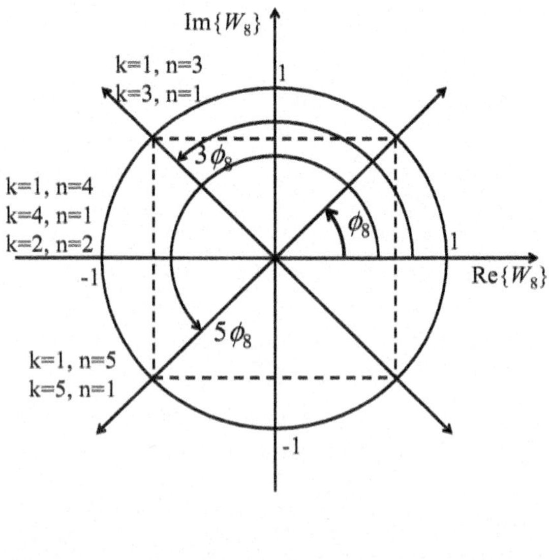

Fig. 4.42 Graphical interpretation of the rotation factor W_N according to Eq. 4.36 for N = 8. The unit circle is divided into eight angles. Because of the rotational symmetry, there are only two different values for all angles. For $k.n$ = const., the angles are identical

this procedure is shown in Fig. 4.43. The division can be realized recursively for all powers of the base sequence. The most favorable configuration for signal processing is according to the root-2 algorithm, as shown in Fig. 4.44. This division can be generalized as a butterfly for $N = 2$, as shown in Fig. 4.45.

According to Eq. 4.43, the algorithm has been frequently improved and optimized, and today numerous implementation variants are available. As a standard signal processing method, it belongs to the essential equipment of DSP, FPGA, and controller-based technical solutions and every programming package of signal analysis. The above platforms usually use a hardware multiplier (MUL) and an ALU. Therefore, the algorithmic effort for calculating the DFT according to Eq. 4.34 can be estimated with $2.N^2$ (exactly N^2 multiplications and $N(N - 1)$ additions). Composing the computational structure into butterflies (Fig. 4.45) reduces the effort to $2N.\log_2(N)$. That is a more than 99% reduction for a sequence length of 1024 samples and a decrease of 99.99% for $N = 65{,}536$ (2^{16}) (Fig. 4.46).

For reasons of the correct application of terms, it should be noted that the FFT is an *algorithm for* the fast calculation of the DFT but *not a transformation* itself.

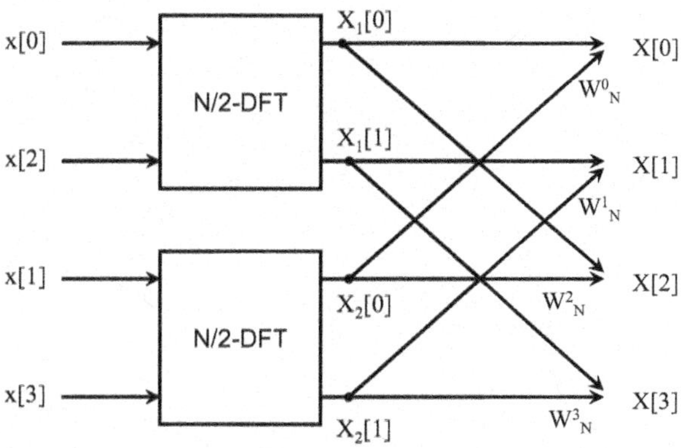

Fig. 4.43 FFT algorithm according to the principle of dividing the input sequence into even and odd indices. It can therefore be cascaded for sequences with 2^N input values. Arrows without a rotation vector have transmission 1; addition occurs at the arrows' ends

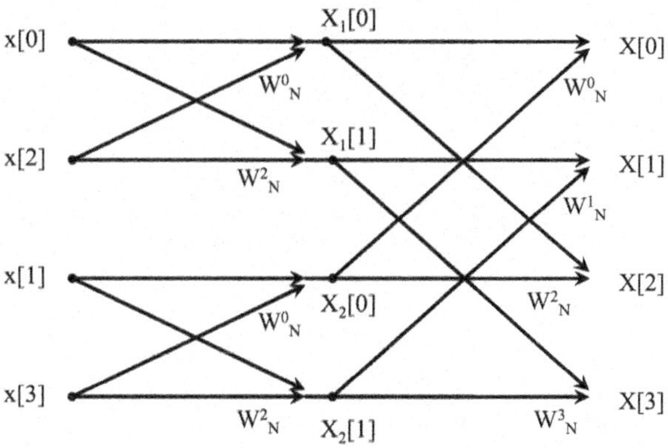

Fig. 4.44 The DFT blocks from Fig. 4.43 were replaced by a further decomposition according to Eq. 4.43. It is done for all powers of N until; finally, the whole structure consists only of butterflies (Fig. 4.45)

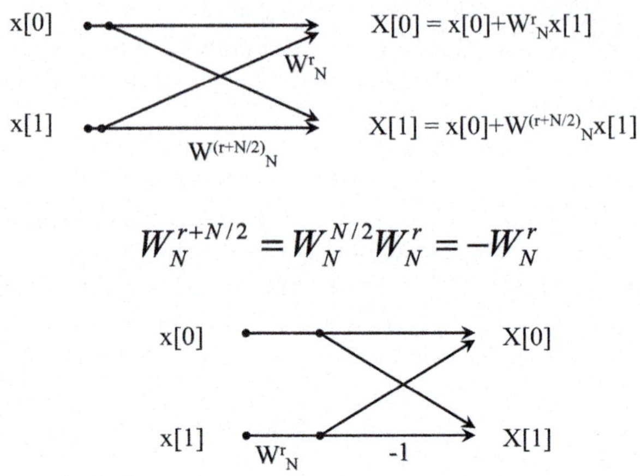

Fig. 4.45 Generalisation of the fundamental element of the FFT to a butterfly (r is the power of the algorithm)

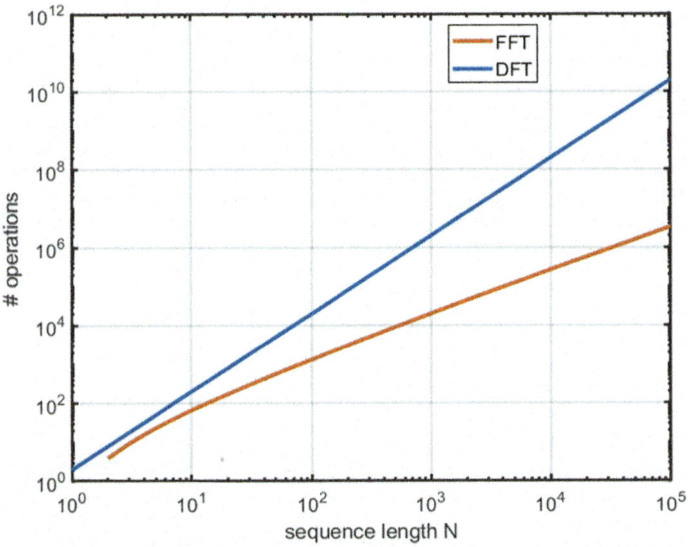

Fig. 4.46 Computational effort for calculating the DFT and FFT as a function of the sequence length as the number of operations multiplication and addition. The data transport was not considered here, as it depends on the concrete porting

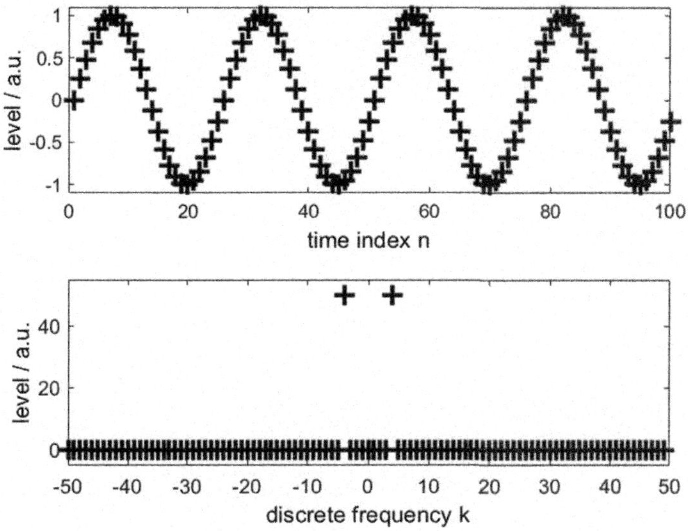

Fig. 4.47 Time course of a harmonic (top) with the period $T = 25T_A$ and a window length of N = 100. The Fourier coefficients on the grid points k = −5, 5 amounts to 50 (bottom)

4.2.2.5 Sources of Error in the DFT

Finite Window Length and Window Function

According to the calculation formula of the DFT according to Eq. 4.34, after the transformation, N discrete Fourier coefficients are present, distributed over the grid points lying at a distance of f_A/N (f_A is the first harmonic of the sampling rate). If a harmonic is transformed whose period corresponds precisely to the multiple of the sampling period, it is correctly interpreted as a spectral needle (Fig. 4.47). However, suppose a different window length is applied to the same signal, which does not equal the integer multiple of the signal period. In that case, none of the discrete frequencies will correspond to the harmonic. Since the DFT is a linear transformation, it approximates the harmonic by a linear combination of several discrete Fourier coefficients. The signal energy is distributed over several support points in the spectrum (Fig. 4.48). In signal processing, this is called the leakage effect (*leakage effect*). The DFT simulates a continuous spectrum with a maximum that does not correspond to reality, and the actual amplitude of the harmonics is reduced.

In biosignal processing, harmonics are particularly interesting when one stimulates humans' sensory and motor systems with periodic sequences (visual stimulation of the eye with pattern reversal, acoustic stimulation with sinusoidal oscillations) or examines periodic biosignals (heart rate and its variability). Especially in the case of stimulation, the period duration is known in advance so that for processing, one can use a corresponding window length can be chosen. In the case of endogenous signals, the period before the analysis is unknown. If possible,

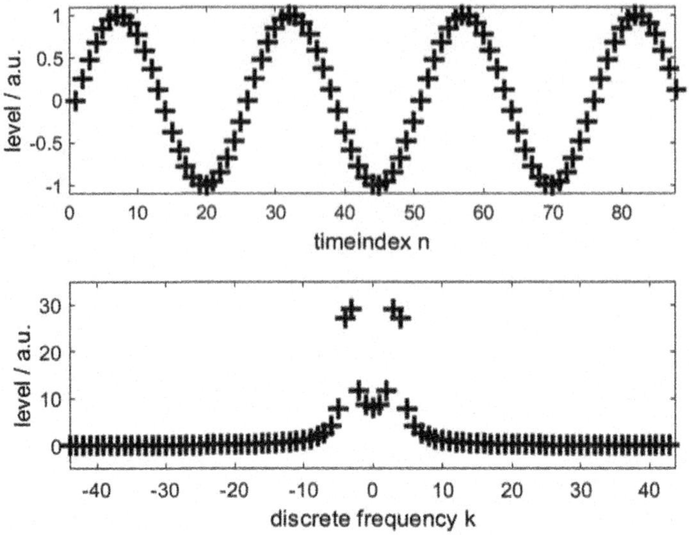

Fig. 4.48 Time course of a harmonic with the period of $T = 25 * T_A$ (as in Fig. 4.47, top) and a window length of N = 88. Since the window length does not correspond to the integer multiple of the signal period, the harmonic is distributed over several grid points (bottom)

a very long window should be selected because the leakage effect decreases with the window length. With a long window, however, the following effect is added:

Padding with Zeros (Zero Padding)

As described in the previous chapter, the FFT can only be used to calculate DFT if the input sequence has a certain length, which takes on a power of 2, 4, and so on, according to the FFT algorithm. In practical BSV, one rarely influences the exact signal length. In practice, one most often uses the power of 2 algorithms, where the following window lengths are applied:

$$r = 2 \rightarrow N = 2^n = \{2, 4, ...128, 256, 512, 1024, ..., 16384, 65536...\} \tag{4.44}$$

One way to adjust the length of the biosignal of a power of two is to choose the next lower power of two (Eq. 4.45)

$$N_W = 2^n \leq N_S \tag{4.45}$$

In Eq. 4.45, N_W is the window length, n *is* the next lower power concerning N_S, and N_S is the signal length.

The solution for adjusting the window length according to Eq. 4.45 is analytically correct, but practically it means a possibly significant loss of information. It is unfavorable for biosignals with strong dynamics and is not recommended. An alternative approach is to use the next higher power of two (Eq. 4.46).

$$N_W = 2^n \geq N_S \tag{4.46}$$

To use this approach practically, one fills the missing data with zeros (zero padding, Eq. 4.47).

$$\{x_r[n]\} = \{x[1], ..., x[N_S], 0, ..., 0\} \tag{4.47}$$

If the phase frequency response is essential, symmetrical padding should be used (Eq. 4.48).

$$\{x_r[n]\} = \{0, ..., 0, x[1], ..., x[N_s], 0..., 0\} \tag{4.48}$$

The following example demonstrates the effect of filling with zeros. If one transforms the sinusoidal oscillation from Fig. 4.47 with 100 coefficients, one correctly obtains a level of 50 in the imaginary part alone at the points $k = -4, 4$ (Fig. 4.49). For applying an FFT, one fills the harmonic with 28 zeros to reach the next higher power of two of 128. It changes the real and imaginary parts of the coefficients, but the magnitude remains the same (Fig. 4.49). Contrary to the indications in several books that the FFT algorithm is indifferent to the number of zeros, a change is clearly present and verifiable here (also theoretically). The symmetrical or asymmetrical filling with zeros changes the Fourier coefficients. The additional zeros and their symmetrical or asymmetrical arrangement can be interpreted as a time shift of the original signal. Then Eq. 4.18 (shift theorem) comes into play (Fig. 4.50). Since the phase or latency of biosignals is a diagnostically crucial parameter, appropriate precautions must be taken when using this method. Usually, the number and distribution of the zeros are known so that the correct phase or latency can be corrected retrospectively using the relationship according to Eq. 4.18. See also Sect. 3.2.3 Sampling in multi-channel systems.

One can reduce the influence of the padded zeros (Fig. 4.50) by choosing a much higher power of two of the FFT than is necessary.

For the signal in Fig. 4.47 with $N_S = 100$, one can use an FFT with $N_W = 1024$, i.e., more than ten times the original length (Fig. 4.51). It improves the purely optical resolution. However, the actual resolution (or bandwidth of the Fourier coefficient at the relative frequency of 0.04) resulting from the original signal length of $N_S = 100$ is retained. That is a fundamental insight for DFT-based analysis after padding with zeros:

> The characteristic properties of the original analysis window and the analyzed signal, such as spectral composition, bandwidth, and statistical properties, change when applying an FFT after filling in zeros.

Increasing the window length can reduce errors but not improve the information in the windowed signal. In terms of statistical signal analysis, increasing window length increases data redundancy but not content.

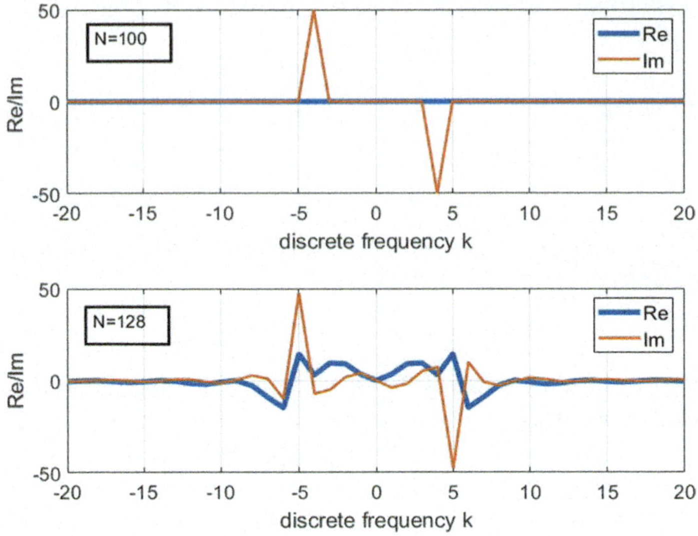

Fig. 4.49 Effect of padding with zeros in a signal to obtain a window length corresponding to a power of two. Real and imaginary parts of the DFT of the harmonics from Fig. 4.47 at k = 4 and N = 100 (top) and N = 128 (bottom). The discrete frequency of k = 5 at the bottom results from the window length 128

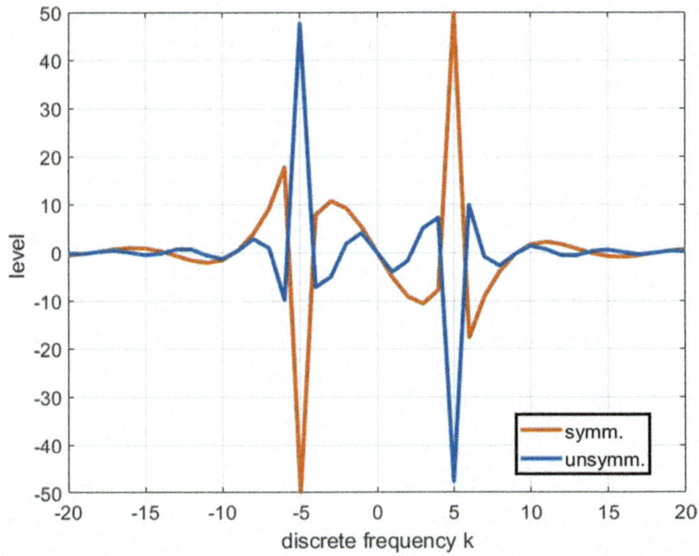

Fig. 4.50 Imaginary part of DFT coefficients of the sinusoidal oscillation from Fig. 4.47 with symmetrical (red, 14 zeros each before and after the signal) and unsymmetrical (blue, all zeros after the signal) padding with 28 zeros

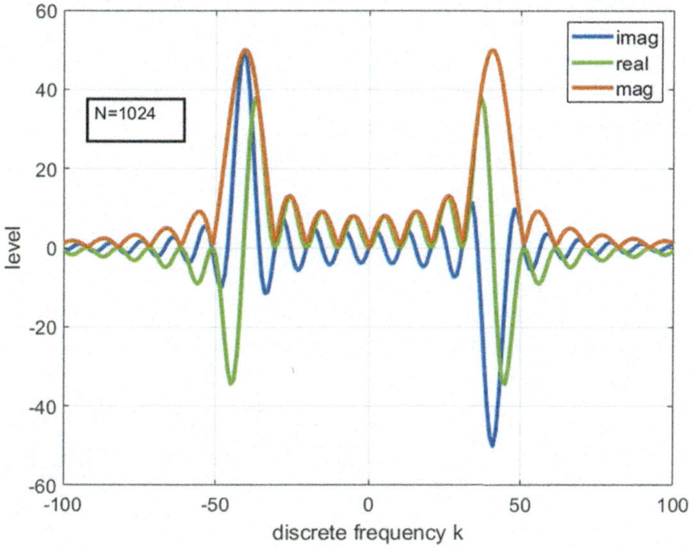

Fig. 4.51 Magnitude, real and imaginary parts of Fourier signal coefficients from Fig. 4.47 for a window length of N = 1024, with 924 zeros padded

4.3 Signal Analysis in the Time–Frequency Composite Range

Indeed, biosignal analysis in the time domain is still the essential method in clinical routine for measuring diagnostic features. Some critical parameters in the time domain alone are either not measurable (heart rate variability) or not visible (stimulus responses in the EEG). To some extent, spectral analysis methods can determine the parameters of interest. The spectral analysis only succeeds if the signal under investigation is stationary in its spectral composition or the spectral components are constant without needing a temporal reference as in time series analysis. It is the case, for example, when the sensory or motor system of the human being is stimulated with a periodic sequence of impulses. In typical cases, however, one does not influence the form of the biosignals, which are strongly transient even with periodicities and whose spectrum changes all the time dynamically. A composite of both domains is suitable for analyzing such signals—the composite representation of time and frequency (TFD Time-Frequency Distribution). The following chapters deal with methods that enable dynamic spectral analysis. The methodological spectrum of such methods is extensive today. The selection for this book was based on relevance to biosignal analysis. The signal theory was adapted to the selection and reduced to a necessary minimum.

4.3.1 Introduction to Time-Frequency Distributions

If the temporal course of a spectrum is to be examined, the window used for spectral analysis (Fig. 4.52) must be shortened and shifted equidistantly along the time axis (Fig. 4.53). Since the window length is shortened, the spectral resolution also changes according to Eq. 4.49.

$$\Delta f = \frac{1}{L \cdot T_A} \tag{4.49}$$

In Eq. 4.49, f is the spectral resolution of an analysis window in the time domain with the length of L samples and the sampling period T_{A}.

> The analysis window must be shortened for the transition from spectral analysis to time-frequency analysis, whereby the temporal resolution increases, but the spectral resolution decreases.

This relationship is shown in Figs. 4.52, 4.53, 4.54 and 4.55. Based on the representation in Fig. 4.54, one can observe the generally poorer temporal resolution due to the windowing compared to the time course. Especially for critical short-time processes (R-wave of the ECG), the smearing along the time axis is unfavorable. For the spectral resolution, it is irrelevant which length the DFT used for the transformation had (in Figs. 4.54 and 4.55 $N = 1000$), but only the length of the

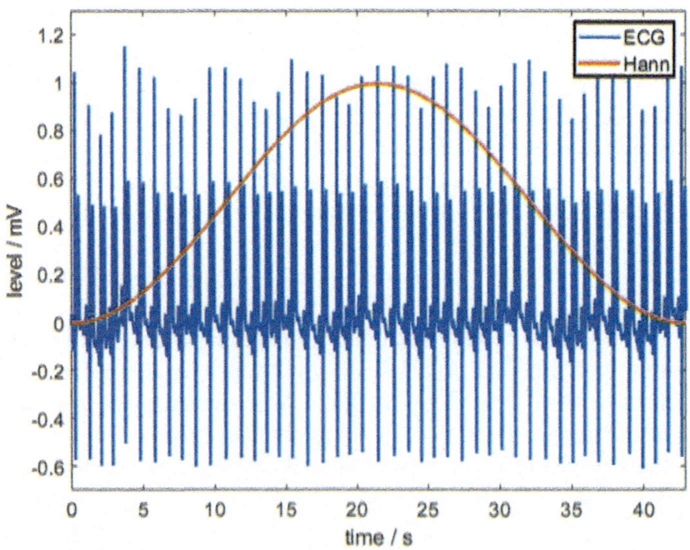

Fig. 4.52 An ECG recording over 43 s with a Hann window. Spectral analysis is therefore carried out with a resolution of 0.023 Hz

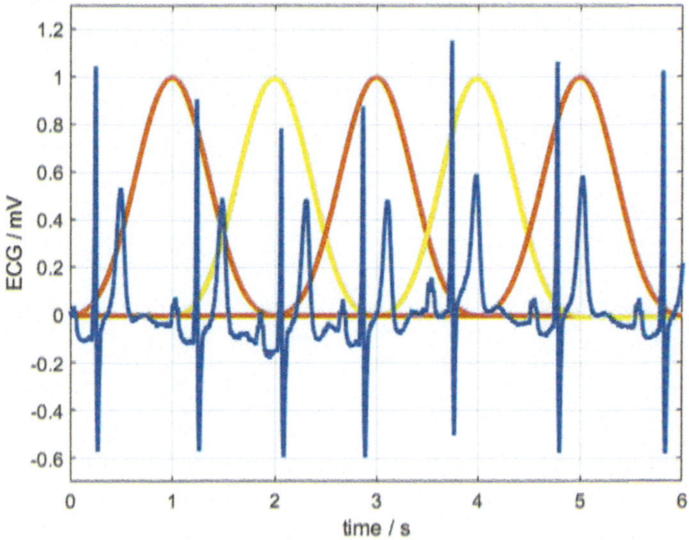

Fig. 4.53 Segmentation of an ECG recording. The segment length is 2 s, and the adjacent segments overlap 50% or 1 s. The spectral resolution corresponds to the segment length of 0.5 Hz ($L = 1000$, $T_A = 2$ ms)

analysis window (in Fig. 4.55 $L = 101$ above, $L = 51$ below). The effect of halving the length of the window can be observed in the diagram in Fig. 4.55: The temporal resolution is much better with the shorter window (double resolution), but the spectral resolution is halved (half resolution). That is particularly visible in the T-wave. Therefore, defining the temporal and spectral resolution independently would be advantageous. The class of Wigner-based time–frequency distributions offers this possibility. One can bring the two resolutions up to the theoretical limit, given by the signal-analytical interpretation of the Heisenberg uncertainty relation (Eq. 4.50).

$$\Delta t \cdot \Delta f \geq \frac{1}{4\pi} \tag{4.50}$$

In Eq. 4.50, t is the length of the analysis window along the time axis or the observation time, and f is the spectral resolution or the bandwidth.

The effect of the uncertainty principle can be observed in the example of HRV in Fig. 4.56. Like any physiological parameter, HRV is a time-varying variable whose dynamic changes are of diagnostic importance. HRV can also be a frequency-modulated signal according to Eq. 4.13, so one would like to measure its instantaneous frequency at specific points in time. The length of the analysis window in Fig. 4.56 was 20 s, so according to the relationship according to Eq. 4.50, the frequency cannot be determined more precisely than with 4 MHz uncertainty (width of the spectral components shown). It applies, although, with

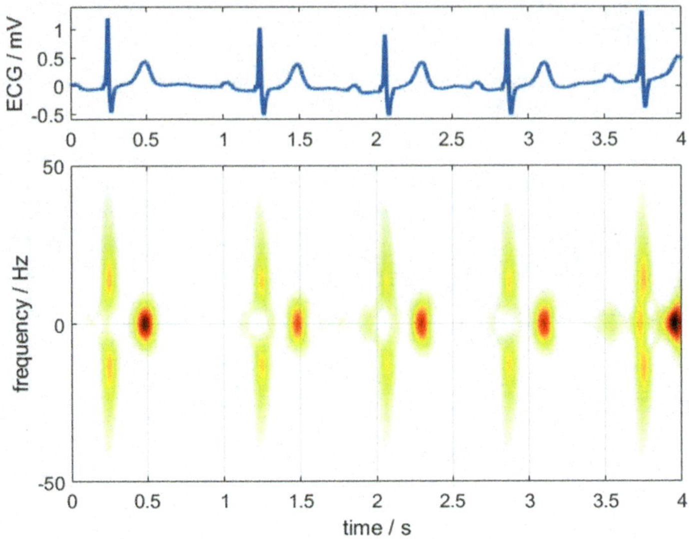

Fig. 4.54 Time course of an ECG (top) and its time–frequency distribution (bottom). Window length $L = 101$, FFT-length $N = 1000$, window function Hann, color scale of the energy, window shift in each case by one sampling period $T_A = 2$ ms

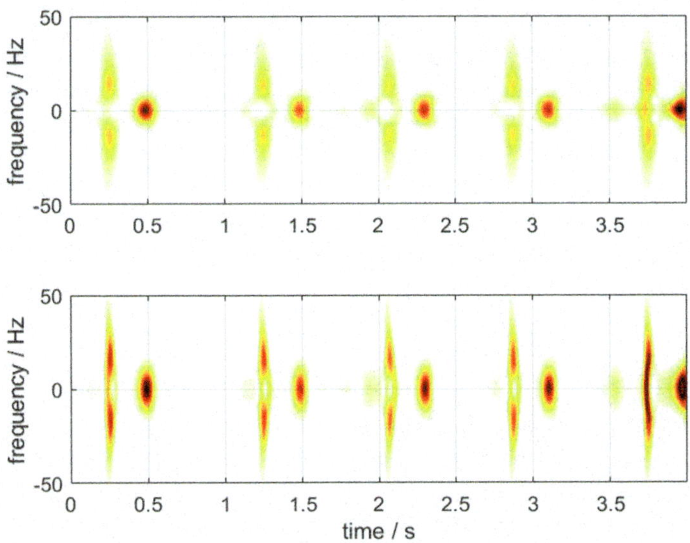

Fig. 4.55 Time–frequency distribution of an ECG with window length L = 101 (top) and L = 51 (bottom), FFT length N = 1000. The bottom temporal resolution is better because of the half-window length, but the spectral resolution is worse than on top

a length of the DFT of $N = 10^6$, a resolution of $f = f_A/10^6 = 0.5$ MHz would have been theoretically possible. Another consequence of the uncertainty principle is that while the HF and LF spectral bands of HRV are captured for dynamic spectral analysis (yellow bars around the HR in Fig. 4.56), the analysis window is too short for the VLF band. Therefore, the HR appears as a time-varying curve with a period of about 30 s. This analysis concludes that although the computational means allow extreme resolutions or accuracies, the uncertainty principle sets a clear physical limit to the signal analysis. Furthermore, based on the example in Fig. 4.56 for the dynamic spectral analysis of HRV, another consequence is clear:

Suppose the deepest spectral components of a biosignal are to be mapped not only as a time course but also as part of the dynamic spectrum. In that case, the analysis window must have a sufficient minimum length corresponding to the necessary resolution.

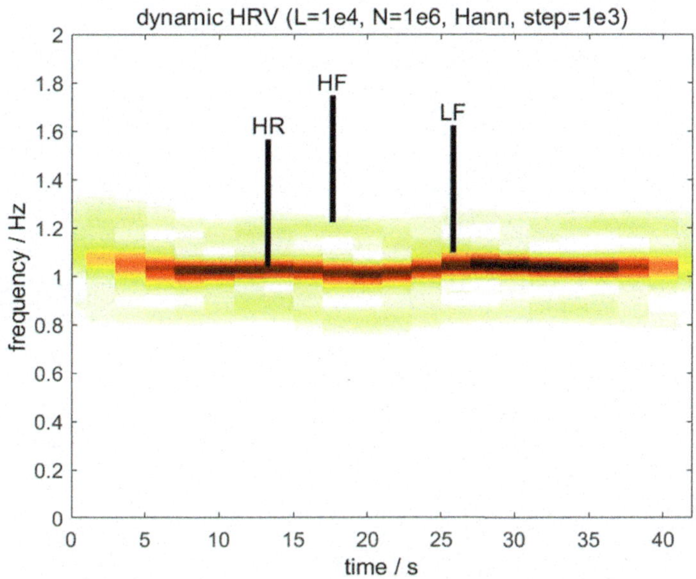

Fig. 4.56 Dynamic spectrum of HRV from Fig. 4.23 (Sect. 4.1.2.3). The Hann window length is $L = 10^4$, the length of the DFT $N = 10^6$, and the distance between the windows is 2 s. HR is heart rate, HF is the high-frequency band, and LF is the low-frequency band

Classification of time–frequency distributions according to the exponent of the analyzed signal (TFD), time–frequency-distribution:

1. Linear: Short-Time Fourier Transform (STFT), wavelet transform (WT)
2. Square: Spectrogram (SG), Wigner-distributions (WD)
3. Higher orders: Bispectrum, Trispectrum

Classification of time–frequency distributions according to the analysis function:

1. Complex harmonics: STFT, spectrogram, Wigner-distributions
2. Basic function: Wavelet

4.3.2 Fourier-Based Time–Frequency Distributions

4.3.2.1 Short-Time Fourier Transform

To represent the dynamics of the spectrum in the time-continuous range, the extension of the conventional Fourier transformation by another time variable is necessary. Following the relationship according to Eq. 4.51, an analysis window $w(\tau)$ is applied to the signal $x(t)$ at time t. Then, the spectrum is calculated with the Fourier transformation via the time variable. Since the spectrum calculation is carried out with a shorter window than pure spectral analysis, this time–frequency distribution is also called *Short-Time Fourier Transform* (STFT).

$$STFT_x^{(w)}(t, f) = \int_{\tau=-\infty}^{\infty} x(\tau) \cdot w^*(\tau - t) \cdot e^{-j2\pi f \tau} d\tau \tag{4.51}$$

The STFT is a linear, Fourier-based transformation so that all properties of the Fourier transformation (Sect. 4.2.1) can also be transferred to the composite time–frequency range accordingly. For example, the time shift according to Eq. 4.18 and the frequency shift according to Eq. 4.19 are combined with the time–frequency shift in the composite range (Eq. 4.52).

$$STFT_x^{(w)}(t, f) = e^{-j2\pi(f-\nu)\tau} \cdot STFT_x^{(w)}(t - \tau, f - \nu) \tag{4.52}$$

In Eq. 4.52, w is a window function, x is the signal to be analyzed, t and f *are* independent variables in the TFD for time and frequency, and independent shifts in time and frequency.

Analogous to the conventional Fourier transformation, the inverse transformation of the STFT is also possible because of the linearity (Eq. 4.53).

$$x(t) = \int_{-\infty}^{\infty} STFT_x^{(w)}(t, f) \cdot H(f) e^{2\pi f t} df \tag{4.53}$$

The window function $h()$ (synthesis window) is the reciprocal function (inverse Fourier transform of $H(f)$) of the analysis window in the time domain $w(\tau)$ so that (Eq. 4.54) must apply:

$$\int_{-\infty}^{\infty} w(\tau) \cdot h(\tau) d\tau = 1. \tag{4.54}$$

According to Eqs. 4.51–4.54, the relations form the theoretical basis for several practicable signal analysis and synthesis methods. As already explained, many features can only be represented in the composite time–frequency plane, especially with biosignals. Therefore, they can be emphasized with corresponding time-variable filters or undesired parts can be explicitly suppressed and then transformed back to the usual time course (time-variable filters are dealt with in the chapter Digital Filtering). From the point of view of practical analysis, implementing the condition according to Eq. 4.54 will cause the most significant problems. Most analysis windows converge towards zero at their edges, so the synthesis window theoretically diverges towards infinity at these points. The practical aspects are dealt with in the following.

After the transition into the discrete-time range, the following applies to the STFT relationship according to Eq. 4.55 applies.

$$STFT_x^{(w)}(n,k) = \sum_{m=0}^{M-1} x(n - m - (M-1)/2) \cdot w(m) \cdot e^{-j2\pi km/M} \tag{4.55}$$

In Eq. 4.55, n is the time index, k is the frequency index, m is the index of the auxiliary variable in time, and M is the odd window length.

Accordingly, for the discrete inverse STFT, the relationship according to Eq. 4.56 applies.

$$x(n) = \sum_{k=0}^{M-1} STFT_x^{(w)}(n,k) \cdot H(k) \cdot e^{j2\pi km/M} \tag{4.56}$$

According to Eq. 4.54, the condition also applies in the discrete-time range (Eq. 4.57).

$$\sum_{n=1}^{M} w(n) \cdot h(n) = 1 \tag{4.57}$$

As explained in Sect. 4.2.1.2 Windowing and Sect. 4.2.2.3 Discrete Fourier Transform as Filter Bank, the analysis window's spectral resolution or bandwidth is inversely proportional to the window length. An important parameter is added in the composite time–frequency range—the time resolution. It is directly proportional to the window length, so the frequency and time resolution are coupled via the window length (Eq. 4.58).

$$\Delta t \sim M \sim \frac{1}{\Delta f} \tag{4.58}$$

The effect of the window length is shown in Fig. 4.54 for an ECG section (magnitude of the Fourier coefficients). For practical analysis, this dependency means that for each biosignal or its features, one has to find a compromise between the desired spectral and temporal resolution.

A typical analysis task is shown in Fig. 4.57. The eye of a test subject was stimulated with 8 pps (pulse-per-second; short flashes of light 125 ms apart), and the visual evoked potential (VEP) was obtained from the EEG with stimulus-synchronous averaging. With a window length of M = 255 or 1020 ms, the individual spectral components are easily identifiable (Fig. 4.57 top): The sought-after, time-varying amplitude of the VEP at 8 Hz, α-wave (11 Hz) between t = 0.5 s…1 s, a transient component at approx. 1 Hz (smeared between 0 s…1.2 s). In comparison, a window half as long with M = 127 or 508 ms (Fig. 4.57) shows a poorer spectral resolution but a significantly better time resolution. It is evident in the short transient component of the VEP at t = 0.2 s…0.6 s. The transient component ranges from 0.1 to 0.6 s, so the lower representation corresponds better to reality concerning this signal feature. However, the spectral resolution is so poor here that the stimulus-response could be confused with the α-wave.

4.3.2.2 Spectrogram

In many areas of spectral signal analysis, it is not the (complex) Fourier coefficients in the foreground but rather the physical nature of the signal source, power, or energy. These signals are produced by energy sources (sound, light, heat, radiation), but biosignals can also be considered power signals for specific questions. For example, the RMS value (RMS—root-mean-square or standard deviation) of an EMG, for instance, measures the current mechanical force of the muscle under examination, or the instantaneous power in an EEG band is an expression of specific brain activity. Of course, one would like to know how the muscle force and the spectral EEG power change over time. Dynamic spectra of power or energy are suitable for this. Voltages and currents in their second power go into the calculation of power. Such quantities are called power signals (e.g., periodic signals of constant power) or energy signals (signals limited in time) (Eq. 4.59).

$$\begin{aligned} p(t) &= \tfrac{u^2(t)}{R} = i^2(t)R && \rightarrow x^2(t) \\ E &= \tfrac{1}{R}\int_T u^2(t)dt = R\int_T i^2(t)dt && \rightarrow \int_T x^2(t)dt \end{aligned} \tag{4.59}$$

In Eq. 4.59, $u(t)$ is the voltage across resistor R, $i(t)$ is the current through resistor R, $p(t)$ is the instantaneous power, E is the energy, $x(t)$ is a signal that abstracts the voltage $u(t)$ or current $i(t)$ for signal processing. The resistor R is

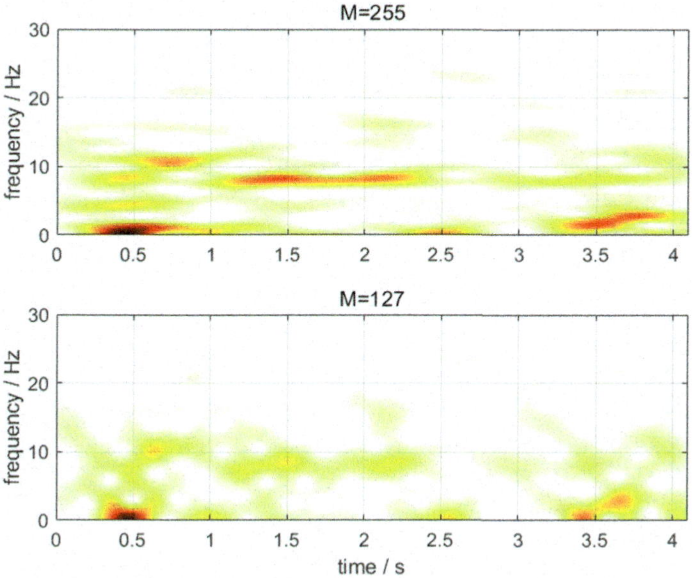

Fig. 4.57 STFT of a visual evoked potential after an optical stimulation with 8 pps and a window length of M = 255 (top) and M = 127 (bottom). The DFT length was N = 1000, shown as the magnitude of the Fourier coefficients (red and black represent the highest energy)

a constant factor for signal processing that can be inserted at any time and is, therefore, unimportant for signal processing.

The energetic equivalent of the STFT is the spectrogram (Eq. 4.60). The magnitude of the Fourier coefficients is squared, which yields a dynamic power spectrum. However, the phase information is lost by this operation (from the formulation of the complex Fourier coefficients, only the magnitude remains). Hence, a reconstruction of the signal is no longer possible. Calculating a spectrogram according to Eq. 4.60 is particularly advantageous for the computational technique since it can be carried out with the help of the FFT. One can also calculate the SG based on the Wiener-Khinchine-theorem via the AKF (autocorrelation function), which is particularly suitable for theoretical analyses.

$$SG_x^{(w)}(t, f) = \left| STFT_x^{(w)}(t, f) \right|^2 \tag{4.60}$$

The spectrogram of the VEP from Fig. 4.57 is shown in Fig. 4.58. Here the distribution of the signal energy in the time–frequency plane is visible. The highest signal energy is found in the transient part of the VEP around 0.5 s after the stimulus. The representation is logarithmic (in dB), as is usual for the representation of signal power.

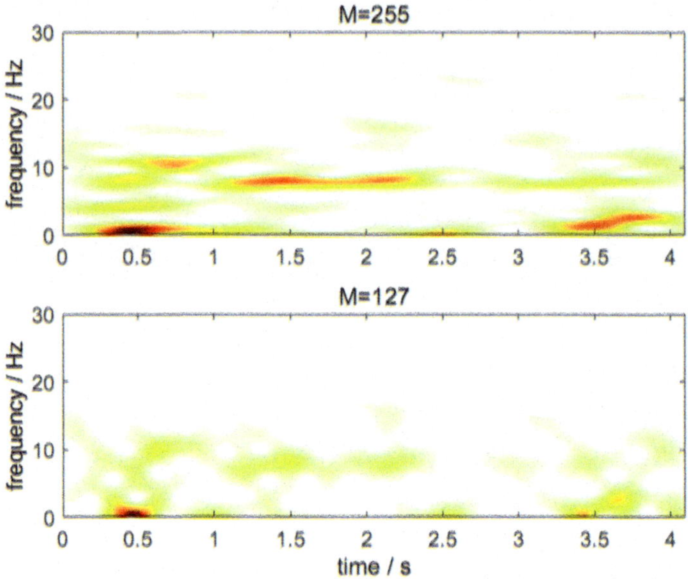

Fig. 4.58 Spectrogram of the VEP from Fig. 4.56. The grey values correspond to the logarithm of the signal power (the color scale represents the energy). The analysis window (top) has been halved (bottom)

> Since biosignals and disturbances always have a stochastic component in addition to periodic components, the generalized representation with statistical measures is used from here on where it makes sense. That implicitly includes deterministic signals.

The signal power in the time domain is determined by the autocorrelation function (ACF) or the cross-correlation function (Eq. 4.61).

$$\rho_{xx}(\tau) = \int_{-\infty}^{\infty} x(t) \cdot x(t+\tau)dt$$

$$\rho_{xy}(\tau) = \int_{-\infty}^{\infty} x(t) \cdot y(t+\tau)dt \tag{4.61}$$

In Eq. 4.61, ρ_{xx} is the ACF (autocorrelation function), ρ_{xy} is the CCF (cross-correlation function), τ the independent time shift between signal components, $x(t)$ and $y(t)$ are the signals to be analyzed, t is the independent time variable.

Since the correlation functions according to Eq. 4.61 only differ in the combination of the signals, for simplicity's sake, only the ACF will be treated in the following. The ACF is an even function, so temporal mirroring is possible (Eq. 4.62).

$$\int_{-\infty}^{\infty} x(t) \cdot x(t+\tau)dt = \int_{-\infty}^{\infty} x(t) \cdot x(\tau - t)dt \tag{4.62}$$

The right part of the formula, according to Eq. 4.62, corresponds to the convolution of the signal $x(t)$ with itself. It corresponds to the multiplication of the spectrum with itself (Eq. 4.63).

$$S_{xx}(f) = S_x(f) \cdot S_x^*(f) \tag{4.63}$$

Transforming the correlation function into the frequency domain yields the auto or cross-power density (Eq. 4.64).

$$\begin{aligned} S_{xx}(f) &= \int_{-\infty}^{\infty} \rho_{xx}(\tau) \cdot e^{-j2\pi f\tau} d\tau \\ S_{xy}(f) &= \int_{-\infty}^{\infty} \rho_{xy}(\tau) \cdot e^{-j2\pi f\tau} d\tau \end{aligned} \tag{4.64}$$

According to Eq. 4.64, the relationship is known as the Wiener-Khinchine theorem. For a time–frequency representation, as with the STFT, another auxiliary variable in time is necessary (Eq. 4.65). It is called the temporal (instantaneous) correlation function (TCF).

$$\rho_{xx}(t,\tau) = x(t) \cdot x(t+\tau) \tag{4.65}$$

Of course, one can use different windows described in Sect. 4.2.1.2 Windowing. Then the relation according to Eq. 4.65 can be generalized to Eq. 4.66.

$$\rho_{xx}^{(w)}(t,\tau) = \int_{-\infty}^{\infty} x(t) \cdot x(t+\tau) \cdot w(t'-t)dt' \tag{4.66}$$

In Eq. 4.66, t is the current time, the time shift of the AKF, and t' is the time variable for integration over the window function $w(t)$.

Starting from the local ACF, the dynamic power spectral density can be formulated according to Eq. 4.67.

$$S_{xx}(t,f) = \int_{-\infty}^{\infty} \rho_{xx}(t,\tau) \cdot e^{-j2\pi f\tau} d\tau \tag{4.67}$$

Although both direct and indirect paths theoretically lead to the same energy distribution in the time–frequency plane, the practical calculation yields different results, as expected, as shown in Fig. 4.59. While the spectral resolution is necessarily the same for the same window length, the temporal resolution of the indirect path is slightly better. However, the indirect path's time–frequency distribution (TFD) shows local nonstationarities, which will be evaluated as disturbances.

In the discrete-time range, one can calculate the AKF and the power density (discrete power spectrum) with the help of the formulae Eqs. 4.68 and 4.69.

$$\rho_{xx}(n, m) = x(n) \cdot x(n + m) \tag{4.68}$$

In Eq. 4.68, n is the current time index, m *is* the index of the AKFM is the odd window length.

$$S_{xx}(n, k) = \sum_{m=0}^{M} \rho_{xx}(n, m) \cdot w(m) \cdot e^{-j2\pi mk} \tag{4.69}$$

In Eq. 4.69, $w(m)$ is a window function, in the simplest case, a rectangle (unwindowed signal section). Another window function reduces the local nonstationarities (Fig. 4.59) by smoothing, which degrades the spectral resolution.

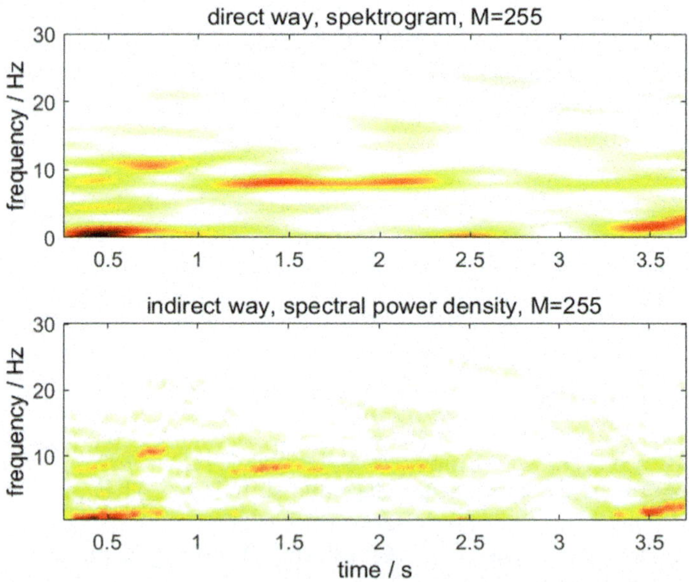

Fig. 4.59 Spectrogram of a VEP (top) on the direct path (spectrogram using STFT) and spectral power density on the indirect path (bottom, spectrogram using ACF). In both distributions, the window length equals $M = 255$ (the color scale represents the energy)

4.3.2.3 Wigner-Distribution

The ACF, according to Eq. 4.65, can be modified so that the shift time is halved (or distributed completely arbitrarily on both sides) and varies symmetrically around the time t (Eq. 4.70). In addition, a so-called *analytical* signal is used here introduced by Ville (Boashash, 2003).

$$\rho(t,\tau) = x\left(t - \frac{\tau}{2}\right) \cdot x^*\left(t + \frac{\tau}{2}\right) \tag{4.70}$$

If one transforms over the time variable into the frequency domain, one obtains the Wigner-Ville spectrum (Boashash, 2003, p. 33) according to Eq. 4.71:

$$W_x(t,f) = \int_{-\infty}^{\infty} \rho(t,\tau) \cdot e^{-j2\pi\tau \cdot f} d\tau \tag{4.71}$$

There is a difference between the ACF according to Eqs. 4.65 and 4.70 regarding the time shift τ around the current time t. One can generalize it with the help of a parameter α (Eq. 4.72).

$$\rho^{(\alpha)}(t,\tau) = x\left(t - \left(\frac{1}{2} - \alpha\right)\tau\right) \cdot x^*\left(t - \left(\frac{1}{2} + \alpha\right)\tau\right) \tag{4.72}$$

After the Fourier transform of the generalized ACF over the time shift, one obtains the generalized Wigner-Ville spectrum (Eq. 4.73).

$$W_x^{(\alpha)}(t,f) = \int_{-\infty}^{\infty} \rho^{(\alpha)}(t,\tau) \cdot e^{-j2\pi\tau f} d\tau \tag{4.73}$$

For $\alpha = 0$, one obtains the ordinary Wigner-Ville spectrum (Eqs. 4.71 and 4.74), and for $\alpha = 1/2$, the Rihaczek spectrum (Eq. 4.74).

$$\begin{aligned} W_x^{(0)}(t,f) &= \int_{-\infty}^{\infty} \rho^{(0)}\left(t - \frac{\tau}{2}, t + \frac{\tau}{2}\right) \cdot e^{-j2\pi\tau f} d\tau \\ W_x^{(1/2)}(t,f) &= \int_{-\infty}^{\infty} \rho^{(1/2)}(t, t-\tau) \cdot e^{-j2\pi\tau f} d\tau \end{aligned} \tag{4.74}$$

The methodological goal in developing the Wigner distribution was originally to formulate a signal kernel (analysis function) so that, with its help, the instantaneous frequency can be determined as precisely as possible (Boashash, 2003, p. 9). This approach was developed by Wigner in 1932 in quantum mechanics and is used in signal analysis under his name.

The Wigner distribution can be formulated (according to the original approach) for a one-component signal (e.g., an FM signal) as follows (Eq. 4.75).

$$WD_s(t,f) = \int_{-\infty}^{\infty} s\left(t+\frac{\tau}{2}\right)\cdot s^*\left(t-\frac{\tau}{2}\right)e^{-j2\pi f\tau}d\tau$$
$$= \int_{-\infty}^{\infty} S\left(f+\frac{\nu}{2}\right)S^*\left(f-\frac{\nu}{2}\right)e^{j2\pi f\nu}d\nu \quad (4.75)$$

In Eq. 4.75, note that a *local* correlation function (Eq. 4.76) is calculated so that (also in practical analysis) for a time t or a frequency f, the entire signal length must be integrated.

$$\rho(t,\tau) = s\left(t+\frac{\tau}{2}\right)\cdot s^*\left(t-\frac{\tau}{2}\right) \quad (4.76)$$

The specific feature of the local (TCF, Temporal Correlation Function) correlation function (Eq. 4.76) compared to the ACF is that it is not integrated over time, i.e., it represents an instantaneous correlation recording. Since the goal is to determine the instantaneous frequency precisely, the effect of the WD on a linearly modulated harmonic signal is examined according to Eq. 4.77.

$$s(t) = A\cdot\cos\left(2\pi t\left(f_0+\frac{\alpha}{2}t\right)\right) \quad (4.77)$$

In Eq. 4.77, f_0 is the initial frequency, and $\alpha/2$ is the linear frequency increase.

If we substitute the relationship according to Eq. 4.77 into Eq. 4.75, we obtain Eq. 4.78 as a result.

$$WD_s(t,f) = \frac{A^2}{4}(\delta(f-f_i(t))+\delta(f+f_i(t)))$$
$$+\frac{A^2}{2}\mathrm{F}\left\{\cos 2\pi\left(\frac{\alpha\tau^2}{4}+2f_0t+\alpha t^2\right)\right\} \quad (4.78)$$

In Eq. 4.78, F is the operator of the Fourier transform, and $f_i(t)$ is the instantaneous frequency of the signal according to Eq. 4.77.

The relationship in Eq. 4.78 shows that the exact determination of the instantaneous frequency $f_i(t)$ has succeeded mathematically. The terms $(f - f_i(t))$ and $(f + f_i(t))$ describe, as expected, the frequency course in mirror image in the range of the positive and negative frequencies with infinitely narrow time and frequency resolution. The last term, however, is undesirable; it is an artifact (cross-term interference), which is additionally caused by the non-linearity of the quadratic signal representation. The artifacts are symmetrically arranged around the frequency axis as high-frequency disturbances, as shown in Fig. 4.60. One can solve this problem by expanding the real signal $s(t)$ into an analytical signal $x(t)$. The analytical signal $x(t)$ is complex, whereby the imaginary part is formed from the real part, which contains the real signal, so all frequencies are shifted by $\pi/2$. This operation corresponds to the Hilbert transformation (Eq. 4.79) regarding signal theory.

$$\begin{aligned} S(f) &= \mathbf{F}\{s(t)\} \\ \mathbf{F}\{\mathbf{H}\{s(t)\}\} &= -j \cdot \operatorname{sgn}(f) \cdot S(f) \end{aligned} \tag{4.79}$$

The complex analytical signal $x(t)$ is composed of the real signal $s(t)$ in the real part and its Hilbert transform in the imaginary part (Eq. 4.80).

$$x(t) = s(t) + j \cdot \mathbf{H}\{s(t)\} \tag{4.80}$$

The distribution in the negative frequency range disappears through the Hilbert transformation, which Ville carried out for the first time (Eqs. 4.70 and 4.71), and the power of the positive components doubles (Fig. 4.61). The Wigner-Ville transformation (WVT) is thus the ideal analysis method for harmonic one-component signals with linear frequency rise or fall. The instantaneous frequency can be read exactly at any moment since the formula, according to Eq. 4.78 for the analytical signal, simplifies to a single Dirac pulse train (Eq. 4.81).

$$WD_x(t, f) = \frac{A^2}{2}\delta(f - f_i(t)) \tag{4.81}$$

Test signals with linearly varying frequency offer almost ideal analysis properties for the functional test of the human sensory systems, not only in terms of signal theory (Eq. 4.81) but also electrophysiologically. However, human biosignals are not monocomponent. Due to the always-existing non-linearities and technical and biological disturbances, the measurement signal contains several disturbed harmonics. The following examines the Wigner-Ville distribution (WVV) of a three-component ramp. For example, a three-component ramp can arise from a fundamental chirp due to a third-order nonlinearity, as shown in Fig. 4.62. The spectrogram represents the harmonics in the measurement signal. The spectrogram represents the three chirps correctly, but the time–frequency resolution is insufficient. The spectrogram chirps are smeared in both directions—frequency and time—and the amount of smeared depends on the length of the analysis window. The WVV represents the chirps with an ideal time–frequency resolution, but interference occurs between the chirps. In principle, each pair of components (auto terms) generates an interference (cross term). This example leads to the first and third chirps causing interference in their middle, which directly affects the second chirp. In addition, the second chirp generates further interference with the first and the third. Mathematically, the interferences can be explained analogously to Eqs. 4.75–4.78. This representation shows the fundamental problem of Wigner-based distributions:

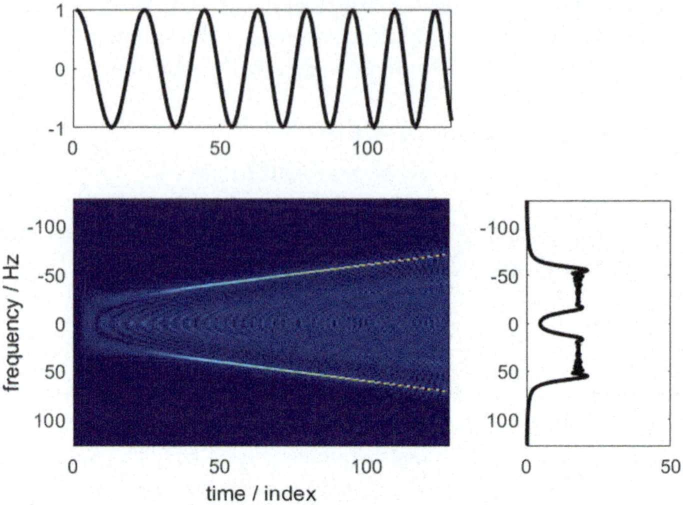

Fig. 4.60 Wigner distribution (WD) of a harmonic signal with linearly increasing frequency. The time course (top) and the amplitude spectrum (right) show the one-dimensional representations of the signal. In the composite area (bottom left), the near-delta sequences correctly show the frequency progression in time. Between the ramps are the interfering artifacts

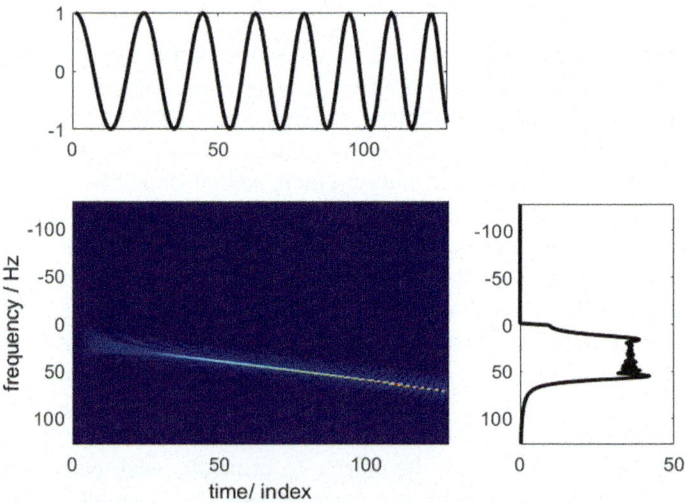

Fig. 4.61 Wigner distribution (WD) of an analytical signal formed from the real signal in Fig. 4.60 with the help of the Hilbert transformation. After the transformation, the component in the negative frequency range is omitted, thus, the artifacts. Note that the spectral amplitude is twice as large as in Fig. 4.60

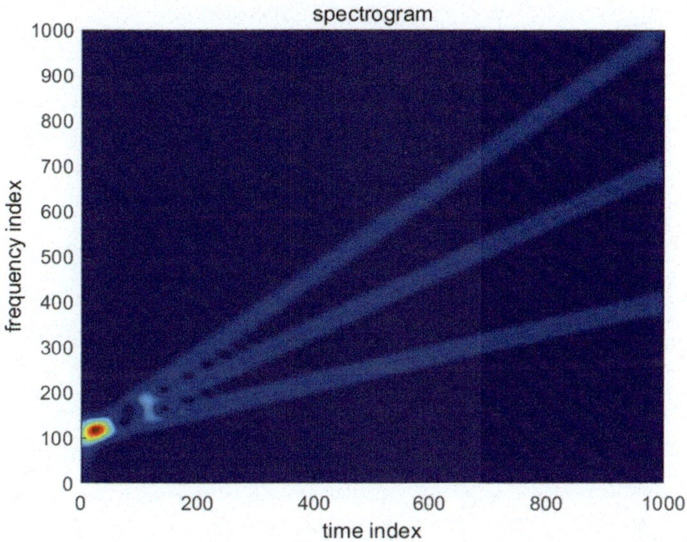

Fig. 4.62 Spectrogram of a three-component ramp (linear chirps) as three harmonics of a fundamental chirp. The interferences are weak (near the chirp junction), and the time–frequency resolution is poor (smearing in time and frequency direction), cf. Fig. 4.61

In the Wigner distribution, each pair of components produces interference. These interferences complicate the interpretation; they even make it impossible in extreme cases.

A component pair is understood to be two chirps isolated from each other and parts of a non-linear chirp with a common projection. This situation is shown in Fig. 4.64: The legs of the sinusoidal chirp are for the different WVD components, so each pair produces interference. This example shows that (non-linear) single-component signals also produce interference. It further complicates the interpretation of biosignals, e.g., heart rate and heart rate variability, where the instantaneous frequency is of fundamental interest (see Sect. 4.1.2.3 Determination of frequency and rate). For a reliable determination of the instantaneous frequency, possibilities for eliminating or at least suppressing the interferences are necessary. As expected, the correlation functions of the chirps and the interferences are different. From the one-dimensional signal analysis, the dual relationship between the spectral power density and the shift operator of the correlation function is known (Eq. 4.29).

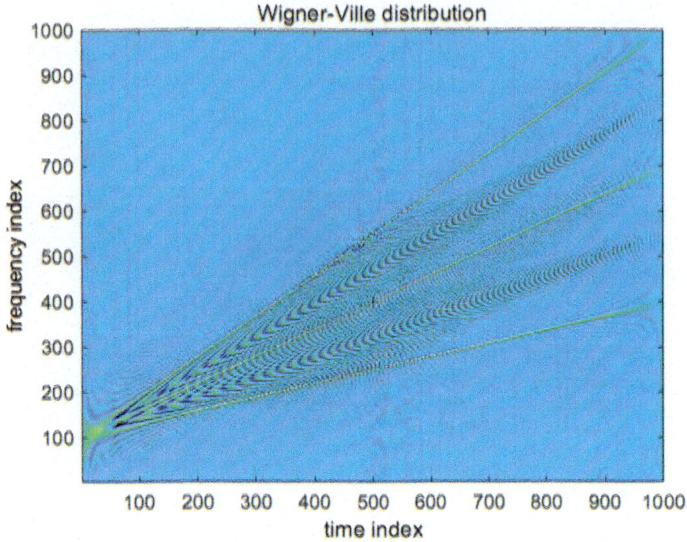

Fig. 4.63 WVD, Wigner-Ville distribution of a three-component ramp (see Fig. 4.62). The basis is the local correlation function according to Eq. 4.76, which is transferred one-dimensionally into the time–frequency domain using the Fourier transformation (Eq. 4.75)

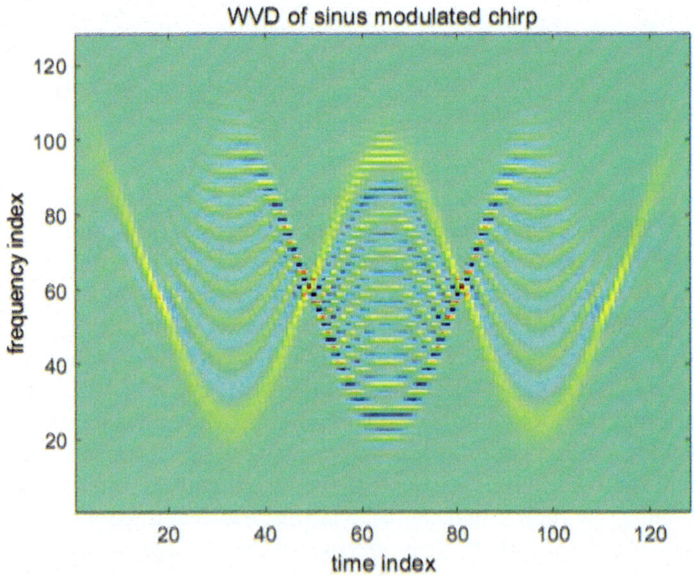

Fig. 4.64 Wigner-Ville distribution of a sinusoidally modulated chirp. The legs of the sinusoidal chirp produce paired interferences as if they were different components (internal interferences between parts of the auto terms)

Since the two-dimensional correlation function $A_x(\tau, v)$ is functional depending on the time and frequency shifts, it can be calculated according to Eq. 4.82.

$$A_x(\tau, \nu) = \iint_{t,f} W_x(t, f) \cdot e^{-j2\pi(\nu t - \tau f)} df\, dt \tag{4.82}$$

It is also referred to as the *similarity function* (Ambiguity Function). Its meaning is demonstrated in the following example. From the chirps (first and third), according to Fig. 4.62, two isolated mono-components (auto terms) are formed in the time–frequency plane (Fig. 4.65). These mono-components produce interference in their center (cross-term). Their time–frequency distribution is transformed two-dimensionally via the Fourier transformation (Eq. 4.82) into the similarity plane (Fig. 4.66). In the similarity plane, the auto-terms (desired components) are gathered around the origin, while the cross-terms (interferences) are distant from the origin and the auto-terms. In this case, one can create the auto-terms (a two-dimensional filter function) around the auto-terms so that the auto-terms are preserved, and the cross-terms are eliminated or at least suppressed. A binary mask in the form of a circle would already be applicable here (inside the circle filled with ones, outside with zeros). This low-pass filter operation degrades the resolution in the time–frequency plane (low-pass smoothes, therefore, poorer resolution).

The masks that can be used to separate auto and cross terms are also called *signal kernels* or simply kernels. Analogous to analysis windows in spectral analysis (see Sect. 4.2.1.2 Windowing), there are various possibilities for kernel design.

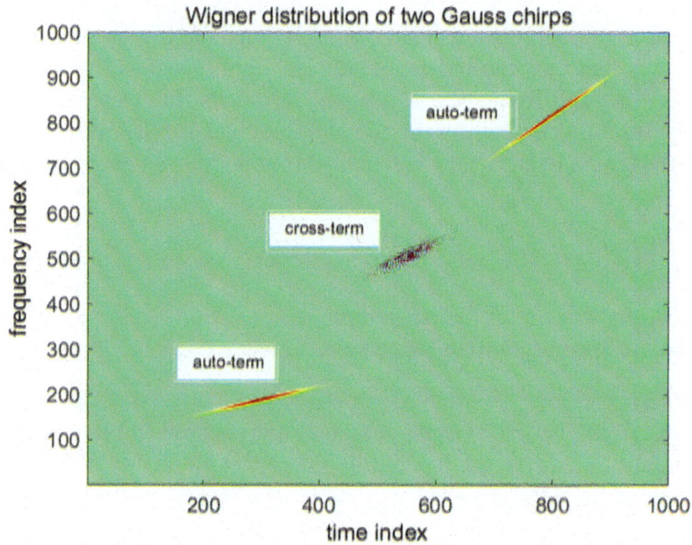

Fig. 4.65 Wigner-Ville distribution of two Gaussian-modulated linear chirps (auto-terms). Note the interference lying in the middle between the chirps (cross-term)

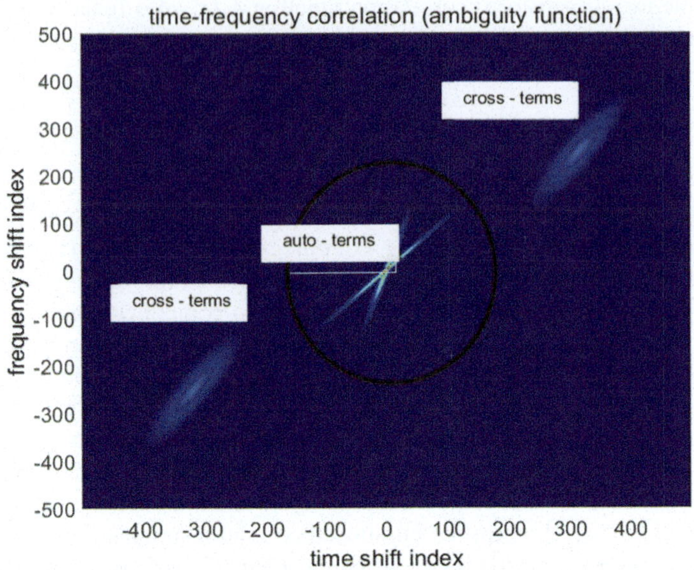

Fig. 4.66 Ambiguity function of the Gaussian-modulated chirps from Fig. 4.65. The auto-terms (desired parts) are arranged around the origin. The cross terms (interferences) are located away from the origin and the auto terms. The circle around auto-terms represents a (binary) mask with which the auto-terms are preserved, and the cross-terms are eliminated.

Since the kernels are two-dimensional, they can be constructed in each dimension separately, independent of the other dimension. They can then be merged into a two-dimensional core by fusion in the composite representation. However, this also means that one can independently define the analysis windows for both dimensions—time and frequency. Moreover, this is the fundamental difference to the spectrogram. In the spectrogram, the temporal and spectral resolution depends directly on the length of the analysis window M and, thus, on each other (Eq. 4.83).

$$\Delta t \sim M \sim \frac{1}{\Delta f} \tag{4.83}$$

As shown in Fig. 4.62, good temporal resolution leads to poor spectral resolution in the spectrogram and vice versa.

> In Wigner-based distributions, the temporal and spectral resolution can be set independently. However, the resolution limit is determined by the Heisenberg uncertainty principle.

That is the most significant advantage of the Wigner distributions compared to all other time–frequency analysis methods. In practical analysis, the limit given by Heisenberg's uncertainty principle (Eq. 4.50) is hardly reached, as previously known limitations of digital signal analysis become effective (bandwidth, spectral resolution, edge effects).

A kernel is applied to the similarity function (Eq. 4.84) in the same way as a one-dimensional window (see Sect. 4.2.1.2 Windowing).

$$A_x^K(\tau, \nu) = A_x(\tau, \nu) \cdot K(\tau, \nu) \tag{4.84}$$

After the Fourier transformation, the filtered similarity function yields the WVV adjusted for cross terms (Eq. 4.85, Fig. 4.65).

$$W_x^K(t, f) = \iint\limits_{t,f} A_x^K(\tau, \nu) e^{j2\pi(\nu t - \tau f)} d\nu d\tau \tag{4.85}$$

After the Fourier transformation (Eq. 4.85) of the masked similarity function (Eq. 4.84), the cross terms disappear, and the desired auto-terms remain (Fig. 4.67).

Note: Equation 4.85 transfers the similarity function via backward and forward transformations into the Wigner planes.

In real signals, the filter effect of the kernel is not as practical as in this simulated case. Eliminating the cross terms becomes more difficult in the practical

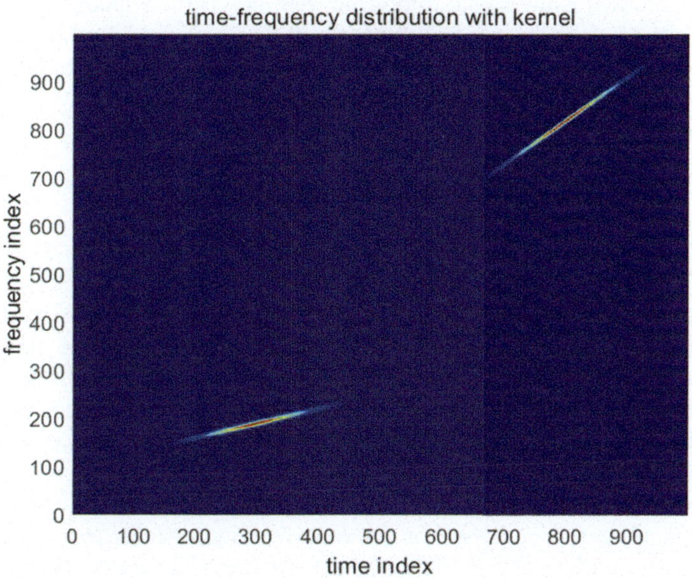

Fig. 4.67 WVV of the signal from Fig. 4.65 after applying a filter mask (binary circular kernel) to the similarity function in Fig. 4.66

analysis since the auto terms overlap in time, frequency, or both. Therefore, specific kernels have been developed, suitable for a particular group of signals, and named after their inventors.

The classical spectrogram (see Sect. 4.3.2.2 Spectrogram, Eq. 4.60) can be formulated with the help of the WV as follows (Eq. 4.86).

$$SG_x^{(w)}(t,f) = \int_{t'}\int_{f'} W_w(t'-t, f'-f) \cdot W_x(t', f') df' dt' \tag{4.86}$$

In Eq. 4.86, W_W is the Wigner distribution of a conventional analysis window. Since there is only one window w with a single degree of freedom (window length), on whose size the spectral resolution also depends (Eq. 4.83), compliance with the uncertainty condition is ensured in advance. However, a window has a soothing effect (like any window), effectively suppresses the interference, and results in a smearing of the auto-terms (Fig. 4.62). Regarding the two-dimensional kernel, one can decouple the time and frequency resolution by introducing a separate window for each dimension. This approach leads to the (SPWD; Smoothed Pseudo-Wigner Distribution) smoothed pseudo-Wigner-distribution (Eq. 4.87).

$$SPWD_x^{(g,H)}(t,f) = \int_{t'}\int_{f'} g(t-t') H(f-f') W_x(t', f') df' dt' \tag{4.87}$$

In Eq. 4.87, x(t) is the signal to be analyzed, $g(t)$ is the window function in the time domain, $H(f)$ is the window function in the frequency domain, $W_X\ (t, f)$ is the Wigner distribution of the signal $x(t)$, t' and f' are auxiliary variables for the integration in the respective domain along the window functions.

The smoothing kernel in the time–frequency plane thus results in (Eq. 4.88)

$$K_{SPWD}(t,f) = g(t) \cdot H(f), \tag{4.88}$$

where $g(t)$ and $H(f)$ are two independent, freely selectable window functions. In Eq. 4.88, note further that the kernel is defined in the time–frequency plane.

Here it can be used according to Eq. 4.86 or transformed as a mask into the similarity plane and applied there to the similarity function of the chirps can be applied. A particular case of the SPWD is the PWD, i.e., an unsmoothed WD where $g(t) = (t)$. The PWD is particularly suitable where the fundamental properties of the Wigner distribution are to be preserved, but the (considerable) signal length does not allow its calculation (if the online estimate is necessary). Therefore, as with the spectrogram, an analysis window is slid over the signal so that only the signal section required for the analysis window must ever be present. In reality, the auto-terms and cross-terms of the biosignals are not distributed as favorably, as shown in Figs. 4.65 and 4.66. On the contrary, the biosignals' cross-terms are not distributed as favorably. On the contrary, the cross terms overlap strongly in time and frequency with the auto terms, so their separation also poses a significant challenge to empiricism. Currently, about 20 different smoothing kernels are known. Some typical distributions with a triple chirp as a test signal are shown in Figs. 4.68, 4.69, 4.70 and 4.71. The PWD does not differ significantly from the conventional WD concerning the unwanted cross terms. Because only one window is applied in the time direction for the analysis, it has a more practical than analytical significance. Due to the relatively short window length (compared to the signal length), the PWD (Fig. 4.68) is also applicable in practical analysis.

While for the real WD, the whole signal before calculating it, the PWD can be calculated slidingly, similar to the STFT or the spectrogram. However, it is impossible to effectively suppress the interfering cross terms with the help of the analysis window. Another window along the frequency axis is necessary to suppress cross-terms or interferences. That is realized with the smoothed PWD (SPWD) (Eq. 4.88, Fig. 4.69). The additional window has the effect that the cross terms are partially reduced, whereby the original time–frequency resolution does not change significantly. The WVV with a cone core and the RID with a Hamming window was used to compare the two distributions examined. The WVV with a cone core (Fig. 4.70) suppresses mainly the frequency-based interferences in the case of harmonic linear chirps, but the time resolution is improved. The RID shows the best suppression of the cross terms with a Hamming window (Fig. 4.71) with few orthogonal residuals. However, good suppression comes at the expense of resolution. The time and frequency resolution is worse than the previous distributions but better than the spectrogram.

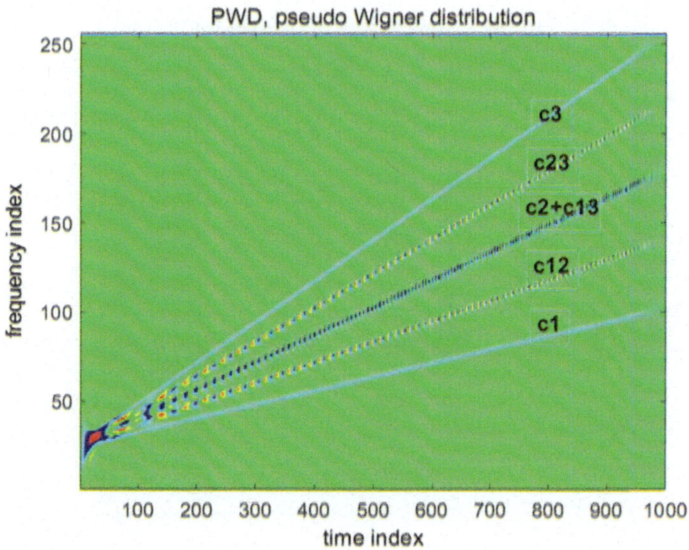

Fig. 4.68 PWD, pseudo-Wigner-distribution of three harmonic chirps: c1, c2, c3 are autoterms c12, c13, c23 are cross terms or interferences. The interferences and their effect on the second chirp are much stronger than in the conventional Wigner distribution (Fig. 4.63). Here the analysis window Hann(255) was used

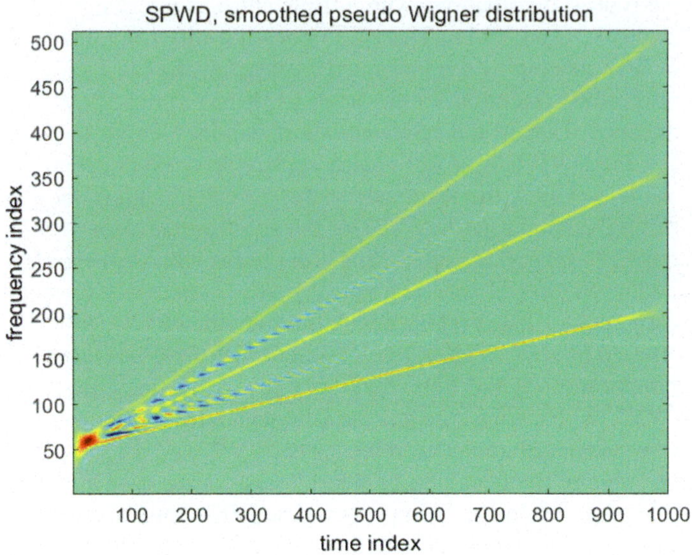

Fig. 4.69 SPWD, smoothed pseudo-Wigner-distribution of three harmonic chirps. Compared to the PWD, the interference is damped, but the resolution is worsened. The window in time direction was Hann(511), and the frequency direction was Hann(31). The interferences are comparable to the spectrogram's (Fig. 4.62), but the resolution is enhanced

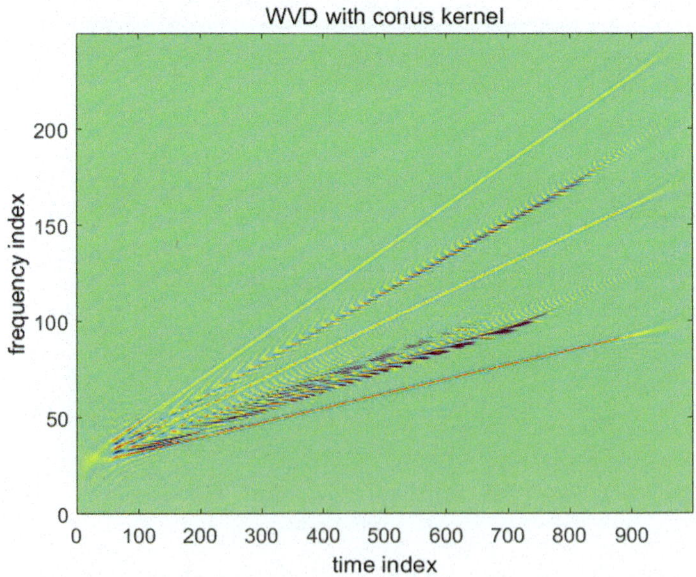

Fig. 4.70 Wigner-Ville distribution of three harmonic chirps with a cone core. Vertical components of the interference are more effectively suppressed than horizontal ones, but the horizontal resolution is better than the vertical

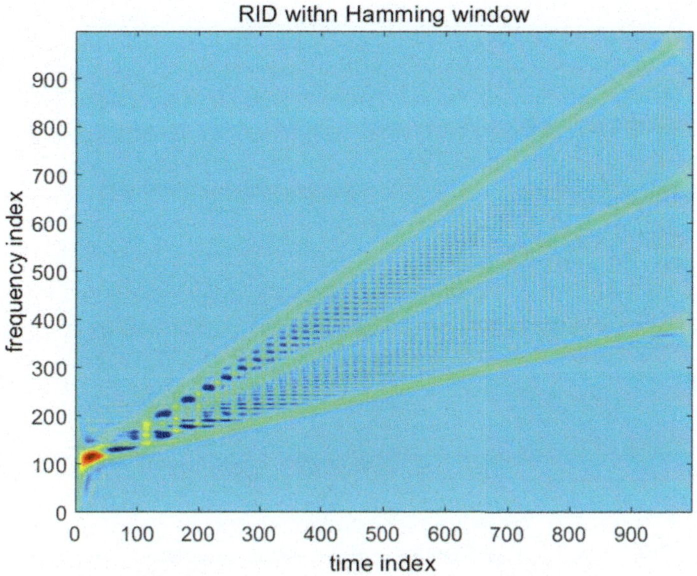

Fig. 4.71 Reduced-interference distribution (RID) of three harmonic chirps with a Hamming window. The interferences are strongly damped with orthogonal residuals. The resolution is worse than for other distributions but still better than for the spectrogram

The conclusion of these comparisons is: Searching for a one-dimensional or a two-dimensional smoothing kernel is, in principle, the compromise between a good time–frequency resolution and the smearing of the auto-terms in the time–frequency domain (Fig. 4.74).

Based on our investigations into suitable biosignal cores, the following tested alternatives can be recommended (Henning et al., 1996):

- The core of the SPWD: $K(\tau, \upsilon) = \eta\left(\frac{\tau}{2}\right) \cdot \eta^*\left(-\frac{\tau}{2}\right) \cdot G(\upsilon)$
- Cone core: $K(\tau, \upsilon) = g(\tau).|\tau|.\frac{\sin(\pi\tau\upsilon)}{\pi\tau\upsilon}$
- Choi Williams Core: $K(\tau, \upsilon) = \exp\left(-\frac{(2\pi\tau\upsilon)^2}{\sigma}\right)$
- Reduced-interference core: $K(\tau, \nu) = S(\tau\nu)$
- Optimal core according to Baraniuk-Jones Adaptive core (Baraniuk & Jones, 1994).

Examples of the application of different nuclei to VEP after burst stimulation are shown in Figs. 4.72, 4.73, 4.74 and 4.75. The VEP after burst stimulation with 8 pps (pulse per second) was examined for 4 s. The EEG recording started 4 s before the burst onset, so 8 s (8192 ms) recordings were averaged (averaging order $M = 16$). The spectrogram (Fig. 4.72) shows good frequency resolution but poor time resolution. The transient part of the VEP is smeared over three seconds ($t = 3$ s…6 s). The SPWD shows very good time and frequency resolution

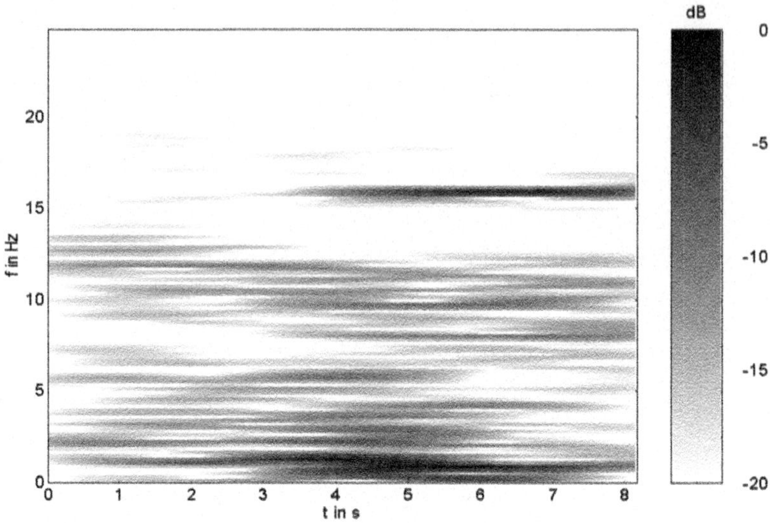

Fig. 4.72 Spectrogram of a visual stimulus-response. Stimulus onset at $t = 4$ s, stimulus rate 8 pps or 16 alternations/s. The spectral resolution is very good; the temporal resolution is insufficient

(Fig. 4.73) but is subject to substantial interference. The typical hyperbolic shape of the transient component of the VEP is well pronounced (t = 4.5 s…6 s). The cone-core WVV suppresses the vertical details of the interference very well, but also those of the transient part of the VEP. For this signal type, the RID with a Choi-Williams core (Fig. 4.75) shows optimal properties: The interferences are effectively suppressed, and the searched components of the VEP (transient and periodic) are well visible and sharply defined. The time and frequency resolution is sufficiently good.

4.3.3 Wavelets

4.3.3.1 Scaling and Fourier-Based Time–Frequency Distributions

In the chapters Short-Time Fourier Transform (STFT) and Spectrogram, it was explained how important the choice of the window function and the window length is for the time and frequency resolution and the spectral properties of the window for reliable signal analysis. The fundamental problem with the window length is that the time and frequency resolution are inversely proportional. If there is a good time resolution, there is a poor frequency resolution, and vice versa. This problem could be partly solved with the help of the Wigner-distributions (independent adjustment of time and frequency resolution, e.g., in the SPWD), but at the expense of disturbing cross-terms or interference. The following consideration opens up another way: For the spectral analysis of a signal, a few—in extreme cases, a single—periods of the frequencies to be determined are sufficient. If, for

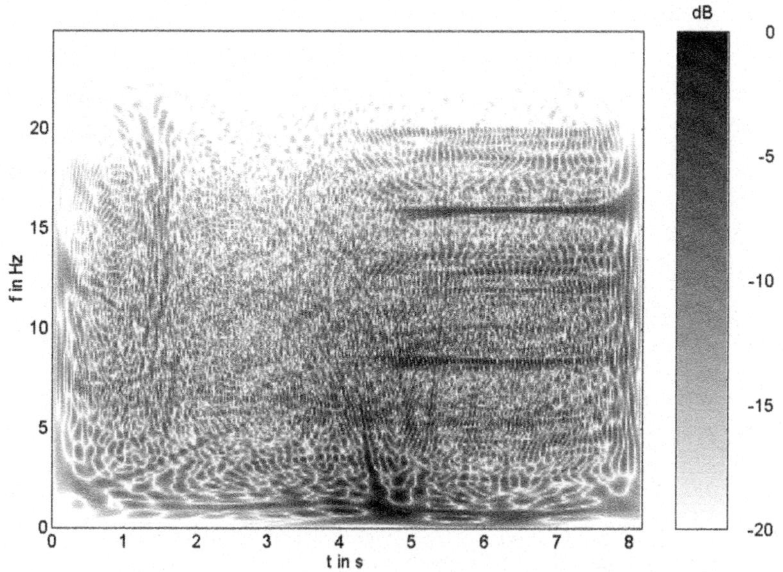

Fig. 4.73 Smoothed pseudo-Wigner-distribution (SPWD) of a visual stimulus-response as in Fig. 4.72. Good time and frequency resolution. The transient part of the VEP is (hyperbolic arc between t = 4 s and 5 s) very well pronounced. Strong interference is also found on fundamental wave 8 Hz

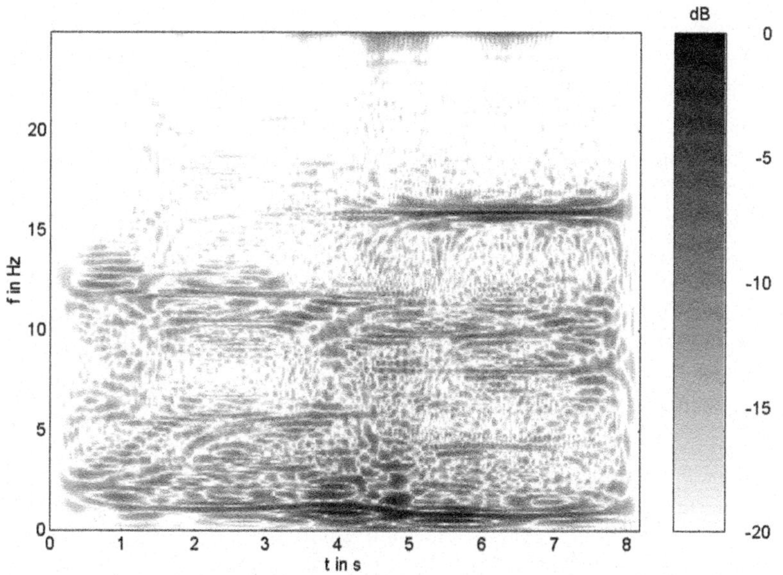

Fig. 4.74 Cone kernel distribution (CKR) of a visual stimulus-response as in Fig. 4.72. Vertical interferences are well attenuated, but horizontal residues distort the real signal—good spectral resolution

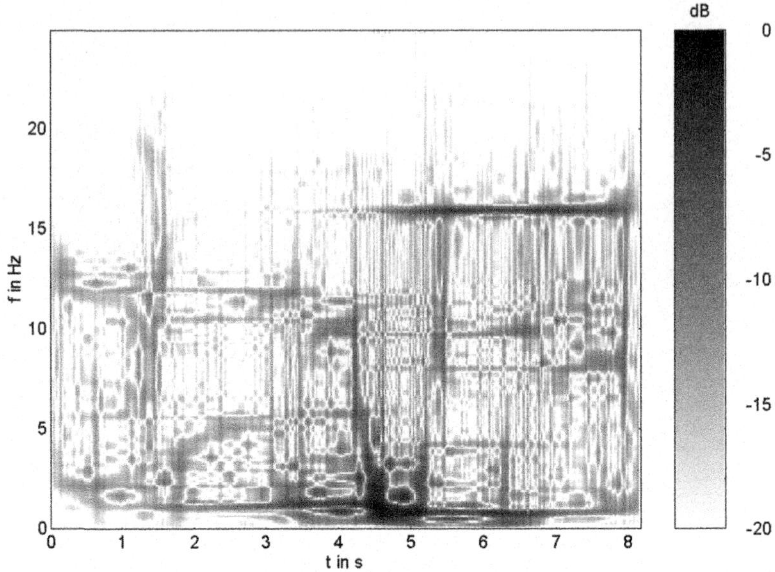

Fig. 4.75 Reduced interference distribution (RID) with Choi-Williams kernel of a visual stimulus-response as in Fig. 4.72. The best distribution for this signal type is a good time and frequency resolution; important components are very well visible, with few interference residues

example, the average respiration period is 5 s, one needs an analysis window of at least 5 s in length to determine the respiration rate. With such a window length, one obtains a relatively good frequency resolution of $f = 0.2$ Hz but a very poor time resolution of $t = 5$ s.

If the ECG and the EEG are also recorded in parallel with the respiration (usual, e.g., in polygraphy), window lengths of about $T_{\mathrm{EKG}} = 0.2$ s would be sufficient for the ECG (resolution of 5 Hz) and $T_{\mathrm{EEG}} = 1$ s for the EEG (range 1–100 Hz). However, since only a window length of $T_{\mathrm{A}} = 5$ s has been assumed so far, it would be an unacceptably poor time resolution for the ECG and the EEG. The approach to solving this problem results from the fact that for higher frequencies, correspondingly shorter analysis windows (with an equal number of periods) are sufficient for frequency determination, which offer a better time resolution. This way, five windows for the EEG ($T_{\mathrm{A}} = 5\mathrm{T}_{\mathrm{EEG}}$) and 25 windows for the ECG ($T_{\mathrm{A}} = 25\mathrm{T}_{\mathrm{EKG}}$) could be accommodated in the considered window for breath analysis of $T_{\mathrm{A}} = 5$ s length. However, the shortening of the windows is accompanied by a deteriorating spectral resolution ($f \sim 1/t$). This behavior is shown qualitatively in Fig. 4.76. Such structures, which are characterized by shorter windows at higher frequencies, can, of course, also be applied to single signals. Qualitatively, this approach corresponds to the STFT with a variable window length that becomes shorter towards higher frequencies. Therefore, such a window structure is particularly suitable for signals where good frequency resolution is desired in the low-frequency range and good time resolution in the high-frequency range, e.g., ECG or EMG. The direct comparison of the STFT with constant and scaled windows is shown in Fig. 4.77.

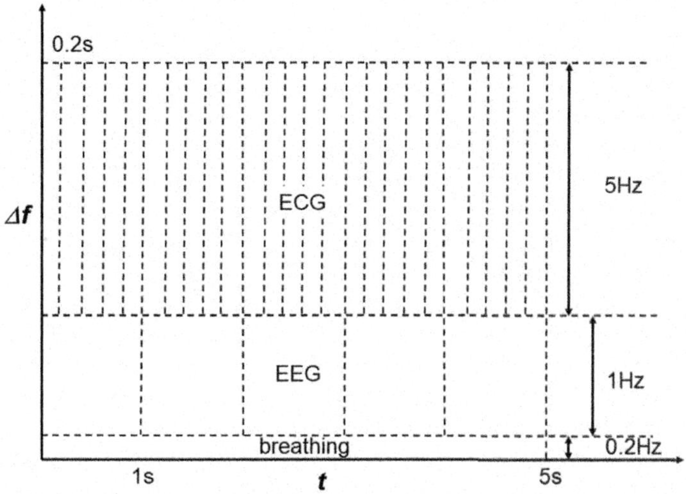

Fig. 4.76 Different window lengths for signal analysis and their bandwidths. The longer the window, the better the frequency resolution, but the worse the time resolution and vice versa (the range $f = 5$ Hz is graphically scaled compressed due to the expansion in the image)

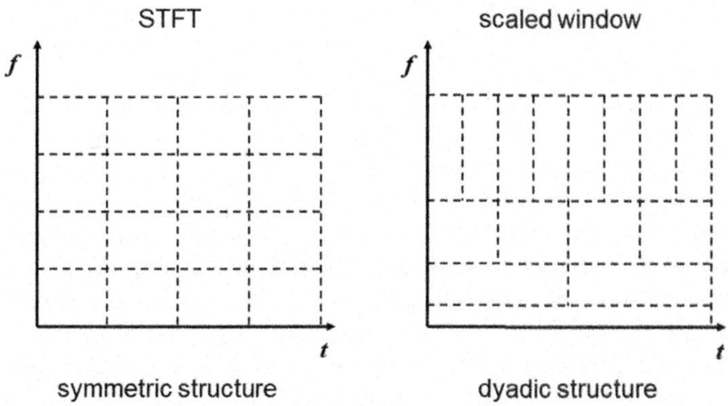

Fig. 4.77 Constant window length with the STFT (left) and the window length halving towards higher frequencies in the scaled (dyadic) structure. The scaled window is particularly suitable for signals with short-time broadband components in the upper-frequency range and with slow, spectrally low-frequency but interesting components (e.g., ECG, EMG, EOG, ERG). *Note* The dyadic structure represents a practical example of window scaling here, as it can be implemented algorithmically effectively, especially for binary computing. In principle, the scaling can be carried out in gradations of integers (discrete WT) or real numbers (continuous WT)

With the STFT, the total spectral bandwidth is constant and independent of the frequency (Eq. 4.89).

$$\left.\begin{array}{l}\Delta t = T_F \\ f_2 > f_1\end{array}\right\} \Rightarrow \Delta f_2 = \Delta f_1 \sim \frac{1}{\Delta t} \tag{4.89}$$

In Eq. 4.89, T_F is the window length, t is the time resolution, f is the absolute frequency, and f is the bandwidth.

For a scaled window, the absolute bandwidth increases with increasing frequency in inverse proportion to the window length, so the relative spectral bandwidth is constant at all frequencies (Eq. 4.90).

$$\left.\begin{array}{l}\Delta t_2 = \frac{\Delta t_1}{s}, \Delta f \sim \frac{1}{\Delta t} \\ f_2 > f_1, \Delta f_2 = s \cdot \Delta f_1\end{array}\right\} \Rightarrow \frac{\Delta f_2}{f_2} = \frac{\Delta f_1}{f_1} \tag{4.90}$$

In Eq. 4.90, t is the time resolution, f is the frequency resolution, f is the absolute frequency, and s is the scaling factor.

The Gaussian window is well-suited as a window function for transitioning from constant to scaled window length. This window has several very favorable properties from the point of view of signal theory (infinitely often differentiable identity in the uncertainty relation). The dynamic spectral analysis can be described analogously to the STFT (Eq. 4.51) and can be formulated as follows (Eq. 4.89):

$$X^{(Gauss)}(t, f) = \int_{-\infty}^{\infty} x(\tau) \cdot e^{-\frac{(t-\tau)^2}{2}} \cdot e^{-j2\pi f \tau} d\tau \tag{4.91}$$

The product of the (Gaussian) window function and the spectral analysis function in Eq. 4.91 is summarised as (Eq. 4.92):

$$\psi(\tau) = e^{-\frac{1}{2}(t-\tau)^2 - j2\pi f \tau} \tag{4.92}$$

In Eq. 4.92, concerning Eq. 4.91, t is the constant time considered and the auxiliary variable of the window function for integration. For further consideration, the roles of the variables are reversed: represents a time shift concerning the starting time $t = 0$, and t is the time variable. The frequency f is not more the independent variable but the current frequency. It results from the scaling factor (Eq. 4.90). To vary the window length, one introduces a scaling factor s so that starting with the shortest window at the highest frequencies, it becomes longer and longer towards lower frequencies (Fig. 4.77). After these considerations, the relationship Eq. 4.92 can be modified (Eq. 4.93):

$$\psi_{s,\tau}(t) = e^{-\frac{1}{2}\left(\frac{t-\tau}{s}\right)^2 - j2\pi f_m \frac{\tau}{s}} \tag{4.93}$$

In Eq. 4.93, t is the independent time variable; the time shift with respect to $t = 0$, s is the scaling factor, and f_{m} is the (constant) highest frequency to be investigated.

4.3.3.2 Wavelet Transform

The time–frequency distribution with a scaled window can be formulated from Eqs. 4.91 and 4.93 as follows (Eq. 4.94):

$$X(s,\tau) = \int_{-\infty}^{\infty} x(t) \cdot \psi_{s,\tau}(t)dt \tag{4.94}$$

The interpretative relationship between the conventional time–frequency distributions and the distributions with a scaled window according to Eq. 4.94 is shown in Fig. 4.78. In this example, there are seven periods of the fundamental frequency f in a Gaussian window (Eq. 4.93, Fig. 4.78). This window fits four times in the time course of the signal ($\tau = 0, 1, 2, 3$), starting with the shortest window ($s = 1$). With the following doubling of the scaling factor ($s = 2$), the window length also doubles so that the window can only be segmented twice on the signal section ($\tau = 0, 1$). At the same time, scaling the time halves the original frequency f and the absolute bandwidth f. It is the principle of the *wavelet transform* (WT). We obtain a valuable and interpretative reference to the time–frequency distributions based on the analysis function by Gaussian window and harmonic (complex) oscillation according to Eq. 4.93. The WT with an analysis function, according to Eq. 4.93, can be interpreted as a dynamic spectral analysis with scaled windows. According to Eq. 4.93, the function is interpreted as a *Morlet-wavelet.* It is the only wavelet for which the term "spectral analysis" is correct in the sense of a harmonic decomposition of the analyzed signal. The formulation for wavelets can be generalized from Eq. 4.93 concerning the analytical function to (Eq. 4.95):

$$\psi_{s,\tau}(t) = \frac{1}{\sqrt{s}}\psi\left(\frac{t-\tau}{s}\right) \tag{4.95}$$

In the formulation according to Eq. 4.95, the factor $1/s^{0.5}$ is necessary for weighting the wavelet for the following reason: Similar to the spectrogram, energy distribution in the scale-time shift plane (s, -plane) is also common in the wavelet analysis, which is called a scalogram. The spectrogram and the scalogram are related since both represent the energy distribution of the analyzed signal spectrotemporally (Eq. 4.96). However, with the spectrogram, the length of the analysis window is constant so that the representation of the spectral-temporal energy distribution can also maintain the qualitative relationships between spectral bands in a time-variant manner and reproduce them correctly (Eq. 4.60 or Eq. 4.66). It is somewhat different from the scaling in the wavelet transform: The window length is not constant; it depends on the scaling s. In the case of the general wavelet (Eq. 4.95), without the correction factor $1/s^{0.5}$, this would lead to the energy of one, and the same signal component increases with the scaling by the scaling factor s (Eq. 4.96). It is, of course, not permissible, which is why the correction factor was introduced.

$$SG_x^{(w)}(t, f) = \left|STFT_x^{(w)}(t, f)\right|^2, \quad T_W = konst.$$

$$SC_x^{(w)}(s, \tau) = \left|WT_x^{(w)}(s, \tau)\right|^2, \quad T_S = sT_W \tag{4.96}$$

In Eq. 4.96, x is the analyzed signal, w the analysis window or wavelet, *STFT* the short-time Fourier transform, *SG* the spectrogram, *WT* the wavelet transform, *SC* the scalogram, T_W and T_S the window lengths at *SG* and *SC*, respectively.

According to Eq. 4.95, the generalization states that any function (including a non-differentiable and a discontinuous one) that can be shifted and scaled can be applied as a wavelet. One can therefore design one's suitable wavelet for every signal-analytical problem.

$$\int_{-\infty}^{\infty} \psi(t)dt = \mathbf{F}(\psi(t))|_{f=0} = 0 \tag{4.97}$$

Since every wavelet also has a spectral decomposition function in a certain sense due to the scaling, the time average or the Fourier coefficient must be zero-valued at $f = 0$ (Eq. 4.97). This property is explained for the wavelet ("Mexican

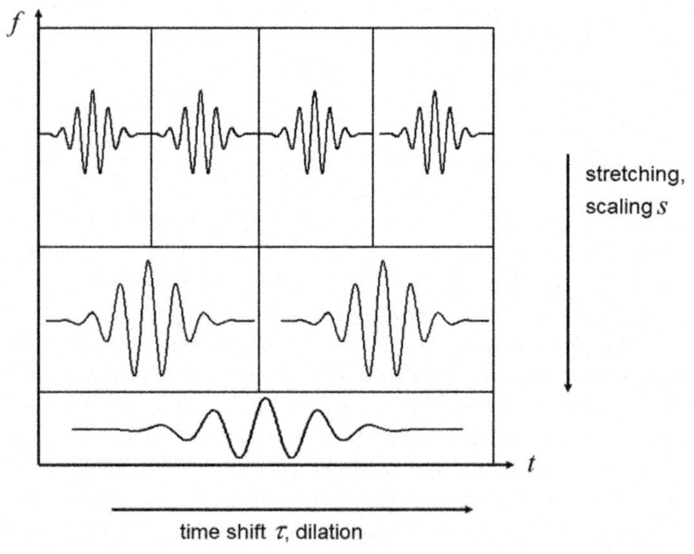

Fig. 4.78 Relationship between interpreting a time–frequency distribution (time and frequency axes, t and f) and the distribution with scaled windows (time shift τ, and scaling and s) for the Morlet-wavelet. Note that the scaling and the frequency are inversely proportional to each other

hat") considered in the following. For reasons already mentioned, the Gaussian bell is particularly well suited for analyses. Its negative second derivative is shown in Fig. 4.79. The wavelet "Mexico hat" explicitly has no harmonic signal component. However, this was not to be expected concerning the mathematical formulation of the negative second derivative of the Gaussian function (Eq. 4.98).

$$\psi^{(mexhat)}(t) = \left(1 - t^2\right) \cdot e^{-\frac{t}{2}^2} \tag{4.98}$$

Now the question arises of how the wavelet decomposition with time and scale can be interpreted concerning previously treated methods of harmonic time–frequency analysis. For this purpose, one can compare the filter functions of the Morlet-wavelet, which can still be interpreted as a spectral decomposition, and the Mexican hat (Figs. 4.80 and 4.81). The Morlet wavelet shows a bandpass behavior (Fig. 4.80), as discussed in the Discrete Fourier Transform chapter. With increasing scaling s (and a decreasing frequency), the analysis window becomes longer and the bandwidth narrower; nevertheless, the character of a harmonic analysis remains because of the harmonic fundamental under the Gaussian window. Unlike the Morlet wavelet, the Mexicohut wavelet (and all others) also shows a bandpass character (Fig. 4.81). However, the lower cut-off frequency of each bandpass is at or near zero.

For this reason, wavelets, except for the Morlet wavelet, are no longer a harmonic decomposition, so the interpretation of the wavelet transform as a spectral-temporal representation is no longer correct. It is important to note that although the bandwidth decreases with increasing scaling in both cases, in the case of the Mexican hat (and all others), the band always starts at zero. This effect can mean that when the scaling is doubled, the upper half of the signal spectrum is

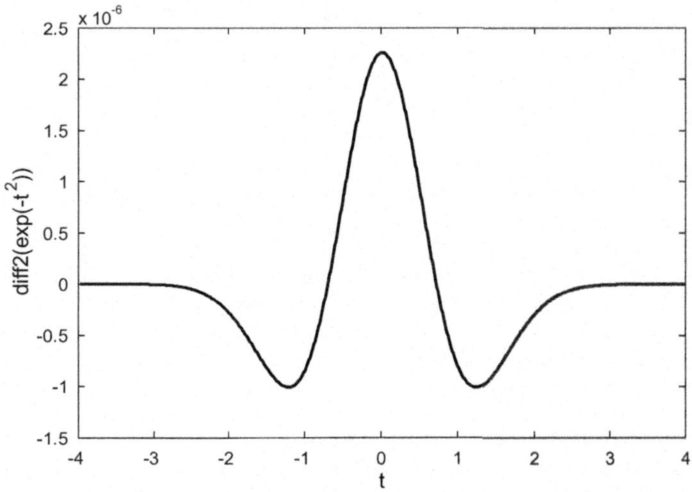

Fig. 4.79 Negative second derivative of the Gaussian bell-wavelet "Mexican hat"

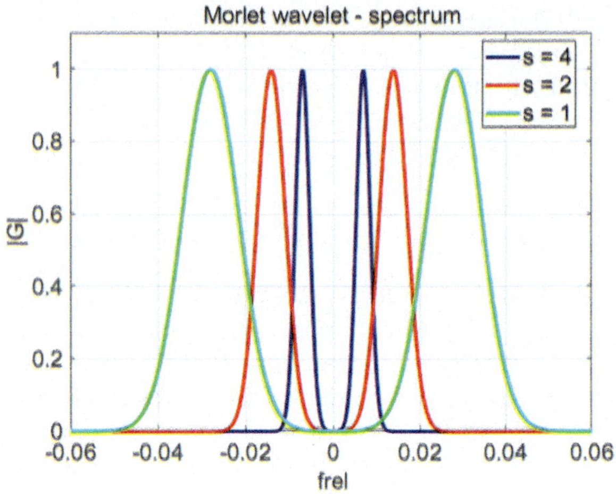

Fig. 4.80 Filter characteristics of Morlet-wavelets with dyadic (binary) scales. Note the band-pass character around a mean normalized frequency of $f_{\text{rel}}/\text{s} = 0.28/\text{s}$ in the sense of harmonic analysis

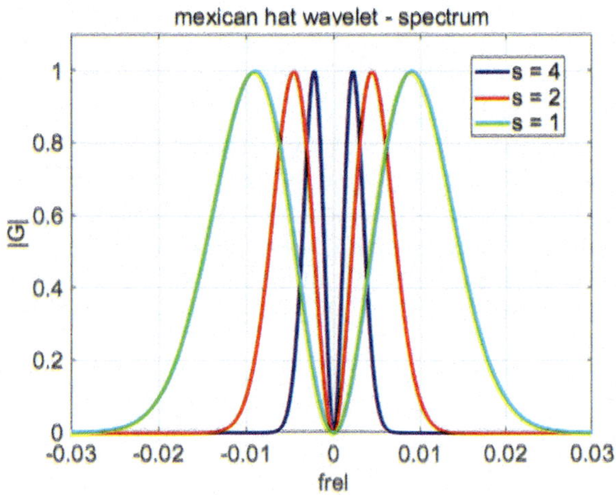

Fig. 4.81 Filter characteristics of Mexican hat wavelets with dyadic scales. Note the band-pass character of the filters, which always starts at zero frequency

suppressed. That is a fundamental insight into how to interpret wavelet spectral properties.

Since this property is known in advance, the spectral composition can still be determined with the help of linear algebra. Nevertheless, this is unnecessary because the time–frequency analysis methods are available. Subsequently, the question arises as to what is unique about wavelets that time–frequency analysis has not been able to capture so far.

The unique feature of wavelets is the possibility to construct a specific wavelet for every signal analytical problem and every conceivable signal character.

With any wavelet, there is only a signal-analytical advantage over the previous methods of time–frequency analysis if the signal character roughly corresponds to that of the ECG or EMG (fast high-frequency or broadband as well as narrowband low-frequency signal components). At low frequencies, a good spectral resolution is required (e.g., heart rate, respiration, heart rate variability); at high frequencies, a good time resolution and not necessarily a good spectral resolution (e.g., R-waves, action potentials). If specific frequencies with a good spectral resolution are to be specifically searched for in the high-frequency range, e.g., for visual or acoustic stimulation of the sensory systems with defined stimulus sequences, wavelets are unsuitable. Two typical examples of the calculation according to Eq. 4.94 and the presentation of the wavelet coefficients are shown in Figs. 4.82 and 4.83. The scalogram of an ECG in Fig. 4.82 offers the advantages of wavelets compared to previous methods of spectral-temporal analysis. While in the low-frequency range, a good spectral resolution for the analysis of slow cardiological processes prevails, a good temporal resolution for the exact localization of QRS complexes is available at high frequencies. To also make the different spectral widths of the associated scales clear, the height of the bars corresponds to the actual absolute bandwidth of the windows. The following example with a VEP after periodic stimulation with 8 pps and spontaneous α-waves are shown in the scalogram in Fig. 4.83. The sought harmonics of the stimulation rate ($n \times 8$ Hz) and the α-wave trains are not identifiable in this representation; even the transient component is hardly more distinct.

At this point, it becomes clear that the wavelet transform is also not the ultimate method of spectral-temporal analysis but—like any other method—offers advantages only in some regions of the signal properties.

4.3.3.3 Wavelets as a Filter Bank

While the Morlet-wavelet (Fig. 4.80) allows a spectral-temporal interpretation from the harmonic decomposition point of view, this interpretation is no longer permissible for all other wavelets (e.g., Mexico hat, Fig. 4.81). Spectral decomposition of signals is needed in signal analysis despite the theoretically arbitrary shape of a wavelet. Wavelets may be specifically adapted to the current analysis problem. Not at least for reference reasons, a suitable spectral decomposition is sought. The qualitative spectral behavior of a wavelet, as shown in Fig. 4.81, can be interpreted as a low-pass filter whose DC component has been eliminated. Eliminating the DC component (of the first Fourier coefficient) is necessary for signal-theoretical reasons (spectral delimitation of neighboring wavelets). It leads to the interpretation of the wavelet as a bandpass. In the further procedure, the high-pass character of the DC component elimination is not considered.

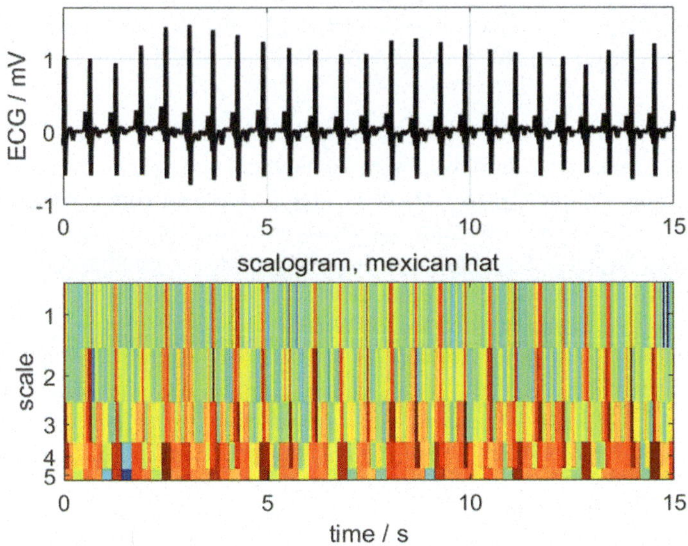

Fig. 4.82 Time course of an ECG (top) and the scalogram with the wavelet "Mexican hat" (bottom). For details, see exercises and *uebung_4_12.m*

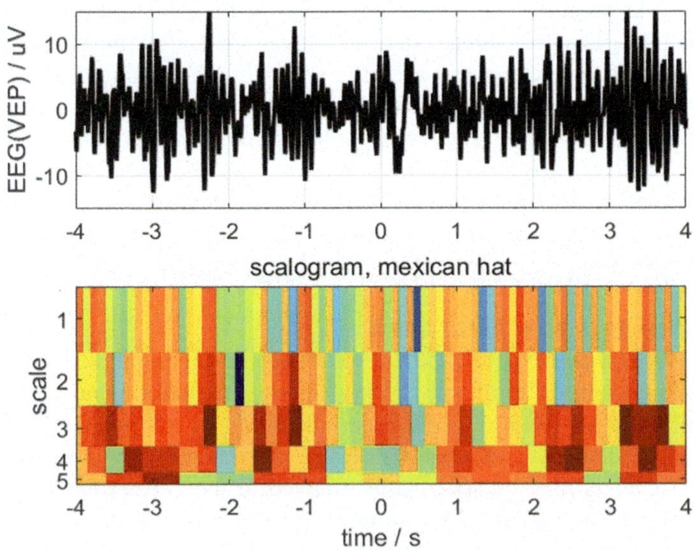

Fig. 4.83 Time course of an EEG/VEP (top) and the scalogram with the wavelet "Mexico hat" (bottom)

Accordingly, one concludes that the spectrum divided by the low-pass filter is sensibly halved (dyadic structure) for pragmatic reasons (practical algorithms). After each halving, one would unnecessarily take the low-frequency part repeatedly, leading to enormous redundancy. It makes sense to halve the spectrum at each doubling of the window length by a low-pass and a high-pass. For this, one needs the spectrally inverse variant of a wavelet—a high pass. The following considerations are helpful before constructing its high-pass variant from a wavelet.

One can divide any spectrum with the help of spectral filters. If one knows in advance that it is to be halved in the next step (dyadic structure), one can construct a corresponding low-pass filter (Fig. 4.85) and an equivalent high-pass filter (Fig. 4.86). Ideally, both filters have a square wave characteristic, which results in the ideal impulse response (Fig. 4.84). The length of both impulse responses in Fig. 4.84 was $L = 1001$, with which the ideal spectral filter characteristics can be approximately realized (full lines in Figs. 4.85 and 4.86). Of course, such filter lengths are difficult to realize in practical signal analysis; the computational effort and the necessary transmission capacity would increase unreasonably. For practical applications, reducing the filter length by about 100 is necessary. The fact that the baseband ($-f_A/2 < f < f_A/2, f_A$ is the fundamental frequency of the sampling rate) is to be halved with the lowpass and highpass filters ($f_G = \pm f_A/4$, f_G is the cut-off frequency of both filters) results in the following computationally very favorable effect. Due to the periodicity of the spectra resulting from the sampling, the half-band high-pass can be interpreted as a half-band low-pass shifted by $f = f_A/2$ to both sides (see chapter Sampling, Eq. 4.99).

$$G_{HP}(f) = G_{TP}\left(f \pm n\frac{f_A}{2}\right) \tag{4.99}$$

According to the frequency shift theorem of the Fourier transform (Eq. 4.19), the expression in Eq. 4.99 corresponds to the multiplication of the impulse response of the low pass with a harmonic frequency $f_A/2$ (Eq. 4.100).

$$g_{HP}(t) = g_{TP}(t) \cdot e^{\pm j2\pi \frac{f_A}{2} t} \tag{4.100}$$

If the relationship Eq. 4.100 is transferred to the discrete-time range, then (Eq. 4.101) applies:

$$g_{HP}(n) = g_{TP}(n) \cdot (-1)^n \tag{4.101}$$

According to Eq. 4.101, the relation offers an effortless computational realization of the half-band high-pass: One calculates the filter coefficients of the half-band low-pass and changes the sign at odd coefficients (Fig. 4.84).

Since after filtering with the half-band filters, the resulting frequency bands are only half as wide as the spectrum before halving; only half the sampling rate would be necessary for their sampling. Since the data is already sampled, the sampling rate can be reduced to half after filtering without loss of information.

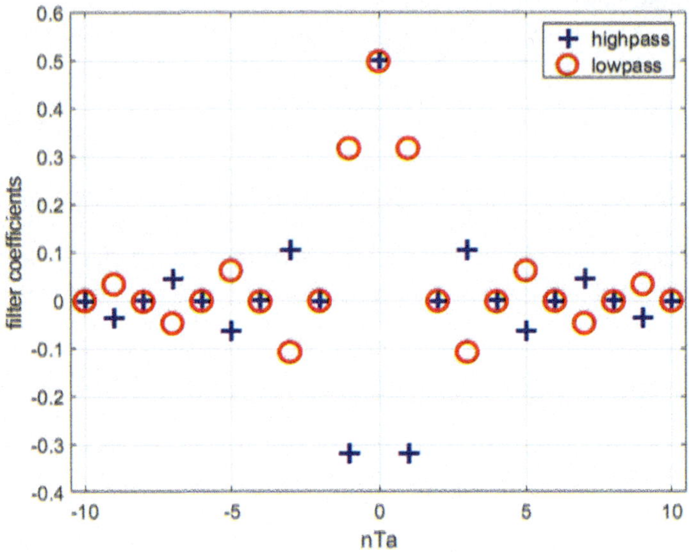

Fig. 4.84 Filter coefficients of an ideal low pass (red rings) and an ideal high pass (blue crosses). The relative cut-off frequency is 0.25 (first halving of the signal spectrum). Shown are discrete values for filters of length $L = 21$

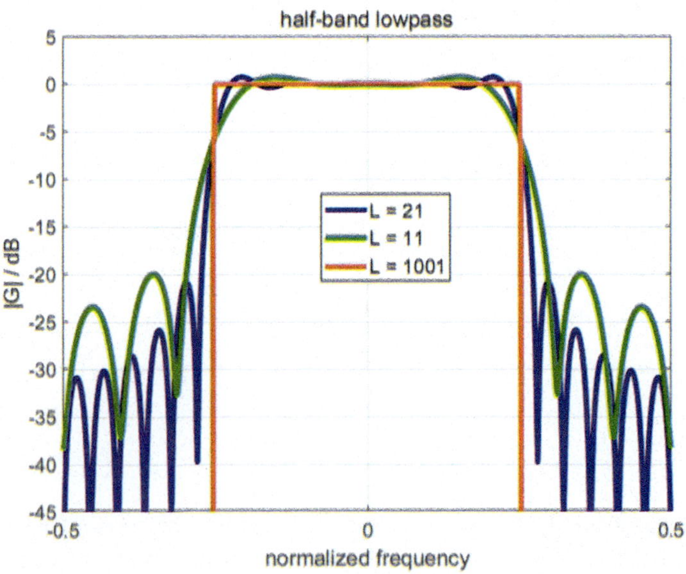

Fig. 4.85 Half-band low-pass filter (cut-off frequency 0.25) with different filter lengths of the discrete-time sinc function (L = 11, 21, 1001). The impulse response is shown in Fig. 4.84 (L = 21)

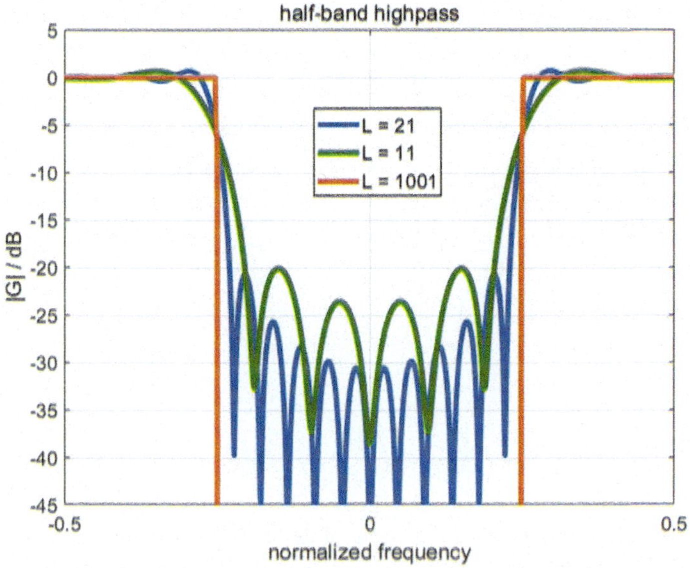

Fig. 4.86 Half-band high-pass (cut-off frequency 0.25) with different discrete-time sinc function filter lengths (L = 11, 21, 1001). The impulse response is shown in Fig. 4.83 (L = 21)

In practical processing, every second digital value of the sequence is omitted. It achieves a spectral decomposition of the analyzed signal without increasing the redundancy (the number of values remains constant) and without losing information (Figs. 4.87 and 4.88). In literature, the impulse response of the high-pass is called a *wavelet*, and that of the low-pass a "scaling function." Both terms return to Meyer, who was the first to use them in this decomposition. In a less strict sense, however, the introduced *wavelet is* used today for all required filters. The terminology of the decomposition into "approximations" and "details" used in the technical literature is based on the characteristics of the low-frequency and high-frequency signal components. The low-frequency components are typical for slow progressions (large-scale structures in images); they reflect the average or smoothed component—hence approximations. The high-frequency components express short-time and local progressions (subtleties in images); therefore, they are called details.

As shown in Fig. 4.88, the spectrum of a signal is described by approximations at the largest scale (A3) and all details (D1, D2, D3). Therefore, a signal reconstruction is possible in the spectral sense by summing all details and the last approximation (Eq. 4.102).

$$S_x(f) = A_N + \sum_{n=1}^{N} D_n \tag{4.102}$$

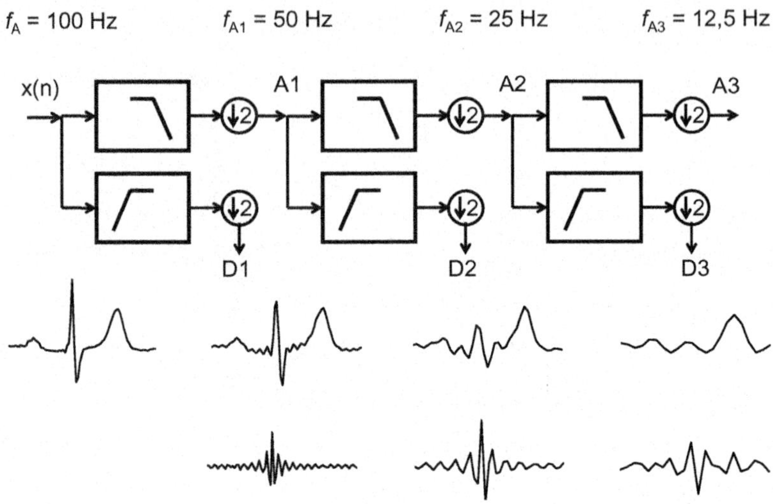

Fig. 4.87 Half-band filtering and sub-sampling of an ECG. Starting with the fundamental frequency of the sampling rate $f_A = 100$ Hz, this is halved in each step. The filter coefficients remain the same in each step. The signal curves belong to the components marked in the block diagram

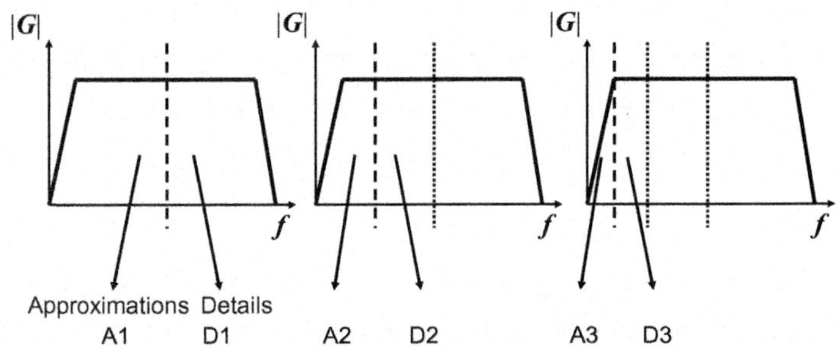

Fig. 4.88 Interpretation of wavelets as a filter bank. With increasing scaling (here, dyadic structure), the current spectrum is halved to the low-frequency range (approximation) and the high-frequency range (details)

It should be noted that each detail D_n spectrally represents a passband resulting from the approximation of the pre-stage A_{n-1} and the high-pass of the current stage *n*. This can be easily checked, especially with the conventional ECG. If the ECG is decomposed according to the scheme in Fig. 4.87, the detail D2 after the second decomposition lies spectrally in the range between 6.25 and 12.5 Hz since half-band filters are used the fundamental frequency of the sampling rate after the reduction is 25 Hz.

As shown in Fig. 4.87, each approximation at level N results from a series connection of N low-passes and each detail from a series connection of N − 1 low-passes and a high-pass (Eq. 4.103).

$$A_N(f) = X(f) \cdot \prod_{n=1}^{N} G_{TP}^{(n)}(f)$$

$$D_N(f) = X(f) \cdot G_{HP}^{(N)}(f) \cdot \prod_{n=1}^{N-1} G_{TP}^{(n)}(f) \tag{4.103}$$

One could therefore calculate a common impulse response for the decomposition at level N via N low-pass filters or N − 1 low-pass filters and one high-pass filter. Since the individual impulse responses theoretically correspond to a Sinc function (Fig. 4.84), one must fold all involved impulse responses accordingly for the resulting Sinc function. Such wavelets are also called Sinc wavelets (Shannon wavelets). As is well known, the Sinc function is theoretically infinitely long; in practice, one has to reckon with about 1000 coefficients. These wavelets are hardly practicable because of their length; one needs much shorter wavelets, especially for real-time applications.

The wavelet transform is linear; therefore, reconstructing the decomposed signal or synthesis from an otherwise generated decomposition is theoretically and practically possible. If details and approximations are available according to the formulation in Eq. 4.102, the structure shown in Fig. 4.87 is processed in reverse order for the synthesis. The details and the approximation are brought to twice the sampling rate by inserting zeros in between (oversampling by a factor of 2) and then filtered with the same high and low passes (with symmetrical impulse response) that were used for the analysis (Fig. 4.89), starting with the largest scale. A transmission or processing system exists between the analysis (Fig. 4.87) and the synthesis (Fig. 4.89, Eq. 4.103). Whether this system serves a data compression for transmission, noise reduction, or feature detection is a question of the concrete application. A complete wavelet system for analysis and synthesis with a general processing system of the decomposed signal is schematically shown in Fig. 4.90. One needs up to four different filters of analysis and synthesis, whereby they can all be derived from a basic filter (analysis half-band low-pass). How an analysis high pass can be constructed has already been shown (Eq. 4.101). Suppose the impulse responses of these filters are not symmetrical (as is often the case with wavelets). In that case, the synthesis lowpass corresponds to a time-mirrored variant of the analysis lowpass (Eq. 4.104).

$$g_{TP}^{(Synth)}(t) = g_{TP}^{(Anal)}(-t) \tag{4.104}$$

In the discrete-time case, applying the formula according to Eq. 4.104 means a reversal of the order of the filter coefficients (Eq. 4.105).

$$g_{TP}^{(Synth)}(n) = g_{TP}^{(Anal)}(-n) \tag{4.105}$$

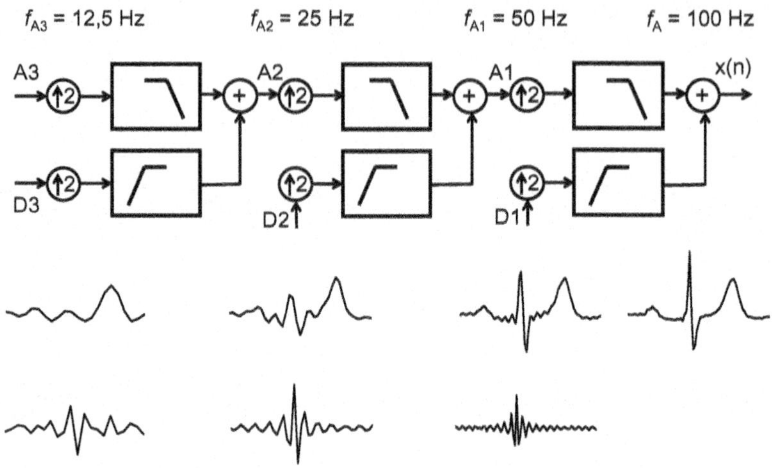

Fig. 4.89 Synthesis from approximations and details derived from the analysis according to Fig. 4.87. The signal curves belong to the components marked in the block diagram

According to Eq. 4.105, the relation is also valid for the other filters in case of an asymmetrical impulse response. From the relations according to Eqs. 4.105 and 4.101 results for the analysis high pass (Eq. 4.106):

$$g_{HP}^{(Anal)}(n) = g_{TP}^{(Anal)}(-n) \cdot (-1)^n \tag{4.106}$$

Finally, the synthesis high-pass (Eq. 4.107) holds:

$$g_{HP}^{(Synth)}(n) = g_{TP}^{(Anal)}(n) \cdot (-1)^n \tag{4.107}$$

From the relations according to Eqs. 4.105–4.107, one single basic wavelet is needed for the complete analysis and synthesis (see Eqs. 4.121 and 4.122). Simple operations can calculate all its spectral, dilated, and scaled variants. Note that in the relations according to Eqs. 4.101, 4.105, 4.106 and 4.107, the index n always starts with a one so that the unambiguous assignment to even and odd indices is ensured. The negative sign in front of the index n formally indicates the reverse order of the indices since they cannot be negative if they begin with a one (see exercises for details).

4.3.3.4 Types and Applications of Wavelets

As shown before, the analysis and synthesis with Sinc wavelets are possible with ideal spectral filters. However, the Sinc wavelets are not practical because of their length.

For this reason, attempts were made to construct much shorter wavelets. The trade-off between the spectral filter function and the number of coefficients was that the filter function deviated significantly from the ideal at the expense of the

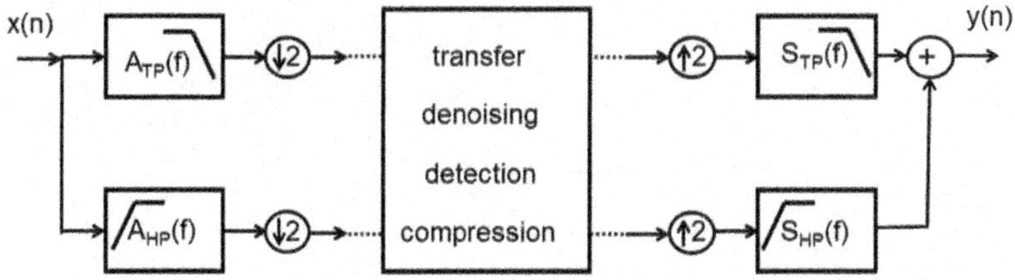

Fig. 4.90 Block diagram for analyzing and synthesizing signals using wavelets. The four different filter types listed here are needed: one high-pass and one low-pass each for the analysis and the synthesis. These filters are cascaded according to the structures in Figs. 4.87 and 4.89. If the impulse response is symmetrical, the low-passes and high-passes are identical for analysis and synthesis. The high pass is derived from the low-pass, according to Eq. 4.96

impulse response length. Meyer rounded the otherwise desired sharp corners of the filter function by piecewise approximations (Eqs. 4.108 and 4.109), which are visible in Fig. 4.91 at locations of the corner frequencies (the angular transitions of the otherwise round courses at the cut-off frequency are a consequence of the Meyer approximation). This apparent weakening of the spectral sharpness leads to Meyer's wavelets managing with few coefficients. That is a significant simplification compared to Sinc wavelets, which require many more coefficients. The very low relative frequency in Fig. 4.91 results from the high scaling used for this calculation. It was necessary to illustrate the piecewise approximation, according to Meyer. With dyadically scaled wavelets, all filter functions are half-band filters.

Meyers Wavelet

Meyer's wavelet (Bergh et al., 1999) definition in the frequency domain (for reasons of clarity, the frequency f is partly replaced by the angular frequency in the following):

$$\psi^{FT}(\omega) = \begin{cases} \frac{1}{\sqrt{2\pi}} e^{j\omega/2} \sin\left(\frac{\pi}{2}\gamma\left(\frac{3}{2\pi}|\omega|\right) - 1\right), & \frac{2\pi}{3} \le |\omega| \le \frac{4\pi}{3} \\ \frac{1}{\sqrt{2\pi}} e^{j\omega/2} \sin\left(\frac{\pi}{2}\gamma\left(\frac{3}{4\pi}|\omega|\right) - 1\right), & \frac{4\pi}{3} \le |\omega| \le \frac{8\pi}{3} \\ 0, & |\omega| \notin \left[\frac{2\pi}{3}, \frac{8\pi}{3}\right] \end{cases}. \tag{4.108}$$

Meyer's scaling function—definition in the frequency domain:

$$\phi^{FT}(\omega) = \begin{cases} \frac{1}{\sqrt{2\pi}}, & |\omega| \le \frac{2\pi}{3} \\ \frac{1}{\sqrt{2\pi}} \cos\left(\frac{\pi}{2}\gamma\left(\frac{3}{2\pi}|\omega| - 1\right)\right), & \frac{2\pi}{3} \le |\omega| \le \frac{4\pi}{3} \\ 0, & |\omega| > \frac{4\pi}{3} \end{cases}. \tag{4.109}$$

Meyer's auxiliary function (x):

$$\gamma(x) = x^4\left(35 - 84x + 70x^2 - 20x^3\right), \quad x \in [0, 1] \tag{4.110}$$

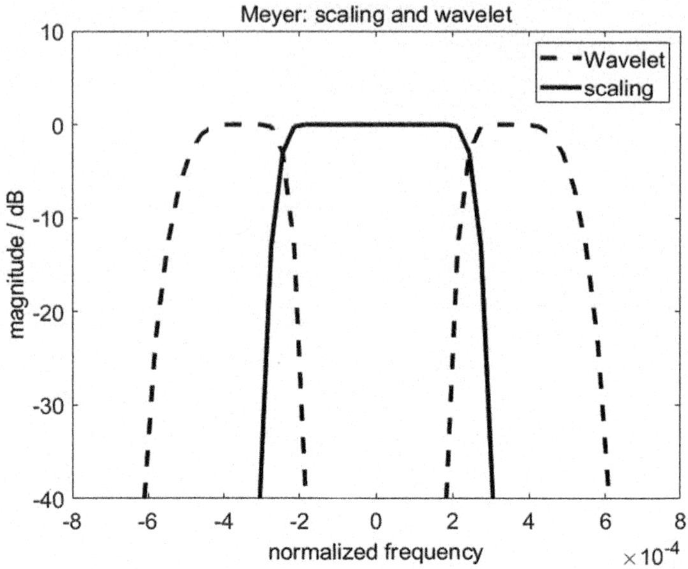

Fig. 4.91 Meyer wavelet as a spectral function: the filter functions for the low pass (scaling), and the high-pass (wavelet) are approximated piecewise, recognizable by corners of the characteristics in the cut-off frequency range

From the definitions for Meyer's scaling function and wavelet in the frequency domain (index FT in the formulae), the inverse Fourier transformation can determine the corresponding functions (Fig. 4.92). One can generate any Meyer wavelet families with the auxiliary function (x) (Eq. 4.110).

Haar's Wavelet

The Haar wavelet (1909) is the simplest and the oldest (Figs. 4.93 and 4.94). Nevertheless, it is still widely used in several variants (e.g., image compression). Its simplicity becomes apparent when considering the number of relevant coefficients (Eqs. 4.111 and 4.112). Two coefficients for both functions, the scaling and the wavelet are sufficient; they are also identical in magnitude. The simplicity of the Haar wavelet is bought by a very flat filter characteristic of the high pass (wavelet) and the low pass (scaling) (Fig. 4.94).

Note: It should be noted that compared to all the analysis functions discussed so far, wavelets do not necessarily have to be continuous and differentiable (Meyer, Haar, Daubechies).

Haar's Wavelet:

$$\begin{aligned} \psi(t) &= 1, \quad t \in [0, 0.5) \\ \psi(t) &= -1, \quad t \in [0.5, 1) \\ \psi(t) &= 0, \quad t \notin [0, 1) \end{aligned} \tag{4.111}$$

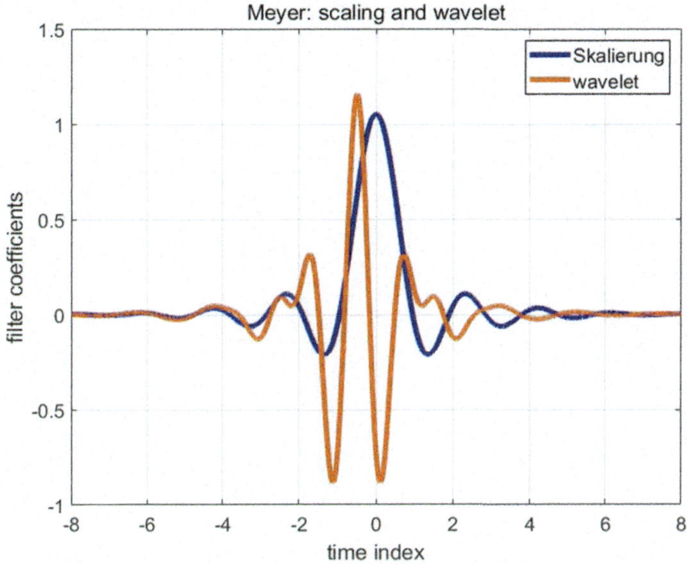

Fig. 4.92 Meyer's scaling function (low pass, blue) and wavelet (high pass, red)

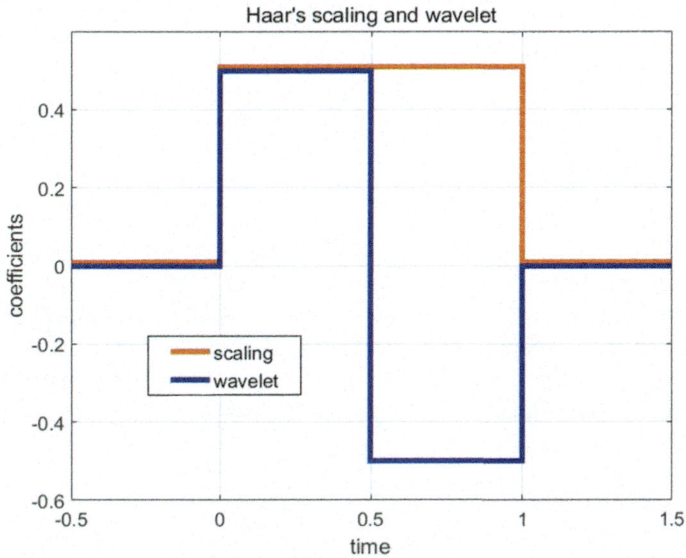

Fig. 4.93 Haar wavelet: scaling function and wavelet. Note that both functions are discontinuous and, therefore, not differentiable. Contrary to the definition according to Eqs. 4.111 and 4.112, the two discrete coefficients are 0.5, so their application as a filter receives a transmission of $|G(f)| = 1$ overall (see Fig. 4.94)

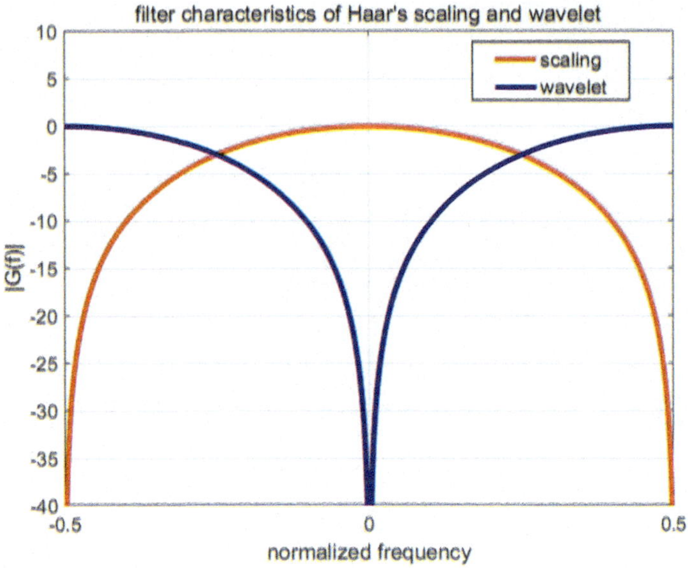

Fig. 4.94 Haar wavelet: half-band filter characteristics of the scaling and the wavelet. The cut-off frequencies are identical, but the overlap range is extensive

Haar's scaling function:

$$\begin{aligned} \phi(t) &= 1,\ t \in [0, 1] \\ \phi(t) &= 0,\ t \notin [0, 1] \end{aligned} \tag{4.112}$$

From the relationships according to Eqs. 4.111 and 4.112 and the representation in Fig. 4.93, it can be seen that the Haar wavelet is discontinuous and, therefore, also not differentiable. Consequently, analyzing this wavelet concerning its properties and comparing it with other wavelets is problematic, especially in theoretical analysis. In developing new wavelet families in the 1990s, the Haar wavelet became a subset of Ingrid Daubechies' new wavelet class. However, this wavelet class has no closed mathematical formulation, such as the wavelets according to Morlet or the Mexican hat.

Daubechies' Wavelet

There is neither an explicit mathematical formulation nor a piecewise approximation for the Daubechies wavelet. Instead, Daubechies defines conditions and properties to be met while generating wavelets (Debnath, 2002). For the calculation, one first needs a polynomial with binomial coefficients (Eq. 4.113)

$$P_N(x) = \sum_{k=0}^{N-1} \binom{N-1+k}{k} \cdot x^k \tag{4.113}$$

and a trigonometric polynomial (Eq. 4.114):

$$L(\omega) = \sum_{k=0}^{n} b_k \cdot e^{-jk\omega} \tag{4.114}$$

The following applies to the generation function (Eq. 4.115)

$$m_0(\omega) = \left(\frac{1 + e^{-j\omega}}{2}\right)^N \cdot L(\omega) \tag{4.115}$$

and for $m()$:

$$m(\omega) = \frac{1}{\sqrt{2}} \sum_{-\infty}^{\infty} c_n e^{-jn\omega}. \tag{4.116}$$

If the relationships Eqs. 4.115 and 4.116 are equal, the coefficients c_n can be calculated.

Daubechies Wavelet for N = 1

From Eq. 4.113, it follows for $P_\mathrm{N}(x) = 1$ and from Eq. 4.114 for $L() = 1$, so that the generating function according to Eq. 4.115 holds (Eq. 4.117):

$$m_0(\omega) = \frac{1}{2}\left(1 + e^{-j\omega}\right) \tag{4.117}$$

According to Eq. 4.117, the generation function corresponds to the Haar wavelet in the frequency domain. That makes it clear that the Haar wavelet is also a Daubechies wavelet for $N = 1$.

Daubechies Wavelet for N = 2

From Eq. 4.113, it follows for $P_\mathrm{N}(x)$ with $N = 2$

$$P_2(x) = \sum_{k=0}^{1} \binom{k+1}{k} x^k = 1 + 2x$$

and from Eq. 4.114 for $L()$

$$L(\omega)L(-\omega) = 2 - \frac{1}{2}\left(e^{j\omega} + e^{-j\omega}\right). \tag{4.118}$$

If one equates the relationships according to Eqs. 4.114 and 4.118, one obtains by the coefficient comparison (Eq. 4.119)

$$b_0 = \frac{1}{2}\left(1+\sqrt{3}\right),\ b_1 = \frac{1}{2}\left(1-\sqrt{3}\right).. \tag{4.119}$$

Substituting Eq. 4.119 into Eq. 4.116, we obtain (Eq. 4.120)

$$\begin{aligned} c_0 &= \frac{1}{4\sqrt{2}}\left(1+\sqrt{3}\right),\ c_1 = \frac{1}{4\sqrt{2}}\left(3+\sqrt{3}\right) \\ c_2 &= \frac{1}{4\sqrt{2}}\left(3-\sqrt{3}\right),\ c_3 = \frac{1}{4\sqrt{2}}\left(1-\sqrt{3}\right) \end{aligned}. \tag{4.120}$$

Using the coefficients c_n from Eq. 4.120, one can formulate the scaling function (Eq. 4.121) and the wavelet (Eq. 4.122).

$$\phi(x) = \sqrt{2}[c_0\phi(2x) + c_1\phi(2x-1) + c_2\phi(2x-2) + c_3\phi(2x-3)] \tag{4.121}$$

$$\psi(x) = \sqrt{2}[-c_3\phi(2x) + c_2\phi(2x-1) - c_1\phi(2x-2) + c_0\phi(2x-3)] \tag{4.122}$$

In practical analysis, of course, one does not have to calculate the Daubechies coefficients in this complicated way every time. The scaling and wavelet coefficients are known for the orders $N = 2, \ldots, 20$ (and calculable for further orders) and tabulated, which is sufficient for the standard case. It should be noted that only even orders are used (the number of coefficients is even, and the signal length is a power of 2) since the theory and the implementation would hardly be manageable with odd orders.

In the practical calculation of the wavelet coefficients, this means that, for example, a Daubechies wavelet with $N = 3$ has a length of 6 coefficients (see Eq. 4.120 for $N = 2$).

Figures 4.95 and 4.96 show two selected, frequently used wavelets, Daubechies with $N = 2$ and $N = 8$. Both examples show that these wavelets are strongly asymmetrical, which can be advantageous in some exceptional cases, e.g., a rapidly changing spectrum. However, this is unsuitable for analyzing the usual biosignals (ECG, EEG, evoked potentials, pulse progression). As is well known, e.g., in the ECG, the QRS complex and the P- and T-wave are largely symmetrical as a time course. Thus, an asymmetrical wavelet can easily lead to misinterpretations. Figure 4.97 (ECG) and Fig. 4.98 (EEG/VEP) show a typical analysis.

Wavelet Application

If one considers the decomposition using wavelets as band-pass filtering, then in this sense, a spectral analysis is carried out, which differs from the STFT or spectrogram by the variable length of the analysis window. The time resolution is better than with STFT because the analysis window becomes shorter and shorter toward the high frequencies. That makes wavelets—especially the Morlet-wavelet—suitable for the dynamic spectral analysis of biosignals, some of which have a pulse character, e.g., ECG, EMG, or EOG.

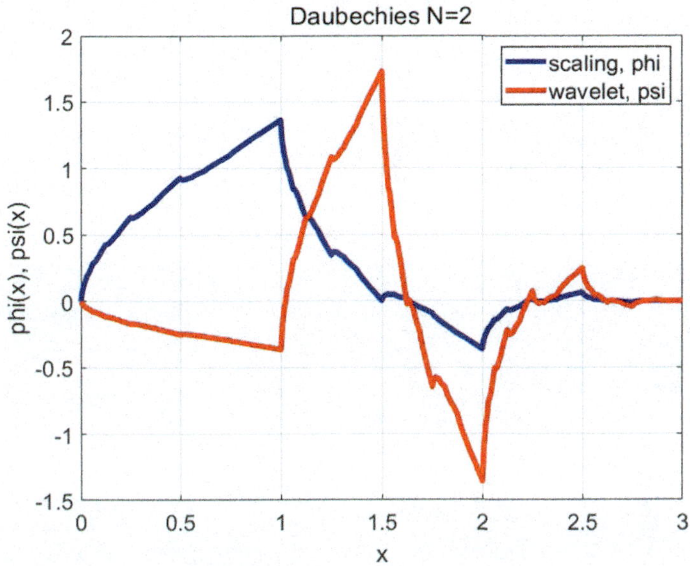

Fig. 4.95 Daubechies wavelet and scaling for N = 2

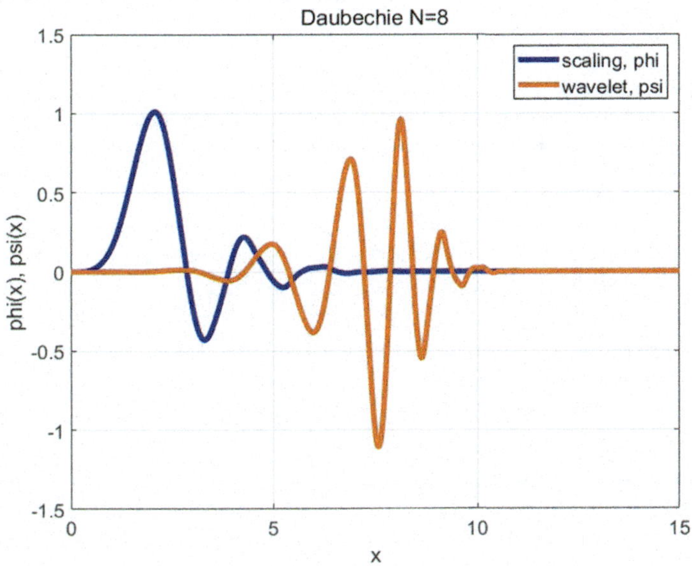

Fig. 4.96 Daubechies wavelet and scaling for N = 8

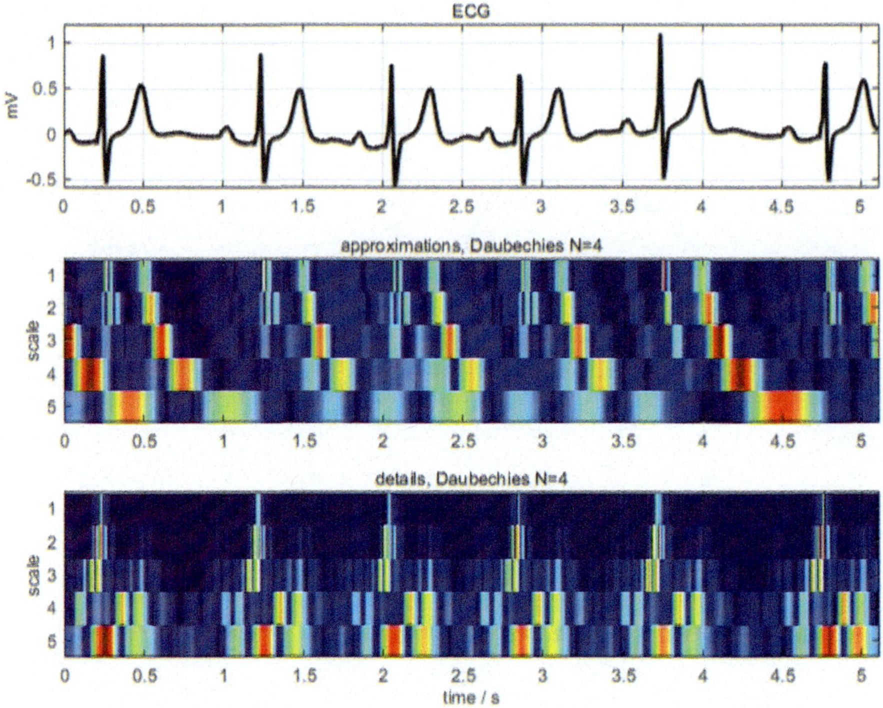

Fig. 4.97 Decomposition of an ECG section with Daubechies wavelet *db4* and decomposition level $S = 5$. The asymmetric character of the Daubechies wavelet is evident in the approximations. The high signal energy is shown in red. It should be noted that high scaling corresponds to low frequencies

However, the shorter analysis window also makes the spectral resolution poorer than at low frequencies. Thus, wavelets are unsuitable for the spectral analysis of high-frequency biosignals with good frequency resolution, e.g., otoacoustic emissions (OAE) in the frequency range of several kilohertz. Due to the rapid development of wavelets, especially in the eighties and nineties, they were given a significance that was not appropriate to real needs.

Wavelets are used, for example, to reduce noise in signals or images and for compression (see also Wavelet Toolbox in Matlab, Mathworks). Both applications have in common that a part of the signal energy, attributed to the noise or the supposedly unnecessary details, is eliminated. By suppressing the noise, better signal quality is achieved, and by compression, the amount of data is reduced, sometimes considerably. However, these advantages are tied to some preconditions. Before noise can be removed from a biosignal, the signal-to-noise ratio (SNR) must already be relatively high (at least 10 dB). If this condition does not meet, a significant portion of the biosignal is eliminated with the noise. It may even result

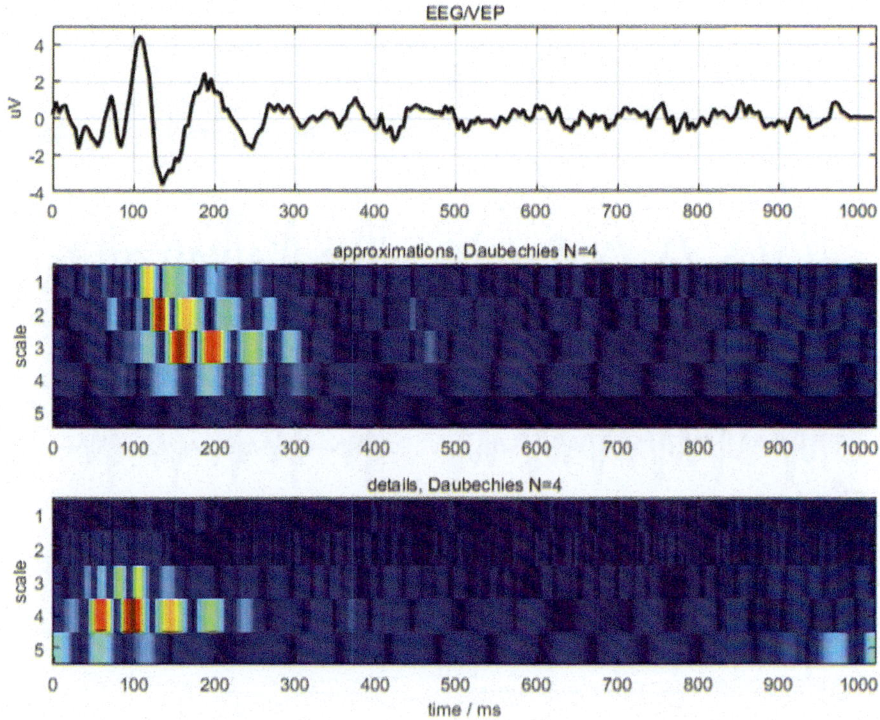

Fig. 4.98 Decomposition of an EEG section with VEP with Daubechies wavelet *db4* and decomposition level $S = 5$. The high signal energy is shown in red. Note that high scaling corresponds to low frequencies. The sampling rate is 250 sps. The bands A5 (0…4 Hz), as well as D1 (62.5 Hz … 125 Hz) and D2 (31.25 Hz … 62.5 Hz), contain very little signal energy because the VEP was filtered with a high-pass filter $f_G = 7$ Hz and a bandstop $f_S = 45$ Hz … 55 Hz

in a deterioration of the SNR. In this sense, the methods of time–frequency analysis (STFT Wigner distribution and wavelets) are equivalent; with all of them, the noise can be reduced by eliminating it below a discriminant threshold (see exercises) (see exercises). A similar situation arises with data compression. They set a threshold above which lower signal energies are eliminated, thus decreasing the data amount. It is basically at the expense of details or short-term processes in the signal, which can certainly be compared with noise in terms of signal character. Such data compression is, therefore, lossy and, in extreme cases, can inadmissibly suppress important details in biosignals or medical images (Fig. 4.99).

The most crucial advantage of wavelets is the possibility of constructing one's wavelet, optimized according to one's criteria, for almost every specific problem in biosignal analysis. For example, one can formulate a corresponding signal pattern as a wavelet to search for epileptic patterns in the EEG (e.g., spike-wave complexes). As is known, such signal patterns are distributed arbitrarily over the time

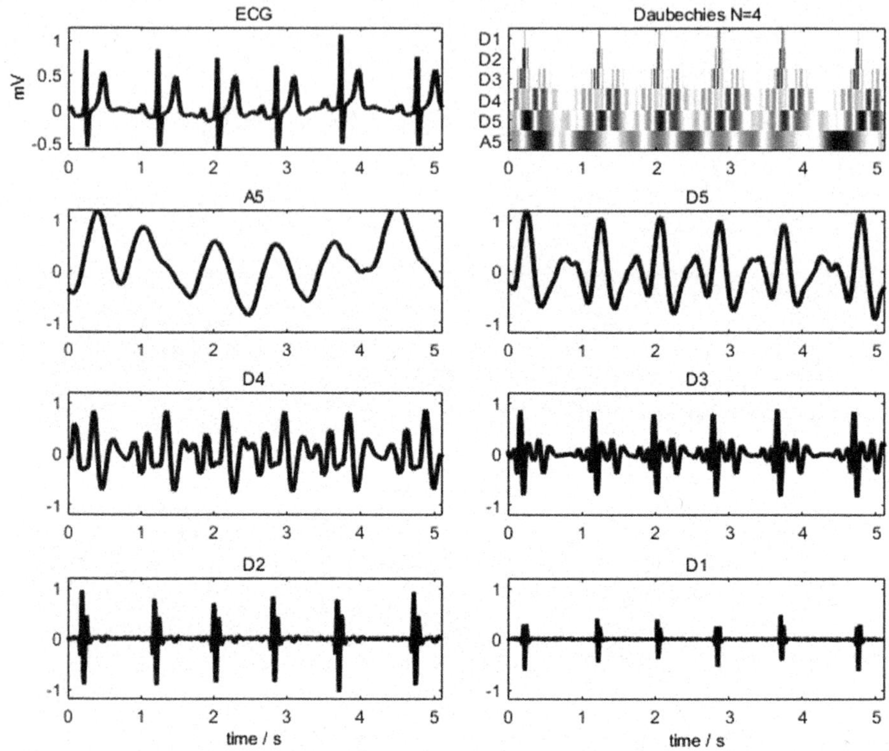

Fig. 4.99 Decomposition of an ECG with Daubechies wavelet *db4*. The time–frequency plot (top right) shows the six components that describe the signal entirely after a decomposition at $S = 5$. The signal reconstruction is possible by adding these components (Eq. 4.102). While a low pass forms A5, all details D1, …, and D5 arise as band-pass signals (Fig. 4.88)

axis and have variable extension and duration. A wavelet with a spike-wave pattern as a template would be very suitable for automatically searching such patterns. In the sense of second-order statistics, this would be a cross-correlation analysis with a template of variable length and an examined EEG section. Therefore, one can interpret the wavelet coefficients as (non-normalized) correlation coefficients between the examined signal (image) and the scaled and shifted pattern (template wavelet).

4.4 Exercises

4.4.1 Tasks

4.4.1.1 Elimination of Slow Settlement Processes in an ECG

Load the saved ECG (*e23.mat*). Display it as a real ECG with $T_A = 4$ ms and amplification factor $V = 100$. In the following, we will work with indices, not physical units, for easier handling. That makes the analysis easier because the units can be added to the result.

- Create a compensation process from two e-functions, whereby the fall time constant active at the beginning should be 1 s, and the rise time constant following it should be 3 s. Add this compensation process to the ECG to begin at $t = 4$ s or time index 1001. Plot the time course; it should correspond to the course in Fig. 4.3.
- Calculate the spectra of the original ECG, the settlement process, and the disturbed ECG and display them (see Fig. 4.4). Decide based on the spectra at which point the cut-off frequency of a high-pass filter can be located with which the settlement process is to be suppressed. Note that the fundamental wave of the ECG with its sidebands (the first harmonic of the heart rate with adjacent spectral peaks of respiration) must not be affected. In what range could the cut-off frequency lie?
- Design a high pass using the Matlab packages SPtool or FDAtool as an FIR and an IIR filter. Compare the effect of the high passes on the artifact and the waveform of the ECG. Which filter causes the shortest delay? Which filter can distort the waveform of the ECG? Which filter with which cut-off frequency would be better suited for this measurement task, taking into account the requirements of the measurement chain?

Non-linearities in the Biosignal

Check how well the signal model is according to Eq. 4.3 corresponds to reality with the signal *ekg.mat*. Start with the additive part of the signal model $A(s(t) + \mathrm{a}(t))$. The desired signal $s(t)$ is the recorded ECG without changes in amplitude, and the artifact $a(t)$ is the respiration, which changes the ECG amplitude. Since the periods of cardiac activity and respiration are different, one can try to separate them spectrally.

- Using the calculated spectrum, construct a filter (Matlab Toolbox Signal Processing: SPTool) with which the additive component of respiration can be eliminated. Estimate the effect of the filter.
- Further investigate the multiplicative part of the signal model $Ms(t)a(t)$, representing an amplitude modulation. For this purpose, note the spectral effect of amplitude modulation as exemplified in Fig. 4.100.
- Check whether a qualitatively similar image (a dominant carrier frequency with symmetrically lying sidebands) is present in the spectrum of the real ECG.

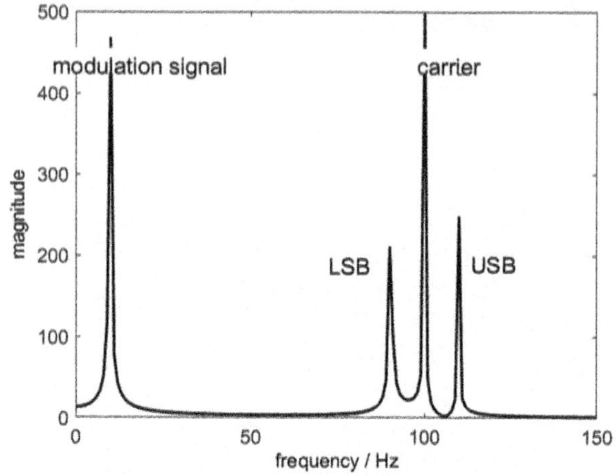

Fig. 4.100 Spectral position of the modulation signal (10 Hz), the carrier (100 Hz), and the sidebands created after amplitude modulation (90, 110 Hz). This figure is intended to illustrate the basic spectral relationships in AM. The frequencies in a real ECG are much lower

Note that in the real ECG, the fundamental frequency of the heart rate is to be interpreted as a carrier (near 1 Hz), and the fundamental frequency of respiration (near 0.2 Hz) as a modulation signal. If such an image is available, one can try to demodulate the ECG so that the carrier and the modulation signal are separated. To do this, use the Matlab function *amdemod.m*, which performs amplitude demodulation. Determine the carrier frequency F_c as the fundamental frequency of the heart rate from the spectrum; the sampling rate F_s was 500 sps. Plot the spectrum of the modulation signal and estimate whether it corresponds to the sidebands of the original ECG. Determine whether the examined signal model corresponds to reality.

4.4.1.2 Threshold-Based Feature Identification

Investigate the reliability of fixed and relative level thresholds in detecting R-waves in a real ECG (*ekg.mat*). Simplify by assuming that the ECG has no jumps and compensation processes and that the R-wave has the highest amplitude of all waves. Plot the detected "threshold exceeded" events as a function of the threshold value. Determine the most reliable amplitude range from this representation. Would a relative detection threshold common in technical measurement technology also apply to biosignals? If so, under what conditions?

4.4.1.3 Pan-Tompkins Algorithm

The Pan-Tompkins algorithm is one of the oldest methods for automatically detecting cardiac actions, yet it is one of the most reliable for regular ECG. It can be interpreted as follows (Fig. 4.17): in the band of maximum signal energy of the QRS complex (5–11 Hz), the slope of the edges in a time window is evaluated and converted to the R-spike detector with a discriminator. Investigate whether this detector works independently of the type of ECG lead and what influence the current sampling rate has. For the test, a regular ECG without artifacts and arrhythmias is first used (*uebung_4_4.m*). The Matlab function contains a resampling to 200 sps, which is necessary for the Pan-Tompkins algorithm. Modify the function so it does not resample to see how the initial sampling rate of 500 sps will affect the result.

4.4.1.4 Determination of Local Extremes

Using the example of a saved VEP (*vep.mat*), automatically determine local extrema and their latencies with the help of your program (Matlab function) using the derivatives and the tangent method. Add a trend (settlement process) to the VEP and examine how this affects the parameters to be determined. Limit the analysis to a time range from 0 to 300 ms. One can use Matlab functions (*polyfit.m*) for linear regression for the tangent calculation on the main shaft P100.

4.4.1.5 Determination of the Instantaneous Heart Rate and Its Variability

The instantaneous heart rate (IHR, Instantaneous Heart Rate) theoretically results from the reciprocal instantaneous heart period (Eq. 4.14, $N = 1$). Practically, the problem is that the length of the current heart period is only known at its end, and the heart rate can be calculated retrospectively. Since the heart periods are not of constant length, one obtains a non-equidistant sequence of different values (Fig. 4.24). This sequence must be converted to an equidistant sequence to investigate the heart rate behavior. Examine the linear and cosine interpolation results to determine the heart rate and variability. Use the stored ECG (*ekg.mat*) and Matlab functions (*interp1.m, hrv_cosinusinterpolation.m*) for this purpose. Use an algorithm to detect R-waves with a binary detection sequence as an output signal.

4.4.1.6 Influence of Analysis Windows on Spectral Resolution

As explained in Sect. 4.2.1.2, the theoretical limits of the Fourier transformation are not realizable, so windowing is necessary. About 20 different windows are available by default. They can be roughly classified into non-differentiable windows (rectangle, triangle, Bartlett, Tukey), cosine windows (Hamming, Hann), Gaussian windows (Gauss, Parzen), and others (Chebyshev, Flat Top). Generate a sum of two harmonics with the relative frequencies 0.1 and 0.11 at a length of $N = 100$. The spectral spacing of the two harmonics corresponds precisely to the theoretical resolution of the DFT ($1/N = 0.01$). Investigate what influence different windows have on the theoretically achievable spectral resolution. From each window group mentioned, select the best window that maintains the resolution and has the lowest side maxima. The Matlab functions of the Signal Processing Tool *wintool.m* and *wvtool.m* can be used to calculate and visualize the window functions.

4.4.1.7 Length of the Analysis Window in the Spectrogram

The time resolution depends directly; the frequency resolution is inversely proportional to the window length. The waves of an ECG, especially the QRS complex, are short-time processes, so a short analysis window is required for good time resolution. If, in addition to the ECG, heart rate and its variability are under examination, an excellent frequency resolution (up to 0.01 Hz) is required. Thus, a lengthy analysis window is needed. Investigate which window lengths are suitable for the two measurement tasks. Is there a compromise length that can answer both questions? Use the stored ECG (*ekg.mat*) as the signal and write your own Matlab function to calculate the spectrogram (Eqs. 4.60 and 4.55). For checking, use the Matlab function from *uebung_4_8.m* and the signal processing toolbox *spectrogram.m*.

4.4.1.8 Comparison of Direct and Indirect Spectrogram

The direct path to the spectrogram via the STFT (Eq. 4.60) and the indirect path via the ACF (Eq. 4.69) are theoretically equivalent in the result. In practical signal analysis, especially for signals with high dynamics (nonstationarity) and only a few or single realizations, the different ways can sometimes lead to considerably different results. Compare the two possibilities on one of the most dynamic biosignals—the ECG. For this purpose, use the ECG already analyzed in previous tasks from the Matlab file *ekg.mat*. The lengths of the analysis windows and the FFT must be identical for comparison. The AKF becomes twice as long as the analyzed signal section (Eq. 4.68). One can use the functions of the Matlab toolbox Signal Processing *xcov.m* or *xcorr.m to* calculate the AKF. Do the two ways lead to comparable results?

4.4.1.9 Discrete Wigner-Distributions

The basis of all Wigner-based distributions is the local correlation function (LKF, Eq. 4.76) or the AKF (Eq. 4.70). In the DSV, which works with discrete-time values, a shift by /2 can, of course, only be realized with integers. Therefore, a step size of at least two samples is necessary. Investigate how the doubling of the minimum step size affects the demands from the sampling theorem or the permissible bandwidth of the analyzed signal. Calculate the Wigner distribution or the Wigner-Ville distribution according to Eq. 4.75 for the discrete case. Use a sum of three linear chirps as a test signal; the third chirp results spectrally from the sum of the first two (Fig. 4.62).

4.4.1.10 Removal of Cross-Terms

As already apparent in the solution of the 10th task (Fig. 4.121), cross-terms (interferences) arise in the WVV; they complicate the interpretation in the time–frequency plane. Under certain conditions, the cross terms can be eliminated relatively quickly. To check this, first, generate two chirps in the time–frequency plane to overlap neither in time nor in frequency. Calculate the WVV and plot it. In the middle between the chirps (autoterms), cross terms occur. Transform the WVV into the similarity plane and filter out the autoterms with a suitable mask. Check the effect of the masking by transforming back into the time–frequency plane. Repeat the process for two chirps that overlap in time or frequency (or both).

4.4.1.11 The Harmonic and Spectral Transfer Function of Wavelets

The conventional harmonic (Morlet) and the spectral decomposition (Mexico hat) with wavelets will be investigated in the spectral range. For this purpose, construct a Morlet wavelet with seven periods of harmonic oscillation and a Mexican hat in an appropriately long window ($L \sim 250...1000$). Plot the transfer function of the wavelets in the frequency domain. Note the central frequency (frequency of the maximum) and the spectral width of the filter characteristics. Scale the wavelets twice dyadically by halving the sampling rate (at each halving, every second value of the sequences is omitted). At each halving, represent the filter characteristics with the same FFT length as in the first scaling. Compare the transfer functions depending on the scaling. What is the main difference between the transfer functions of the Morlet wavelet and the Mexican hat?

4.4.1.12 Sinc Wavelets for ECG Analysis

Based on the transfer functions of an ideal high-pass and an ideal low-pass as a half-band filter, calculate their impulse responses of length $L = 1001$. Plot the transfer functions in the frequency domain for the lengths of the impulse response for $L = 11$, 101, and 1001 and compare the quality of the filter function. Load the stored ECG (*ekg.mat*) and display its spectrum (sampling rate 500 sps). Based on the spectrum and the available impulse responses, decide on a length of the impulse response L and a modified sampling rate of the ECG so that the spectrum is captured as completely as possible after a triple dyadic scaling by the filters. Then carry out a function check by synthesis.

4.4.2 Solutions

4.4.2.1 Elimination of Slow Settlement Processes in an ECG

The individual solution steps are described in the Matlab file *uebung_4_1.m*. The calculation of the spectra results in the following picture (Fig. 4.101). From the comparison of the spectra, it follows that the optimal cut-off frequency is in the range of about 0.7–1.2 Hz. It must not be higher so that the ECG is not affected; if it were reduced, the artifact would be insufficiently suppressed.

The result after filtering with an FIR high-pass filter is shown in Fig. 4.102. The settlement process has been eliminated, and the signal shape of the ECG has been preserved. The time shift between the original and the filtered signal is half a filter length $(L - 1)/2 = 302$ or 1208 ms. Such a time shift would be unacceptable for time-critical problems, e.g., for R-wave detection in the pacemaker, but acceptable for conventional recordings.

With an IIR high-pass filter (Fig. 4.103), a short time delay is achieved because of the low filter order, but most IIR filters cause a shape distortion of the biosignal. How strong the shape distortion depends on the real signal spectrum and instationarities. In Fig. 4.103, shape distortions are already recognizable: reduction of the

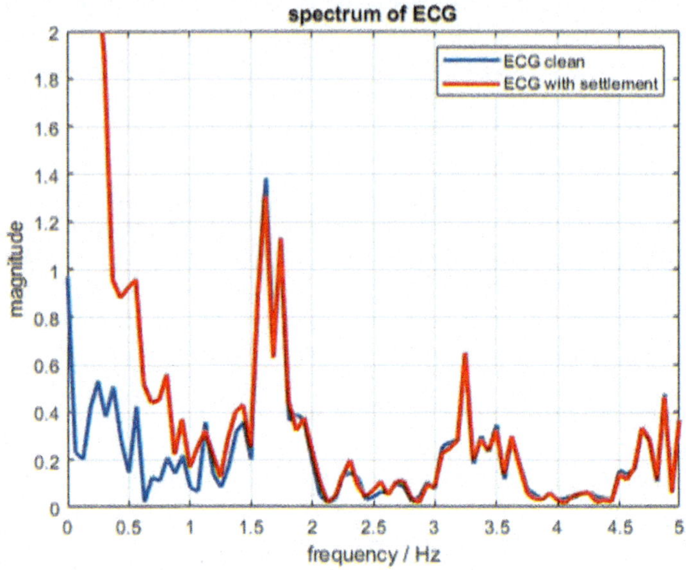

Fig. 4.101 Spectrum of the original ECG and the settlement process. The optimal cut-off frequency of a high-pass filter lies in the range of about 0.7–1.2 Hz

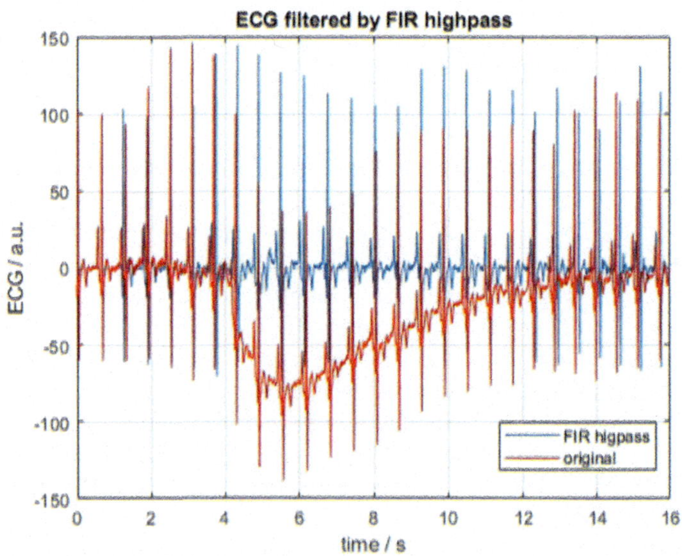

Fig. 4.102 ECG after high-pass filtering with an FIR filter with a cut-off frequency of 1 Hz. There is a time shift between the original and the filtered data due to the filtering (half filter length (L − 1)/2 = 302)

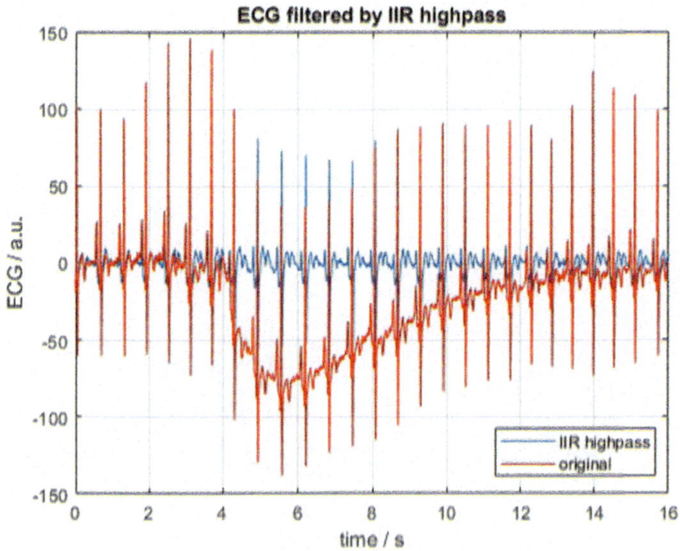

Fig. 4.103 ECG after high-pass filtering with an IIR filter (elliptical, 4th order) with the cut-off frequency of 1 Hz. The time shift due to the filtering is 16 ms (filter order L = 5), which is also acceptable for real-time processing. The signal shape is slightly changed especially its amplitude

amplitude, the shape of the S-wave, and others. Therefore, an IIR high-pass filter is suitable for time-critical tasks but not for recordings where the signal shape must remain undistorted for measurement.

4.4.2.2 Non-linearities in the Biosignal

The spectrum calculated from the stored ECG is shown in Fig. 4.104. The additive component corresponds to the signal model according to Eq. 4.3 can be suppressed with the help of a high-pass filter with a cut-off frequency of about 0.5 Hz. After high-pass filtering, the time course (Fig. 4.105) shows that the suppression of the additive component of the respiration $a(t)$ had no noticeable influence on the variability of the amplitude of the R-spikes. From this, it can be concluded that the often-assumed additive model practically plays a subordinate role. Worsening was the shape distortion because an IIR high-pass filter was used. Since the amplitude of the R-spikes is a diagnostically relevant parameter, one can try the following to free them from amplitude modulation. For this purpose, the Matlab function *amdemod.m is* used, in which the basic frequency of the heart rate $F_c = 1.03$ Hz and the sampling rate Fs = 500 Hz are inserted as parameters.

The spectrum of the modulation signal after amplitude demodulation is shown in Fig. 4.106. The spectrum shows a global maximum at 0.18 Hz, the fundamental frequency of respiration. Further local maxima occur at 0.03 and 0.08 Hz in conformity with the original spectrum in Fig. 4.104. Since these spectral maxima,

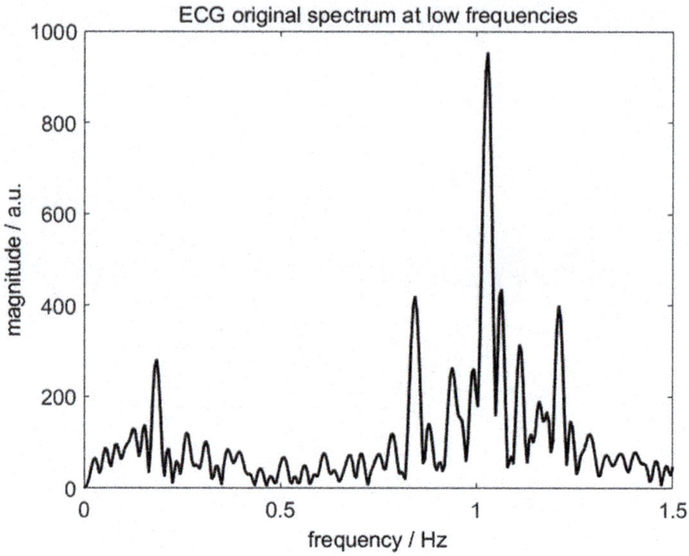

Fig. 4.104 Spectrum of a stored ECG. The respiration shows a spectral needle at 0.18 Hz and, as a sideband after amplitude modulation, is also mirror-symmetrical around the heart rate's fundamental frequency at 1.03 Hz

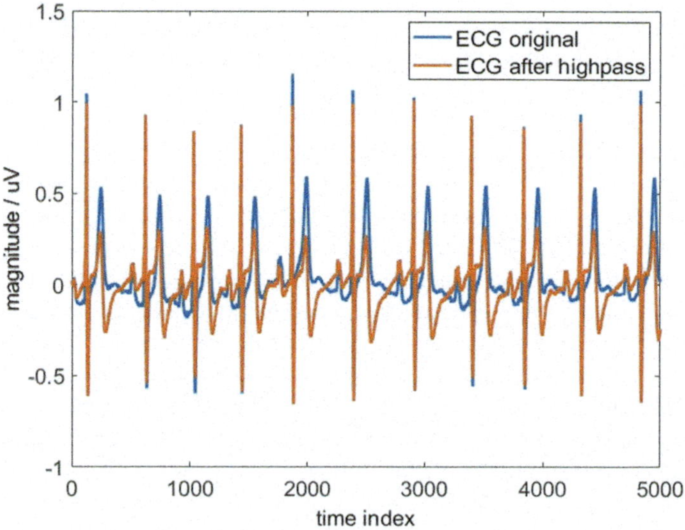

Fig. 4.105 Time course of an ECG after high-pass filtering with a cut-off frequency of 0.5 Hz. The fundamental frequency of respiration is sufficiently suppressed, but the amplitude of the R-spikes hardly changes. Take note of significant changes in the ST stretch, which is diagnostically important but modified after the highpass filtering

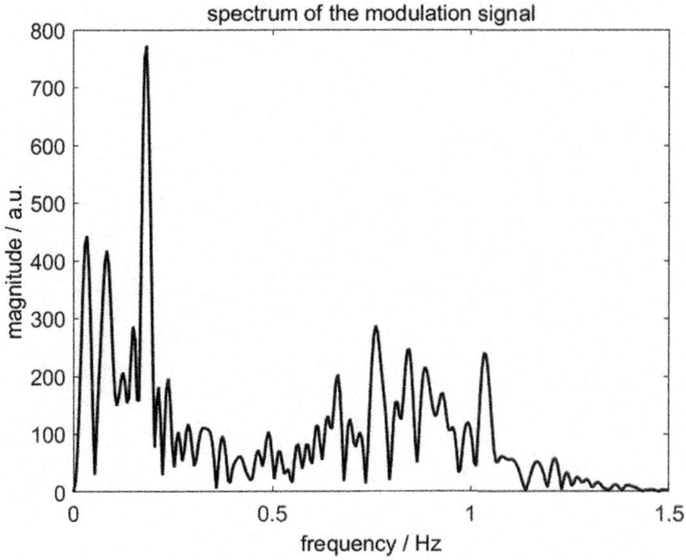

Fig. 4.106 Spectrum of the modulation signal after amplitude demodulation. In addition to the breathing frequency of 0.18 Hz, other components are included at 0.03 Hz (VLF) and 0.08 Hz (LF)

in contrast to the respiratory frequency, are not included as low-frequency components in the original spectrum, it can be assumed that they are solely attributable to the signal model's multiplicative component (modulation) (see Fig. 4.27).

In the spectrum of the modulation signal, one can still observe relatively strong components between 0.7 and 1.2 Hz. No physiological reasons are known for these components, so one must assume they arise as products of further modulation alternatives (phase, frequency modulation). It further follows that the signal model with an additive, multiplicative component and noise is not yet precise enough. As a result of amplitude demodulation, the modulation signal is obtained, as is usual in communications engineering. However, the carrier signal (ECG) would also be interesting, which was implicitly assumed to be harmonic here. Theoretically, two further possibilities exist to obtain both components—the carrier and the modulation signal. On the one hand, via the Hilbert transformation, and on the other hand, with a comb filter. Both possibilities are dealt with in corresponding chapters.

4.4.2.3 Threshold-Based Feature Identification

With the help of the Matlab function *uebung_4_3.m*, the value range of the stored ECG can be examined with large selectable steps between the lowest and the highest level. The lowest threshold, at which false detections disappear, and the highest threshold, at which the first R-spikes are no longer detected, are determined. The lowest threshold (Fig. 4.107) in the concrete example is 0.63 (filtered digitized values), and the highest threshold is 0.8 (Fig. 4.108). Considering the

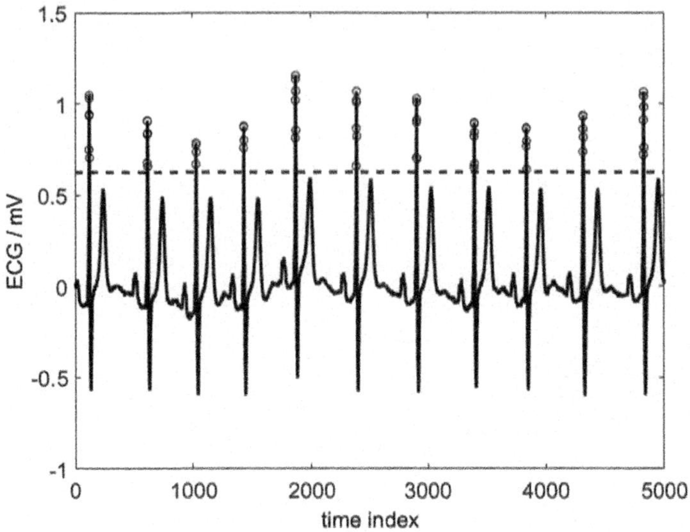

Fig. 4.107 After reaching the lowest detection threshold (dashed) of 0.63, no more T-waves are detected, but all R-spikes are still detected. Circles mark values above the threshold

range of values between − 0.6 and 1.16, the threshold distance of 0.17 for R-spike detection is very tight; it is about 10% of the range. It is very little for reliable detection, as even minor fluctuations in the baseline and slight settlement processes can reach this range. Suppose there are no further requirements regarding risk functions (ratio of false positive to false negative detections). In that case, one will place the discriminant threshold in the middle of the range between the minimum and maximum detection threshold, i.e., at 0.71. Of course, the multiple detections of an R-wave resulting from the sampling rate must be combined into a single detection of cardiac action.

In technical measurement technology, the 10 and 90% thresholds of the range between the baseline and the amplitude are used to detect an edge or wave's start and end times. That is because the lower and upper 10% are usually superimposed by noise, and the start and end phases of the edges are relatively flat. If this principle is applied to the ECG, the following picture emerges (Fig. 4.109). Since there is no baseline for biosignals, the time average can be used. The 10% threshold results from the distance between the mean and the maximum values. It is problematic with the ECG because the P-wave also exceeds this threshold. One could also use the 90% threshold as a criterion to detect the R-waves alone. However, these fluctuate by more than 10% due to respiration. This criterion would therefore be unsuitable without further preprocessing. This criterion is not applicable for biosignals that, in contrast to the ECG, do not show any phases of inactivity, e.g., the EEG or the pulse curve.

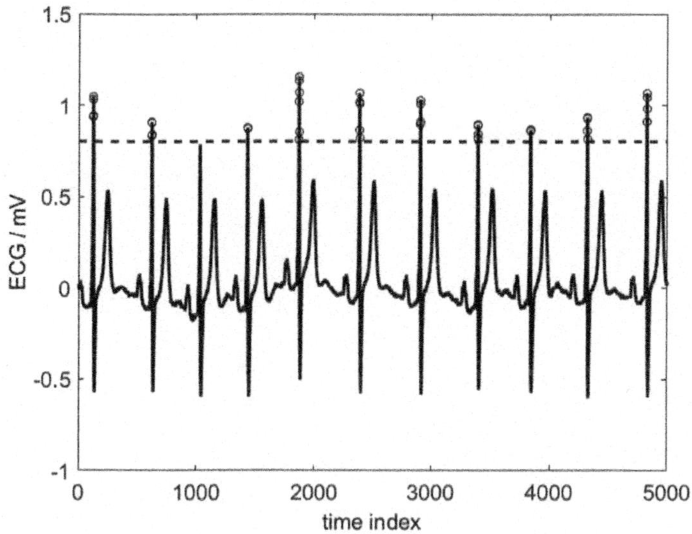

Fig. 4.108 After falling below the upper detection threshold (dashed) of 0.8, the first R-prongs are no longer detected (third heart action)

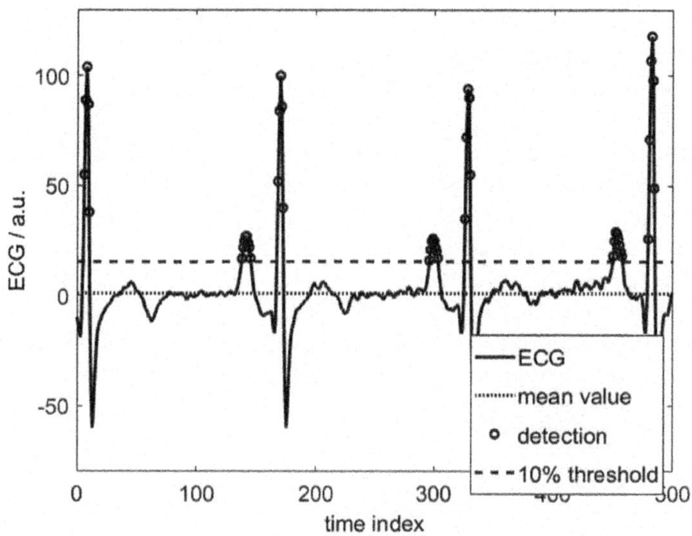

Fig. 4.109 Detections in the ECG after introducing a 10% threshold

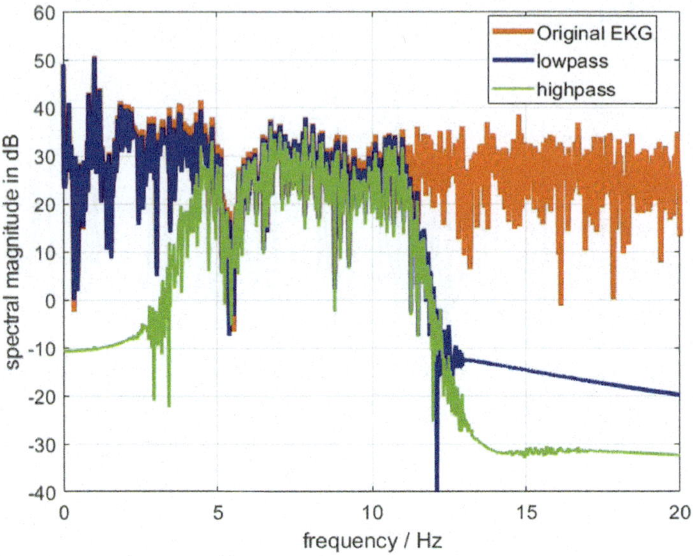

Fig. 4.110 Spectra of the ECG after steps of the Pan-Tompkins algorithm: original ECG (dotted), after low-pass filtering (dashed), and subsequent high-pass filtering (full line)

4.4.2.4 Pan-Tompkins Algorithm

The Pan-Tompkins algorithm is implemented in the Matlab function *uebung_4_4.m*. With the help of a high pass at 5 Hz and a low pass at 11 Hz, the ECG is limited to the spectral range of the maximum signal energy (Figs. 4.110 and 4.111)

Band-pass filtering is followed by band-limited differentiation and squaring and a smoothing filter in the form of a moving average. It should be noted that the differentiator and the moving average are tuned to a sampling rate of 200 sps. Therefore, a resampling to 200 sps is necessary before the algorithm. Alternatively, the two operations can be reconfigured to the current sampling rate. For practical use, the output of the moving average (MA) can be followed by a maximum detector that can accurately determine the timing of the QRS complex in real-time.

4.4.2.5 Determination of Local Extremes

Theoretically, one can search for local extremes by looking for points on the time axis where the first derivative is zero-valent and recognizing whether it is a local maximum or minimum according to the polarity of the second derivative. In practical analysis with digital values, a zero-value first derivative is an exception (Fig. 4.112), so a search must first be done for a zero crossing of the first derivative. Once this is identified, the associated second derivative gives the type of the local extreme (Fig. 4.113).

Of course, this is also the case with noise, so local extremes of the signal and the noise cannot be distinguished. An additional criterion must be introduced, e.g.,

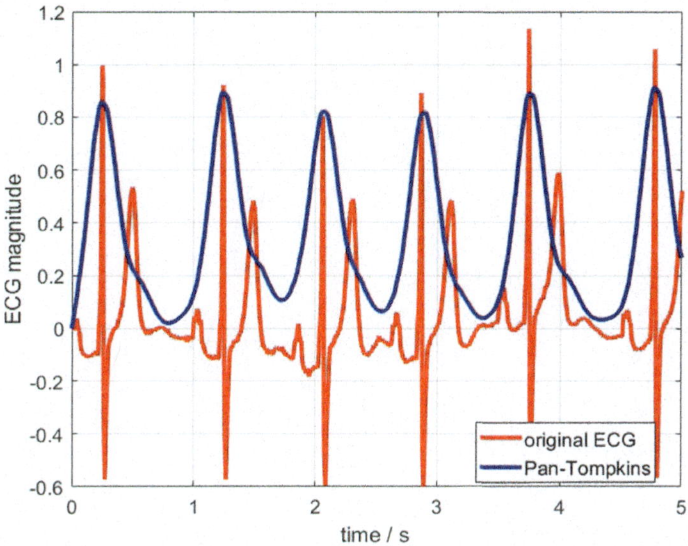

Fig. 4.111 Original ECG (dashed) and the output of the Pan-Tompkins algorithm (full line). The maximum of the algorithm's output signal must be determined for binary R-spike detection

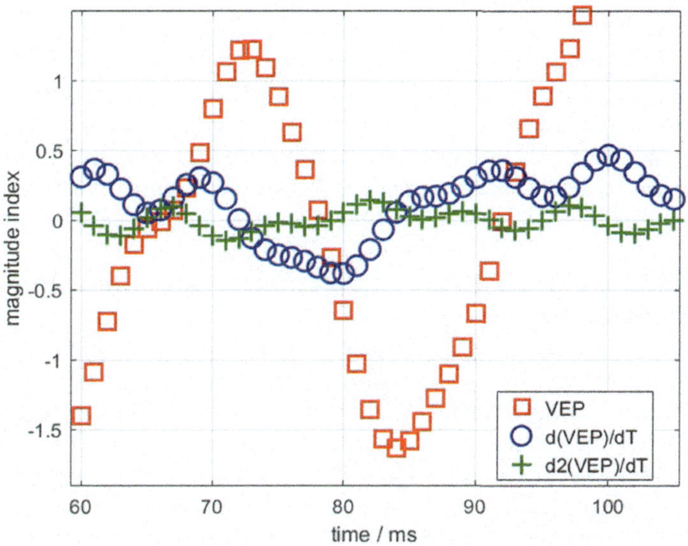

Fig. 4.112 Two waves of a VEP (P70, N85), their first and second difference. An extreme value is present at the zero crossing of the 1st derivative; at the negative 2nd derivative, a local maximum; and at the positive, a local minimum

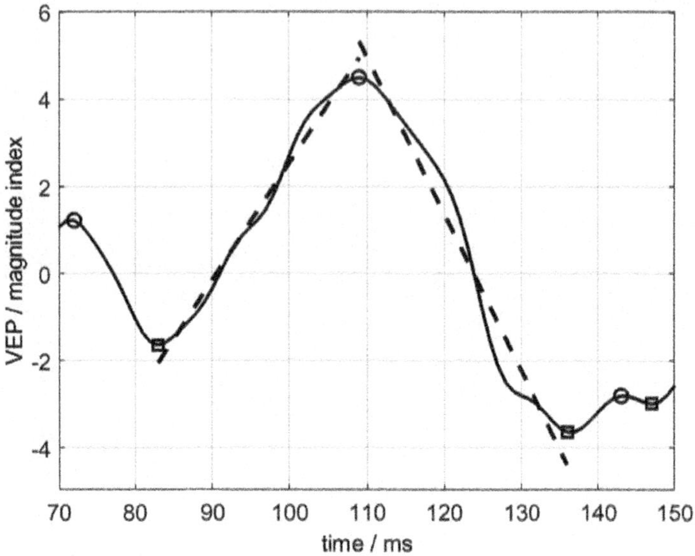

Fig. 4.113 Tangents on the main shaft of the VEP. Local maximums are marked with circles and minimums with rectangles

an amplitude discriminator for the noise (*uebung_4_5.m*). With the help of the tangent method, the extrema can be determined with the same reliability as with derivatives. The point of intersection of the tangents gives the times when the extreme occurs and their value. However, this only applies to undisturbed signals. If the two methods are compared based on disturbed signal courses (compensation process), it can be determined that the tangent method is more robust (see *exercise_4_5.m*).

4.4.2.6 Determination of the Instantaneous Heart Rate and Its Variability

To determine the heart rate, it is first necessary to determine the time points of the heart actions (R-points). One can use the Pan-Tompkins algorithm described here, the Matlab function *MaximumSuche.m*, or one's program. A binary sequence (zero sequences with ones at the positions of the detected heart actions) or an index sequence (indices of the time points of detected heart actions) is required as the output signal. An evaluation shown in Fig. 4.22 serves as a starting point.

The result of the linear interpolation (between the discrete heart rate values) is shown in Fig. 4.114. In this heart rate time, the dominant respiratory rhythm can be seen (period of about 5 s). The spectral analysis of the HR curve will show the modulation signals but not the HR fundamental frequency. That is because the linear interpolation has only captured the envelope of the heart rate, which is equivalent to AM demodulation. The heart rate spectrum can be displayed after cosine interpolation (Fig. 4.27). The HR transfer after cosine interpolation into the

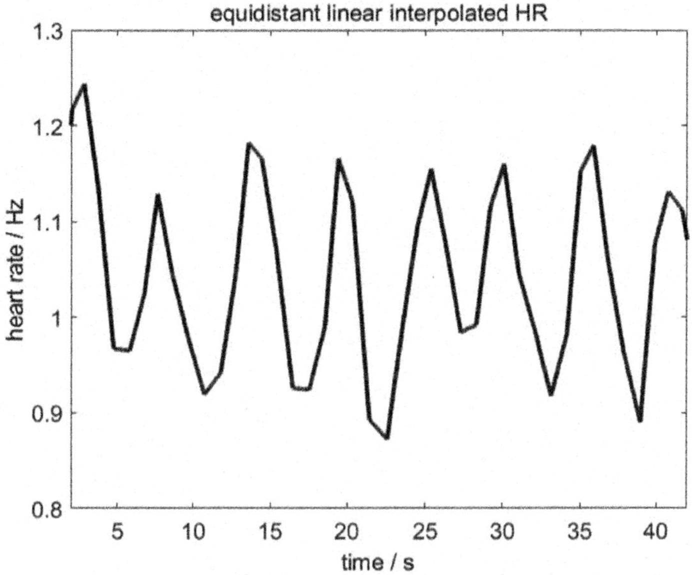

Fig. 4.114 Linear interpolation to determine the instantaneous heart rate. The interpolated values result from the non-equidistantly determined heart rate according to Fig. 4.24

baseband is necessary to examine the informative value of the two methods for the bands of HRV. Since the amplitude is constant after cosine interpolation, but the frequency changes, the interpolated signal is treated with the help of FM demodulation (Matlab function *fmdemod.m*). That is also indirect proof that the additive and multiplicative signal model is not sufficiently accurate; a frequency-changing component must be implemented as a supplement. The linear and cosine interpolation evaluations concerning the spectra are shown in Fig. 4.115. Except for the low-frequency component of the linear interpolation, the spectra are mainly identical. However, cosine interpolation offers the possibility of a direct comparison to the heart rate's fundamental frequency, so it is better suited for the analysis. In addition, cosine interpolation (in the total spectrum) shows considerably less interference, noise, and edge effects than linear interpolation.

4.4.2.7 Influence of Analysis Windows on Spectral Resolution

The desired window functions can be generated for the given length using their definition formulas or comfortably with the help of the Matlab toolbox Signal Processing (wintool, wvtool). After multiplying the window function by the sum of the harmonics, one obtains the amplitude spectrum of the respective harmonics with the help of the DFT (FFT) the amplitude spectrum of the respective windowing is obtained. The spectrum provides information about the resolution, the width of the main maxima (main lobe) at the relative frequencies 0.1 and 0.11,

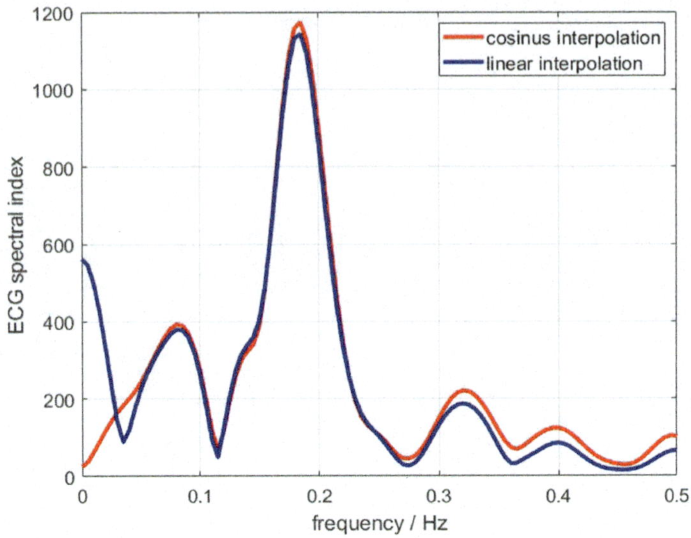

Fig. 4.115 Spectrum after linear and cosine interpolation of the heart rate

and the distance of the unwanted side maxima (side lobes). The essential window functions for these criteria are shown in Fig. 4.116: The rectangular window has the best resolution with the narrowest main maxima (as desired) but with the strongest (unwanted) side maxima. The Chebyshev window has the most vigorous suppression of the side maximum and the broadest main maximum at the same resolution. The Gaussian window shows a compromise of both extremes (as well as all others). In practical analysis, which window function is optimal for the analysis depends on the concrete question. A compromise must always be found between the main lobes' width and the side lobes' suppression. Particular care must be taken with real spectra, as a broad main lobe can cause necessary harmonic spectral needles to merge.

It can lead to an initially discrete multi-component needle spectrum being converted to a continuous curve, resulting in an incorrect interpretation.

4.4.2.8 Length of the Analysis Window in the Spectrogram

It is known that the length of ECG waves is 100 ms (QRS) to 200 ms (T-wave), which gives the necessary time resolution. The total spectrum is first evaluated to estimate the necessary spectral resolution (Fig. 4.117). Although the normal ECG has maximum energy in the range of 5–11 Hz, the heart rate with its sidebands and its harmonics dominate the spectrum, so the range of up to 3 Hz is sufficient for these purposes. The spectrum clearly shows the fundamental respiratory frequency of 0.2 Hz, also contained in the sidebands of the heart rate. This results in the following requirement for HRV analysis, a window length of at least 5 s is required. This requirement is in marked contrast to the necessary time resolution.

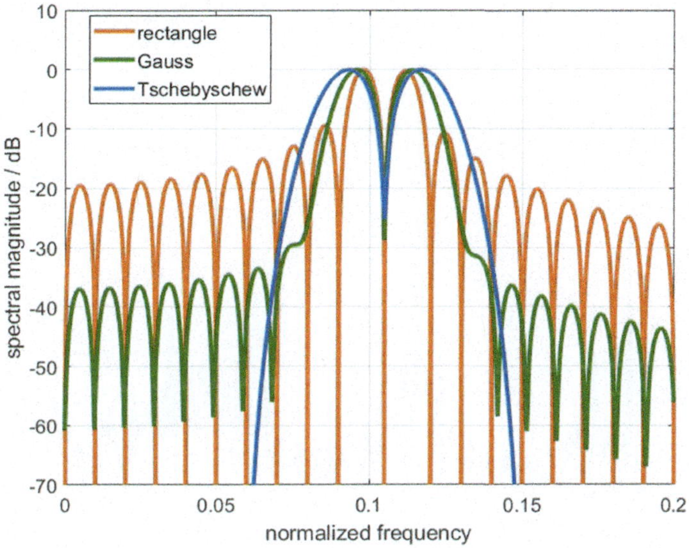

Fig. 4.116 Spectrum of three window functions (rectangle, Gauss, Chebyshev)

Figure 4.118 shows the spectrogram with a window length of 10 s and a frequency resolution of 0.05 Hz. All spectral components identified in the total spectrum are also contained in the spectrogram. In addition, an apparent periodicity of the fundamental frequency of the heart rate can be seen in the spectrogram (respiratory period).

That is direct evidence that the heart rate is amplitude-modulated and frequency-modulated (see cosine interpolation).

The analysis window must be significantly shortened to display the dynamic spectrum of cardiac actions. Figure 4.119 shows the spectrogram of the ECG analyzed so far with a window length of 200 ms. Here the time resolution is sufficiently good for dynamic analysis, but the spectral resolution is no longer sufficient for an HRV analysis (Fig. 4.118). Comparing the spectrograms shows that a compromise for a good spectral and temporal resolution is impossible. Therefore, one must orient oneself according to the primary question when choosing a suitable window length.

4.4.2.9 Comparison of Direct and Indirect Spectrograms

The comparison of the direct method for calculating the spectrogram (Eq. 4.60, Fig. 4.119) and the indirect method (Eq. 4.69, Fig. 4.120) shows that with the indirect method, better signal-to-noise ratio (image contrast) is achievable. Further, a better representation of the existing spectral width and a partly better time resolution (with identical window length) can be achieved.

These are typical properties of indirect methods in biosignal analysis, especially for transient and noisy or disturbed signals.

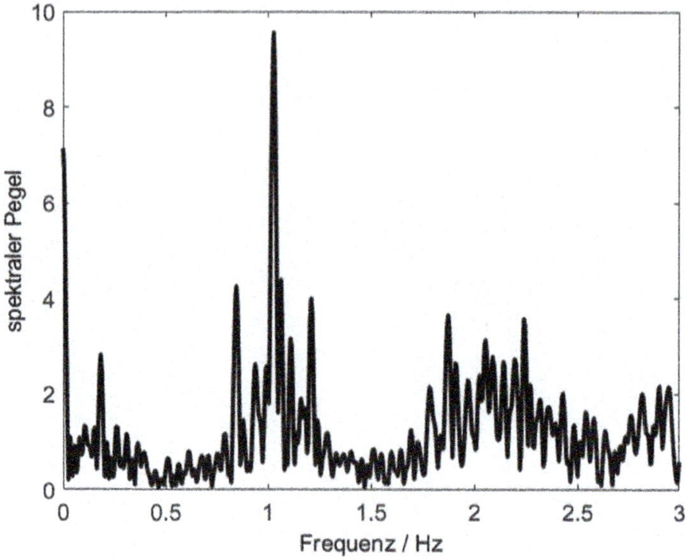

Fig. 4.117 Spectrum of an ECG recording of length 42 s. Note the fundamental frequency of the heart rate (1.025 Hz) with its sidebands and first harmonics. The spectral needle at 0.2 Hz comes from respiration

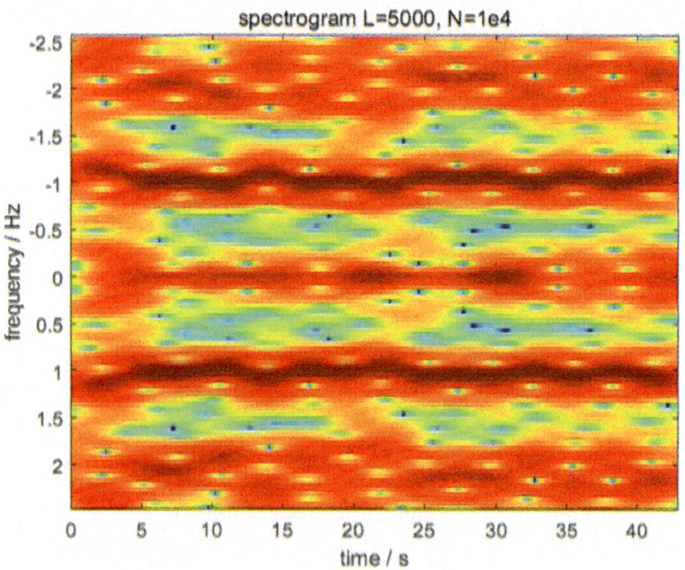

Fig. 4.118 Spectrogram of an ECG using a window length of 5000 samples or 10 s and an FFT length of 10^4 at a sampling rate of 500 sps

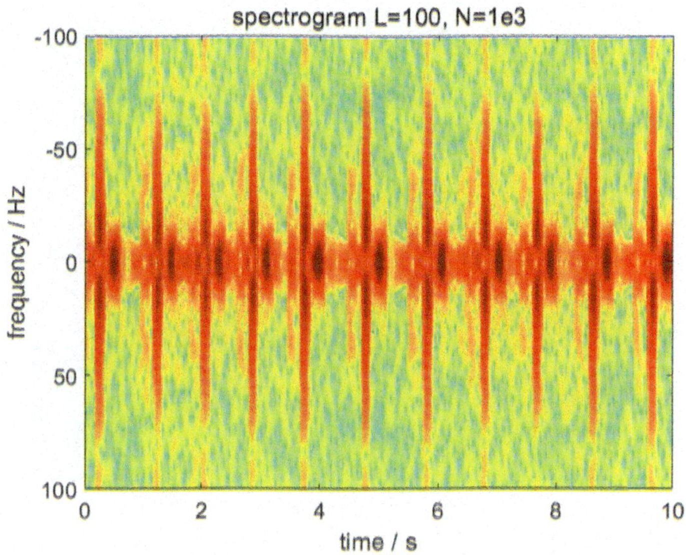

Fig. 4.119 Spectrogram of an ECG (Figs. 4.117 and 4.118) with a window length of 100 samples or 200 ms and an FFT length of 10^3 at a sampling rate of 500 sps

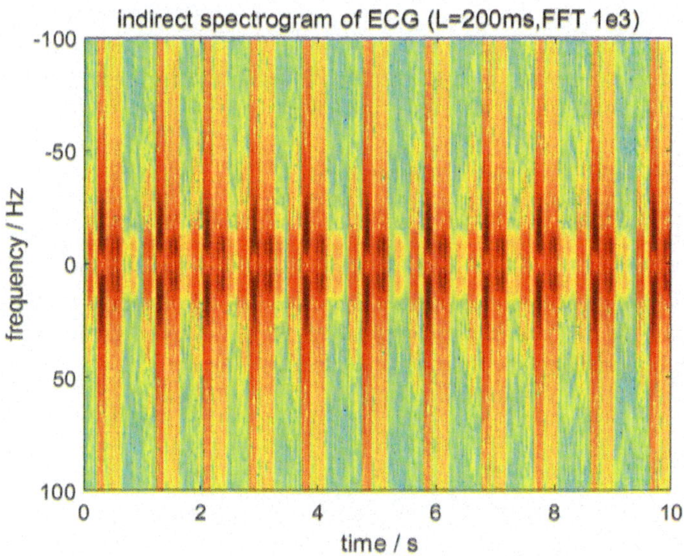

Fig. 4.120 Indirect spectrogram (Eq. 4.69) with otherwise identical parameters as in the direct spectrogram (Fig. 4.119)

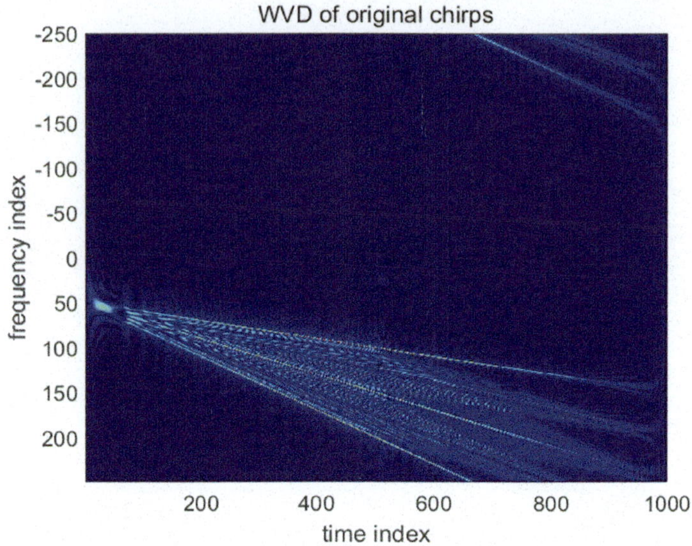

Fig. 4.121 Wigner-Ville distribution of the sum of three linear chirps. The modified sampling theorem was not observed; therefore, aliasing occurs (in the graph top right)

The explanation for the better picture (Fig. 4.120) is the smoothing effect of temporal integration when calculating the ASF in the indirect method. Although the DFT is also a temporal integration, this linearly transforms the time signal into the frequency domain, i.e., without a smoothing effect.

4.4.2.10 Discrete Wigner-Distributions

To implement the formulae according to Eqs. 4.70 and 4.75 in the discrete-time range, doubling the minimum step size for the time shift is necessary. In the general case, according to Eq. 4.72, even multiplication of the minimum step size may be necessary, e.g., with the Wigner-bispectrum. Multiplication of the step size corresponds to a de facto undersampling of the analyzed signal. This results in modifying the sampling theorem for the analysis (Eq. 4.123).

$$T_A \leq \frac{1}{2Q \cdot f_{\max}} \tag{4.123}$$

In Eq. 4.123, $f_{\max}$ is the highest frequency contained in the signal, Q is the multiplication factor to apply Eq. 4.72, T_{A} is the maximum sampling period.

Usually, the biosignals are already so limited in their bandwidth that a further band reduction in the sense of the modified sampling theorem, according to Eq. 4.89, is not possible without impairing the information content of the biosignal. However, a subsequent oversampling is also possible in already stored signals. A separate function or the Matlab function *resample.m* can be used for oversampling (*uebung_4_10.m*). With the double oversampled signal (Fig. 4.122), the spectral

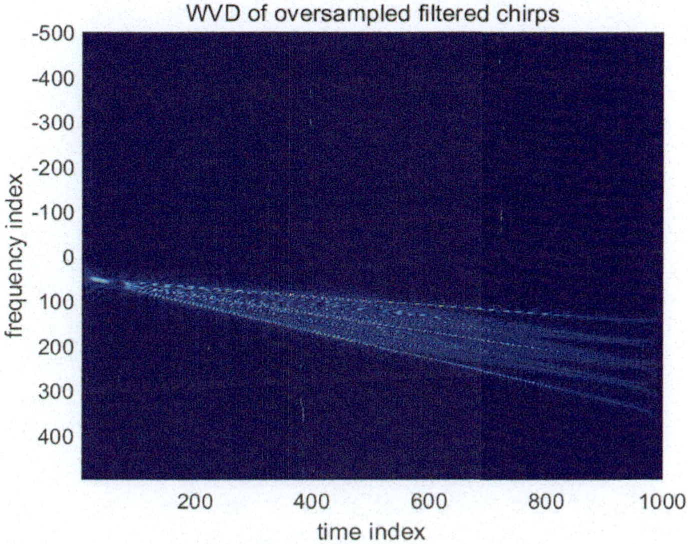

Fig. 4.122 Wigner-Ville distribution of the sum of three linear chirps. The modified sampling theorem was observed after the original signal was double-oversampled and filtered

bandwidth is also doubled, and in the sense of the normalized frequency, there is a compression of the original spectrum. The subsequent undersampling as part of the calculation of the WVV no longer leads to aliasing as with the original signal (Fig. 4.121).

4.4.2.11 Removal of Cross-Terms

Ideally, autoterms and cross terms in the similarity plane can be separated with a two-dimensional filter function (Kernelmask) to separate them. Figure 4.123 shows the similarity function of two linear chirps with a Gaussian window, which overlaps neither time nor frequency. In this simple representation, the autoterms can be filtered out with a filter in the form of a binary rectangular mask (– 50:50, – 150:150) (*exercise_4_11.m*). If the chirps overlap in time, the cross terms in the similarity plane are also arranged accordingly around the time shift (Fig. 4.124).

To filter out the autoterms, a corresponding filter function (skew ellipse, skew rectangle) would be necessary. This example shows that there cannot be a generally valid filter function (kernel); it must be adapted to the current signal character.

4.4.2.12 The Harmonic and Spectral Filter Function of Wavelets

The solution path is described in detail in *uebung_4_12.m*. While calculating the Morlet-wavelet and the Mexican hat should not cause any significant problems, the following must be considered when calculating the spectra and their representation. With dyadic scaling, one spreads the spectrum twice the width after halving

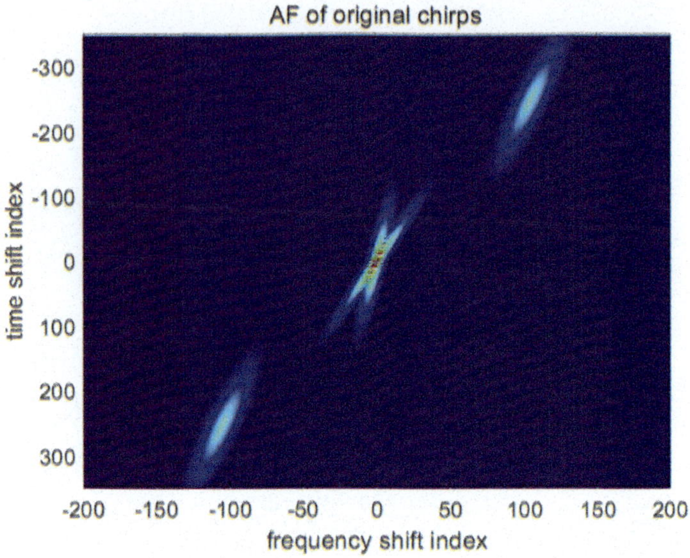

Fig. 4.123 Similarity function (AF, ambiguity function) of two linear chirps with Gaussian windows that overlap neither in time nor frequency. The autoterms are arranged in the center, the cross terms on the outside

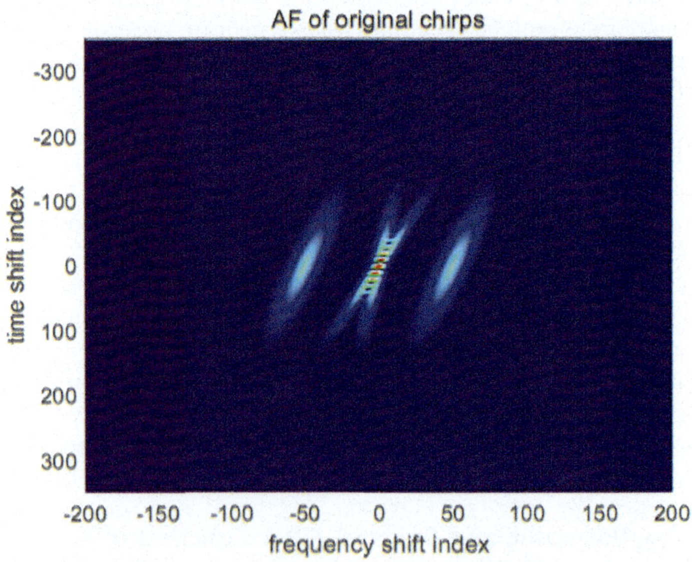

Fig. 4.124 Similarity function (AF, ambiguity function) of two linear chirps with Gaussian windows that overlap entirely in time. The autoterms are arranged in the center, the cross terms on the outside symmetrically to the zero-value time shift

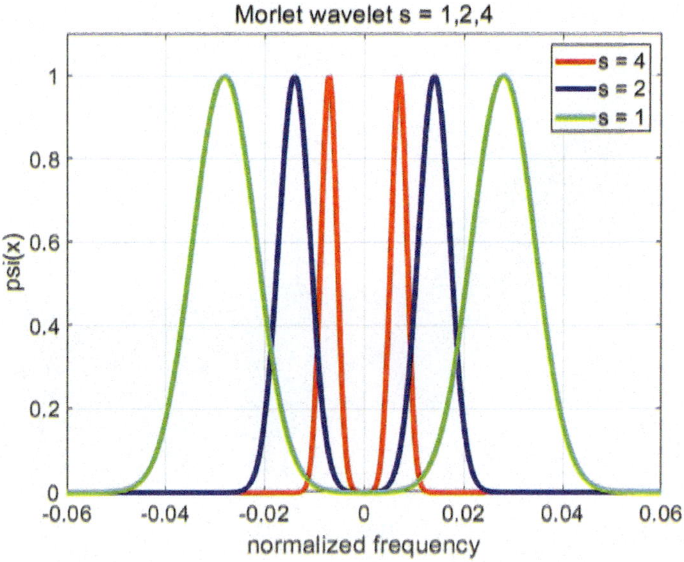

Fig. 4.125 Transfer function of the Morletwavelet in the frequency domain for the scalings $s = 1, 2, 4$. The length of the FFT was $N_{\mathrm{FFT}} = 10^4$. The filters have identical coefficients and transfer functions at their current sampling rate

the sampling rate. In reverse order, one obtains so-called half-band filters. However, each halving has identical filter characteristics and filter coefficients. It is an excellent advantage for filter applications because only one calculation of the coefficients (Morlet or Mexican Hat) is necessary. However, why are the spectra the same after each halving? It is because the filter coefficients have not changed.

All spectra to the original signal must be related to show the effect of the bisection. The only way to do this is to calculate all spectra with the same length of the FFT, which should be considerably longer than the original signal length to be safe. In this concrete example, $N_{\mathrm{FFT}} = 10^4$ (Fig. 4.125).

4.4.2.13 Sinc Wavelets for ECG Analysis

The ideal curves of the half-band filters for the low-pass and the high-pass are defined as square-wave functions with a cut-off frequency of 0.25 on the relative frequency scale (Figs. 4.85 and 4.86). From this ideal curve, impulse response coefficients can be calculated via the IFFT for the desired length (Fig. 4.84). For details see *uebung_4_13.m*. From the courses of the transfer functions in the frequency range, it can be concluded that the longest impulse response with $L = 1001$ comes closest to the ideal course, as expected. However, a long impulse response also means a poor time resolution. The original spectrum of the ECG is shown in Fig. 4.126.

If one analyses the spectrum of the measured ECG, one can already draw some conclusions. A filter with a low cut-off frequency (about 20–30 Hz) has been

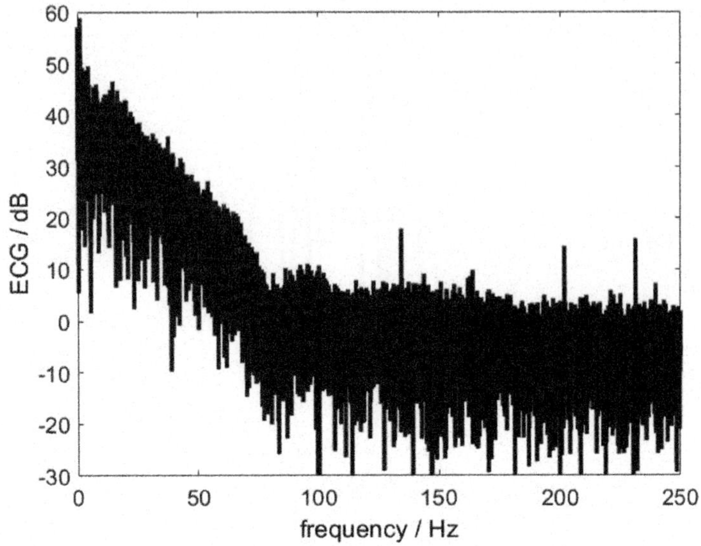

Fig. 4.126 Spectrum of an ECG section of length 45 s, sampling rate 500 sps

used, with an attenuation of 40 dB at 50 Hz. It leads to the conclusion that the spectral components above 50 Hz contribute nothing to the information except noise. Therefore the sampling rate can be reduced to a fifth (100 sps) (the Nyquist frequency corresponds to 50 Hz). That makes sense because the undersampling spreads the information-carrying part of the spectrum over the entire baseband (0…50 Hz). The half-band filters can already be applied to the total bandwidth. Then the question arises about a suitable length of the impulse responses of the half-band filters. The impulse response with the length $L = 1001$ comes close to the ideal filter but is unsuitable for real-time processing. With a length of the impulse response of $L = 11$, spectral filters are obtained that can be called low-pass and high-pass filters and that are not worse than usual wavelets. However, the task here is a harmonic decomposition or a sinc wavelet construction; therefore, this short impulse response is unacceptable. A compromise is offered by the length of $L = 101$, with which the ECG was filtered. The results are shown in Fig. 4.127. The gradation of the lengths has been done on purpose in decades of tens so that the orders of magnitude become visible. In practice, one can choose one's length as a compromise solution.

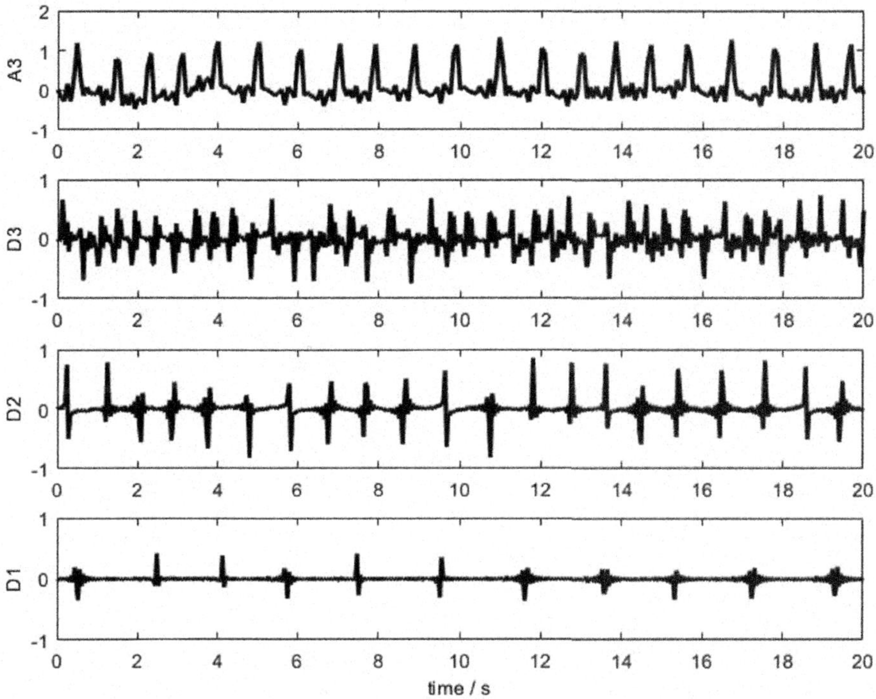

Fig. 4.127 Decomposition of an ECG with Sinc wavelets as half-band filters in three stages at a sampling rate of 100 sps. The band ranges are accordingly: D1 (25 Hz … 50 Hz), D2 (12.5 Hz … 25 Hz), D3 (6.25 Hz … 12.5 Hz), and A3 (0 Hz … 6.25 Hz)

References

Baraniuk, R., & Jones, D. (1994, January). A signal-dependent time-frequency representation: fast algorithm for optimal kernel design. *IEEE Transactions on Signal Processing*, 134–146.

Berger, R., Akselrod, S., Gordon, D., & Cohen, R. (1986, September). An efficient algorithm for spectral analysis of heart rate variability. *IEEE Transactions on Biomedical Engineering*, 900–904.

Bergh, J., Ekstedt, F., & Lindberg, M. (1999). *Wavelets with applications in signal and image processing*. Springer.

Boashash, B. (2003). *Time frequency signal analysis and processing*. Elsevier.

Debnath, L. (2002). *Wavelet transforms and their applications*. Birkhäuser.

Henning, G., Hoenecke, O., Husar, P., & Schellhorn, K. (1996, March). Time-frequency analysis in objective perimetry. *Applied Signal Processing*, 95–103.

Pan, J., & Tompkins, W. J. (1985, March). A real-time QRS detection algorithm. *IEEE Transactions on Biomedical Engineering*, 230–236.

Digital Filtering 5

5.1 Introduction to Digital Filtering

In biosignal processing, digital filters play a central role. On the one hand, their importance has increased due to the long-term progress of digitalization. On the other hand, with digital filters, it is possible to realize many special functions that do not exist with analog filters. Section 2.3 deals with the basics of the theory of analog filters and their design and realization. It becomes clear that only a limited number of functions can be realized in the analog world. In addition to the conventional spectral filters, such as low pass, high pass, band pass, and band stop, some other functions, such as differentiator or integrator, can be interpreted as filters. Thus, the range of analog filtering can be limited to the spectrum. However, this area will retain its functional significance as long as signal processing is necessary before digitization. Furthermore, since it is not foreseeable when and if there will be sensors with a direct digital output, analog signal conditioning must always precede a digital filter. After digitalization, however, a library of digital functions and algorithms opens up that can hardly be overlooked. Here, the term "filter" loses its original meaning of a specific spectral function. Filters are also used, for example, to describe blocks for transformation (e.g., Hilbert filters) or phase correctors (e.g., all-pass filters) that have no spectral characteristic at all. Now the question arises, what leads to such an enormous increase in possibilities in the digital world? Through temporal discretization (see Sect. 3.2), analog signals become sequences of values that can also be interpreted as vectors or matrices. In this way, one loses the reference to the analog time axis; the knowledge of the

---

Supplementary Information The online version contains supplementary material available at https://doi.org/10.1007/978-3-662-67998-2_5.

P. Husar and G. Gašpar, *Electrical Biosignals in Biomedical Engineering*,
https://doi.org/10.1007/978-3-662-67998-2_5

sampling period is only necessary for the correct scaling of representations or signal reconstruction. Moreover, since, in principle, every mathematical formulation can be discretized with the help of numerical mathematics and converted into a matrix form, it can also be realized in discrete time. For the sake of completeness, however, it must also be noted that mathematical formulas could also be realized in principle in the analog world (e.g., with analog computers in the 1970s), but at a barely acceptable cost. The transition from continuous to discrete time is made possible by the *Z-transformation.* The *Z-transformation* of a signal sequence $x(n)$ results from (Eq. 5.1):

$$X(z) = Z\{x(n)\} = \sum_{n=-\infty}^{\infty} x(n) \cdot z^{-n} \tag{5.1}$$

In Eq. 5.1, z is a complex variable.

Several conditions must be met for calculating the Z-transformation (ZT), particularly the convergence condition (Ingle & Proakis, 2007). Then one can establish a direct relation to the FT, where z lies on the unit circle (Eq. 5.2).

$$\left.\begin{array}{l} |z| = 1 \\ z = e^{j\omega} \end{array}\right\} \rightarrow X(z) = X(j\omega) = \mathrm{F}\{x(nT_A)\} = \sum x(nT_A) \cdot e^{-j\omega nT_A} \tag{5.2}$$

The ZT has the following essential properties:

Linearity (compliance with the superposition principle)

$$ax(n) + by(n) \overset{Z}{\longleftrightarrow} aX(z) + bY(z) \tag{5.3}$$

Timeshift

$$x(n-N) \overset{Z}{\longleftrightarrow} z^{-N} X(z) \tag{5.4}$$

Frequency shift (multiplication with an exponential series)

$$z_0^n x(n) \overset{Z}{\longleftrightarrow} X\left(\frac{z}{z_0}\right) \tag{5.5}$$

Time mirror

$$x(-n) \overset{Z}{\longleftrightarrow} X\left(\frac{1}{z}\right) \tag{5.6}$$

Folding

$$x(n) * y(n) \overset{Z}{\longleftrightarrow} X(z)Y(z) \tag{5.7}$$

Complex conjugation

$$x^*(n) \overset{Z}{\longleftrightarrow} X^*(z^*) \tag{5.8}$$

There is a long series of methods and algorithms for designing digital filters. Selected methods are presented here as examples to illustrate the primary procedure. The relevant literature is recommended for studying further possibilities.

5.2 LTI-Systems: FIR and IIR

The classification between FIR (finite impulse response) and IIR (infinite impulse response) is based on the actual properties of the digital filters. Theoretically, all impulse responses in the analog domain are infinite. FIR filters have a finite impulse response, which is necessary for practical processing because they do not use recursion. Impulse responses of IIR filters are also practically infinitely long because they are based on recursion.

5.2.1 Introduction to Impulse Response and Filter Structure

Filter classification can be carried out in digital filtering according to various characteristics. Here, the characteristics of *realizable filter structure* and *impulse response length are* dealt with first. According to the filter structure, a distinction is made between filters with feedback—*recursive filters*—and filters without feedback—*non-recursive* or *transversal filters. Filter.* Figure 5.1 shows both filter structures from the system analysis point of view in the continuous-time range. That makes it clear that system stability must also be considered, especially in the case of recursive filters. It is known from system analysis that a weight function (impulse response) is generally infinitely long since it describes dynamic processes characterized by the e-function.

However, generating a finite impulse response is often necessary, especially from a practical point of view. For this purpose, one can use the possibility of windowing, especially in signal processing. These explanations show that the length of an impulse response (finite or infinite) is not dependent on the filter structure. With both structures, both types of impulse response can be realized in the analog range. In practical digital signal processing, however, using transversal filters to realize impulse responses of finite length makes sense, and recursive filters to realize impulse responses of infinite length. This division results from pragmatic considerations: A recursion is an infinitely repeatable process, so it is also suitable for realizing infinite impulse responses. A transversal filter has no feedback, thus has only a limited temporal length available, and is, therefore, suitable for realizing finite impulse responses. In the following, two filters will be discussed—filters with infinite impulse response realizable with recursive filters and filters with finite impulse response realizable with transversal filters.

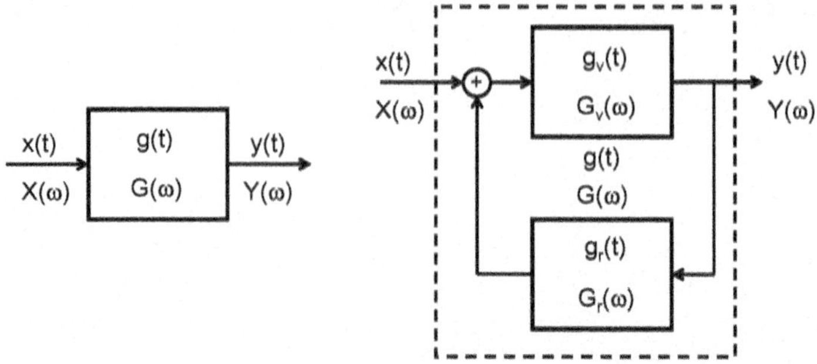

Fig. 5.1 Filter structures without feedback (left) for the realization of transversal and with feedback (right) for the realization of recursive filters. It should be noted that recursion in the discrete-time domain (Fig. 5.3) corresponds system-analytically to a feedback system (right)

5.2.2 Infinite Impulse Response Filter, IIR

5.2.2.1 Analog and Digital Filtering—Pulse Invariant Technique

For a continuous-time filter (system) with time-constant system parameters (LTI—Linear Time Invariant), the following applies for the input–output relationship and the impulse response (system function, Eq. 5.9)

$$y(t) = x(t) * g(t), \tag{5.9}$$

and in the frequency range (Eq. 5.10)

$$Y(j\omega) = X(j\omega) \cdot G(j\omega). \tag{5.10}$$

In Eqs. 5.9 and 5.10, x(t) or X() is the input signal, y(t) or Y() is the output signal and g(t) or G() is the impulse response or transfer function of the filter.

The transition into the discrete-time range and a design method for digital filters will be explained using the example of a low-pass filter (Fig. 5.1) with the time constant = RC = 0.2 s. The impulse response of the low pass is an exponential function according to Eq. 5.11, which is obtained from the reverse transformation of the transfer function in the frequency domain.

$$G(j\omega) = \frac{1}{1 + j\omega\tau} \xrightarrow{IFT} g(t) = \frac{1}{\tau} e^{-\frac{t}{\tau}} \tag{5.11}$$

For the transition into the discrete-time range, the impulse response is sampled like an analog signal (in compliance with the sampling theorem). Analogous to sampling any signal, the sampled sequence can be formulated as a Dirac pulse series, here as an exponential sequence with $p = \exp(-T/)$ because of the exponential course (Eq. 5.12).

$$g(t) = p_0 \Big(\delta(t) + p\delta(t - T) + p^2\delta(t - 2T) + \cdots + p^N \delta(t - NT) \Big) \tag{5.12}$$

If one transfers the impulse response of a low-pass filter according to Fig. 5.2 into the frequency range, one obtains, starting from the equation (Eq. 5.13) for the impulse response, the expression according to Eq. 5.14.

$$F(\delta(t-T)) = e^{-j\omega T} \tag{5.13}$$

$$G(j\omega) = p_0\left(1 + pe^{-j\omega T} + p^2 e^{-j\omega 2T} + \cdots + p^N e^{-j\omega NT}\right) \tag{5.14}$$

Corresponding to the formula for the sum of a geometric sequence, Eq. 5.14 can also be written as follows for a large N (Eq. 5.15):

$$G(j\omega) = p_0 \frac{1}{1 - pe^{-j\omega T}}. \tag{5.15}$$

Assuming that all prerequisites for the Z-transformation are fulfilled, one can write according to the relationship according to Eq. 5.2 for the transfer function of the discrete-time filter in the z-domain (Eq. 5.16)

$$G(z) = p_0 \frac{1}{1 - pz^{-1}}. \tag{5.16}$$

In digital signal processing, the formula Eq. 5.16 can already be implemented algorithmically or in terms of circuitry. For practical implementation, z^{-1} means a delay of one sampling or clock period. Thus, the low-pass, according to Eq. 5.11 (Fig. 5.2), can be directly transferred into the algorithm of a simple recursion formula (Eq. 5.17) or a circuit (Fig. 5.4, digital circuits, FPGA; Fig. 5.3).

$$\begin{aligned} y_0 &= p_0 x_0 \\ y_{n+1} &= p_0 x_{n+1} + p y_n \end{aligned} \tag{5.17}$$

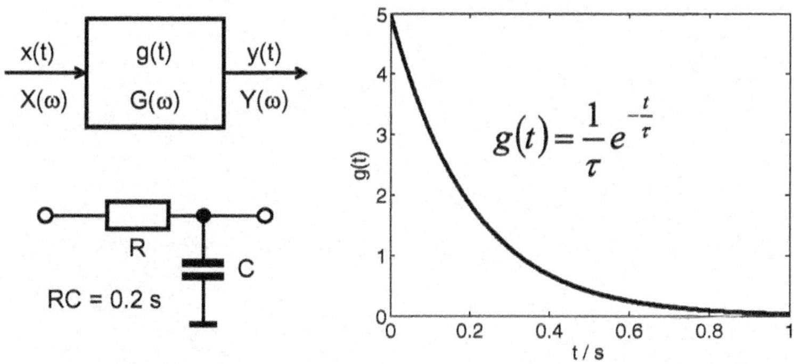

Fig. 5.2 An analog filter as an LTI-system (linear time invariant). Example of a 1st order low pass filter and its impulse response (weight function)

The question arises about how to apply the previous relationships to an arbitrary, generally formulated filter. Without proof, one can assume that every impulse response can be represented as a sum of infinitely many e-functions. Therefore, according to Eq. 5.16, the relationship can be generalized for $N + 1$ summands (Eq. 5.18).

$$G(z) = \sum_{i=0}^{N} \frac{k_i}{\left(1 - p_i z^{-1}\right)} \tag{5.18}$$

The formulation in the z-range with products or quotients is necessary. According to Eq. 5.18, the sum leads to a common denominator and is formulated as a quotient of two polynomials. Here the quotient is equal to the ratio of the output signal to the input signal (Eq. 5.19).

$$G(z) = \frac{Y(z)}{X(z)} = \frac{\sum_{i=0}^{M} b_i z^{-i}}{1 + \sum_{j=1}^{N} a_j z^{-j}} \tag{5.19}$$

The larger of the two orders of the polynomials M and N is called the order of the filter. Through the inverse z-transformation of the relationship according to

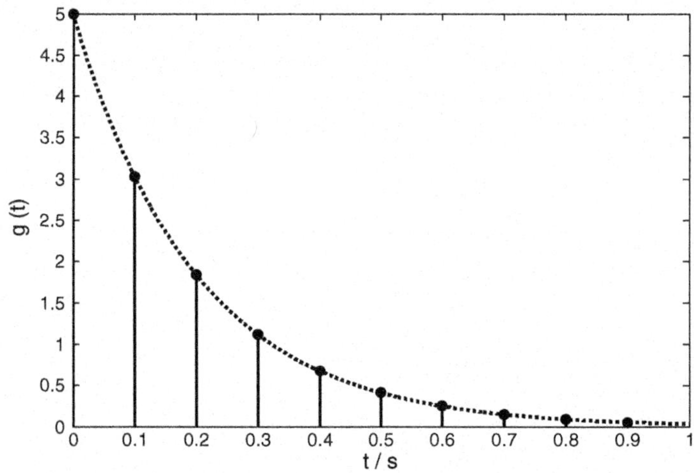

Fig. 5.3 Sampled impulse response of the low pass from Fig. 5.1. The sampling rate is 10 sps. According to Eq. 5.14, the parameters for this low pass are $p_0 = 5$ and $p = \exp(-T/\tau) = 0.61$

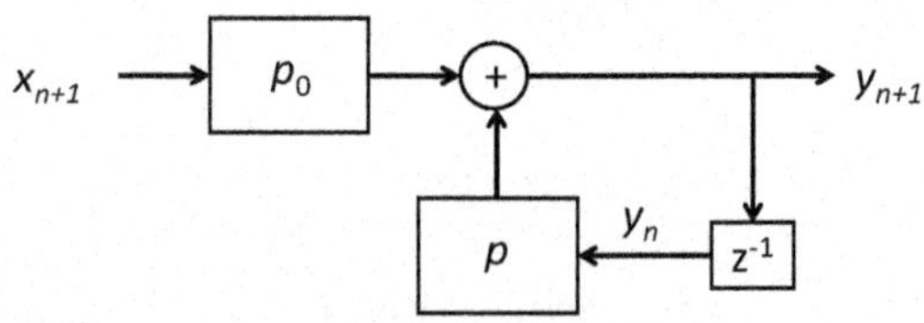

Fig. 5.4 Algorithmic or circuit realization of a low pass according to Eqs. 5.16 and 5.17. The block with z^{-1} represents the delay by one sample/clock period

Eq. 5.19 and conversion, one can express the current output value of a filter as follows (Eq. 5.20):

$$y(n) = \sum_{i=0}^{N} b_i x(n-i) - \sum_{j=1}^{M} a_j y(n-j). \tag{5.20}$$

According to Eq. 5.20, the recursion formula provides the filter coefficients b_i and a_j and a prescription for the realization of arbitrary filters, which can be described in the continuous-time domain with the help of the transfer function $G(j)$.

It follows that every filter that exists in the continuous-time domain also has a discrete-time equivalent. It is an important finding from the filter theory and practice perspective: a direct functional equivalence exists between the analog and the digital domain. However, digital filters differ from analog filters in their periodization in the frequency domain.

Since with the method presented here, the impulse response remains identical (invariant) both in the time-analog and in the time-discrete domain and is merely sampled during the transition from the analog to the discrete domain; this method is called *Impulse invariant Technique*.

5.2.2.2 Recursive Filters

In principle, for analog filters, indeed, their impulse responses are theoretically infinitely long. Since digital IIR filters have a recursive component (Eq. 5.20), the discrete-time impulse response can also become infinite. Because of the recursion, such filters are also called *recursive filters*. The direct structure of a recursive filter is shown in Fig. 5.5. This structure is explicit but unsuitable for realization. For realization with a DSP or an FPGA, the direct canonical form (Fig. 5.6) is better suited since it can be realized effectively in the shift register (DSP) or by clocking (FPGA) and is very fast in data processing. Recursion is a feedback loop in the sense of system analysis, so it must be investigated from the point of view of both system and numerical stability and constructed with corresponding safety.

While the system analysis uses proven tools to calculate the parameters (filter coefficients) for stability (e.g., stability criterion, according to Nyquist), ensuring numerical stability is problematic. It can hardly be put into formulas or recommendations because it depends decisively on the concrete temporal and value discretizations and the values of the calculated filter coefficients. When designing recursive digital filters, it is not uncommon for the calculated filter to be system-analytically stable (according to stability criteria) but diverge in its (quantized) numerical variant. Practice teaches that numerical stability is best achieved with relatively few coefficients of the feedback (low order of the denominator polynomial in Eq. 5.19) (see exercises for the chapter).

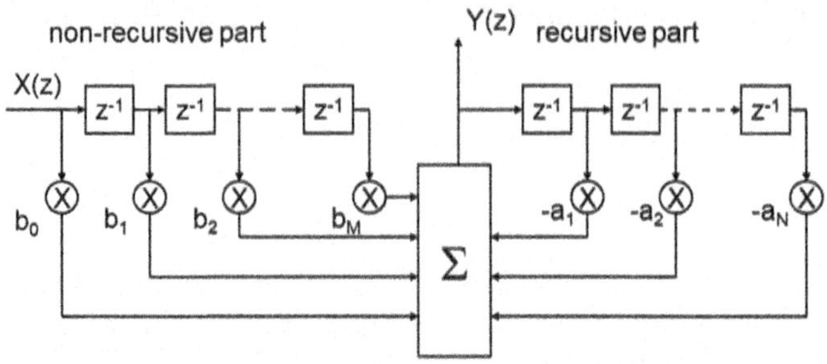

Fig. 5.5 Direct structure for the realization of a recursive filter according to Eq. 5.20

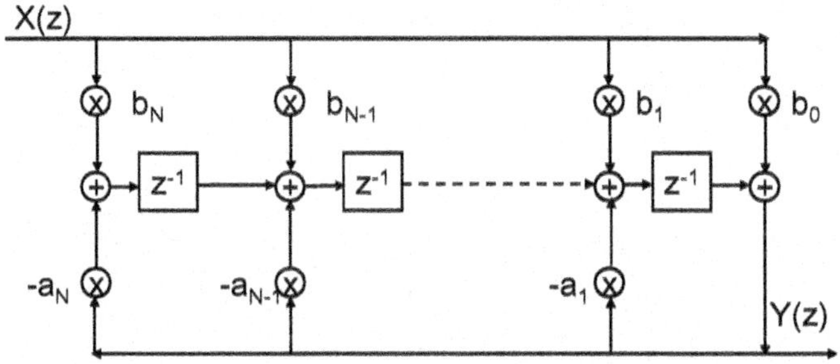

Fig. 5.6 Canonical direct form of a recursive filter according to Eq. 5.20. Compared to the direct structure according to Fig. 5.5, one saves half of the delay elements and a global summation. This structure suits realizations with shift registers (DSP) and FPGA (parallelizable)

Recursive IIR filters are relatively short (Fig. 5.6), i.e., they get by with only a few coefficients or low orders of the polynomials (Eq. 5.19). In practice, the orders are in the range N, $M = 4\ldots20$. This makes them well suited for real-time processing with the abovementioned limitations regarding stability. However, the phase frequency response of the filters is generally non-linear. For biosignals, however, the linearity of the phase is a fundamental requirement, as already discussed in Sect. 2.3.3. A bandstop with a center frequency of 50 Hz was constructed with the same objective as in the mentioned chapter. The bandstop's amplitude and phase frequency response is shown in Fig. 5.7. The amplitude frequency response fulfills the requirements almost ideally (bandstop between 45 and 55 Hz with 60 dB attenuation).

However, the phase frequency response is an order of magnitude worse compared to the analog stopband (Fig. 2.68). Even if the requirement for a linear phase is limited to the passband of the amplitude-frequency response alone, the IIR filter delivers an unacceptable result. The bandstop was tested with a square wave signal; the filtered signal is shown in Fig. 5.8. Here, the typical behavior of the IIR filters are visible: Since the phase drops sharply near the cut-off frequency, there is a correspondingly long delay of the high frequencies.

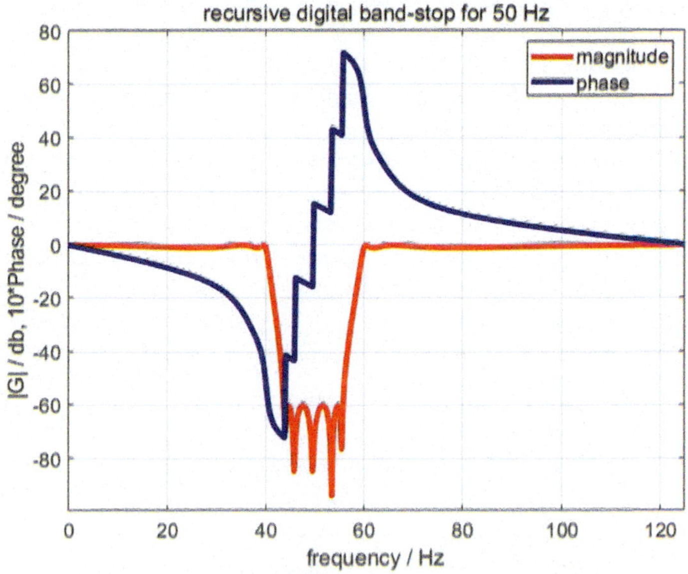

Fig. 5.7 A recursive digital bandstop's amplitude and phase frequency response between 45 and 55 Hz. Filter type IIR, elliptical, 10th order. The phase jumps result from the 2π periodicities

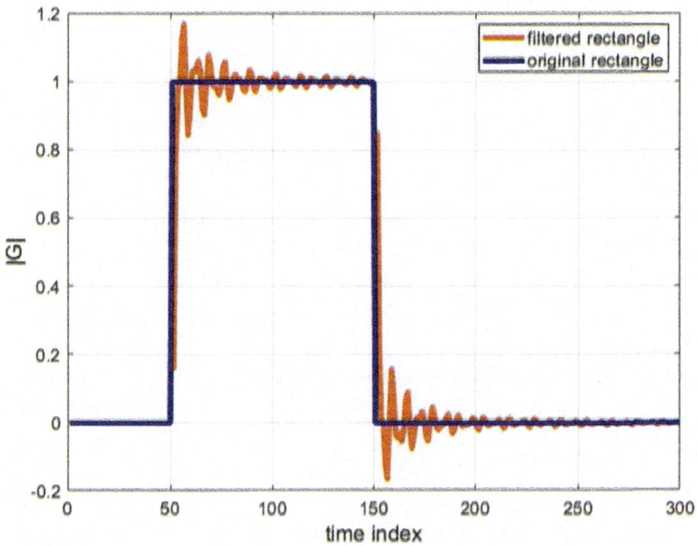

Fig. 5.8 Time course of a filtered rectangle. The digital filter is the IIR bandstop, according to Fig. 5.7

As a result, impulsive, jumpy, and otherwise high-frequency signal components slip backward on the time axis and significantly impair the signal shape. There are two ways to correct the unfavorable phase frequency response. Firstly, it can be linearized with an inverse phase frequency response. It is done by constructing an all-pass filter, which has the function of a phase shifter. It can be realized with an FIR filter (see next chapter). The second possibility can be used for offline processing. The recorded signal is filtered as usual in the direction of the time axis and then again in the opposite direction (mirrored time axis). According to the time mirroring Eq. 4.21, the filter's phase is altogether canceled, and a *zero-phase* filter is obtained (Eq. 5.21). The zero phase leads to the fact that the filtered and the original signal do not show any shift in the time domain. The amplitude-frequency response must be multiplied accordingly or added to the logarithmic representation.

$$\begin{aligned}
&y(n) = x(n) * h(n) \overset{DFT}{\longleftrightarrow} Y(j\omega) = X(j\omega)H(j\omega) \\
&z(n) = y(n) * h(-n) \overset{DFT}{\longleftrightarrow} Z(j\omega) = Y(j\omega)H(-j\omega) \\
&\varphi_Z(j\omega) = \varphi_X(j\omega) + \varphi_H(j\omega) - \varphi_H(j\omega) = \varphi_X(j\omega) \\
&|Z(j\omega)| = |Y(j\omega)||H(j\omega)| = |X(j\omega)||H(j\omega)|^2
\end{aligned} \tag{5.21}$$

In Eq. 5.21, * stands for the convolution of two vectors representing the sampled signal and the impulse response of the filter, n is the time index of the signal sequence, h(n) is the impulse response of a filter, $x(n)$ is the input signal, $y(n)$ is the output signal of the filter after the first filtering, $z(n)$ is the output signal of the filter after the time-mirrored filtering.

For the zero phase, it does not matter whether the signal $y(n)$ or the impulse response of the filter $h(n)$ are mirrored on the time axis. It must be considered by a possibly necessary additional mirroring of the output signal $z(n)$. In practical signal processing, one proceeds as follows:

1. Filtering of the input signal $y(n) = x(n) * h(n)$
2. Filtering of the mirrored output signal $z(-n) = y(-n) * h(n)$
3. Mirroring of the output signal $z(n) := z(-n)$

This algorithm is mathematically equivalent to Eq. 5.21, but for practical reasons, it is easier to mirror the filtered signal than the filter's impulse response (see exercises).

5.2.3 Finite Impulse Response Filter, FIR

5.2.3.1 Basics

The introduction explains that FIR filters are usually realized with a transversal structure. A transversal filter is obtained from a recursion filter by omitting the feedback (Fig. 5.1) or the recursion (Fig. 5.5), resulting in a simplified structure shown in Fig. 5.9. Accordingly, Eq. 5.19 simplifies the relationship according to Eq. 5.22.

$$G(z) = \frac{Y(z)}{X(z)} = \sum_{i=0}^{M} b_i z^{-i} \tag{5.22}$$

After the z-back transformation, Eq. 5.23 results for the discrete-time domain to simplify the relationship according to Eq. 5.20.

$$y(n) = \sum_{i=0}^{N} b_i x(n-i) \tag{5.23}$$

If the formulation, according to Eq. 5.23, is transferred into the continuous-time domain (Eq. 5.24), it becomes clear that the filter coefficients are identical to the sampled impulse response.

$$g(t) = b_0\delta(t) + b_1\delta(t-T) + \cdots + b_N\delta(t-NT) \tag{5.24}$$

From the relations according to Eqs. 5.22–5.24, the following consequence results for the construction of transversal filters:

> The filter coefficients of a transversal filter are identical to the impulse response samples. On the one hand, one can construct a transversal filter directly from an impulse response. On the other hand—and this is the essential advantage of FIR filters—one can calculate an impulse response for any function in the frequency domain, which does not exist in the continuous-time domain.

The following example shows the realization of a transversal filter based on a desired spectral function. An ideal low pass is to be realized digitally. The transfer function and the impulse response of an ideal low pass are shown in Fig. 5.10. The impulse response corresponds to the sinc function, which is theoretically infinitely

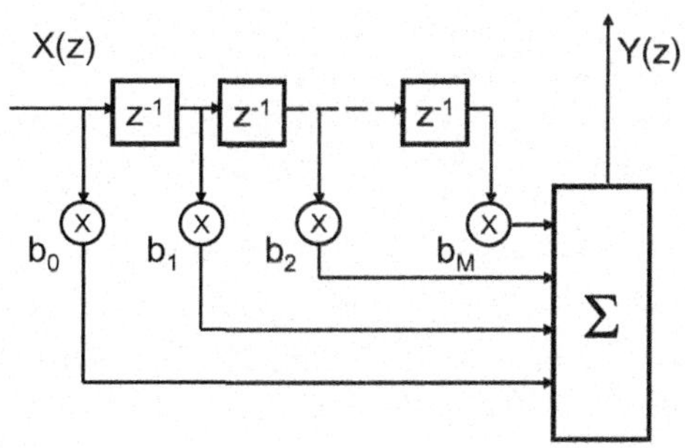

Fig. 5.9 Structure for the realization of a transversal FIR filter

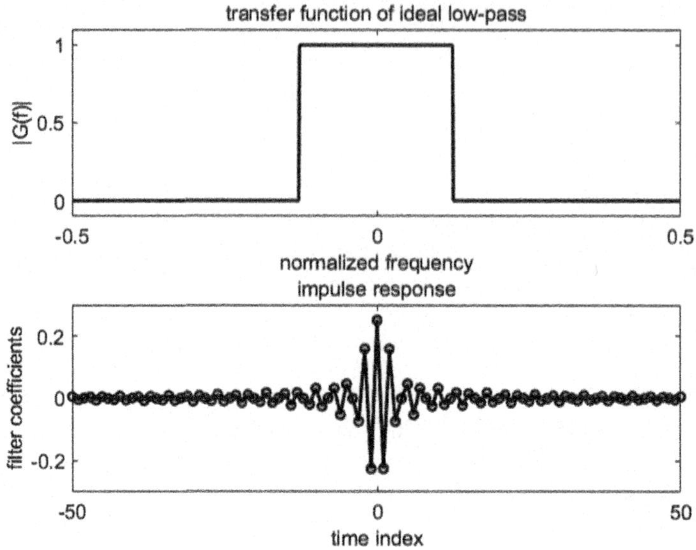

Fig. 5.10 Transfer function of an ideal low pass with the relative cut-off frequency of 0.125 (top) and its impulse response in the form of the sinc function (bottom, zoomed to 101 of 1001 coefficients)

long. It is necessary to realize its limitation with a finite time window, e.g., a rectangle, a triangle, the cosine, and similar. An example of windowing the approximately ideal impulse response ($N = 1001$) with a rectangle of length $N = 11$ and $N = 101$ is shown in Fig. 5.11. The curves of the amplitude-frequency response show that with the decreasing size of the impulse response, the transfer function moves further and further away from the ideal. This example demonstrates a fundamental property of transversal filters:

> With the increasing order of a transversal filter, one approaches the ideal course of a filter function.

In practice, however, ideal filters are not necessary. Usually, a filter is required to comply with specified limits or tolerance ranges of attenuation and frequency, as already explained in Sect. 2.3.1 (Fig. 2.55). Almost any function can be realized with an FIR filter, whereby the limiting parameter is usually the filter length or the filter order. For details on filter design, please refer to the relevant literature.

5.2.3.2 Causality of FIR Filters

The previous formulae (Eqs. 5.22–5.24) and filter structures (Fig. 5.9) assumed that all the data to be processed was available. It became apparent in the example of designing an ideal low-pass filter (Fig. 5.10) that the impulse response on the

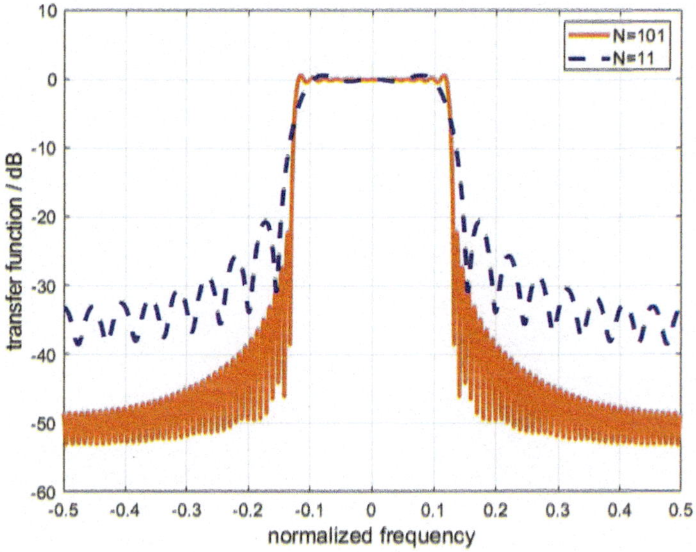

Fig. 5.11 Transfer functions of two realizations of a digital low pass whose impulse response ($N = 1001$) was windowed with a rectangle of length $N = 101$ and $N = 11$

time axis is zero-symmetrical (even). That means that half the impulse response is in the negative time domain. In a practical application, one would therefore have to have $N/2$ digitized signal values from the future, which would amount to a safe prediction. Of course, this is not feasible, so measures to ensure causality must follow. One possible measure could be calculating the impulse response to only differ from zero for non-negative times. It is theoretically feasible but very time-consuming and signal-analytically highly unfavorable. The temporally zero-symmetrical impulse response has important properties: Since it is an even function, its Fourier coefficients are real (special functions, however, can have complex coefficients, e.g., Hilbert filters). For causal filtering of signals, it must be shifted by half the filter length into the positive time domain (Fig. 5.12).

Practically, this means that before the first value appears at the output of an FIR filter, $(N-1)/2$ values must first enter its input (N is the odd filter order). The same effect occurs at the end of the input signal: Because $(N-1)/2$ values from the "future" are needed, the output signal ends half a filter length earlier (Fig. 5.13). It follows that the filtered signal is $N-1$ values shorter than the original. It is impractical for the analysis because the signal lengths would change constantly, and uncontrolled time shifts can occur (index 1 of the output signal in Fig. 5.12 occurs half the filter length later than index 1 of the input signal at index 3 of the input signal). This deficiency can be remedied by extending the input signal in advance to the left and right with zeros corresponding to half the filter length so that the output signal has the same length as the original input signal.

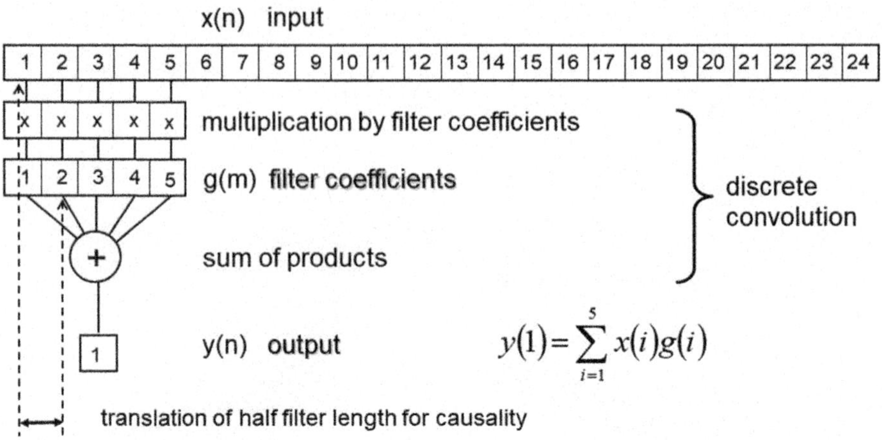

Fig. 5.12 Shifting the filter coefficients (here, their indices) by half a filter length to establish causality. With an odd filter length, the shift is precisely $(N-1)/2$ sampling periods (N is the filter order)

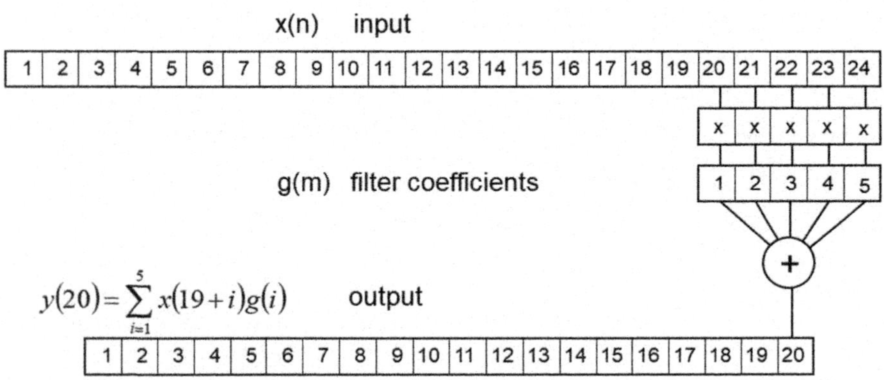

Fig. 5.13 Due to the discrete convolution, the output signal is shorter by $N-1$ values for odd filter lengths (N is the filter order, and shown are the indices)

However, it must be taken into account that even after this, the time shift of the output signal by half a filter length remains.

5.2.3.3 Phase Frequency Response of FIR Filters

The previous section shows that the time shift between an FIR filter's output and input signals with a temporally zero-symmetrical impulse response is half the filter length. This fact is independent of the frequencies contained in the signal, so the group delay is constant and corresponds to half the filter length (Eq. 5.25).

$$d(\omega) = \frac{N-1}{2} T_A \tag{5.25}$$

The phase frequency response then applies accordingly (Eq. 2.26):

$$\varphi(\omega) = -\int_{-\infty}^{\omega} d(\omega) d\omega = -\frac{N-1}{2} T_A \omega \tag{5.26}$$

In Eqs. 5.25 and 5.26, N is the odd filter length, and T_A is the sampling period, the angular frequency.

The relationship, according to Eq. 5.26, shows that the phase frequency response of an FIR filter with a temporally zero-symmetrical impulse response is always *linear*. It applies independently of the amplitude-frequency response. This fact represents the most significant advantage over IIR filters. However, FIR filters have no function equivalents in time-analog signal processing (analog electronics), as with IIR filters. One must also reckon with the fact that the filter lengths of FIR filters—especially with high demands on the amplitude-frequency response—can become two to four decades larger than the orders of comparable IIR filters. That can become a severe problem, especially in real-time processing (pacemakers, neurosurgery).

As already mentioned, FIR filters can be used to realize almost any function that can be converted into matrix form using methods of numerical mathematics. For the phase frequency response, the possibility of constructing an all-pass or a phase shifter is particularly interesting. The following construction specifications apply to the all-pass filter (Eq. 5.27):

$$G(j\omega) = |G(j\omega)| \cdot e^{j\varphi(\omega)} \quad |G(j\omega)| = 1 \quad \varphi(\omega) = f(\omega) \tag{5.27}$$

According to the relationship in Eq. 5.27, the goal is to construct an FIR filter whose amplitude frequency response equals 1 and has a given phase frequency response $\phi(\omega)$. If the all-pass is to be used to correct the non-linear phase of an IIR filter, the inverted phase of the IIR filter is used as the requirement. Figure 5.14 shows an example of phase correction.

The phase frequency response of the bandstop shown in Fig. 5.7 is to be corrected. The impulse response of the bandstop (Fig. 5.14, top left) is causal. Still, the group delay (top right) is strongly decreasing in the passband near the cut-off frequencies (The group delay peaks in the bandstop area are caused by numerical rounding and do not exist in reality). For the construction of the phase shifter, the phase of the bandstop was inverted, and according to Eq. 5.27, the impulse response of the all-pass (center left) as well as the group delay (center right) was calculated (Eq. 5.28). In this representation, half the filter length of all-pass was not yet considered; therefore, the group delay is positive. The bandstop and the all-pass are connected in series, so the resulting impulse response is obtained by

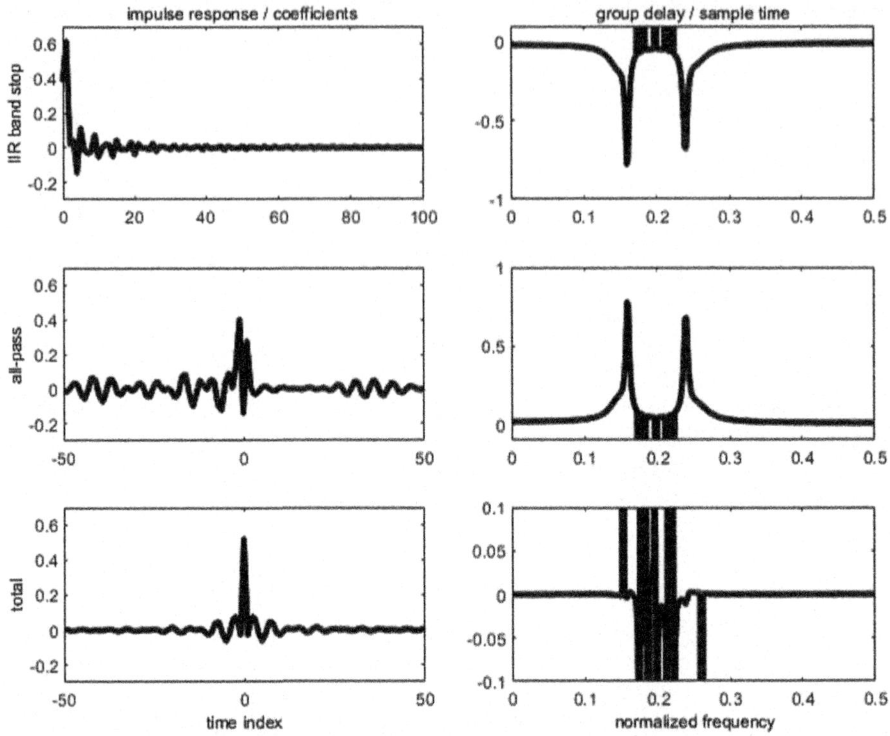

Fig. 5.14 Impulse responses (left column) and group delay times (right column) of an IIR bandstop (top row, see Fig. 5.7), an FIR all-pass or a phase shifter (middle row), and overall behavior of the two filters connected in series (bottom row). The spurious pulses in the group delay (right column) representation are caused by numerical singularity and do not exist

convolution (Fig. 5.14, bottom left, Eq. 5.29). From this, the resulting group delay is calculated (Fig. 5.14, bottom right, Eq. 5.30). Note that neither the impulse response of the all-pass nor the impulse response of the bandstop are symmetrical, but the total impulse response has symmetry. The resulting group delay is constant (without considering the delay of the all-pass even zero).

$$G_{BS}(k) \rightarrow |G_{AP}(k)| = 1, \quad \varphi_{AP}(k) = -\varphi_{BS}(k) \tag{5.28}$$

In Eq. 5.28, k is the DFT's discrete frequency index, BS is the bandstop, and AP is the all-pass.

$$g_{Gesamt}(n) = g_{BS}(n) * g_{AP}(n) \tag{5.29}$$

In Eq. 5.29, $g(n)$ is the impulse response, BS is the bandstop, and AP is the all-pass.

$$d_{total}(k) = -\frac{\Delta\phi_{total}(k)}{\Delta k} = d_{BS}(k) + d_{AP}(k) = const \tag{5.30}$$

In Eq. 5.30, $d(k)$ is the group delay; BS is the bandstop, and AP is the all-pass.

The all-pass filter in this example requires a filter length of $N = 500\ldots1000$ for phase correction. Here the correction problem becomes clear: an IIR filter of order 10 is needed for the desired amplitude frequency response, and for phase correction, an FIR filter with a hundred times the length. The consequence of this comparison would be that one can immediately construct an FIR bandstop that fulfills both requirements and is of comparable length.

5.3 LTV Systems: Time-Variable and Adaptive Filters

5.3.1 Basics of Time-Variable Filtering

In contrast to the technical fields (communications engineering, telecommunications, mechanical engineering, engineering sciences), in which one can assume stationary signals at least in a manageable time range, biosignals are highly dynamic. This section deals with signal dynamics concerning the time-varying spectrum. As already explained in Sect. 2.1, to separate the desired biosignal from the external (technical) and internal (body's biosignals) disturbances, it is necessary to know at least a part of the components involved completely. Since the biosignals must usually be assumed to be unknown, this approach only works for periodic or known interference signals. If both classes—desired and interfering signals—are variable or spectrally dynamic, the methods of the LTI analysis are no longer sufficient. In this case, a filter is needed whose spectral characteristics change according to the signal properties. Therefore, the transfer function of such a filter is a time variable according to the formulation in Eq. 5.31 (transition from an LTI to an LTV system).

$$G(\omega) \rightarrow G(\omega, t) \; g(\tau) \rightarrow g(\tau, t) \tag{5.31}$$

The output signal of an LTV filter then results from (Eq. 5.32)

$$y(t) = x(t) \overset{\tau}{*} g(\tau, t) = \int_{-\infty}^{\infty} x(t - \tau) \cdot g(\tau, t) d\tau. \tag{5.32}$$

Since such filters are only practicable in the discrete-time domain, one needs the formulation's discrete variant, according to Eq. 5.32, to calculate the output signal (Eq. 5.33).

$$y(n) = \sum_{m=-(M-1)/2}^{(M-1)/2} x(n - m) \cdot g((M - 1)/2 + m + 1, n) \tag{5.33}$$

According to Eq. 5.33, the relationship can be expressed in simplified matrix form (Eq. 5.34).

$$\mathbf{y} = \mathbf{G}^T \cdot \mathbf{x} \tag{5.34}$$

In Eq. 5.34, **x** and **y** are column vectors, and **G** is the matrix of discrete impulse responses (in columns).

Calculating the impulse responses g(n, m) has several possibilities. One can start from a known signal or a known disturbance and determine its time-varying transfer function using a mask (ROI, Region Of Interest) to determine the time-varying transfer function. This approach is followed in Sect. 5.3.2. However, it is also possible—especially in real-time processing—to adapt the transfer function to the dynamically changing signal properties, as described in Sect. 5.3.3.

5.3.2 Time Variable Filters

The approach for separating desired and undesired signal components is based on the fact that they can be separated in the time–frequency plane by specifying an ROI can be specified. The procedure is illustrated in the following example. A realistic situation from objective functional diagnostics of the auditory system was simulated with synthetic data (Fig. 5.15). The auditory system is stimulated with a linear chirp (harmonic tone with linearly increasing frequency, Fig. 5.15), whereby the AEP (auditory evoked potentials) are evaluated from the EEG (hyperbolic chirp in Fig. 5.15).

Note: All sensory systems (vision, hearing, touch, smell, taste) show qualitatively identical behavior in the time–frequency plane with transient stimuli: After a trigger, the stimulus–response in the EEG proceeds hyperbolically: It starts at high frequencies, then the period length of the waves increases linearly with time—the frequency decreases hyperbolically.

If the loudspeaker is very close to the auditory canal, the linear chirp can interfere with the EEG recording system, and a mixture of both chirps is obtained, as shown in Fig. 5.15. One can preset (interactively, automatically) an ROI so that the desired signal components are inside the ROI and the undesired signal components are outside the ROI. The simplest (but from a signal analysis point of view not necessarily the optimal) approach is to set a binary mask in such a way that the desired signal components (hyperbolic chirp) are evaluated (masked) with a one, the unwanted signal components (linear chirp) with a zero. Such a mask is shown in Fig. 5.16 (light fields correspond to a weighting of 0.0, dark areas to a weighting of 1.0). Based on an additive signal model, the masking in the time–frequency domain can be formulated as follows (Eq. 5.35). From the representation in Fig. 5.16, it follows that this simple masking involves two basic filters: Between time indices 1 to 80, there is a high-pass filter with a cut-off frequency of $f_{gHP} = 0.11$, and between time indices 81 and 256, there is a low-pass filter with a cut-off frequency of $f_{gTP} = 0.13$. The result of masking in the time–frequency plane is

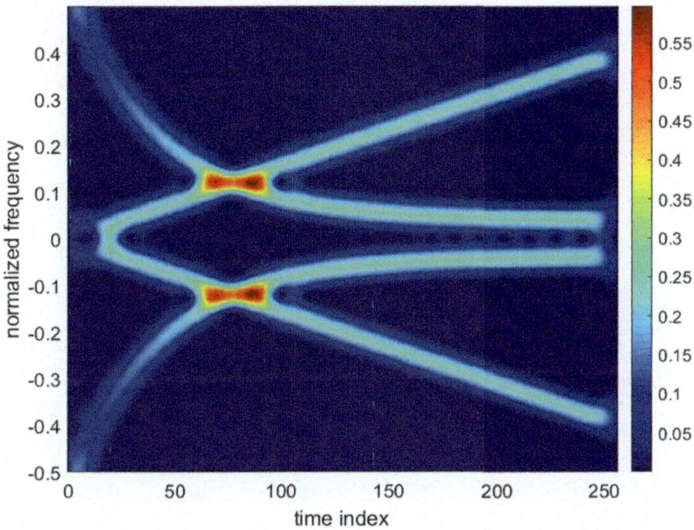

Fig. 5.15 Linear and hyperbolic chirp in the time–frequency plane. The linear increasing chirp was introduced into the measurement signal by induction from the headphones. The hyperbolic chirp is the EEG's stimulus response to the cortical auditory center

shown in Fig. 5.17.

$$\begin{aligned} \mathbf{x} &= \mathbf{s} + \mathbf{n} \\ \mathbf{G}^{\mathbf{T}} \cdot \mathbf{s} &= \mathbf{s} \\ \mathbf{G}^{\mathbf{T}} \cdot \mathbf{n} &= \mathbf{0} \end{aligned} \tag{5.35}$$

The matrix of discrete impulse responses $\mathbf{G}(m, n)$ is obtained from the time–frequency plane by the inverse DFT of the ROIs (Eq. 5.36).

$$\mathbf{g}(m, n) = IDFT_k(\mathbf{G}(k, n)) \tag{5.36}$$

In Eq. 5.36, $\mathbf{G}(k, n)$ is the binary mask in the time–frequency plane with discrete frequency k and time index n, $\mathbf{G}(m, n)$ is the matrix of the discrete impulse responses (column-wise) with displacement time index m and time index n.

The matrix of impulse responses calculated according to Eq. 5.36 can be used directly in the calculation formulas for Eq. 5.33 or 5.34. Figure 5.18 shows a section of the matrix of impulse responses.

From the previous explanations, it follows that applying an LTV filter, at least one of the signal components must be known and largely stable in its form. This assumption applies to some biosignals (evoked potentials of the sensory and motor systems, alpha waves in the EEG, normal ECG, respiration) and many disturbances (stimulation signals of functional diagnostics, technical disturbances). Thus, LTV filters offer an effective alternative to LTI filters.

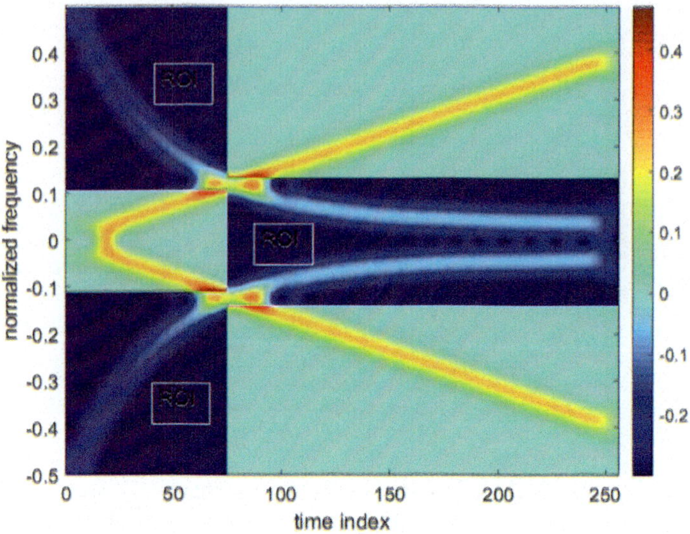

Fig. 5.16 Masking of the desired signal components as ROI: light-marked components (ROI) are masked with a weight of 0.0, dark marked with 1.0

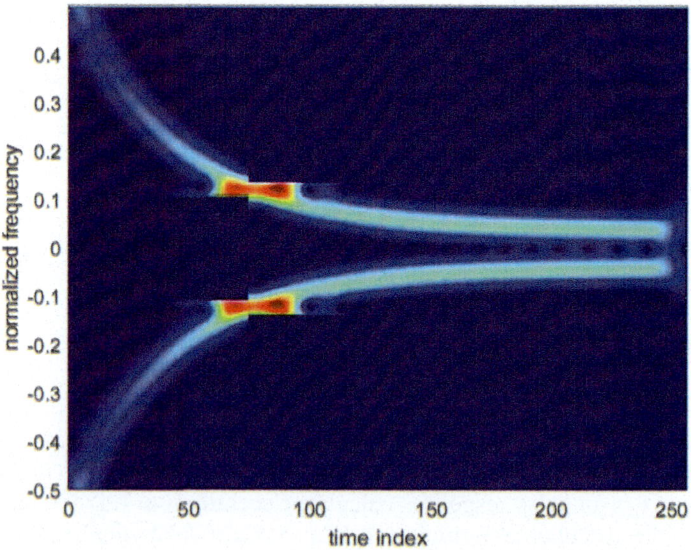

Fig. 5.17 Time–frequency distribution of the simulated signal from Fig. 5.15 after masking with binary ROIs from Fig. 5.16. Separation is not possible in the overlapping area of the components (time indices 50…100)

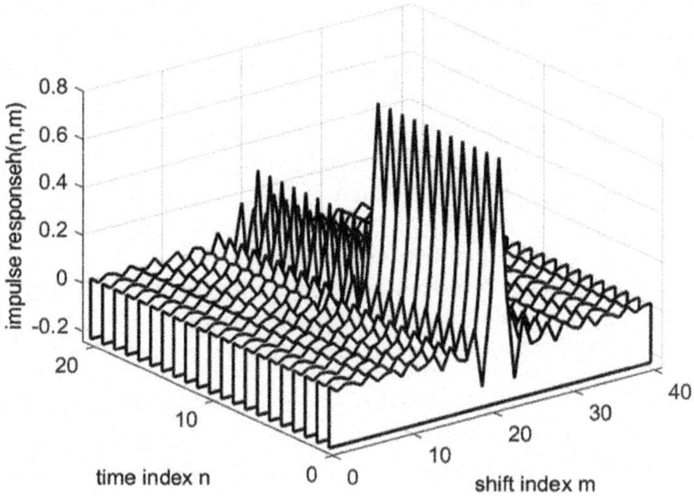

Fig. 5.18 Impulse responses of the mask from Fig. 5.16 at the transition point from high pass to low pass (time index between 65 and 85). Shown are ± 20 coefficients around the center of the impulse responses

In practical signal analysis, some specific features of LTV filtering must be considered (see exercises). Firstly, if possible, the ROIs or the masks must not be too narrow-banded but cover a wide field of the time–frequency plane. LTV filters that are too narrow would lead to generating desired signal components even from white noise, although in reality, none would be present. Furthermore, a too-narrow ROI would lead to the loss of desired components of the biosignals, which are subject to natural, relatively strong fluctuations. From these considerations, it follows for the mask design that the ROIs should capture the desired signal and separate it from the unwanted components, but (as shown in Fig. 5.16) must remain as wide and open as possible in the time–frequency plane. The practical application of LTV filters goes beyond this framework. If interested, please refer to the work.

5.3.3 Adaptive Filters

5.3.3.1 Theoretical Foundations of Adaptive Filtering

In many problems of biosignal analysis, the sought signal or the disturbance to be eliminated is not known in advance. Hence, determining ROIs according to Eq. 5.35 is not possible at all. Since the properties of signals and disturbances change in such cases, an adaptation of the filter (Eq. 5.34) to the current signal parameters is necessary (Widrow et al., 1975). For an adaptation, one needs a target function (*desired* value, model function, *desired function*) to which the filter should adapt. An indicator is required for the quality assessment of an adaptation (residual error, adaptation speed); in signal processing, this is usually the mean squared error (*mse*, mean squared error). In the following, it is assumed that a transversal filter will be used as the filter structure. The system structure of an adaptive filter is shown in Fig. 5.19. This architecture is structurally similar to a control loop. Still, the methodical system approach is different: In both cases, a target function (setpoint) $d(n)$ is given, with which the filter output $y(n)$ (actual value) forms an error signal $e(n)$, which is to be minimized. While in control theory, these signals are treated as dynamic processes, in signal processing, they are initially assumed to be stationary processes.

It initially simplifies the derivation of the algorithms but leads to problems with transient processes. Assuming that the transversal filter **w**—and thus the analysis window—has a length of $M = 2\mathrm{L} + 1$, the following applies to the output signal (Eq. 5.37):

$$y(n) = w_n(-L) \cdot x(n-L) + \cdots + w_n(L) \cdot x(n+L) = \mathbf{x}^T \cdot \mathbf{w}_n \tag{5.37}$$

In Eq. 5.37, $\mathbf{w}_\mathrm{n}$ is a vector of filter coefficients of length $2\mathrm{L} + 1$ at time n, and **x** is a vector of input data at time n.

For the error signal to be minimized (Eq. 5.38) applies:

$$e(n) = d(n) - y(n). \tag{5.38}$$

For simplicity, the time index n is omitted in the following. Furthermore, the following applies to the error (Eq. 5.39):

$$e = d - y = d - \mathbf{x}^T \cdot \mathbf{w}. \tag{5.39}$$

Note: No physical units are needed to develop algorithms and numerical calculations in signal processing. Therefore, signals are interpreted as voltages or currents, and their second power is power or energy after a temporal summation.

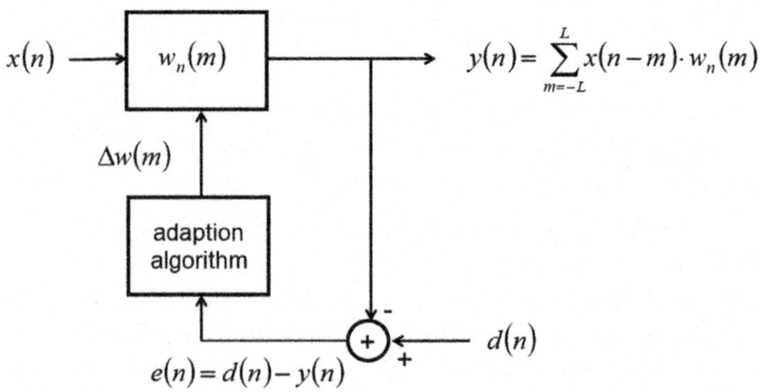

Fig. 5.19 Architecture of an adaptive filter with the basic structure of a transversal filter $w(m)$. The input signal $x(n)$ contains components that correlate with the model signal $d(n)$ and additive disturbances

One must assume that at least the disturbances in the biosignals are stochastic processes (see section "Stochastic processes") so that the error can be considered (without the restriction of generality) as a zero-mean signal. Therefore, minimizing the error would not be sufficient since it can assume both polarities and is already zero-mean. Thus, the error square, which represents the instantaneous signal power, is used to measure the adaptation quality (Eq. 5.40).

$$e^2 = d^2 - 2d \cdot \mathbf{x}^T \mathbf{w} + \mathbf{w}^T \mathbf{x}\mathbf{x}^T \mathbf{w} \tag{5.40}$$

The instantaneous signal power of the error, according to Eq. 5.40, is also a stochastic quantity due to the stochastic character of the disturbances in the input signal $x(n)$. For safe convergence to minimal error power, it is necessary to determine the mean square error *mse*. One introduces the expectation operator E{.} (Eq. 5.41). The expectation operator is usually replaced numerically in the numerical calculation by the arithmetic time average (see section "Stochastic processes").

$$F = E\{e^2\} = E\{d^2\} - 2E\left\{d \cdot \mathbf{x}^T\right\}\mathbf{w} + \mathbf{w}^T E\left\{\mathbf{x} \cdot \mathbf{x}^T\right\}\mathbf{w} \tag{5.41}$$

Based on the product of the model function $d(n)$ and the input data vector $\mathbf{x}_n$, one can interpret the expected value of the vector dx^T as a cross-correlation (Eq. 5.42).

$$\mathbf{p} = E\{d(n) \cdot \mathbf{x}_n\} = E\begin{Bmatrix} d(n) \cdot x_{o,n} \\ \vdots \\ d(n) \cdot x_{2L,n} \end{Bmatrix} \tag{5.42}$$

One interprets the third term in Eq. 5.41 as an autocorrelation (Eq. 5.43).

$$\mathbf{R} = E\left\{\mathbf{x}_n \cdot \mathbf{x}_n^T\right\} = E\begin{Bmatrix} x_{0,n}x_{0,n} & x_{0,n}x_{1,n} & \cdots & x_{0,n}x_{2L,n} \\ x_{1,n}x_{0,n} & x_{1,n}x_{1,n} & & \vdots \\ \vdots & & & \vdots \\ x_{2L,n}x_{0,n} & \cdots & \cdots & x_{2L,n}x_{2L,n} \end{Bmatrix} \tag{5.43}$$

By correlation matrices, the error functional can be expressed as follows (Eq. 5.44):

$$F = E\{e^2\} = E\{d^2\} - 2\mathbf{p}\mathbf{w}^T + \mathbf{w}^T \mathbf{R}\mathbf{w} \tag{5.44}$$

According to Eq. 5.44, the error function F is to be minimized. Therefore, all partial derivatives of the error functional after the filter coefficients are equal to zero (Eq. 5.45).

$$\frac{\partial F(\mathbf{w})}{\partial w_i} = -2\mathbf{p} + 2\mathbf{R}\mathbf{w}_{opt} = 0 \tag{5.45}$$

According to Eq. 5.45, the relationship results in the condition for the minimum *mse* (Eq. 5.46) and the optimal filter coefficients. A check of the extreme using the second derivative is unnecessary since the error functional has a single (global) extreme value—the minimum of a parabola.

$$\mathbf{w}_{opt} = \mathbf{R}^{-1} \cdot \mathbf{p} \tag{5.46}$$

A filter whose coefficients fulfill the condition according to Eq. 5.46 is called a *Wiener filter* or an *optimal filter*. Using the Wiener–Hopf formula, one obtains an equivalent formulation in the spectral domain with the help of the spectral power densities (Eq. 5.47).

$$W(f) = \frac{S_{xd}(f)}{S_{xx}(f)} \tag{5.47}$$

Which of the two formulas is used according to Eq. 5.46 or Eq. 5.47 depends on the concrete problem. However, both formulas are unsuitable for practical use. One needs an analysis window with a fixed and predetermined length to calculate the correlations (Eq. 5.46) and the spectral power densities (Eq. 5.47). Therefore, these two alternatives can only be used in temporally uncritical analyses where the data are static. Stationarity for the calculation of the correlations or the power densities is necessary. It is not fulfilled in biosignals; hence both solution approaches are not practicable. An approach is needed with which it is possible to adapt the filter coefficients $\mathbf{w}$ to the changing signal properties. The gradient method offers one possibility (Widrow et al., 1975).

Gradient Method, LMS Algorithm

The gradient at a point $\mathbf{w}_0$ on the multidimensional error parabola (Fig. 5.20) can be determined directly from Eq. 5.45 (Eq. 5.48):

$$\nabla_0 = \begin{Bmatrix} \frac{\partial E\{e^2\}}{\partial w_1} \\ \vdots \\ \frac{\partial E\{e^2\}}{\partial w_{2L+1}} \end{Bmatrix} \tag{5.48}$$

With the gradient, however, the problem of a sufficient analysis window length for calculating the expected value (mean value) is not yet solved. Assuming *the stationarity* of the error signal $e(n)$, one can interpret the current instantaneous value of the error power $e^2(n)$ as an estimate of the gradient (Eq. 5.49).

$$\hat{\nabla}_0 = \begin{Bmatrix} \frac{\partial e^2}{\partial w_1} \\ \vdots \\ \frac{\partial e^2}{\partial w_{2L+1}} \end{Bmatrix} = 2e \begin{Bmatrix} \frac{\partial e}{\partial w_1} \\ \vdots \\ \frac{\partial e}{\partial w_{2L+1}} \end{Bmatrix} = -2e\mathbf{x} \tag{5.49}$$

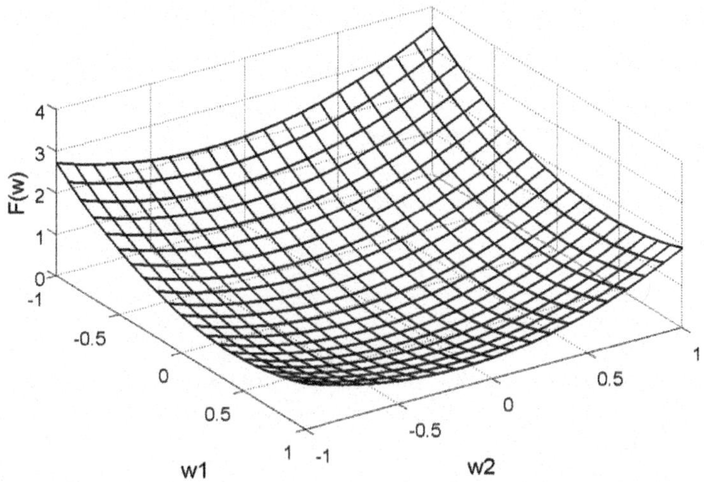

Fig. 5.20 Error parabola for two filter coefficients ($w_2 = -0.2, w_1 = 0.4$)

According to the steepest slope method, one iteratively approaches the error minimum by the following recursion (Eq. 5.50).

$$\mathbf{w}_{n+1} = \mathbf{w}_n - \mu \nabla_n \tag{5.50}$$

Equation 5.50 is the adaptation constant, determining the error's iteration rate and residual variance. For the gradient in Eq. 5.50, substitute its estimate from Eq. 5.49 (Eq. 5.51).

$$\mathbf{w}_{n+1} = \mathbf{w}_n + 2\mu e(n)\mathbf{x}(n) \tag{5.51}$$

According to Eq. 5.51, the recursion is known as the Widrow-Hoff LMS algorithm. It is easy to realize in real-time. Since the recursion represents negative feedback in system theory, a stability analysis is necessary, from which the following convergence and stability condition results (Eq. 5.52):

$$0 < \mu < \frac{1}{\lambda_{\max}} \tag{5.52}$$

In Eq. 5.52, $\lambda_{\max}$ is the largest eigenvalue of the autocorrelation matrix **R**. According to Eq. 5.52, the stability condition is exactly given, but the matrix **R** is not available for recursion. It can be shown that the following relationship holds (Eq. 5.53)

$$\lambda_R = \frac{Sp(\mathbf{R})}{n+1} = \frac{\sum_{n-L}^{n+L} x^2(n)}{2(L+1)}. \tag{5.53}$$

In Eq. 5.53, $Sp(\mathbf{R})$ is the trace of the autocorrelation matrix $\mathbf{R}$ (sum of the elements on the main diagonal), $x(n)$ is the input signal, $2L + 1$ is the length of the transversal filter $\mathbf{w}$ as well as of the input vector $\mathbf{x}$ at location n ($\mathbf{x} = \{x(n-L), \ldots, x(n+L)\}$), λ_R is the upper limit for the adaptation constant corresponding to the average power of the input signal.

The input vector $\mathbf{x}$ is available to the recursion in real-time to check the stability condition or reset the adaptation constant in each iteration step (Eq. 5.54).

$$0 < \mu < \frac{1}{\lambda_R} \tag{5.54}$$

Since $\lambda_R > \lambda_{max}$, the stability condition according to Eq. 5.52 can be safely met even without calculating the autocorrelation matrix $\mathbf{R}$. For practical analysis, normalizing the upper limit for the adaptation constant is helpful to be independent on the current signal (Eq. 5.55).

$$0 < \mu_N < 1,\ \mu = \mu_N \cdot \lambda_R \tag{5.55}$$

An LMS algorithm with a normalized adaptation constant is called the NLMS algorithm (Normed Least Mean Squares). Specifying an exact value for the adaptation constant λ_N is impossible, but only to represent qualitative correlations. The adaptation constant directly influences the adaptation speed and the residual variance of the error. Although the LMS algorithm converges towards the minimum squared error, it is inconsistent. It is because, near the error minimum (global minimum of the $2L + 1$-dimensional parabola), the gradient becomes numerically very small so that it follows the ever-present stochastic fluctuations.

The function of an LMS filter is explained in the following example. An input signal $x(n)$ of the length $L_N = 1000$ was simulated as white, normally distributed noise. It is an ideal, stationary, stochastic process in second-order statistics. The autocorrelation matrix $\mathbf{R}$ is a unit matrix (Eq. 5.56).

$$\left.\begin{array}{r} x(n)|_{n=1..1000} \in N(0,1) \\ S_{xx}(f) = 1 \end{array}\right\} \Rightarrow \mathbf{R} = \mathbf{I} \tag{5.56}$$

The input signal $\mathbf{x}$ was filtered with a filter with the coefficients $\mathbf{w}_X = (-0.2, 0.4, -0.2)$ and fed to the adaptation algorithm as a model signal. This filter has a high-pass character (Fig. 5.21). The initial condition for the filter coefficients is not critical for the iteration and, therefore, arbitrary (here $\mathbf{w}_{start} = (0, 0)$) since the error functional has a single minimum. Depending on the starting point and the adaptation constant, the LMS filter needs a different number of iteration steps to reach the error minimum (Fig. 5.22). In principle, a higher adaptation constant allows the filter to adapt faster, but at the cost of significantly higher fluctuations, which remain even near the global minimum (Fig. 5.23). In this example, both signals of the adaptive filter are stationary so that one can accurately determine the real adaptation time. These empirical values can be used to construct adaptive filters for biosignals if the always-present nonstationarities are to be filtered out or

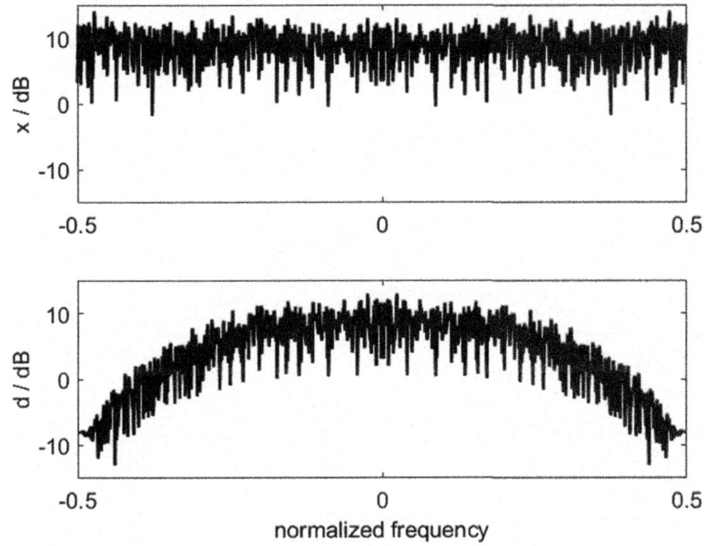

Fig. 5.21 Spectrum of the white normally distributed input signal $\mathbf{x}$ (top) and the model signal $\mathbf{d}$ (bottom) after filtering with $\mathbf{w} = (-0.2, 0.4, -0.2)$

Fig. 5.22 Iteration of the filter coefficients $\mathbf{w}$ from the starting point $\mathbf{w}_{start} = (0, 0)$ to the error minimum $\mathbf{w}_{min} = (0.4, -0.2)$ in the contour plot of the error parabola from Fig. 5.20. The first 500 recursion steps are shown, the normalized adaptation constant was $\lambda_N = 0.02$

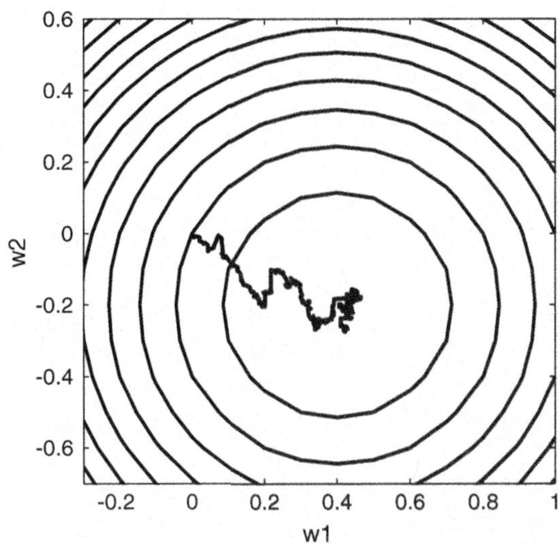

suppressed (e.g., separate R-spikes in the ECG from a network disturbance). The illustration in Fig. 5.23 shows that the original filter coefficients $(0.4, -0.2)$ are perfectly achieved with the adaptation algorithm.

Because of the presentation, the adaptation curve was only shown for two coefficients. It is also unproblematic since after the adaptation is complete, the first

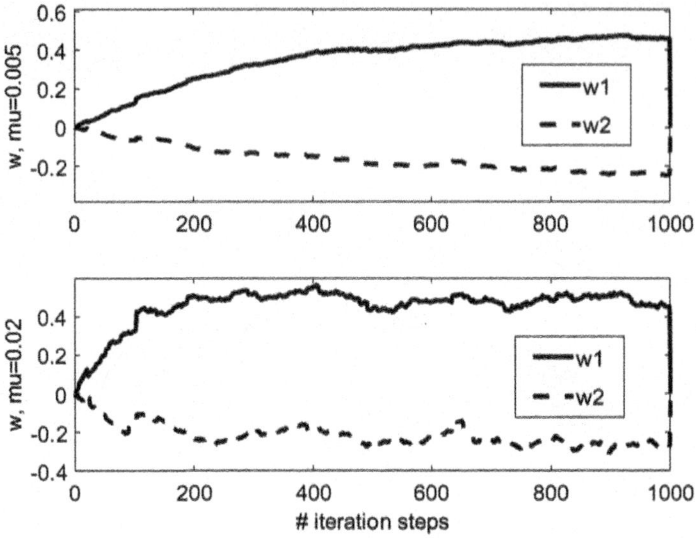

Fig. 5.23 Adaptation of the filter coefficients from Fig. 5.22 with a normalized adaptation constant of $\lambda_N = 0.005$ (top) and $\lambda_N = 0.02$ (bottom). With a larger adaptation constant (bottom), considerably fewer steps are needed for the iteration than with a lower adaptation constant (top). However, the fluctuations are also significantly higher in the optimum

and third coefficients are identical, as with the original filter. However, one must ask *why* an adaptive filter should be used if the model signal $d(n)$ or the filter coefficients are already known before the filtering. It follows from this:

> The optimal filter is practically not applicable in its classical form.

There are signal-analytical and system-structural variants and additions with which optimal filters can be applied in system identification, parameter estimation, interference suppression, or optimization. A practical example follows in the next chapter.

5.3.3.2 Adaptive Noise Cancellation, ANC

Assuming the primary task of a filter is to separate desired and interfering signal components from each other, knowledge of at least one of the components is necessary. With biosignals, one cannot assume that the desired component is known. In many cases, however, one knows the interference that does not contain any components of the biosignal. Therefore, one can change the solution approach of the optimal filter to filter out a known signal from a mixture with disturbances. The new approach is to suppress a known disturbance so that only the desired but unknown signal remains after filtering. At this point, the functioning of an optimal filter in terms of statistics will be explained in more detail (see Fig. 5.19). Suppose one assumes that the adaptation is completed. In that case, the filter coefficients **w** correspond to a Wiener filter (Eq. 5.46), and the optimum in the sense of a linear separation of signal and disturbance is reached.

In the output signal $y(n)$, those parts of the input signal $x(n)$ appear which have a high correlation with the model signal $d(n)$. The error signal $e(n)$ is formed from the disturbance that does not correlate with the model signal $d(n)$.

Here it also becomes clear why one must require stationarity of the error signal $e(n)$ for estimating the gradient according to Eq. 5.49: At least for the adaptation time, the statement about correlations only applies to stationary processes. The solution approach of an ANC (Adaptive Noise Canceller) consists of setting the adaptive filter to a known disturbance (Fig. 5.24). The input of the transversal filter has the function of a noise reference $r_R(n)$.

The noise reference must correlate well with the disturbance $r_S(n)$ in the input signal $x(n)$ and must not correlate with the sought signal $s(n)$ in $x(n)$ (Eq. 5.57).

$$R_{r_R s} = 0, \ \left| R_{r_R r_S} \right| \gg 0 \tag{5.57}$$

In Eq. 5.57, R_{ij} is the process-related correlation coefficient of signals i and j.

According to the correlation relations mentioned above, only the disturbance $r_S(n)$ or its part contained in $r_R(n)$ appears at the filter output $y(n)$ (Fig. 5.24). Suppose the condition, according to Eq. 5.57, is fulfilled. In that case, the filter y(n) output signal does not contain any parts of the searched signal s(n), so this remains after the completed adaptation as an error signal at the ANC output. In the statistical—but not in the algebraic—sense, the interference $r_S(n)$ *is* subtracted

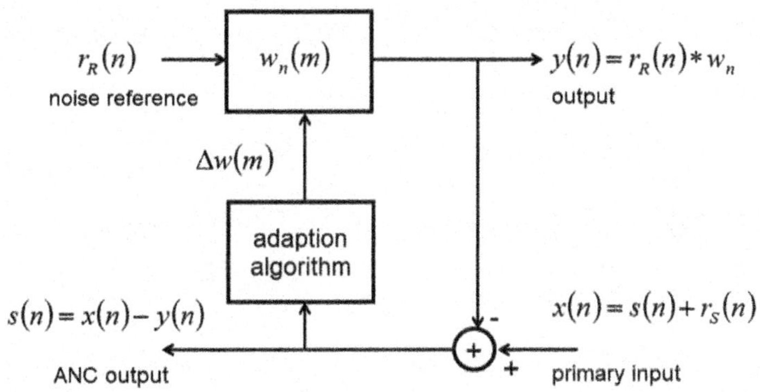

Fig. 5.24 System structure of an ANC (adaptive noise canceller). The input of the transversal filter has the function of a noise reference $r_R(n)$; the model signal here is the noisy sought signal $s(n)$ in $x(n)$, and the ANC output is equal to the error signal

from the input signal $x(n)$ so that the sought signal $s(n)$ is available as the result of the ANC filtering. The following example demonstrates how ANC works:

An impedance plethysmograph (IPG) is strongly disturbed by the mains (Fig. 5.25, primary input $d(n)$). One channel is free at the existing amplifier. That is used to capture a disturbance reference (mains disturbance) which must not contain any components of the IPG.

It is technically straightforward to realize in a practical measurement: Connect an electrode cable and leave it free (without patient contact) in the measurement setup. Thus the mains interference is capacitively coupled in (a static electric field), and the noise reference is formed (Fig. 5.25, Noise reference $x(n)$). Since the exposed electrode cable has no connection to the patient, this reference contains no plethysmographic components. After the start of adaptation (adaptation constant $\lambda_N = 0.01$), the interference appears stable at the filter output after only one second (Fig. 5.25, Filter output $y(n)$). The desired and unknown IPG is entirely free of interference from about 1.5 s after the start (Fig. 5.25, ANC output). With an

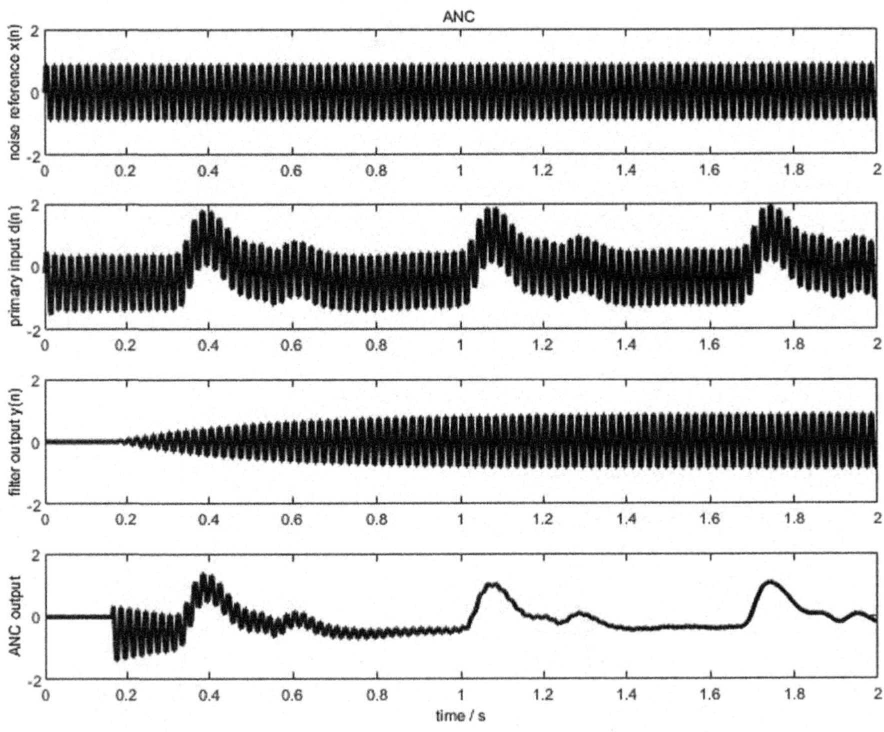

Fig. 5.25 Progressions in the ANC during the suppression of the mains interference in an impedance plethysmograph. The filtering took place offline, so there is no time shift between the ANC output and the disturbed signal $d(n)$. The dead time of 167 ms at the beginning of the adaptation in the filter output $y(n)$ and the ANC output corresponds to half the filter length of 50 sampling periods

adaptation constant of $\lambda_N = 0.1$, the adaptation would already be completed after 100 ms, and the first wave of the IPG would already be almost interference-free.

Obtaining a noise reference (offline evaluation, no suitable or free metrological resources) is not always possible. The following solution approach helps against periodic disturbances in such cases: As explained above, the functioning of an LMS filter is based on the correlation between the input signal of the transversal filter $x(n)$ and the model signal $d(n)$. If the correlation for the disturbance component is high, the disturbance appears at the filter output $y(n)$ and is missing, as desired, at the ANC-output $s(n)$. If one assumes that the disturbance is periodic, its autocorrelation function (AKF) is also periodic. Conversely, biosignals are only very rarely periodic, and if they are, then with intense fluctuations. In addition, the periods of the biosignals are much longer than those of the disturbances (Fig. 5.26). The representation in Fig. 5.26 shows that the disturbance's autocorrelations are periodic with 20 ms (mains disturbance) and those of the pulse curve with about 1 s (heart period). If one shifts the disturbed signal $d(n)$ by a time of, for example, $\tau_V = 0.25$ s, the correlation of the pulse wave is approximately zero. At the same time, that of the periodic disturbance reaches high values (> 0.9). Of course, adjusting the suitable displacement time τ_V by hand and to the current signal each time is impossible. One must decide on a shift beforehand or design it adaptively for a specific measurement task.

It is challenging with biosignals due to their variability. Therefore, it cannot be avoided that a correlation exists between the noise reference and the primary input of the ANC. It leads to the suppression of the interference and the impairment of

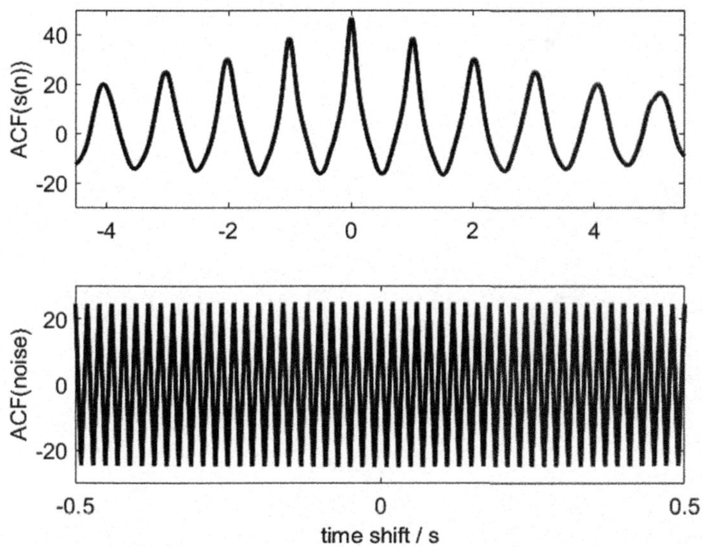

Fig. 5.26 Autocorrelation (ACF) function of an impedance plethysmographic pulse curve (top) and a network disturbance (bottom). Note the different time scales and the period differences

the desired signal (Fig. 5.28). Therefore, in such cases, whether distortion of the wanted signal is acceptable with an ANC without noise reference (Fig. 5.27).

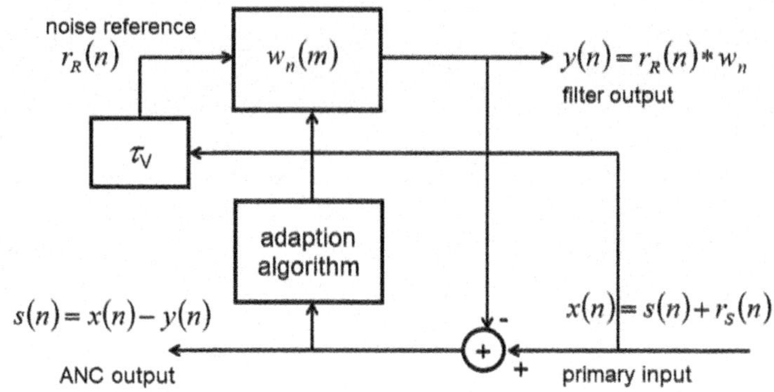

Fig. 5.27 ANC-filter with a delayed input signal as noise reference. The delay τ_V must be dimensioned so that the periodic disturbance has a high correlation and the signal sought has a low correlation

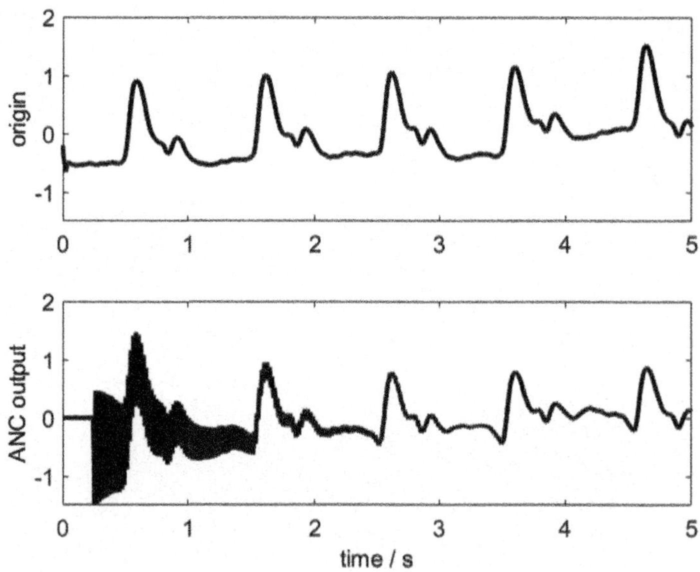

Fig. 5.28 Original signal (top) and output of the ANC filter (bottom) according to Fig. 5.27. The interference is eliminated after 3 s, but the desired signal shows changes in shape

5.4 Spatiotemporal Filtering

5.4.1 Fundamentals of Spatiotemporal Filtering

For many medical questions, multi-channel recording systems are used, e.g., for measuring the EEG or for ECG and EMG mapping. Therefore, using spatial filters to search for specific spatiotemporal signal patterns makes sense. Retaining filter structures based on previous mathematical apparatus and the available algorithms for analyzing, designing, and constructing filters would be advantageous. It is feasible with the structure of the transversal filters (Fig. 5.29): While in the time domain, the input data consist of temporally delayed signal values (Fig. 5.29, top), in space, they are simultaneous signal values along the spatial coordinates (Fig. 5.29, bottom). Thus, the output signal results from the discrete convolution of the filter coefficients with the input data or from the scalar product of the corresponding vectors, independent of the domain (Eq. 5.58).

$$y(k) = \sum_{i=1}^{M} x_i(k) w_i^* \leftrightarrow \mathbf{y} = \mathbf{w}^T \mathbf{x} \tag{5.58}$$

In Eq. 5.58, the index i in the time domain denotes the successive temporal samples concerning time k; in the space domain, the samples in the sensor channels at the same time k (see Fig. 5.29).

Between the filters in the time domain and space, a structural connection exists via the relationship, according to Eq. 5.58. The question arises as to which parameter in space is equivalent to the spectrum for time signals.

To derive the analytical relationships, one can start from an initially simple signal model (Fig. 5.30). M sensors are arranged equidistantly at a distance of d on a straight line. From an infinitely distant source, waves with the wavelength are incident on the sensor line (the propagation speed of the signal is c), where are the wavefront and the sensor line angles. The following applies to the output signal of the filter (Eq. 5.59)

$$y(k) = \frac{1}{M} \sum_{m=1}^{M} x_m(k + \tau_m), \tag{5.59}$$

where the signal delay τ_{m} in channel m results from Eq. 5.60.

$$\tau_m = \frac{(m-1) \cdot d \cos(\theta)}{c} = \frac{d_m}{c} \cos(\theta) \tag{5.60}$$

The transfer function of a channel in the spatial frequency range thus corresponds to (Eq. 5.61)

$$G_m(\omega, \theta) = e^{-j\omega \frac{d_m}{c} \cos(\theta)} w_m^*. \tag{5.61}$$

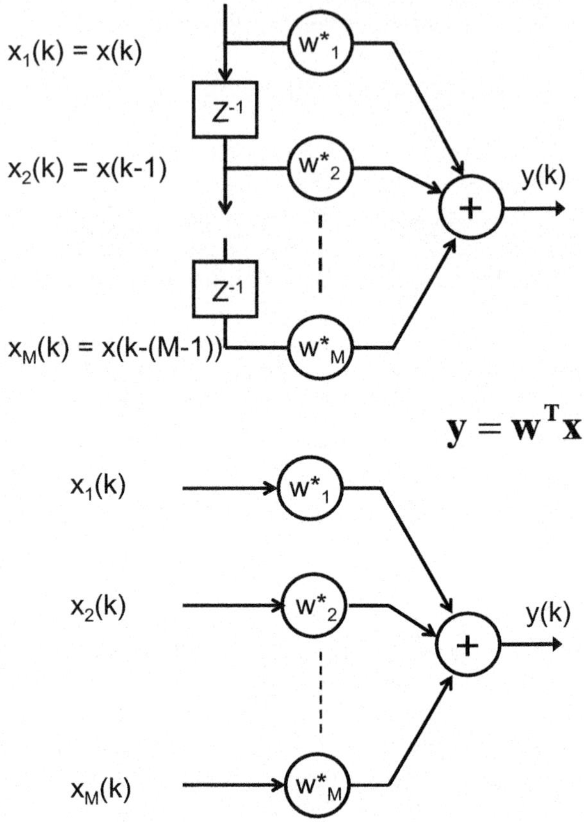

Fig. 5.29 Structural equality of an FIR transversal filter (top) and a spatial filter (bottom). A * marks complex coefficients

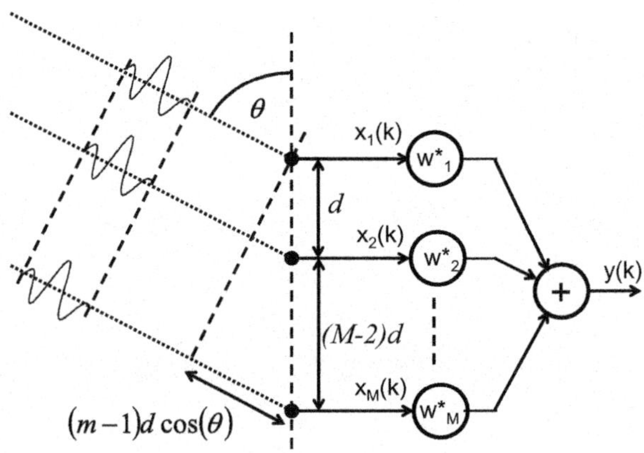

Fig. 5.30 Spatial filter when receiving a wave. The sensors lie equidistantly on a line at a distance of d (sensor line)

Starting from Eq. 5.59, the transfer function of the filter (Eq. 5.62) is as follows

$$G(\omega, \theta) = \frac{1}{M} \sum_{m=1}^{M} e^{-j\omega \frac{dm}{c} \cos(\theta)} w_m^*. \tag{5.62}$$

If one assumes a harmonic signal, the wavelength in Eq. 5.62 is constant, and the transfer function *G*() only depends on the angle. Therefore, the relation according to Eq. 5.62 can also be interpreted as an angular transformation formulated according to Eq. 5.63. According to Eq. 5.62, the relationship corresponds to a discrete-time Fourier transformation in the time domain. Therefore, it can be interpreted as a discrete directional characteristic for the spatially sampled signal (see main and side lobes in DFT).

$$G(\theta) = \frac{1}{M} \sum_{m=1}^{M} e^{-j2\pi \frac{dm}{\lambda} \cos(\theta)} w_m^* \tag{5.63}$$

If the transfer function is to have its main maximum in a certain direction, the relationship according to Eq. 5.62 modifies to (Eq. 5.64):

$$G(\omega, \theta) = \frac{1}{M} \sum_{m=1}^{M} e^{-j\omega \frac{dm}{c} (\cos(\theta) - \cos(\vartheta))} w_m^*. \tag{5.64}$$

Let us make an analogy between the spatial filter and an FIR filter according to Fig. 5.28 (where it is constant here). According to Eq. 5.63, the formula corresponds to a rectangular window at zero frequency (angle 90°). According to Eq. 5.64, the formula corresponds to a rectangular window at a non-zero frequency (angle ϑ).

5.4.2 Beamforming

The following example illustrates the function of a filter according to Eq. 5.61: In the case that all filter coefficients are real and identical, the filter has the maximum transmission perpendicular to the sensor line (= 90°) with a sinc function depending on the angle of incidence (Fig. 5.31). The width of the directivity's main lobe decreases with the harmonic oscillation frequency. The spatial filter function of a sensor array with four microphones at a distance of $d = 2$ cm each is shown as a function of the frequency f and the angle of incidence, which pick up the sound of a very distant acoustic source standing in the air ($c = 330$ m/s). Based on the relationship according to Eq. 5.60, the formula Eq. 5.64 can also be written as follows (Eq. 5.65):

$$G(\omega, \theta) = \frac{1}{M} \sum_{m=1}^{M} e^{-j\omega(\tau_m - \delta_m)}. \tag{5.65}$$

In Eq. 5.65, $_{\mathrm{m}}$ is the difference in travel time when the wave train arrives at the sensors (Fig. 5.30), δ_{m} is the delay to be installed behind the sensors to achieve a preferred spatial direction at the angle. A filter structure corresponding to this requirement is shown in Fig. 5.33. Depending on the desired angle of incidence, one calculates for each channel its delay τ_{m}. If a wave train is incident on the sensor line at the angle ϑ, the delays at the sensors τ_{m} and those of the channels δ_{m} cancel each other out (Eq. 5.65), and the transmission reaches its maximum $|G| = 1$ (Fig. 5.32).

With the help of the channel delays δ_{m} one can—without changing the sensors' spatial positions- change the sensor line's directional characteristic electronically or by calculation. A *beam* is formed; therefore, such structures are called *beamformers*. The simplest filter structure (Fig. 5.33) uses only channel delays and a summing beamformer and is consequently called a *delay-and-sum beamformer* (DSBF). In the practical application of the DSBF, two approaches are conceivable: On the one hand, one can adjust the channel delays for a direction of incidence if it is known for a source and suppress the interfering waves coming from other directions. On the other hand, by changing the channel delays, one can scan the space for unknown sources and thus obtain an angular spectrum (Eq. 5.63).

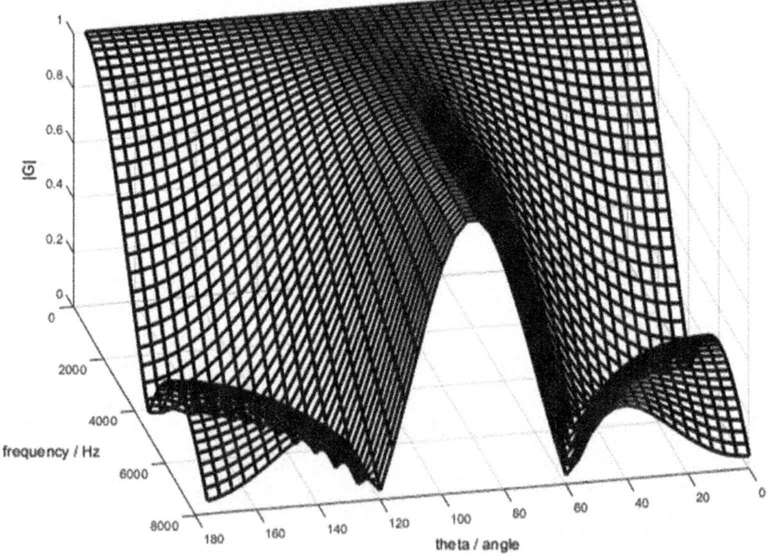

Fig. 5.31 Magnitude of the transfer function of a spatial filter according to Eq. 5.64 (Fig. 5.30) with four identical filter coefficients (rectangular window)

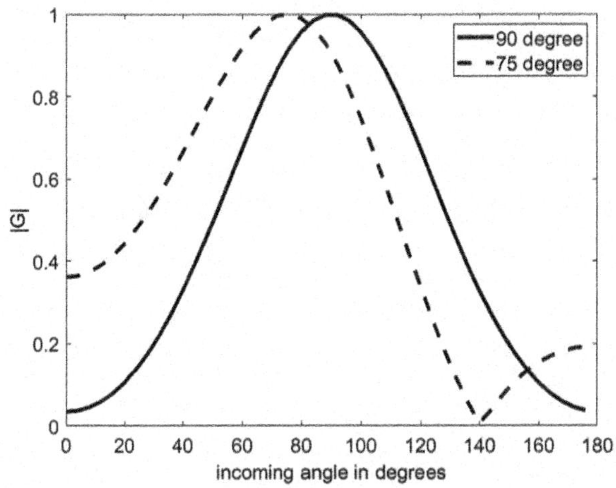

Fig. 5.32 Transfer function (f = 4000 Hz) of a spatial filter with four sensors and identical coefficients (full line) and with additional channel delays (dashed)

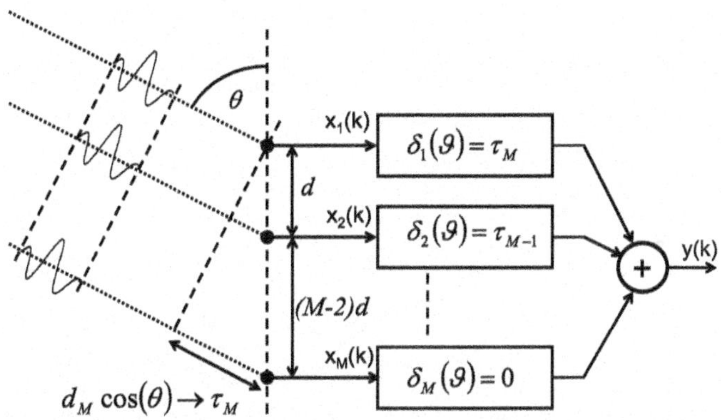

Fig. 5.33 Instead of filter coefficients, each channel in the DSBF receives a defined delay (). This results from the desired angle of incidence of the incoming wave. The transmission reaches its maximum at the incidence angle corresponding to the set channel delays

5.4.3 Spatial Filter

5.4.3.1 Specificity of the Biosignal Sources

The initial conditions for acquiring multi-channel biosignals are much more complex than shown in the model in Fig. 5.33. In Fig. 5.34, the main problem is shown. A sensor array is arranged on the body surface, enabling three-dimensional signal acquisition. The sensor signals are processed by an array processor, which, depending on the structure and technology, is a *beamformer* (BF) or a *Spatial Filter* (SF). From the electrophysiological point of view, the signal source $\mathbf{q}(t, \mathbf{r})$ is a local electrical negativity (signal and current direction **r** variable in time *t*) that acts outwardly like an electrical current source. It is often modeled as a local electric dipole oriented toward the most substantial current. This source moves along the sensory and motor nerve pathways with velocity **v**(**r**), which ranges from 1 to 20 m/s depending on the type of nerve pathway (**v** is the vector of velocity that depends on the current location of the signal source **r**). The electric current generated by the source produces electric and magnetic fields that propagate with the velocity **c**(**r**) (**c** is the vector of the propagation velocity of the electric or magnetic field, which depends on the location **r** because of the anisotropy of the medium). The speed **c**(**r**) is slightly lower than that of light. Still, it is large enough that the speed of movement of the signal source **v**(**r**) can be neglected in connection with the source frequency (< 10 kHz) (the difference in speed is about eight decades, while the period of the source oscillation is over 100 s).

Theoretically, one could track the source movement with known technologies (radar, laser). However, medical measurement technology is not designed for this but for frequencies of 10 kHz maximum. Therefore, the numerous methods and algorithms of beamforming, as they are common in mobile radio, directional radio, sonar, radar, source location, and tracking, are out of the question in medical signal processing because they build on the following essential prerequisites:

- Known and technically controllable propagation velocity c,
- homogeneous and isotropic medium in which waves propagate,
- known properties of the signal source,
- defined and exact sensor arrangement and
- an infinite or sufficiently large expansion of the medium.

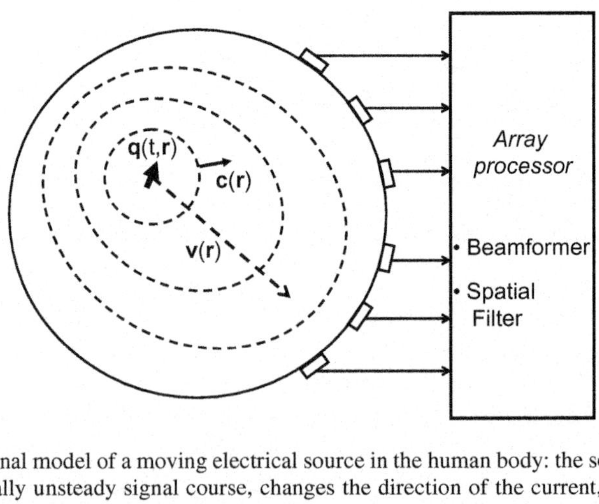

Fig. 5.34 A signal model of a moving electrical source in the human body: the source **q**(*t*, **r**) generates a temporally unsteady signal course, changes the direction of the current, and moves with the speed **v**(**r**). The electric and magnetic field generated by the source (the electric field is shown) propagates with the velocity **c**(**r**) (**r** is the spatial coordinate, t the time)

Real biosignal sources and the medium (human body) in which their signals propagate do not fulfill these conditions. Therefore, beamforming methods are not applicable in principle.

5.4.3.2 Model-Free Spatial Filters

As an alternative to beamforming, spatial filters not bound to the above-mentioned strict conditions are available. The same statement applies to spatial filters as to transversal FIR filters (Fig. 5.29):

> Every mathematically describable function can be realized, whereby the accuracy of the realization depends on the number of available filter coefficients.

It is naturally limited with spatial filters because of the relatively few sensors available. The example of the DSBF illustrates the problem.

One of the essential tasks of biosignal processing is to ensure the signal-to-noise ratio (SNR) necessary for parameter estimation, or at least for detection. The most effective method in the time domain is synchronous or simultaneous *averaging*. (averaging, see section Stochastic Processes). An additive signal model is assumed for a spatial filter (DSBF) (Eq. 5.66).

$$x_m(k) = s(k) + n_m(k),\ y(k) = \frac{1}{M}\sum_{m=1}^{M} x_m(k) = s(k) + \frac{1}{M}\sum_{m=1}^{M} n_m(k) \tag{5.66}$$

In Eq. 5.66, $s(k)$ is the desired biosignal, $n_{\mathrm{m}}(k)$ is the white noise in channel m, where the noise components of the channels are spatially uncorrelated with each other, $y(k)$ is the result of ensemble averaging.

Assuming that the desired signal $s(k)$ is present in the same form in all channels and that the noise power $\mathrm{var}(n_{\mathrm{m}})$ is the same, it decreases inversely proportional with the averaging order M in the output signal y(k), so that for the SNR in the averaged signal $y(k)$ applies (Eq. 5.67)

$$snr(x_m) = \frac{\mathrm{var}(s)}{\mathrm{var}(n_m)},\ snr(y) = M\frac{\mathrm{var}(s)}{\mathrm{var}(n_m)} = M \cdot snr(x_m). \tag{5.67}$$

However, the improvement of the SNR according to Eq. 5.67 only occurs if the signal model according to Eq. 5.66 corresponds to reality, i.e., only if the sought signal $s(k)$ is *simultaneously* present in all channels. Concerning the signal $s(k)$, there must be no time shifts between the channels. In reality, however, one must reckon with time shifts, as already shown in signal models in Figs. 5.30 and 5.33. Therefore, the effect of a time shift between channels is investigated.

If we first assume two channels containing a time-shifted signal $s(k)$, the following relationship applies (Eq. 5.68)

$$\begin{aligned} x_2(k) &= x_1(k+\tau) \\ y(k) &= \tfrac{1}{2}(x_1(k) + x_2(k)) = \tfrac{1}{2}(x_1(k) + x_1(k+\tau)) \end{aligned}. \tag{5.68}$$

In Eq. 5.68, $x_1(k)$ is the first channel of a two-sensor array, and $x_2(k)$, the second channel of the same array, is the discrete-time shift between the two channels, an integer index.

Transforming Eq. 5.68 into the frequency domain, we obtain (Eq. 5.69)

$$Y(\omega) = \frac{1}{2}(X_1(\omega) + X_2(\omega)) = \frac{1}{2}X_1(\omega) \cdot \left(1 + e^{-j\omega\tau}\right). \tag{5.69}$$

The phase shift in Eq. 5.69 can assume two extremes (Eq. 5.70):

$$\left|1 + e^{-j\omega\tau}\right| = \sqrt{2(1 + \cos(\omega\tau))} = \begin{cases} 2, \ \tau = 0 \\ 0, \ \tau = T/2 \end{cases}. \tag{5.70}$$

Equation 5.70 is the time shift from Eq. 5.68, and T is the signal s(k) period assumed to be harmonic.

From Eq. 5.70 follows that a received signal $s(k)$ can be preserved entirely (Eq. 5.66) but also wholly extinguished (Eq. 5.70) depending on its spectral composition in dependence on the time shift between the sensors or channels. If one assumes for the sake of simplicity that the sought signal $s(k)$ is composed of a sum of harmonics (Fourier series), then a time shift of more than half a period to the combined low-pass/high-pass behavior or to a bandstop (Fig. 5.35). A realistic model of a multi-channel VEP is shown in Fig. 5.36. It is known that in EP, the frequency is relatively high at the beginning (up to 30 Hz) and decreases to about 5 Hz. The amplitude of the waves reaches its maximum approximately in the middle of their time course. Accordingly, the amplitudes of the strongest negative and positive waves in Fig. 5.36 are comparable. Due to the time offset between the (EEG) channels, the sum potential is initially weaker than the individual channel signals. As can be seen from the illustration, the amplitudes of the sum wave increase with time and with decreasing frequency when the amplitudes of the channel signals are comparable. The time shift in the channels in the sum signal has a low-pass character (Fig. 5.35). It follows that to improve the SNR, the channel delays must be compensated to fulfill the condition of simultaneity (Eq. 5.66). In principle, the DSBF is suitable for compensating the channel delays (Eq. 5.65, Fig. 5.33). The DSBF does not apply to the electric or magnetic field of the source for the reasons mentioned above but can be used in the range of the source's velocity of motion $\mathbf{v}(\mathbf{r})$ and the time course of the sought signal $s(k)$ (Fig. 5.34). However, the properties of the human body medium and the body's signal sources are hardly usable for an analytical solution approach.

The following overview describes the characteristics of biosignal sources and the human body for signal processing:

- The human body is electrically *inhomogeneous*; the differences in the electrical conductivity of different tissue types amount to several decades,
- The conduction velocity $\mathbf{v}(\mathbf{r})$ of nerve excitation varies by a factor of up to 20 depending on the type of nerve fiber,

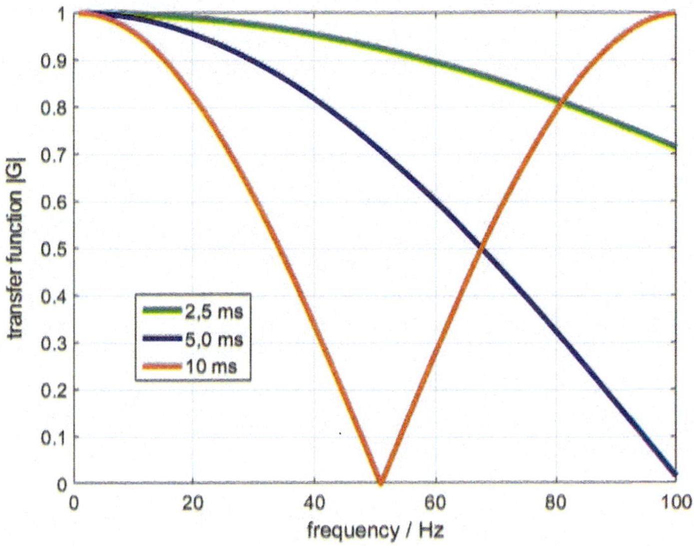

Fig. 5.35 Spectral effect of a time shift between two signals $x_m(k)$, in which the searched signal $s(k)$ is shifted, and the noise $n_m(k)$ is included. If the shift is up to half a period, it has the effect of a low-pass filter. If the shift is over half a period, it behaves like a bandstop

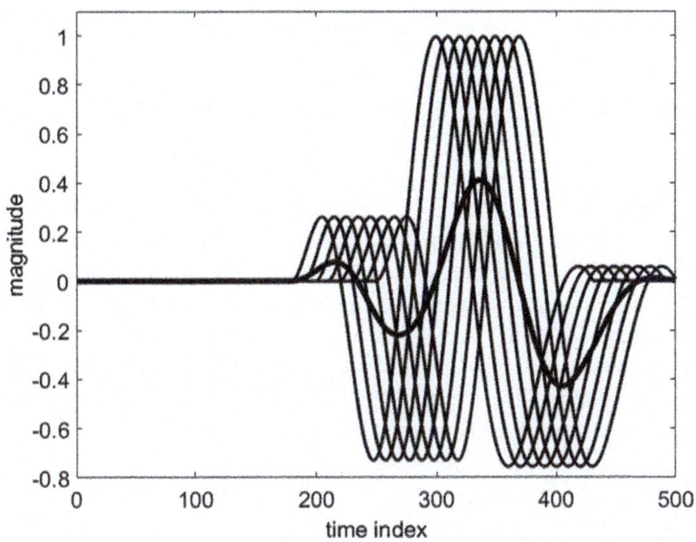

Fig. 5.36 Sum of eight shifted channel signals whose frequency decreases with time: Because of the time shift, the sum as a whole (thick line) is smaller than each channel signal. Its amplitude increases with time and decreasing frequency. That is attributed to the low-pass behavior of the time shift (the period of the waves increases linearly; the instantaneous frequency decreases hyperbolically. Due to the time shift's low-pass effect, the wave damping decreases with increasing time)

- The anatomical variability is so high that the body description is analytically impossible (no spherical or cylindrical symmetry),
- An exact assignment of the sensors to anatomical structures and the reproducibility of the sensor positioning are impossible.

5.4.3.3 Adaptive Spatial Filtering

The projection of the fields on the body surface results from the multimodal activity in the body (movement of the source, time course of the source signal, orientation of the source current). For the above reasons, an analytically determined adjustment of the channel delays δ_m is impossible. Adaptive methods allow tracking the source dynamics and the transient source signal s(k) utilizing the surface projection (see Sect. 5.3.3). However, the problem is complicated because neither a model function nor a noise reference is available. For the adaptation algorithm, an additive signal model is assumed (Eq. 5.71):

$$x_m(k) = A_m s(k + \tau_m) + n_m(k). \tag{5.71}$$

In Eq. 5.71, $x_m(k)$ is a sensor/channel signal, A_m is the amplitude of the sought signal $s(k)$ in channel m, and $n_m(k)$ is the sensor/channel noise.

The output signal of the *Averagers is given by* (Eq. 5.72):

$$y(k) = \frac{1}{M} \sum_{m=1}^{M} x_m(k + \delta_m). \tag{5.72}$$

The SNR of the averager is to be maximized so that applies:

$$snr(y) \to \max|_{\tau}, \; \frac{\partial snr(y)}{\partial \tau_m} = 0, \; \frac{\partial^2 snr(y)}{\partial \tau_m^2} < 0. \tag{5.73}$$

In the search for the maximum SNR, the condition of the second partial derivatives must also be fulfilled because, in contrast to conventional adaptive filters, the shape of the multidimensional error functional is unknown here. From the conditions in Eq. 5.73, all-time shifts τ_m must be zero-valued. It is practically not feasible because the individual sensor signals already have different travel times when the waves arrive. Therefore, an additional channel delay is necessary to compensate for the differences between the channels (Eqs. 5.72 and 5.74, Fig. 5.33):

$$\delta_m = -\tau_m \to \boldsymbol{\delta} = -\boldsymbol{\tau}. \tag{5.74}$$

If one implements the requirement according to Eq. 5.74, one obtains the DSBF already described but with variable channel delays, which are analytically unknown. The error functional is also unknown, but following the LMS algorithm (Eq. 5.50), one can proceed here according to the gradient method (Eq. 5.75):

$$\boldsymbol{\delta}(k+1) = \boldsymbol{\delta}(k) + \mu \nabla_k. \tag{5.75}$$

Since the gradient cannot be described analytically, it must be estimated (Eq. 5.76):

$$\nabla_k = \begin{Bmatrix} \frac{\partial snr(y)}{\partial \delta_1} \\ \vdots \\ \frac{\partial snr(y)}{\partial \delta_M} \end{Bmatrix}, \quad \hat{\nabla}_k = \begin{Bmatrix} \frac{\Delta snr(y)}{\Delta \delta_1} \\ \vdots \\ \frac{\Delta snr(y)}{\Delta \delta_M} \end{Bmatrix}. \tag{5.76}$$

In practical analysis, the difficulty lies in determining the current SNR. The following two approaches can be used for this: On the one hand, one can estimate the current SNR if periods are known in which the sought signal $s(k)$ is not contained, i.e., a signal-free noise reference is available (Eq. 5.77).

$$snr(y) = \frac{\mathrm{var}(s)}{\mathrm{var}(n)} = \frac{\mathrm{var}(x)}{\mathrm{var}(n)} - 1 \tag{5.77}$$

In Eq. 5.77. s is the searched signal $s(k)$, n *is* a section of the sum signal $x(k)$ in which only the noise $n(k)$ *is* located, x is the sum signal of $s(k)$, and $n(k)$, y is the arithmetic mean over the channel signals $x_{\mathrm{m}}(k)$. The variance is calculated in an appropriate time window. The snr may become negative due to natural stochastic fluctuations if the window is relatively short. It must be considered in practical application, especially if the *snr* is given logarithmically (in dB).

Suppose a noise reference is unavailable or signal-free sections are unknown. In that case, the total signal power or energy is used for the SNR, i.e., one does not normalize to the noise reference (Eq. 5.78).

$$snr(y) = \mathrm{var}(x) \tag{5.78}$$

The individual components (j is the channel index) of the gradient according to Eq. 5.76 can be estimated as follows (Eq. 5.79)

$$\nabla_k^j = \frac{\Delta snr(y(k))}{\Delta \delta_j} = \frac{snr\left(y_j(k)\right) - snr(y(k))}{\Delta \delta_j}, \tag{5.79}$$

where one calculates the current signal components according to Eq. 5.80:

$$y_j(k) = \frac{1}{M}\left(\sum_{i=1, i \neq j}^{M} x_i(k) + x_j\left(k + \Delta\delta_j\right)\right), \quad y(k) = \frac{1}{M}\sum_{i=1}^{M} x_i(k). \tag{5.80}$$

In Eqs. 5.79 and 5.80, $_\mathrm{j}$ corresponds to at least one sampling period, where nT_A $= _\mathrm{j}$ applies in principle. One can, therefore, also use the following equation for the time index (Eq. 5.81):

$$x_j\left(k + \Delta\delta_j\right) = x_j(k + n)\big|_{n \geq 1}. \tag{5.81}$$

In the relationships for determining the SNR (Eqs. 5.77–5.79), the window length for calculating the variance should be large enough for a reliable estimate but small enough for the signal dynamics (empirical estimation of the window length).

The procedure is illustrated on simulated data in the following example. The graph in Fig. 5.37 shows a realistic model of a 16-channel VEP from the occipital EEG with different channel delays and amplitudes according to the formulation in Eq. 5.71. Note that the amplitudes of the averaged signal occur significantly later than the individual components with the most significant levels. After the completed adaptation (Fig. 5.38), there are the signal waves are entirely in phase, so the simultaneity of the sought signal $s(k)$, according to Eq. 5.66, is ensured. The number of iteration steps necessary for the adaptation (Fig. 5.40) is relatively high to achieve an acceptable result. The variable course can also be observed in the built-in channel delays (Fig. 5.39): Not all channels adapt simultaneously, but delayed or sequentially. This effect is related to the caution to choose the constant so that the adaptation remains stable (Eq. 5.75). As in the LMS algorithm, an upper limit does not exist in this method since none of the signal components can be described analytically. Therefore, this part of the method is left to empiricism and can lead to long adaptations or instabilities. In practice, however, a suitable adaptation constant or range of the constant for a measuring arrangement and a signal type only has to be determined once (see exercises).

With the help of the adaptation algorithm (Eq. 5.75), the maximum SNR can be reached, and the conditions according to Eq. 5.73 can be fulfilled. After the

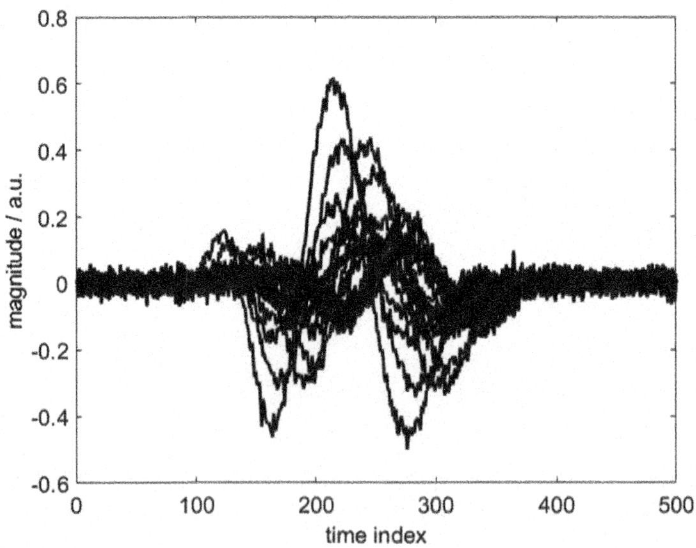

Fig. 5.37 Simulated VEP with a white, normally distributed, additively superimposed noise in 16 channels before adaptation. The averaged signal is marked in bold

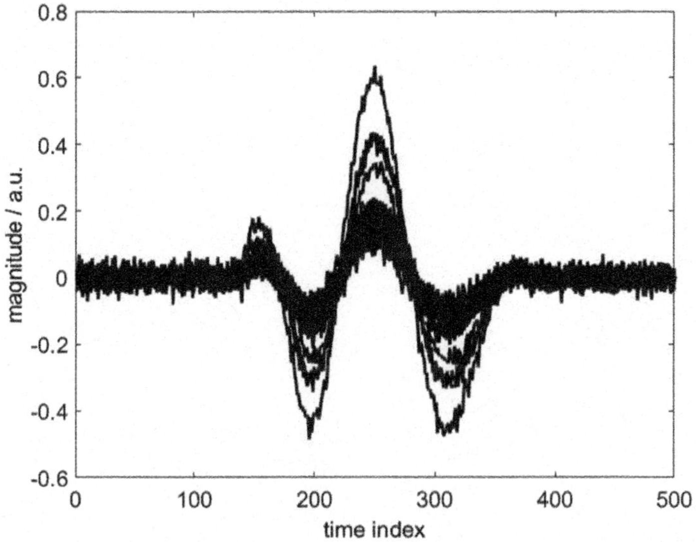

Fig. 5.38 Simulated VEP from Fig. 5.36 after completed adaptation. The averaged signal is marked in bold

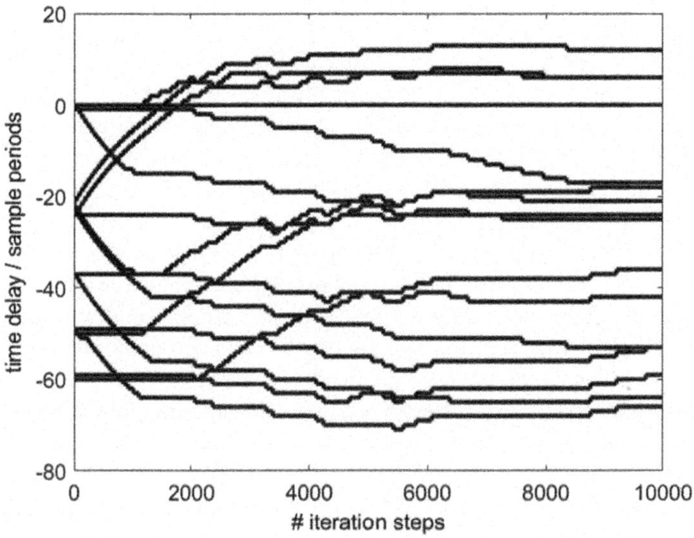

Fig. 5.39 Course of the channel delays during adaptation. Different signs refer to the reference channel. A positive/negative sign indicates a lead/lag of the channel under consideration concerning the reference channel

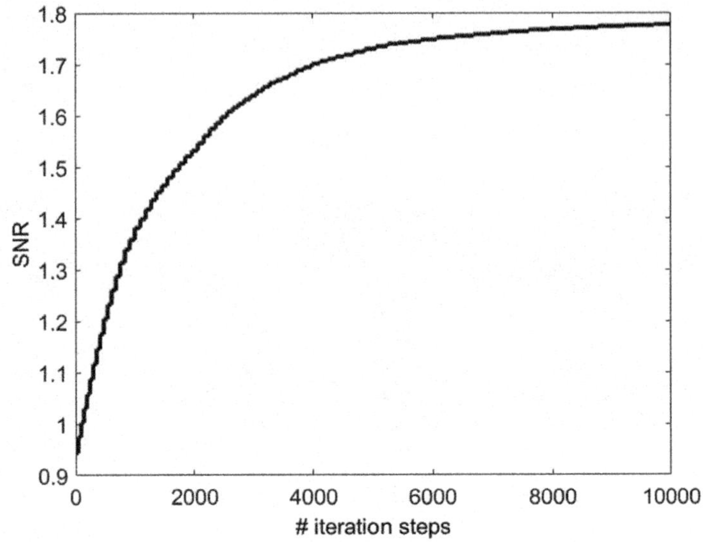

Fig. 5.40 Course of the SNR during adaptation. In this case, an almost doubling of the SNR or an increase of 6 dB could be achieved

adaptation has been completed, it is not sure that the global maximum of the SNR has been reached. Suppose the desired signal s(k) consists of several waves (Fig. 5.37), with an unfavorable choice of the reference channel (in the reference channel $\tau_m = \delta_m = 0$ is set) and with a large distance d between the sensors. In that case, the phase difference between the channels is more than $\pi/2$, or the delay difference is more than T/4. In such a case, only a local but no global maximum of the SNR may be reached.

Therefore, during adaptation, it must always be checked whether the achieved results of the SNR and the channel delays δ_m are plausible. The plausibility check is unproblematic in intact sensory and motor nerve pathways and centers. Since the individual waves of the signals can be assigned anatomically, the transit time differences can be estimated in advance, at least in their magnitude. In the case of disturbed conduction, the waves take on unpredictable forms. Thus, the results of adaptive spatial filtering should be consulted electrophysiologically and neurophysiologically.

5.4.4 Average Reference

The mean reference (CAR) has been described in Sect. 3.1.2; its effect is explained concerning medical measurement technology. Here it will be treated under the aspect of spatial filtering.

5.4.4.1 Theory of the Mean Reference

The theoretical approach to CAR is based on the fact that the area integral of the electric potential over a sphere must be zero-valued if the sphere has only internal current sources and there is no exchange of charge with the environment (Eq. 5.82):

$$\oiint_{\phi,\theta} \varepsilon(r,\phi,\theta) \equiv 0. \tag{5.82}$$

In Eq. 5.82, electrical surface potential, r, ϕ, θ are spherical coordinates of the spatial biosignal source and the electrical potential it generates, respectively.

Equation Eq. 5.82 applies to a sensor arrangement for detecting the electrical potential only under the following conditions:

- The sensors are evenly (equidistantly) distributed over the entire surface of the sphere,
- The sensor density is higher than the source density (spatial sampling theorem),
- The sphere is homogeneous and isotropic in terms of electrical conductivity.

The CAR would be an ideal (virtual) reference according to the formulation in Eq. 5.82 if the above conditions could be fulfilled in practice. Unfortunately, the conditions in a real biosignal derivation are far from ideal:

- No anatomical structure of a biosignal source corresponds to the sphere model, but the brain is still the best,
- No area projection of the electrical activity of a biosignal source can be recorded over the entire surface; in the case of the EEG, it is no more than about 30%,
- Even taking into account the spatial low-pass effect of the human body on the superficial biosignals, the sensor density usually is not higher than the source density, the spatial sampling theorem is not observed,
- The electrical conductivity in the human body is inhomogeneous and anisotropic in the range of several decades.

CAR must be rejected in principle for its application as a virtual reference for the reasons mentioned. However, if one considers it an alternative to spatial filtering, usable approaches arise for practical signal processing.

The Mean Reference as a Spatial Filter

The CAR is calculated with the same formula as the output signal of an average (Eq. 5.72). However, the intention here is fundamentally different—suppressing signals from distant sources. By average, we assume that the signals coming from the source of interest are in phase, and the SNR can be improved in this way. This assumption is only fulfilled sufficiently if the sensor array is placed locally over an accessible area while the signal source is relatively far away. For example, such a situation is given when one wants to detect brainstem potentials in the EEG

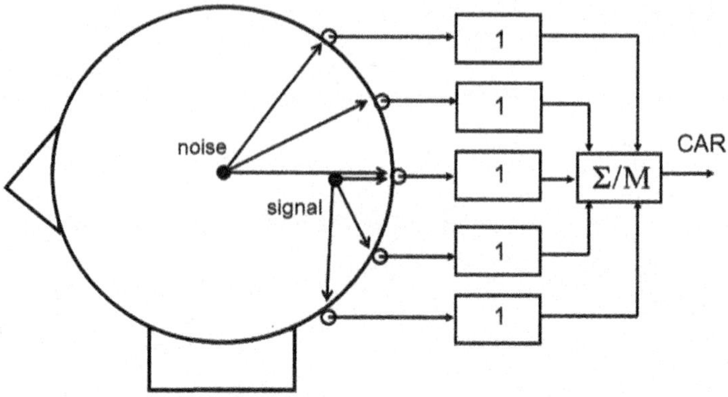

Fig. 5.41 Spatial signal model for the effect of the common average reference (CAR)

or MEG. Often, based on the theoretical assumption according to Eq. 5.82, the mean value signal is used as CAR to generate a supposedly indifferent reference. The channel signals are then related to this reference and interpreted as a unipolar derivative (Eq. 5.83).

$$car(k) = \frac{1}{M}\sum_{m=1}^{M} x_m(k),\ x_m^{car}(k) = x_m(k) - car(k) \tag{5.83}$$

The effect of CAR (Eq. 5.83, Fig. 5.41) on multi-channel biosignals initially presents itself as the elimination of in-phase components in the sum signal (electronically, it corresponds to common mode). The in-phase nature essentially results from the significant distance of the signal sources to the sensor array. One can therefore interpret the referencing of the multi-channel derivative to the CAR as a spatial filter that suppresses distant sources and emphasizes nearby sources. That is advantageous when near-surface potentials are to be recorded, e.g., potentials of a sensory system.

If the effect of the CAR is interpreted vectorially, it suppresses the *radial* components and preserves the *tangential* components of the biosignals.

The effect of the CAR on tangential and radial signal components can be controlled by inserting additional delays into the channels—similar to the DSBF (Fig. 5.42). However, these channel delays are not designed adaptively but are fixed as required. In the highly simplified spatial model shown in Fig. 5.42, the effect of the channel delays can be interpreted as creating virtual sensors that shift the focus closer to or further from the body surface, depending on the setting. The effects of the targeted channel delay are shown in Figs. 5.43 and 5.44. A VEP derived by conventional means (Fig. 5.43, continuous curve) was first averaged

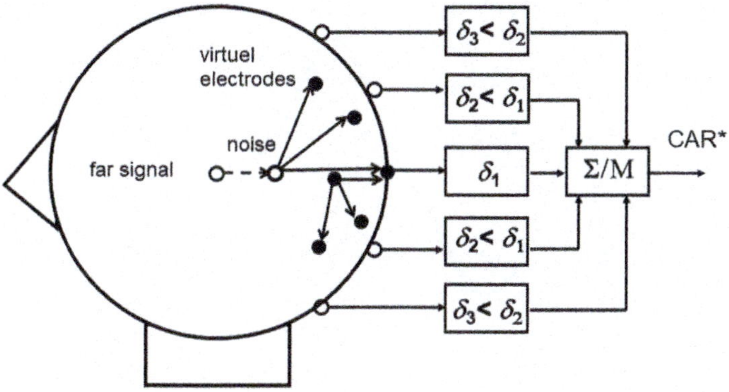

Fig. 5.42 Spatial model for mean reference with additional channel delays

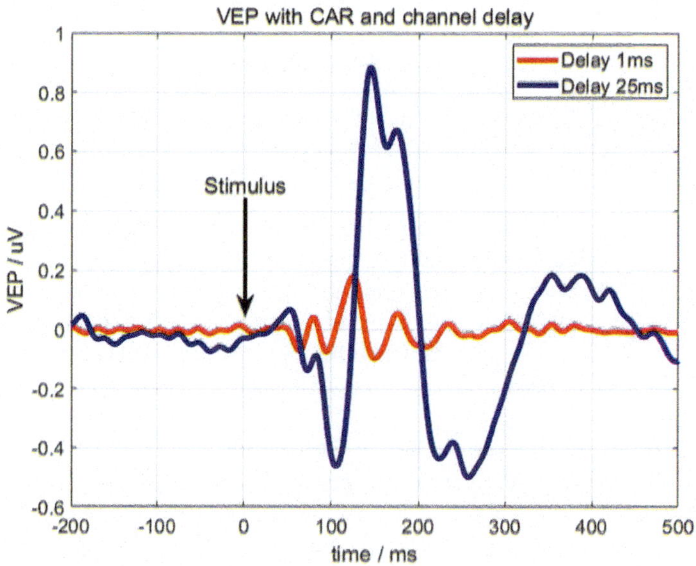

Fig. 5.43 Time course of a VEP derived from 16 channels over the occipital cortex. The information on the delay refers to the second outermost row of electrodes according to Fig. 5.42; the first row has no delay. Each further row of electrodes has a linearly incremented delay

over 50 individual responses in a stimulus-synchronous manner and then simultaneously applied to the Averager according to Eq. 5.72. For a local light stimulus, the relatively weak response with an amplitude of 0.2 μV at 120 ms after the stimulus is typical but can hardly be separated from the noise.

The CAR acts on the SNR or the VEP as a function of channel delays. It shows that—as already mentioned above—the SNR can have several maxima so that the complete (plausible) range must be scanned in the search for the global maximum.

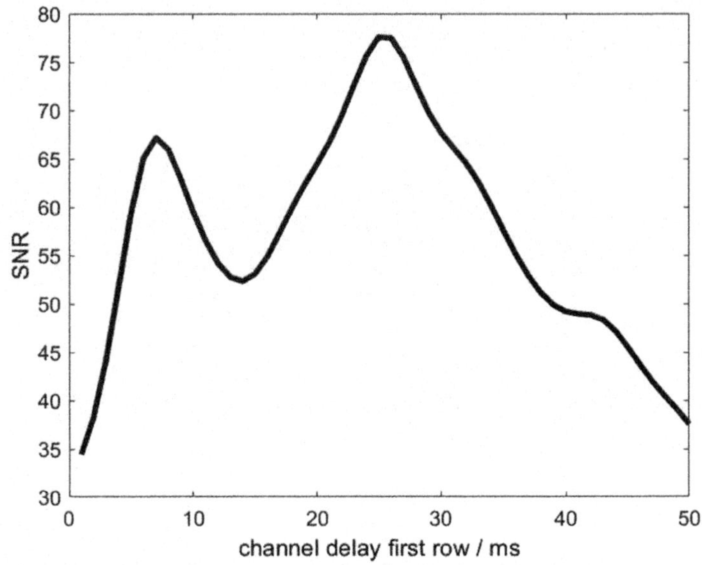

Fig. 5.44 Course of the SNR as a function of the additional channel delay of the second sensor row. Through the targeted control, the SNR could be more than doubled

This concrete experiment finds the global maximum at a channel delay of 25 ms (Fig. 5.44). This value is individually constant under constant examination conditions, so it must only be determined once. If one sets the determined optimal delay for the CAR, one obtains an approximately fivefold VEP amplitude in this specific test (Fig. 5.43, dashed curve). Once the optimal setting has been determined, evoked potentials can be displayed in real-time with the help of the CAR.

5.4.4.2 Effects of the Mean Value Reference

Section 5.4.4.1 discusses the theoretical prerequisites for applying the CAR as an indifferent reference electrode and the real conditions that prevent this approach. Here, the most common errors in the practical analysis that can lead to erroneous conclusions in the medical interpretation will be explained.

Disturbances at the Physical and the Physiological Reference Electrode

The physical reference electrode (Fig. 3.1, electrical ground) of the medical amplifier and the physiological reference electrode (Fig. 3.3, declared physiological reference) are the most sensitive points of the measurement arrangement from the point of view of possible disturbances of the sensor array. All technical (e.g., mains interference) and biological disturbances (e.g., reference positioned over an arterial vessel) appear directly in the measurement signal. The disturbances enter all channel signals via the built-in, electronically realized, or CAR calculated at the DSV. This effect can be explained mathematically. The channel signals $x_m(k)$ used in the formula according to Eq. 5.83 are theoretically monomodal, but in practice, they are always potential differences to a physical reference (see Sect. 5.2) or differences between two measuring points (Eq. 5.84).

$$x_m(k) = \varphi_m(k) - \varphi_R(k) \tag{5.84}$$

In Eq. 5.84, $x_\mathrm{m}(k)$ is the signal in channel m,$_\mathrm{m}(k)$ is the potential at sensor m, and φ_m (k) is the potential at the reference sensor.

For a unipolar derivative, it follows from Eq. 5.84 that any disturbance at the reference sensor or the ground connection is directly transmitted into all channels of the sensor array. The CAR can initially remedy this situation since the following applies when Eqs. 5.83 and 5.84 are taken into account (Eq. 5.85):

$$x_m^{car} = (\varphi_m - \varphi_R) - \left(-\varphi_R + \frac{1}{M}\sum_{i=1}^{M}\varphi_i\right) = \varphi_m - \frac{1}{M}\sum_{i=1}^{M}\varphi_i. \tag{5.85}$$

(Eq. 5.85 and other formulations omitted the time index k for better clarity).

According to Eq. 5.85, the CAR contributes—at least theoretically with the caveats mentioned—to the fact that the measurement data referenced on the CAR are independent of the actual reference $x_\mathrm{R}^*(k)$. However, this approach leads to problems that will be addressed subsequently. With the more frequently applied bipolar derivation, the following applies (Eq. 5.86):

$$x_{mn} = x_m - x_n = (\varphi_m - \varphi_R - car) - (\varphi_n - \varphi_R - car) = \varphi_m - \varphi_n. \tag{5.86}$$

From the relationship, according to Eq. 5.86, it follows that on a bipolar biosignal, neither the actual reference (precondition is a prior unipolar derivative according to Eq. 5.84) nor the CAR (if present) has any influence. The previous statements can be summarised in the following theses:

- A (physical or physiological) reference influences a unipolar biosignal to the same extent as the signal itself (Eq. 5.84).
- The mean reference CAR eliminates the influence of physical and physiological references (Eq. 5.85).
- A bipolar biosignal is independent of physical, physiological, and virtual (CAR) references (if any) and is, therefore, the most robust against interference (Eq. 5.86).

Transfer of Local Activities to the Sensor Array

The application of CAR for referencing multi-channel biosignals, according to Eq. 5.85, can lead to problems in interpreting the spatial distribution of the examined signals in addition to positive effects concerning the disturbances. If one initially interprets the CAR referencing only in terms of signal analysis, one finds that

- apart (1/M) of all other sensor potentials is additively superimposed (with reversed sign) on each sensor potential,
- thus the influence of the unwanted individual channels increases reciprocally with the number of channels,
- mixing increases the cross-correlation between the channels,
- in the sum of all the effects mentioned, each sensor potential simulates the presence of signal components that are not present.

These effects are demonstrated in the following realistic simulation example. A 4 × 3 sensor array is assumed to capture a projection of electrical neuronal activity, as is common in the mapping of muscle activity (EMG), cardiac activity (ECG), or cortical activity (EEG). Here, the alpha wave (frequency 10 Hz) arising in the occipital cortex was simulated, additively embedded in a normally distributed stationary white noise (Fig. 5.45).

All other areas Frontal (F1, Fz, F2), Central (C1, Cz, C2), and Parietal (P1, Pz, P2), initially contain only noise. When channels are referenced to the CAR according to Eq. 5.85, the spectra on all sensor positions also change (Fig. 5.46). The amplitudes of the alpha waves in their area of origin (occipital) are reduced. Still, alpha waves in all other regions appear with a correspondingly lower amplitude (the signal energy of the alpha waves was distributed from three to twelve channels). The consequence of CAR referencing is that signal components from a local area of the examined structure have been carried into all other areas where they were not initially contained, thus electrophysiologically merely feigning their presence. As in this example, the problem can still be recognized early and clearly with simulated signals. With real data, however, it is impossible to check the truth of the signal analysis since neither the signal sought nor the disturbances are known in advance. The consequence of such an incorrect interpretation can sometimes be bizarre conclusions or even fatal diagnoses in the medical field (e.g., "frontal mirroring of the alpha rhythm," which was simulated with this simulation).

An essential part of biosignal processing is the statistical analysis of data and results (see the section on stochastic processes). Correlation analysis is fundamental in BSV, as it can be used to investigate and prove statistically reliable correlations. The initial orientation is provided by the well-known correlation coefficient (KK), according to Pearson. Figure 5.47 shows the KK matrices of the channel signals simulated in the example.

Grey values correspond to the amount of KK, where black (main diagonal) represents the value of KK = 1.0 and white represents the value of KK = 0.0. In the original signal, the occipital channels (channel numbers 10–12) contain noisy alpha waves, which correlate very well (high similarity). The remaining channels contain only noise, which theoretically has a zero-value correlation, which follows the noise's natural fluctuations. After CAR referencing, the correlation between the previous noise channels (channels 1–9) and the occipital channels increases strongly regarding its absolute value and becomes negative (the occipital harmonics contained in the CAR change their sign because of subtraction in referencing to CAR). At the same time, it decreases between the occipital channels (due to

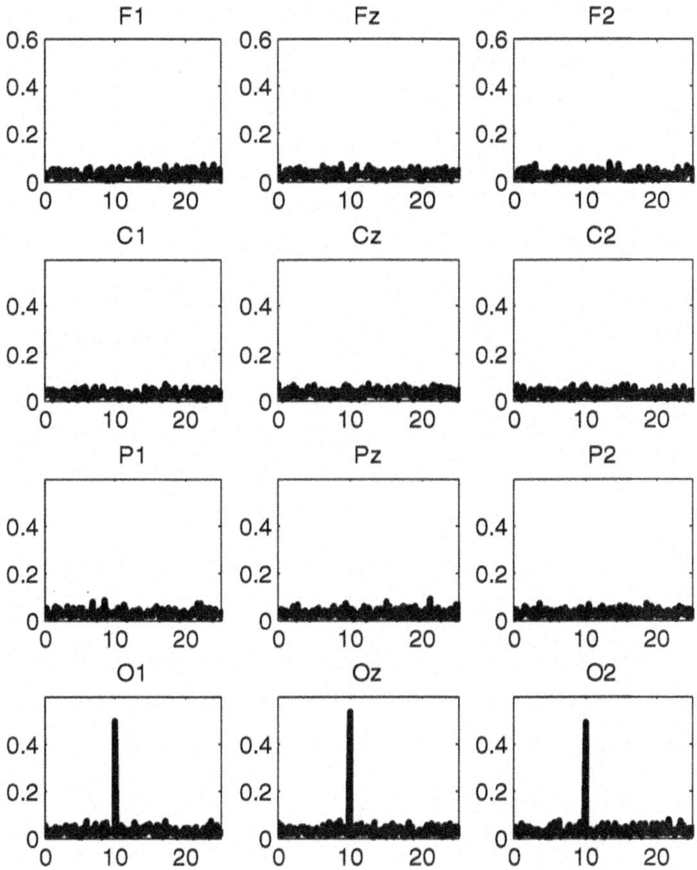

Fig. 5.45 Spectrum (relative amplitude vs. frequency in Hz) of a simulated EEG with alpha waves in the occipital area (O1, Oz, O2). The designation of the electrode positions corresponds to the EEG standard

SNR becoming worsened). It fakes a statistical correlation that was not present in the original data. The possible consequences of such a false conclusion were mentioned above.

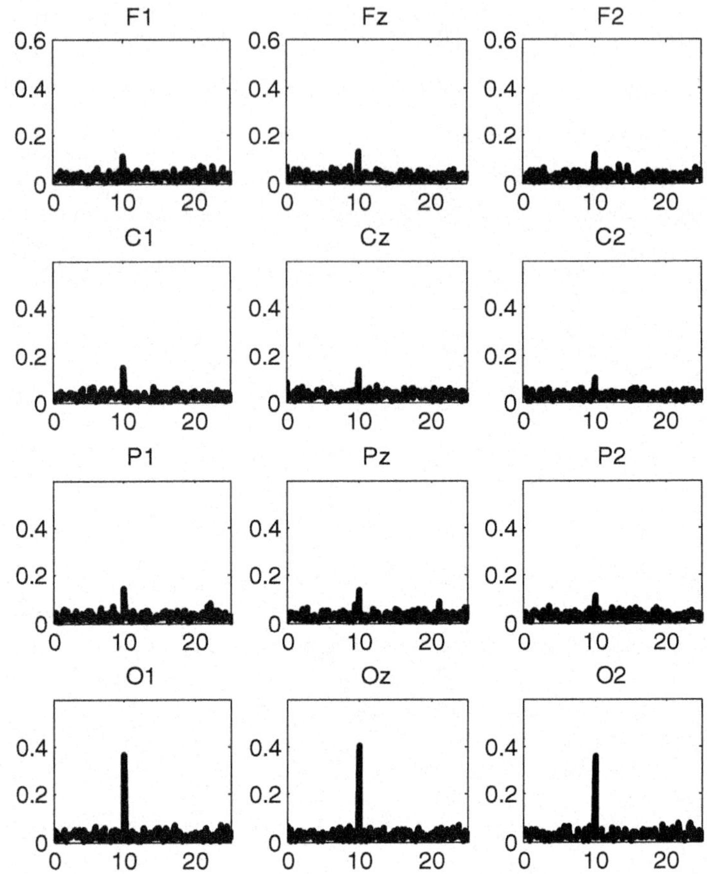

Fig. 5.46 Spectrum (relative amplitude vs. frequency in Hz) of a simulated EEG with alpha waves in the occipital area (O1, Oz, O2) after referencing the CAR of the data from Fig. 5.45. The designation of the electrode positions corresponds to the EEG standard. Amplitudes of the alpha waves in the occipital area are 75%; in the other regions, 25% of the original value or 8.3% in the respective series

5.5 Exercises

5.5.1 Tasks

5.5.1.1 Analog and Digital Filters—Pulse Invariant Technique

Design a 3rd-order analog low pass with a cut-off frequency of 45 Hz. Compare the calculated analog impulse response with the digital impulse response and the impulse response of a simulated electronic circuit. Calculate the filter coefficients of a Butterworth low pass for a sampling rate of 250 sps using the impulse invariant technique according to the following scheme:

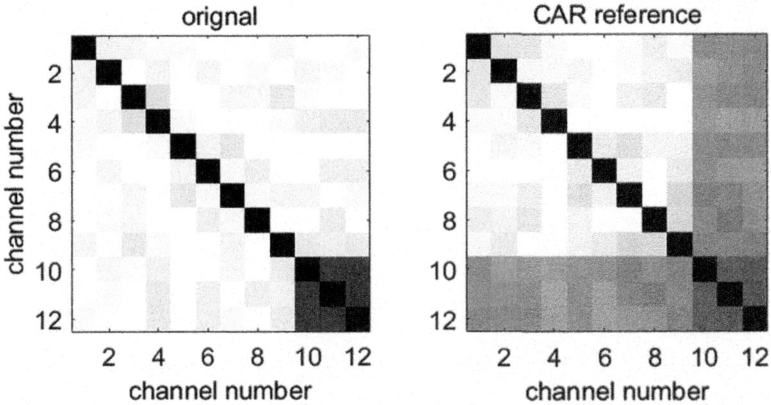

Fig. 5.47 Correlation matrix (magnitude of the correlation coefficient) of the simulated signal (left) from Fig. 5.45 with alpha waves in the occipital region (channel numbers 10–12) and of the same signal (Fig. 5.46) referenced to the CAR (right). The grey values on the main diagonal (black, maximum) correspond to a correlation coefficient 1.0 (autocorrelation), and white corresponds to a correlation coefficient 0.0

1. Analog transfer function

$$G_a(s) = \frac{b(s)}{a(s)} = \frac{b_1 s^N + \cdots + b_{N+1}}{s^N + \cdots + a_{N+1}}$$

2. Partial fraction decomposition

$$G_a(s) = \sum_{i=1}^{N} \frac{k_i}{s - p_i}$$

3. Analog impulse response via the inverse Laplace transform

$$g_a(t) = \sum_{i=1}^{N} k_i e^{p_i t},$$

4. Sampled impulse response (T is the sampling period)

$$g_s(n) = T g_a(nT) = T \sum_{1}^{N} k_i e^{p_i n T},$$

5. Z-transformation from

$$G_s(n)G(z) = \sum_{i=1}^{N} \frac{T k_i}{1 - p_i z^{-1}},$$

6. Polynomial conversion

$$G(z) = \frac{b_0 + b_1 z^{-1} + \cdots + b_m z^{-m}}{1 + a_1 z^{-1} + \cdots + a_n z^{-n}},$$

7. Recursive formula

$$y(n) = b_0 x(n) + b_1 x(n-1) + \cdots - a_1 y(n-1) - a_2 y(n-2) - \cdots$$

5.5.1.2 Design of an FIR Filter According to the Window Method

A digital low-pass filter with a cut-off frequency of 45 Hz for limiting the spectrum of an ECG sampled at 250 sps and, above all, for suppressing the mains interference is to be designed. Investigate the influence of the window function and the window length on the amplitude and phase frequency response or group delay for the window functions rectangle, triangle (Bartlett), Hamming, and Blackman for the window lengths 11, 101, and 1001. Which filter best fulfills the band-limiting requirement? Which filter would be suitable for real-time signal processing (e.g., pacemaker)?

5.5.1.3 LTV, Linear Time-Variable Filters

Simulate the course of an acoustically evoked potential using a hyperbolic chirp. Superimpose this chirp additively with a linear chirp with increasing frequency to simulate the interference from the acoustic stimulator (Fig. 5.15). Plot the superimposed chirps in the time–frequency plane using the spectrogram. Design an arbitrary mask (not necessarily a binary one) so that the linear chirp is suppressed and the ROI includes the hyperbolic chirp (see Fig. 5.16). From this mask, calculate the time-varying impulse response of the LTV filter (Eq. 5.36) and apply it to the time signal (Eq. 5.33). By calculating the spectrogram again, check the effect of the LTV filter.

5.5.1.4 LTV, Adaptive Filters

A stored IPG (impedance plethysmograph) with disturbances from the mains, a tube monitor, and a noise reference is stored (*ipgnois.mat*). The values of the IPG are normalized to m and the time axis to the sampling rate of 300 sps. Design an adaptive filter to remove the noise from the IPG. A simple bandstop would be conceivable for the mains frequency but not for the interference from the monitor, as this can change its frame rate. Investigate the influence of the filter length and the adaptation constant on the filtered IPG.

5.5.1.5 Adaptive Spatial Filtering

Design a workable adaptive spatial filter. A simulated 16-channel noisy evoked potential with delay differences in the individual channels is available and can be loaded from the file *vepnois16ch.mat*. Note that the travel time differences concerning the primary wave are more than half a period. Dimension the adaptation so that the algorithm is neither divergent nor unstable. Note that the adaptation step size of the channel delay can also be smaller than the sampling period for the sake of stability. By accumulating the small steps, one can achieve the necessary change in the grid of sampling periods. Investigate the influence of the noise power on the adaptation and determine the lowest SNR at which the adaptation still works.

5.5.1.6 Influence of the Mean Reference

Investigate the effect of the mean value reference on the transmission of local activities to the entire sensor network. To do this, simulate a sensor network (sensor matrix) in which the individual sensors produce or receive only white and spatially uncorrelated noise. Simulate a local activity in the sensor network with a few sensors. They should contain a deterministic signal additively superimposed to the noise (e.g., a harmonic oscillation or a transient course). This simulation represents an ideal unipolar derivative. The noise and the deterministic signal were generated independently of any reference. Reference the sensor network to the mean reference by calculating it according to Eq. 5.85 and subtracting it from the sensor signals according to Eq. 5.84. Investigate the effect of the average reference in time and frequency domains and the sensor correlation (Figs. 5.48 and 5.49).

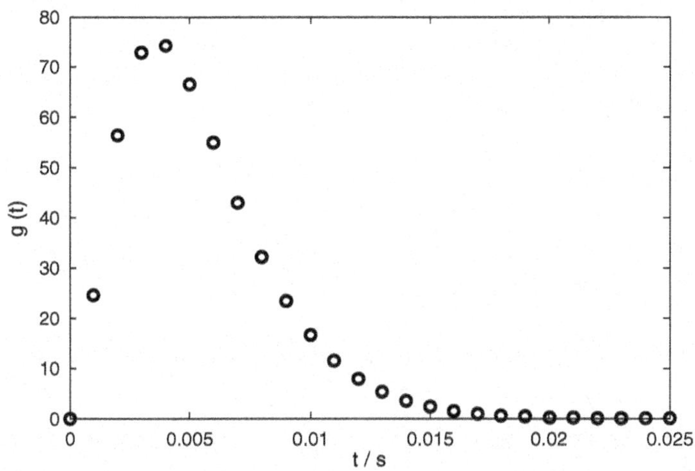

Fig. 5.48 Sampled impulse response of a 3rd-order low pass consisting of three identical first-order low-passes. The sampling rate is 1000 sps. The analog and discrete-time impulse responses are calculated in the Matlab function *uebung_5_1.m*

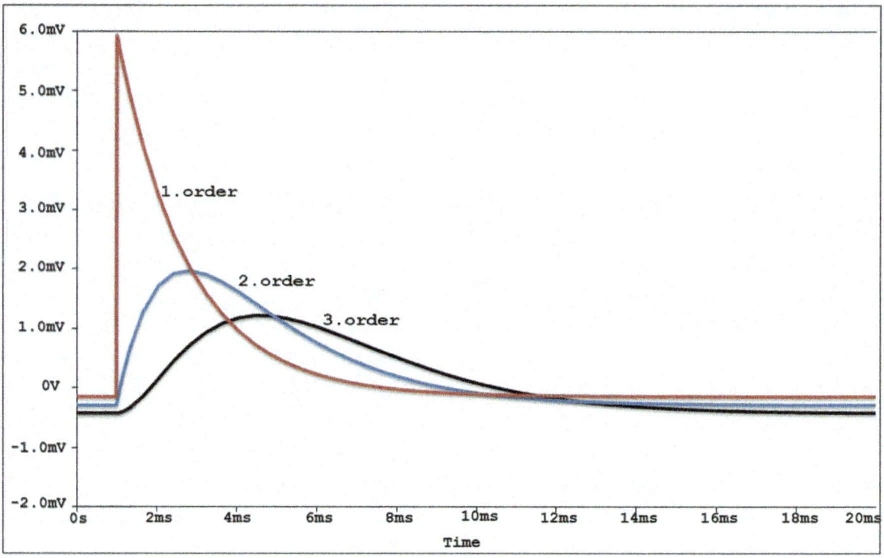

Fig. 5.49 Impulse responses of individual stages of the 3rd order low pass in the analog range (PSpice simulation BSV_5.1)

5.5.2 Solutions

5.5.2.1 Analog and Digital Filters—Pulse Invariant Technique

In the simplest case, a 3rd-order low pass consists of a series connection of five identical 1st-order low passes. The transfer function of such a series circuit is (T = RC)

$$G(j\omega) = \frac{1}{(1 + j\omega T)^3}.$$

After exponentiation, one obtains the transfer function

$$G(j\omega) = \frac{1}{1 - 3\omega^2 T^2 + j\left(3\omega T - \omega^3 T^3\right)}$$

and with the help of the inverse Fourier transform, the analog impulse response

$$g(t) = \frac{t^2}{2T^3} e^{-\frac{t}{T}}.$$

At the cut-off frequency ω_g, the amplitude-frequency response drops by 3 dB so that the following applies

$$\left|G\left(j\omega_g\right)\right| = \frac{1}{\sqrt{2}}.$$

If one applies this requirement to the transfer function, one obtains the time constant

$$T = \frac{\sqrt{\sqrt[3]{2} - 1}}{\omega_g}.$$

With the required cut-off frequency of 45 Hz, the time constant $T = 1.8$ ms, and the cut-off frequency of the individual 1st-order low-pass filters is $f_g = 88$ Hz. The analog impulse response must be sampled. That first requires the selection of a suitable sampling rate. It is known that the cut-off frequency of the entire filter is 45 Hz and that the filter is of order three. It means that at a distance of one decade, i.e., at 450 Hz, the filter has a suppression of 60 dB. Setting the Nyquist frequency here is sufficient, resulting in a sampling rate of about 1000 sps. The sampled impulse response is shown in Fig. 5.50 (see *uebung_5_1.m*). For comparison, this low pass can be simulated in PSpice (see *BSV_5.1.sch*).

According to the given scheme, one can calculate the filter coefficients using the impulse invariant technique as follows:

1. With the help of the Matlab function *butter.m*, one obtains the coefficients of the analog transfer function

$$G_a(s) = \frac{b(s)}{a(s)} = \frac{2.26 \times 10^7}{s^3 + 565s^2 + 1.5989 \times 10^5 s + 2.26 \times 10^7},$$

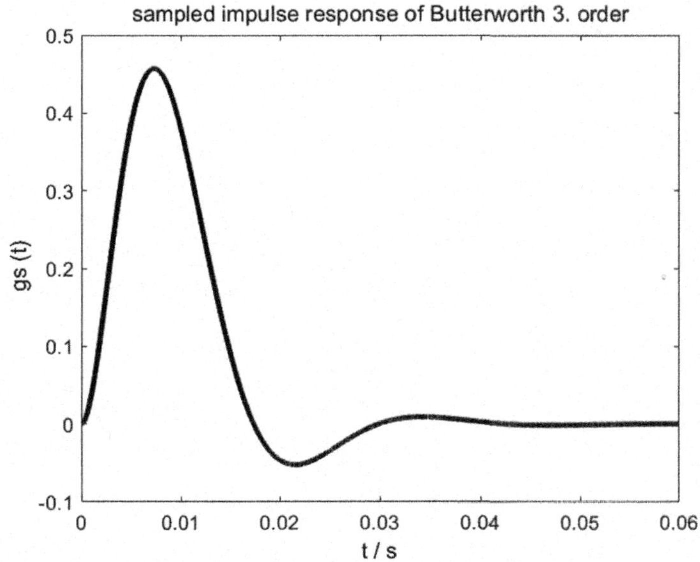

Fig. 5.50 Sampled impulse response of the 3rd order Butterworth low pass with a cut-off frequency of 45 Hz and a sampling rate of 250 sps

2. Partial fraction decomposition

$$G_a(s) = \frac{282.7}{s + 282.7} + \frac{-141.4 - 81.6i}{s + 141.7 - 244.8i} + \frac{-141.4 + 81.6i}{s + 141.7 + 244.8i},$$

3. Analog impulse response via the inverse Laplace transform

$$\begin{aligned} g_a(t) = 282.7e^{-282.7t} &+ (-141.4 - 81.6i)e^{(-141.4+244.8i)t} \\ &+ (141.4 + 81.6i)e^{(-141.4-244.8i)t}, \end{aligned}$$

4. Sampled impulse response (see Fig. 5.50),

$$g_s(n) = 0.004 \cdot \begin{pmatrix} 282.7 \cdot e^{-282.7 \cdot n \cdot 0.004} + \\ (-141.4 - 81.6i) \cdot e^{(-141.4+244.8i) \cdot n \cdot 0.004} + \\ (141.4 + 81.6i) \cdot e^{(-141.4-244.8i) \cdot n \cdot 0.004} \end{pmatrix}$$

5. Z-transformation of $g_s(n)$

$$\begin{aligned} G(z) &= \sum_{i=1}^{N} \frac{Tk_i}{1 - p_i z^{-1}} \\ &= \left(\frac{1.13}{1 - e^{1.13} z^{-1}} + \frac{-0.5656 - 0.3264i}{1 - e^{(-0.5656+0.9792i)} z^{-1}} + \frac{-0.5656 + 0.3264i}{1 - e^{(-0.5656-0.9792i)} z^{-1}} \right), \end{aligned}$$

6. Polynomial formulation

$$G(z) = \frac{-0.0004z^{-2} + 0.3148z^{-1} + 0.1498}{z^{-3} - 0.9563z^{-2} + 0.5271z^{-1} - 0.1041},$$

7. Filter equation

$$\begin{aligned} -0.1041y(n) &= 0.1498x(n) + 0.3148x(n-1) - 0.0004x(n-2) \\ &- 0.527y(n-1) + 0.956y(n-2) - y(n-3). \end{aligned}$$

5.5.2.2 Design of an FIR Filter According to the Window Method

Figure 5.10 shows an ideal low pass's transfer function and its sampled impulse response for the first 1001 sampling points. This impulse response is multiplied (masked) with the given window functions and lengths, and the corresponding amplitude frequency response is shown (see *exercise_5_2.m*). The filter functions for the window length of N = 11 are shown in Fig. 5.51, for the window length of N = 101 in Fig. 5.52. Generally, one can initially state that from a window length of about 1000 coefficients, the desired transfer function is achieved almost ideally in most cases. However, this filter order is unsuitable for fast calculations.

For practical signal processing, significantly shorter filters are necessary. If one compares the window lengths of $N = 11$ and $N = 101$, the better the filter effect is achieved, the higher the filter order of $N = 101$, as expected. If one compares the (given) window functions with this filter order, the clear favorite is the Blackman window (Fig. 5.52).

For a real-time application, however, a window length of $N = 101$ would be completely unsuitable, as it would lead to a time delay of at least $t_d = T_A * (N-1)/2 = 200$ ms (t_d is the group runtime or the delay, T_A is the sampling period, here 4 ms). Therefore, the window length must be significantly shortened once again. If one compares the filter functions for the window length of $N = 11$ (see Fig. 5.51), the favorite is no longer as apparent as with $N = 101$. The Blackman window offers the highest blocking attenuation but has the widest filter edge. The rectangular window has the steepest filter slope but the lowest stopband attenuation. If there is a need for a compromise between these extremes, the Hamming

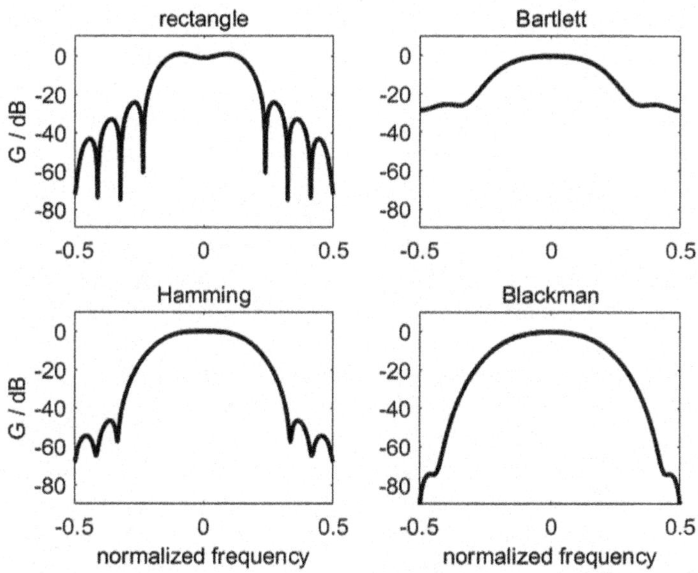

Fig. 5.51 Amplitude-frequency responses with window functions and a window length of $N = 11$

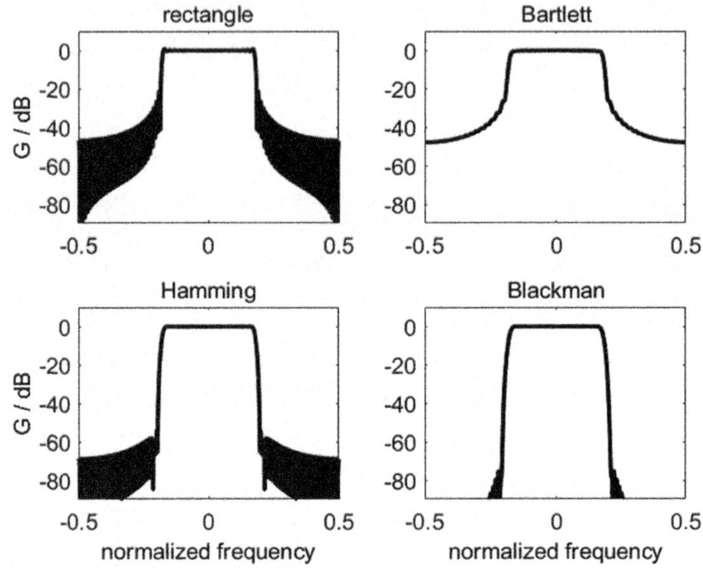

Fig. 5.52 Amplitude frequency responses with different window functions and a window length of $N = 101$

window offers it. In contrast, the Bartlett window is completely unsuitable from both points of view.

5.5.2.3 LTV, Linear Time-Variable Filters

The LTV filter can be calculated in two ways: If the sought signal is known, one can define the ROI can be defined as a pass filter. However, the interference is known more often, as in this case, for which one can formulate a blocking filter.

Assuming an additive superposition of a hyperbolic (nominal signal) and a linear (interference signal) chirp (Fig. 5.15), a time-variable bandstop can be calculated for the linear chirp (Fig. 5.53). The same formula used to generate the chirp can be used to calculate the filter to ensure that the bandstop also covers the chirp; a shorter window is used for its transfer to the time–frequency plane than for the transformation of the linear chirp (Fig. 5.53). Then one inverts the time–frequency distribution with the shorter window and binarises the block (see *uebung_5_3.m*). After applying the mask to the spectrogram, the linear chirp is largely eliminated, and part of the hyperbolic chirp (Fig. 5.54). One can calculate the time-variable impulse response according to Eq. 5.36 for the time-variable bandstop. However, this is impractical because it is the same length as the signal. For a practicable filter, one would have to calculate much shorter time-varying impulse responses with a particular design procedure so that they can also be applied to the time signal in the course. For details on a filter design, please refer to the literature (Boashash, 2003). Adaptive filters offer a variant for calculating the filter coefficients.

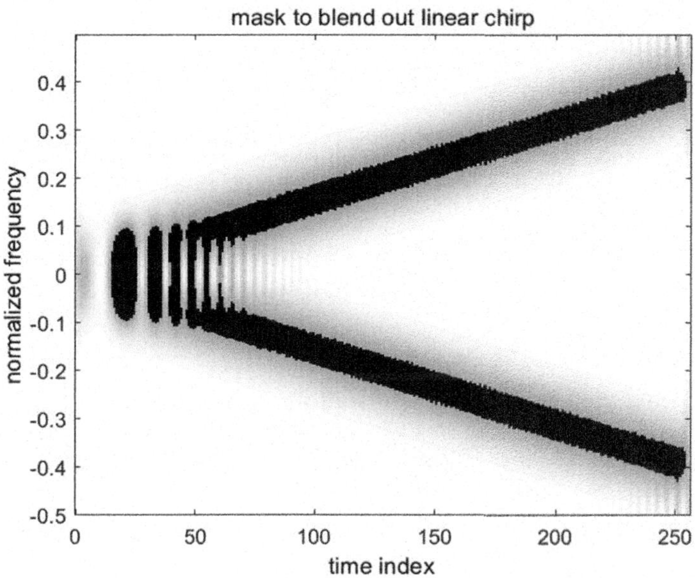

Fig. 5.53 Time variable bandstop to hide the linear chirp in Fig. 5.15

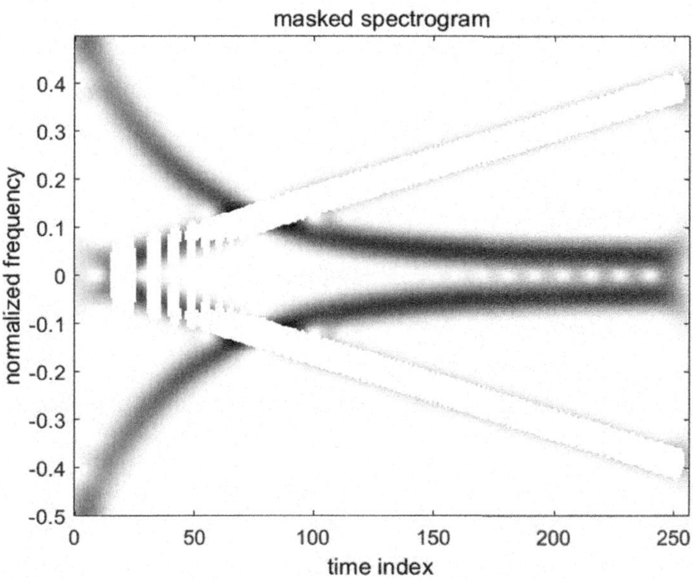

Fig. 5.54 Spectrogram of the additive superposition of a linear and a hyperbolic chirp after applying the time-variable bandstop to the linear chirp

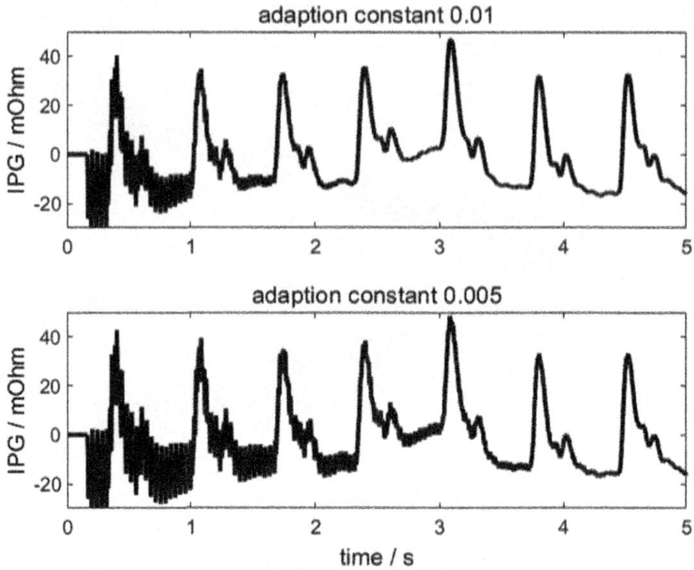

Fig. 5.55 IPG after adaptive filtering with an adaptation constant of 0.01 (top) and 0.005 (bottom)

5.5.2.4 LTV, Adaptive Filters

First, the length of the transversal filter is determined. It makes sense to use odd numbers so that the impulse response is symmetrical around the start value at = 0. Otherwise, the filter length can be chosen freely. When choosing the filter length, considerations of spectral resolution and dynamics play a decisive role. With increasing filter length, the spectral resolution increases, and the dynamics decrease. As an initial value for this task, $N = 101$ is recommended. The normalized adaptation constant must lie in the range $0 \ll 1$; otherwise, its choice is initially free (see *uebung_5_4.m*). A high constant leads to fast adaptation and more significant interference residues in the filtered signal (Fig. 5.55).

5.5.2.5 Adaptive Spatial Filtering

According to Eqs. 5.74–5.81, algorithms will implement the relationships when designing the adaptive spatial filter. In numerical mathematics, the following boundary conditions must be taken into account, which results from the theoretical formulation:

1. To calculate the gradient estimate according to Eq. 5.79 using Eq. 5.80, the amount of data or the data of the channel-related analysis windows must not change. For this purpose, the analysis windows of the channels can be filled up on both sides with zero sequences of sufficient length. A change in the channel data would simulate a gradient, which would not result from the time shift.
2. The adaptation of the channel delay, according to Eq. 5.81, can, of course, only be done in the least available steps—with the sampling period. However, the

sampling period as a quantization step can already lead to instability. Therefore, it is advisable to choose a significantly smaller step size $\Delta\delta_,$ in the algorithm and accumulate it until it reaches the length of the sampling period. Therefore, one can supplement Eq. 5.81 with the algorithmic solution:

$$\sum_i \mu \cdot \Delta\delta_i > T_A \Rightarrow n := n + 1.$$

The investigation of the adaptation on the SNR shows that the borderline SNR lies at about SNR = 5 dB. Below this value, the gradient can hardly be determined, and the gain achieved by the adaptation is only insignificant. To check the proposed solution, one can use the function *uebung_5.5.m* and change its parameters accordingly.

5.5.2.6 Influence of the Mean Reference

One chooses a sensor matrix with dimensions similar to those used in EEG or mapping techniques, e.g., $S(m, n)$, $m = 4$, $n = 3$. A signal sequence of length L is assigned to each sensor, initially the required white and spatially uncorrelated noise: $s_{\mathrm{m,n}} = \{\mathrm{N}_1\ (0,\ 1),\ \ldots,\ N_{\mathrm{L}}\ (0,\ 1)\}$, where $N(0, 1)$ represents the normal distribution with mean = 0 and scatter = 1. A sequence length of at least L = 1000 is recommended for the correlation analysis. For example, local activity can be simulated by giving the last sensor row ($m = 4$) a harmonic oscillation. This approach would correspond, in reality, to the emergence of -waves in the occipital cortex (see Fig. 5.45). After calculating the CAR and subtracting it from all sensor sequences, we find (see Fig. 5.46):

1. The deterministic signal that was originally only present locally is contained in all sensors.
2. The amplitude of the deterministic signal decreased in the original local activity. Amplitude ratios of active and passive regions behave according to the quotient of the number of sensors in the active to the passive region.
3. The signals in the passive region have an opposite phase or sign compared to those in the active area (a consequence of CAR subtraction).
4. The correlation between sensors in the active region decreases; in the passive area, it increases. The correlation between sensors in the active and passive region lies between the two correlations mentioned (see Fig. 5.47).

The uebung_5_6.m function can be used to check or investigate the influence of various parameters.

References

Ingle, V. K., & Proakis, J. G. (2007). *Digital signal processing using MATLAB*. Thomson.

Widrow, B., Glover, J., McCool, J., Kaunitz, J., Williams, C., Hearn, R., Zeidler, J. R., Dong, J. E., & Goodlin, R. (1975, December). Adaptive noise canceling: Principles and applications. *Proceedings of the IEEE, 63*, 1692–1716.

Part III
Biostatistics and Stochastic Processes

Biostatistics 6

6.1 Introduction

In technology and science, stochastic refers to processes that (at least partially) have a random character or to which a random character is attributed. Strictly speaking—according to the principle of causality—there is no random process in the macroscopic area (biosignals). At this point, the theory of relativity, quantum physics, and all philosophically based theories are excluded from further consideration. The discussed area narrows to biology, electro neurophysiology, and electrically based biosignal analysis. When one cannot describe a process analytically because of too much effort, that does not necessarily mean it is a purely random process. Unknown and verified correlations indirectly prove that supposedly random processes obey specific laws. In signal processing, for example, it is generally accepted that thermal noise is a random process that is also white and normally distributed in an almost ideal way.

Nevertheless, this process has apparent, physically based regularities: The noise power increases with increasing temperature and spectral bandwidth. In extreme cases, these regularities could be followed up to the level of individual electrons or ions. In standard cases, this analytical exactness is optional for signal processing and useless. For theory and practice, a single parameter is sufficient to describe thermal noise: The noise power or the variance in terms of second-order statistics. These considerations conclude that all processes (under the abovementioned restriction) are subject to causal relationships. For the analysis, however, one assumes for reasons of simplification or ignorance of causality that they are random processes. This view of conventional signal theory was initially adopted

Supplementary Information The online version contains supplementary material available at https://doi.org/10.1007/978-3-662-67998-2_6.

P. Husar and G. Gašpar, *Electrical Biosignals in Biomedical Engineering*,
https://doi.org/10.1007/978-3-662-67998-2_6

unchanged in biosignal processing. However, the practical analysis of biosignals quickly revealed the limitations of this approach. No assumption about signal properties valid in technical and scientific fields (stationarity of processes, white or known modified spectrum, Gaussian or known distribution of random variables) also applies to biosignals.

Nevertheless, for lack of acceptable alternatives, one can only do with the comprehensive, cumulative tools of stochastic process analysis. Therefore, the way out may initially be to adapt the established methods to the natural signal properties. The following chapter deals with this approach: Building on conventional analytical statistics and process analysis, ways to apply them to biosignals are shown. Another way out is offered by developing new analytical methods that assume natural signal properties from the outset. This approach is the subject of current research—also at the Institute BMTI of the TU Ilmenau—and can be looked up in corresponding sources if interested.

6.2 Fundamentals of Analytical Statistics

6.2.1 Distributions of Random Variables

6.2.1.1 Continuous and Discrete Distributions

A random variable is called continuous if it can assume all real values. Then the function (Eq. 6.1) is called

$$F(x) := P(X < x), \quad x \in (-\infty, \infty) \tag{6.1}$$

as a *distribution function* of X (*P* is the probability, *X* is the random variable, and *x* is a concrete value of the random variable *X*). The probability of occurrence of values in interval (a, b) is calculated as the area of the distribution function in this interval (Eq. 6.2).

$$P(a \le X \le b) = F(b) - F(a) = \int_a^b f(x)dx \tag{6.2}$$

From Eq. 6.2 follows

$$f(x) = \frac{d}{dx}F(x), \tag{6.3}$$

$f(x)$ is the *distribution density* (probability density function). The best-known and one of the most essential distributions for statistics is the Gaussian distribution (Eq. 6.4), also called the normal distribution.

$$f(x|\mu,\sigma) = \frac{1}{\sigma\sqrt{2\pi}}e^{\frac{-(x-\mu)^2}{2\sigma^2}} \tag{6.4}$$

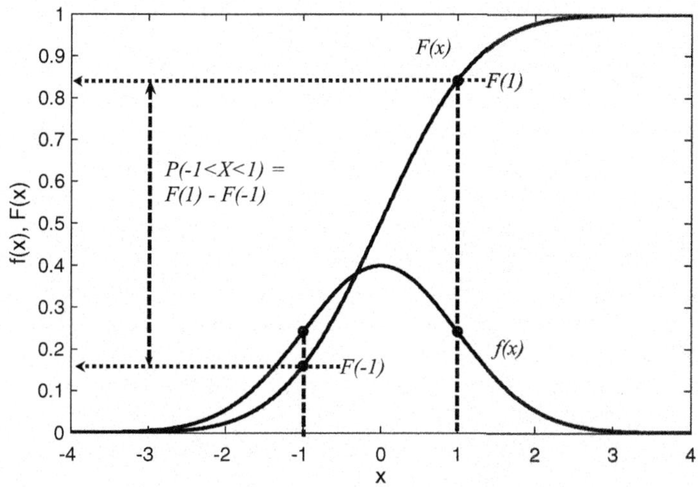

Fig. 6.1 Distribution density $f(x)$ and distribution function $F(x)$ of a Gaussian distribution with $\mu = 0$ and $\sigma = 1$ (standard normal distribution)

In Eq. 6.4 μ is the position parameter, σ the dispersion parameter.

The relationship between the distribution function and the density is shown in Fig. 6.1. From the relationships, Eqs. 6.1–6.3 it follows that for continuous random variables, one cannot state the probability for a single real value since this is asymptotically zero-valued (see exercises). Therefore, the probability of values' occurrence can only be determined for an interval. For a standard normal distribution $N(\mu, \sigma) = N(0,1)$, $F(1) - F(-1) = 68\%$ of the values lie in the range of $-\sigma < X < \sigma$ so that the probability for the occurrence of a value within this interval is $F(1) - \mathrm{F}(-1) = 0.84–0.16 = 0.68$ (Fig. 6.1).

A distribution is called discrete if it can only assume isolated, discrete values. These values can be real numbers, just as with a continuous distribution (Eqs. 6.5 and 6.6)

$$X \in \{x_1, x_2, x_3, \ldots\}, \quad x_j \in R \tag{6.5}$$

$$p_1 := P(X = x_1), \ldots, p_n := P(X = x_n) \tag{6.6}$$

so that the sequence of these probabilities describes the distribution density (Eq. 6.7).

$$f(x_n) = \{p_1, p_2, p_3, \ldots, p_n\} = \frac{\Delta F(x_n)}{\Delta x}, \quad F(x) = \sum_{j=1}^{n} f(x_j) \tag{6.7}$$

As described above, the distinction between continuous and discrete distributions is unambiguous in theory. In practical analysis, however, one quickly encounters the problem of at which point in the data collection and processing one should apply this distinguishing criterion. On the one hand, random variables are already discrete in their origin, such as birth rates. On the other hand, some variables are continuous by nature but become discretized during collection and processing. For example, although it is a continuous quantity, blood pressure is an integer value in mmHg (not an SI unit, but accepted). In this way, an infinitely long real pressure scale is reduced to a few discrete levels (about 50–200 mmHg). An even coarser discretization takes place in the determination of the pulse rate (heart rate), which is popular in practice: One counts the pulse beats (palpation at the wrist) in a relatively short time (10 or 15 s) and multiplies these by the respective factor (6 or 4) to obtain the heart rate per minute. This quantization gives a very coarse grid of discrete values (sixes or fours)—apart from the measurement error. Another dimension of discretization is opened up by automated or computer-assisted data acquisition. Since all patient data are digitized (in a patient monitor, up to 20 parameters), one can only assume discrete values. However, at this point, one must consider the resolution with which the data are digitized. The continuous oxygen saturation determined by a pulse oximeter with a width of 24 bits (16.8 million quantization steps) can still be considered a continuous quantity. On the other hand, quantities digitized in the range of up to 10 bits for speed or capacity reasons (ECG mapping, ultrasound images) can only be regarded as discrete quantities (maximum 1024 quantization steps).

6.2.1.2 Distributions and Descriptive Measures

In biostatistics, one can only very rarely assume a Gaussian distribution. It occurs mainly with quantities of technical origin, such as the noise voltage of a medical amplifier (Fig. 6.2). For the description of normally distributed data, according to the relationship according to Eq. 6.4, the following is sufficient.

Two parameters: the position parameter μ (which is zero-valued in the case of noise) and the dispersion parameter σ. Since the actual parameters are unknown, they are estimated with the help of the empirical mean value m and the standard deviation s or the variance v (Eq. 6.8).

$$m = \overline{x} = \frac{1}{N}\sum_{n=1}^{N} x_n \quad v = \frac{1}{N-1}\sum_{n=1}^{N} (x_n - \overline{x})^2 \quad s = \sqrt{v} \tag{6.8}$$

Second-order statistics are often used in biostatistics and biosignal processing to calculate the correlation (Pearson's correlation coefficient) or the power or variance. For this, the product of two random variables must be formed. The product is no longer normally distributed (Fig. 6.3), even if the random variables initially originate from a normal distribution (Fig. 6.4). As a general rule, products (of second and higher orders) of random variables are doubly exponentially distributed (see exponential distribution). If the random variables are *independent*, the product distributions are symmetrical. If they depend *on each other*, the product distribution is skewed (Fig. 6.3). Then, two parameters (position and dispersion) are no longer sufficient to describe such a distribution. To capture the asymmetry, one introduces

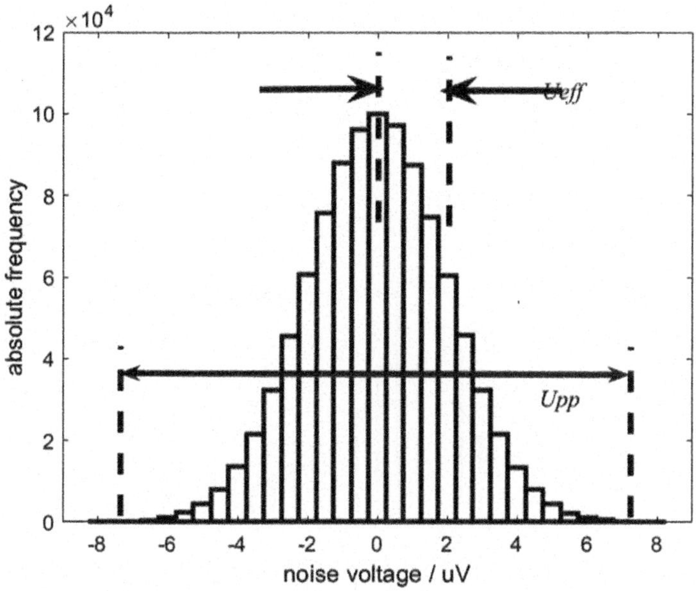

Fig. 6.2 Absolute distribution density of the noise voltage of a medical amplifier from one million measured values. The width of the histogram classes is 0.5 V. The electrical quantities effective noise voltage $u_{\mathrm{eff}} = 2$ V and peak-to-peak voltage $u_{\mathrm{pp}} = 14$ V correspond to the statistical measures standard deviation s *and* range R, respectively

the statistical measure of third-order skewness (Eq. 6.9). To quantify the deviation from the normal distribution, the fourth-order statistical measure Excess (Eq. 6.10) is introduced.

$$c = \frac{\sum_{n=1}^{N} (x_n - \overline{x})^3}{N s^3} \tag{6.9}$$

In Eq. 6.9, s is the standard deviation according to Eq. 6.8. For symmetrical distributions, $c = 0$, right-skewed $c > 0$ (Fig. 6.3), left-skewed $c < 0$.

$$e = \frac{\mu_4}{\sigma^4} - 3 = \frac{\sum_{n=1}^{N} (x_n - \overline{x})^4}{N s^4} - 3 \tag{6.10}$$

Equation 6.10, s is the standard deviation according to Eq. 6.5. The quotient in Eq. 6.10 takes the value 3 for the normal distribution. Since the normal distribution is often used as a reference, 3 is subtracted from the quotient to obtain an excess of $e = 0$ for the normal distribution. Accordingly, narrower or steeper distributions than the normal distribution (Fig. 6.3) have a positive excess $e > 0$, and flatter or wider distributions have a negative excess $e < 0$.

Calculating the skewness and excess for the product distribution according to Fig. 6.3, we obtain values $c = 2.3$ (clearly right-skewed) and $e = 10$ (clearly

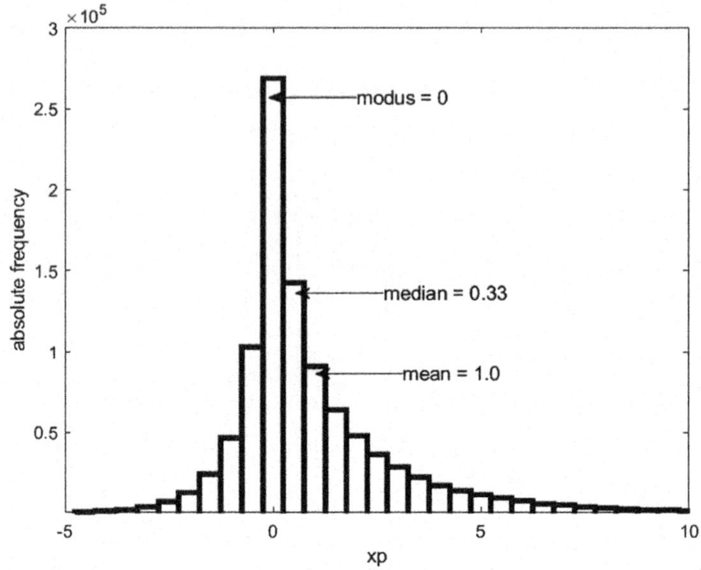

Fig. 6.3 Absolute distribution density of the product of two normally distributed random variables from one million values whose correlation coefficient according to Pearson = 0.5. Comparison of three different position measures for a right-skewed, double exponential distribution

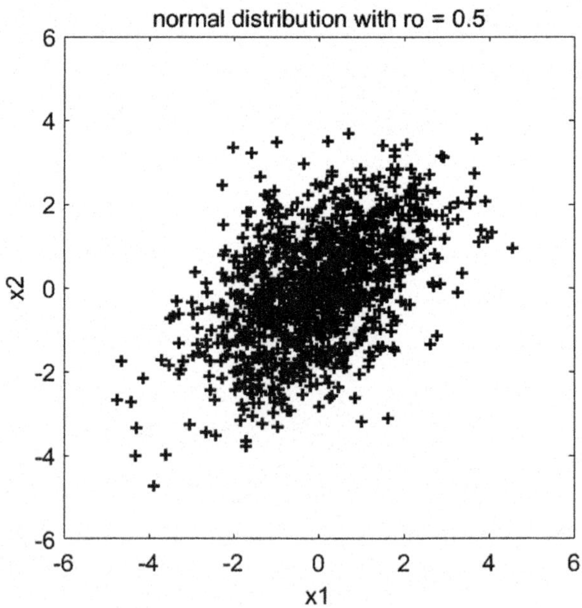

Fig. 6.4 Two normally distributed random variables whose correlation coefficient, according to Pearson = 0.5. Shown are 10^3 pairs of values from a total of 10^6 values

steeper than the normal distribution). Descriptive statistics thus has four measures (first four orders) at its disposal (arithmetic mean, variance or standard deviation, skewness, and excess) to compactly describe a random variable that is not normally distributed. However, this example (Fig. 6.3) shows that the arithmetic mean is unsuitable as a meaningful measure of location. Although it is an essential tool for the subsequent statistical analysis, since it realizes a linear operation on data (linear algebra), it does not describe a prominent distribution point. The mode (or modal value), which represents the most frequent value of the distribution, is more suitable as a location parameter (Eq. 6.11).

$$x_{\text{mod}} := x|_{P(X=x)\to\max} \tag{6.11}$$

An empirical distribution often has several modes (several local maxima), so the position parameter mode is not unique. Another possible location parameter is the median. The median is the 50% quantile (see quantiles) and therefore belongs to the group of rank statistics. Rank statistics have the unique property that statistical measures based on them do not depend on the empirical values but on their rank number (position in the series ordered by values) (Eq. 6.12).

$$\begin{aligned} &\tilde{x} \in \{x_1 \le x_2 \le \cdots \le x_{N-1} \le x_N\}, \\ &\tilde{x} = \begin{cases} x_{M+1}, & N = 2M+1 \\ \frac{x_M + x_{M+1}}{2}, & N = 2M \end{cases} \end{aligned} \tag{6.12}$$

The median can be described similarly to the mode with the help of the distribution function (Eq. 6.13).

$$\tilde{x} := x|_{P(X\le x)=0.5} \tag{6.13}$$

The formula for the median, according to Eq. 6.13, could be used in theoretical analyses. In practical analysis, however, the underlying distribution is usually unknown, so the median is determined according to Eq. 6.12. The median is a non-linear measure of the data. It can, therefore, only be used in the analysis to a limited extent. In contrast to the arithmetic mean, however, it is much more robust against extreme values and outliers. This property is demonstrated in the following example.

Example of the use of median and mean

One generates a random sequence x with 10 values that originate from a uniform distribution (all values equally probable) in the interval (0, 1):

$$x = \{0.95, 0.23, 0.60, 0.48, 0.89, 0.76, 0.45, 0.01, 0.82, 0.44\}$$

The following applies to the arithmetic mean and the median:

$$\overline{x} = 0.56, \quad \tilde{x} = 0.54$$

One simulates an outlier by multiplying the fifth value by a factor of 10:

$$y = \{0.95, 0.23, 0.60, 0.48, 8.91, 0.76, 0.45, 0.01, 0.82, 0.44\}$$

The following applies to the arithmetic mean and the median of the modified series:

$$\overline{y} = 1.37, \quad \tilde{y} = 0.55.$$

This example shows the sensitivity of the position parameters' arithmetic mean and median: while a single outlier causes the mean to rise to 241% of the actual value, the median does not change. Therefore, the median is one of descriptive statistics' most robust position estimators.

From the point of view of statistical analysis, it would make sense to combine the advantages of the arithmetic mean (a linear measure of the empirical data) and the median (robustness against extreme values) in an ordinary parameter. The trimmed mean offers a possible compromise (Eqs. 6.14 and 6.15).

$$\overline{x}_T = \frac{1}{N(1-T)} \sum_{n=NT/2+1}^{N-NT/2-1} \left(x_{NT/2+1} \leq x_n \leq \cdots \leq x_{N-NT/2-1}\right) \tag{6.14}$$

In Eq. 6.14, N is the number of values, and $0 < T < 1$ is the trimming fraction, divided half at the left and half at the right end of the ordered row for symmetrical trimming.

$$\overline{x}_{L,R} = \frac{1}{N(1-L-R)} \sum_{n=LN+1}^{N-NR-1} (x_{NL+1} \leq x_n \leq \cdots \leq x_{N-NR-1}) \tag{6.15}$$

In Eq. 6.15, N is the number of values, and $0 < L < 1$ and $0 < R < 1$ are the proportions at the left and right edges of the ordered row for asymmetric trimming, respectively.

Applying the symmetrically trimmed mean (Eq. 6.14) to the empirical data in Fig. 6.3, we obtain the following values (Fig. 6.5):

$$\tilde{x} = 0.33, \quad \overline{x}_{T=0.8} = 0.35, \quad \overline{x}_{T=0.5} = 0.46, \quad \overline{x}_{T=0.1} = 0.8, \quad \overline{x} = 1.$$

This comparison exemplifies the qualitative behavior of the trimmed mean as a function of the trimming proportion (i.e., the ratio of values excluded from the calculation): With the increase of the trimming ratio, the arithmetic mean converges towards the median. With maximum trimming $T = (N - 1)/N$ (N is the number of all values), where all values except the middle of the ordered series are rejected, the trimmed mean is identical to the median. With minimum trimming $T = 0$, the trimmed mean equals the arithmetic mean.

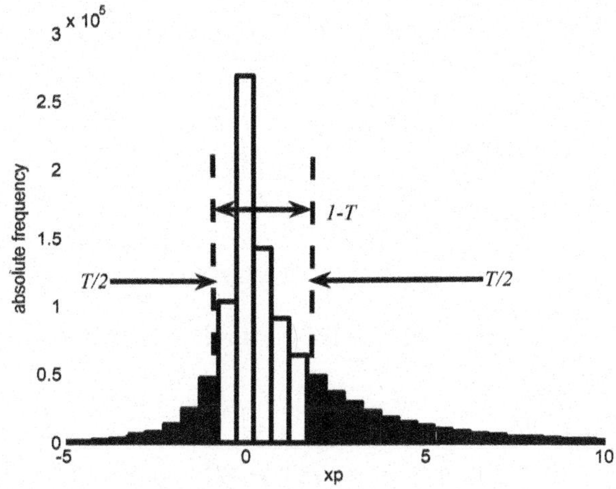

Fig. 6.5 Functioning of the symmetrically trimmed mean: The part of T/2 from the left and T/2 from the right edge of the distribution of the empirically obtained values is rejected from the calculation formula Eq. 6.5 for the mean. Part 1—T of the values left in the middle goes into the calculation (The number of empirical values in this example N = 106, T = 0.25)

6.2.1.3 Quantiles and Graphical Representation

A relatively large amount of data ($N > 10^3 \ldots 10^6$) is required for the representation (Figs. 6.2 and 6.3) and qualitative assessment of the distribution (histogram) underlying a data set (Fig. 6.6 and Table 6.1).

In practical analysis, such a quantity of data is rarely available, so alternative measures are necessary. Quantiles are suitable for the parametric description of distributions. The *p-quantile* is the value below which the proportion $0 < p < 1$ of

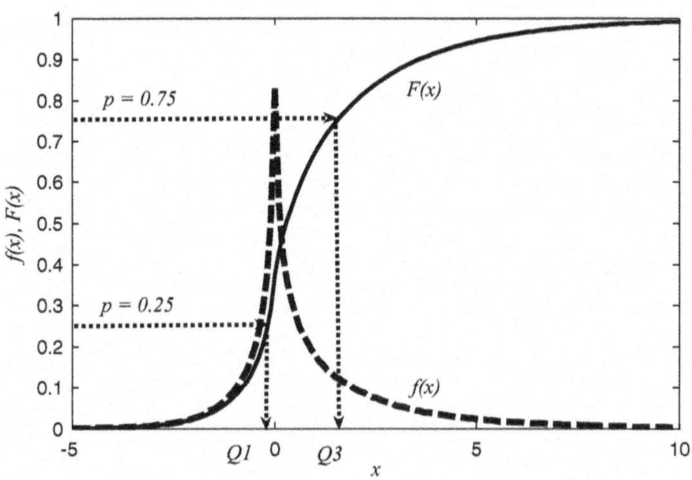

Fig. 6.6 The first Q1 = – 0.13 and the third quartile Q3 = 1.59 of the double exponential, right-skewed distribution from Fig. 6.3

Table. 6.1 Quartiles, 25% quantiles

$p = 0.25$	Lower quartile	$Q1 = Q_{0.25}$
$p = 0.50$	Middle quartile, median	$Q2 = Q_{0.50}$
$p = 0.75$	Upper quartile	$Q3 = Q_{0.75}$

all values lies (Eq. 6.16).

$$Q_P := x|_{P(X<x)=p} \tag{6.16}$$

One uses some typical quantiles, mainly those in 25% increments called quartiles.

The distance between the lower and upper quartiles is called the interquartile range, $IQ = Q_{0.75} - Q_{0.25}$. The interquartile range thus indicates how wide the center of the distribution is with 50% of all values. If we apply these measures to the double exponential right-skewed distribution discussed in Fig. 6.3, we get IQ = 1.72, where 50% of all values are located.

The three quartiles describe compact empirical data, especially distributions with relatively few values ($N < 10^2 \ldots 10^3$). The usual graphical representation is the box-whisker plot, shown for two typical distributions (Gaussian and double exponential, right-skewed) in Fig. 6.7. This gives a clear and concise picture of the empirical distributions to be analyzed.

In this plot, one can optionally also enter the confidence intervals of the median so that a statistical test (see hypothesis testing) is already possible at this stage. If the confidence intervals of the distributions do not overlap, one can reject the null hypothesis with a statistical uncertainty of = 0.05. In practice, one can assume different medians of the distributions. Practically, this means that one can assume different medians of the distributions.

6.2.1.4 *Selected Distributions*

This chapter overviews the most common distributions of biostatistics and biosignal analysis. Here discrete distributions are assumed for the reasons mentioned above (discretization or digitization of parameters and biosignals).

Equal Distribution

The uniform distribution density corresponds to a sequence of identical values (Eq. 6.17).

$$P(X = x) = f(x) = \begin{cases} \frac{1}{N}, & x = x_{1,\ldots,N} \\ 0, & sonst \end{cases} \tag{6.17}$$

The following applies accordingly to the distribution function (Eq. 6.18):

$$P(X \le x) = F(x) = \begin{cases} 0, & x < x_1 \\ \frac{|x|}{N}, & x_1 \le x \le x_N \\ 1, & x_N < x \end{cases} \tag{6.18}$$

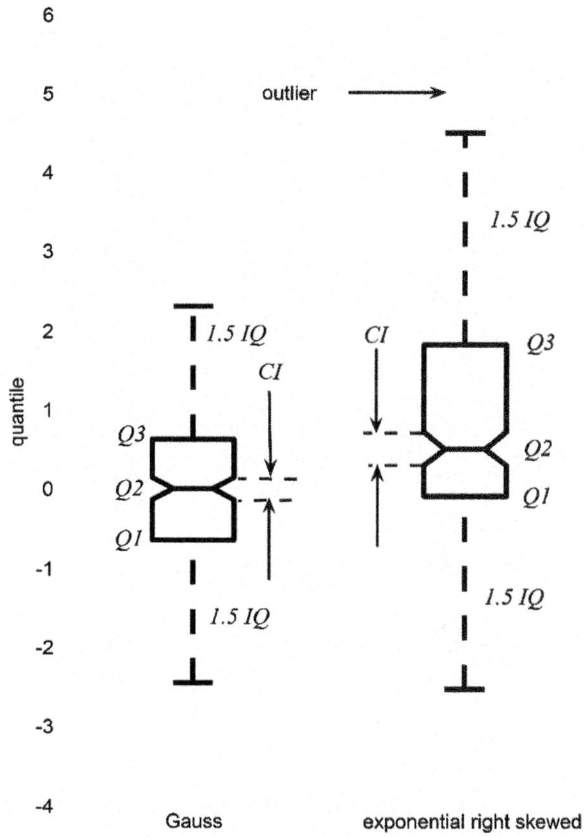

Fig. 6.7 Box whisker plot of two distributions with 200 values each: Gaussian (left) and double exponential right-skewed distribution (right) from Fig. 6.3. Q1, 2, 3 are quartiles (horizontal lines of the boxes), dashed vertical lines with horizontal cross lines ("whiskers") indicate the range for 1.5 times IQ = Q3 − Q1 at the top and bottom. It follows that the whiskers together represent three times the IQ. Therefore, values outside the whiskers are also interpreted as outliers and marked separately. The notches around the median Q2 represent the confidence interval KI of the median for the statistical uncertainty, which is 5% by default

A frequently occurring quantity uniformly distributed is the phase of complex Fourier coefficients in a stochastic process (Fig. 6.8). The discrete probabilities are subject to natural fluctuations; the theoretical value is $f(x) = 2\pi/N$. However, the phase has a problematic property of periodicity around 2 π. With real signals, it is not possible at all to avoid the phase jumps caused by the periodicity. Therefore, despite the statistically good properties, the phase alone is insufficient for the analysis.

Binomial Distribution

The binomial distribution describes the distribution density of a binary random variable, i.e., a variable that can only assume two different states. The theoretical probability for the occurrence of an event is p. The binomial distribution indicates the probability that for n trials that can occur with probability p, x trials will end positively. Its distribution density is described by Eq. 6.19.

$$P(X = x) = f(x|n, p) = \binom{n}{x} p^x (1 - p)^{n-x} \tag{6.19}$$

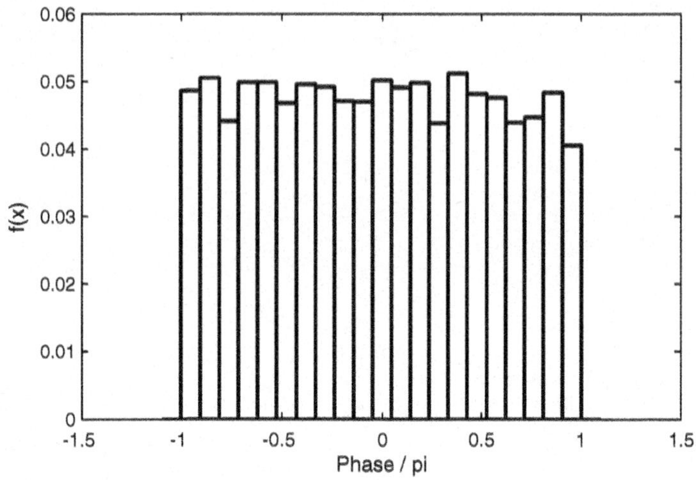

Fig. 6.8 Distribution density of the phase of complex Fourier coefficients of an uncorrelated noise (number of values 10^4, number of histogram classes $N = 21$)

In Eq. 6.19, n is the number of all trials, p *is* the *theoretical* probability for the occurrence of the investigated characteristic, x is the investigated number of trials, and $f(x)$ is the probability that in a total of n *trials,* the characteristic occurs exactly in x *trials.*

The binomial distribution is asymmetric, so the skewness and excess must also be used for its (empirical) description. For large n, it converges to the (symmetrical) normal distribution. The practical significance of the binomial distribution is to be demonstrated in the following example:

Sufficiently significant statistics prove (Germany, 80 million inhabitants) that 30% of the population is overweight and has problems with blood pressure (hypertension). Because of the number of people recorded, one can set the relative frequency of 30% (according to the law of large numbers) equal to the theoretical probability, hence $p = 0.3$. For a practicing internist who treats 18 patients daily, how many hypertensives are most likely to come for treatment? Alternatively, one can ask how high the probability is that the patients have no hypertension. Therefore, one sets $n = 18$ (for the total number of trials) and $x = 0, \ldots, 18$ as the number of trials examined. The results are shown in Fig. 6.9. From the graph; it follows that with the highest probability, five hypertensive patients come for treatment every day. However, it should be noted that the probability of four or six hypertensives is only slightly lower. The probability that no hypertensive patient is present is negligible.

Poisson Distribution

The binomial distribution converges to the Poisson distribution for large n and small p. *The* parameter of the Poisson distribution results from the product $= np$,

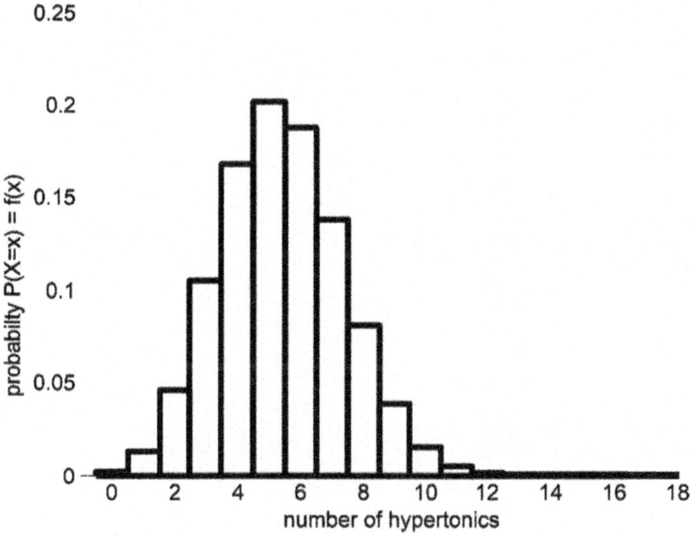

Fig. 6.9 Probability of x hypertensives occurring in practice with 18 patients daily and a portion of 30% of hypertensives in the population

which is independent of n *for* large n and small p ($n > 50$, $p < 0.05$). Therefore, the Poisson distribution can be interpreted as a limiting case of the convergence of the binomial distribution (Eq. 6.20). The Poisson distribution is essential from the point of view of the law of small numbers.

$$P(X = x) = f(x) = \frac{\lambda^x}{x!} e^{-\lambda} \tag{6.20}$$

The following example is intended to illustrate the Poisson distribution: In the maternity ward of a women's clinic, it was determined from long-term data ($n > 10^4$) that, on average, five births are to be expected daily (= 5). It is crucial for staff and resource planning to know how likely deviations from the mean are, first for the extreme cases of "no birth" or "10 births a day". The distribution gives the following picture (Fig. 6.10). While the probability for four and five births per day is the same and amounts to 17.5% (i.e., about 1/3 for four to five births per day), it is 0.6% for $x = 0$ (no birth) and 1.8% for $x = 10$. It follows that extremely low or extremely high birth rates are improbable, yet not impossible.

The following example will demonstrate the Poisson distribution as a theoretical background for the law of small numbers.

Example of the Law of Small Numbers

With an ANN (Artificial Neural Network), which consists of 100 neurons, pattern recognition of graphic structures is to be carried out. The problem with learning an ANN is that the adaptation of the neuronal weights to the learning goal

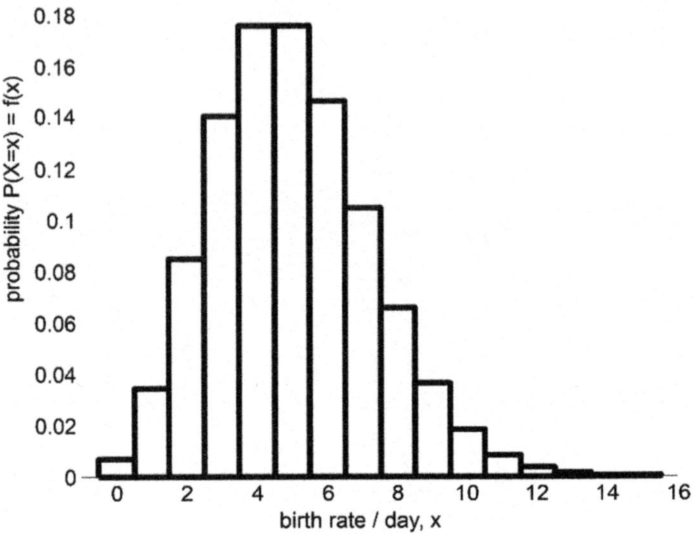

Fig. 6.10 Probability of birth numbers between 0 and 15 with a long-term mean of 5 births per day (Poisson distribution with = 5)

takes place in a multi-dimensional non-linear space. Therefore, one has to look for effective methods to comb this nonlinear space—or at least representatively—for the global minimum of the error functional. The Monte Carlo method (MC) is a well-known random search method in stochastics. It consists of assigning—in this concrete case—an equally probable random number between 1 and 100 to each of the 100 bias inputs of the neurons (and often normalized to the range [− 1,1]). The assigned numbers in the ANN in such an experiment show the Poisson distribution (Fig. 6.11) with parameter = 1 (100 number values for 100 bias values, i.e., on average, one number per bias). Despite the equally probable numbers (equal distribution) between 1 and 100, about one-third of the numbers ($x = 0, f_{Poisson}(0) = 0.37$) do not occur at all, and about one-third exactly once ($x = 1, f_{Poisson}(1) = 0.37$). In other words, only about two-thirds of the numbers ($1 - f(0) = 0.63$) occur at least once. That is why the law of small numbers is also called the *two-thirds law.* At this point, the difference in the interpretation of the laws of large and small numbers becomes clear (Eq. 6.21):

> While the law of large numbers states that for a theoretically known probability of an event, the relative frequency of its occurrence converges to this value with the number of trials, the relative frequency of occurrence for a small number of trials corresponds to the Poisson distribution.

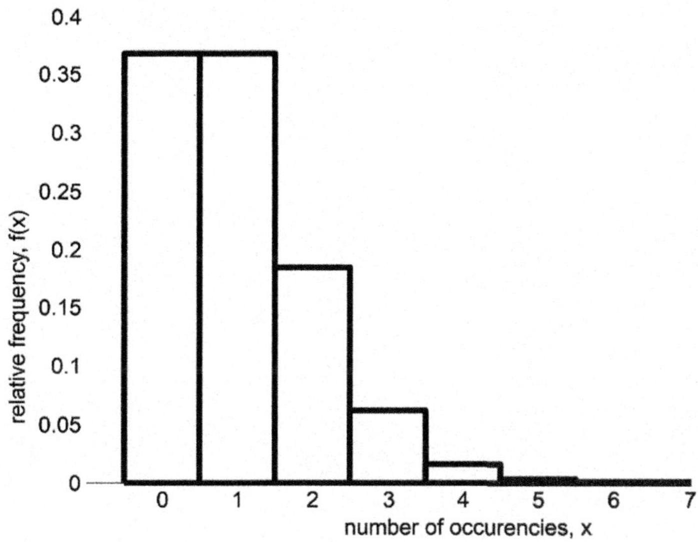

Fig. 6.11 Relative frequency (probability of occurrence) of the occurrence of numbers between 1 and 100 in 100 trials (law of small numbers)

$$\begin{aligned} h_r &= \frac{h}{N} \\ h_r &\overset{N \to \infty}{\longrightarrow} p \\ h_r &\overset{N << \infty}{\longrightarrow} P_{Poisson}(\lambda) \end{aligned} \tag{6.21}$$

In Eq. 6.21, h is the absolute frequency of an event, h_r is the relative frequency, p is the theoretical probability of an event, and $P(\lambda)$ is the distribution density of the Poisson distribution with parameter.

Exponential Distribution

Most time intervals are asymptotically subject to the exponential distribution, e.g., the life span of X-ray tubes, the half-life of the decay of radioactive isotopes, or the life expectancy of humans (life insurance). It is a continuous distribution that can be described with the formula according to Eq. 6.22 (Fig. 6.12).

$$P(X = x) = f(x|\mu) = \frac{1}{\mu} e^{-\frac{x}{\mu}} = \lambda e^{-\lambda x} \tag{6.22}$$

In Eq. 6.22, the service life is the failure rate, where $\lambda = 1/\mu$. The use of the definition is not uniform, so the expected value μ should always be given as well.

The discrete variant of the exponential distribution is the *geometric distribution* that describes the probability distribution of independent Bernoulli experiments (Eq. 6.20).

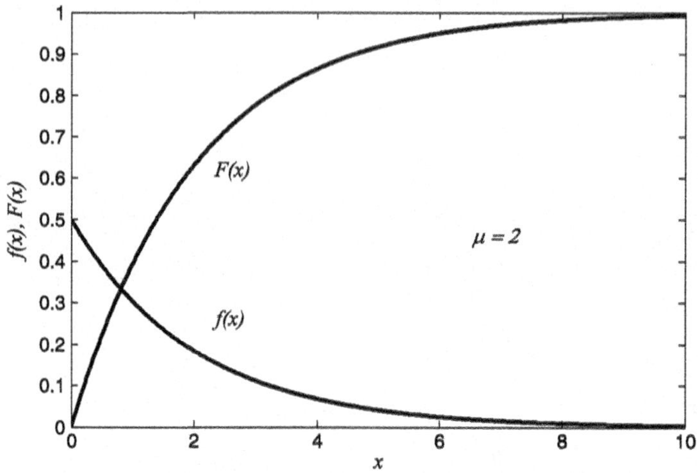

Fig. 6.12 Probability density $f(x)$ and distribution function $F(x)$ of an exponential distribution with parameter $\mu = 2$

$$P(X \leq n) = F(X) = 1 - (1 - p)^n \tag{6.23}$$

In Eq. 6.23, p is the probability of success in independent Bernoulli experiments, n is the number of trials, and $P(X \leq n)$ is the probability that one needs at least n trials to succeed.

The exponential distribution (discretely the geometric distribution) is the only one with a substantial Markov property in process analysis (Eq. 6.24).

$$P(X \geq x + y | X \geq x) = P(X \geq y) \tag{6.24}$$

If one first interprets the relationship according to Eq. 6.24 from the point of view of process analysis, the future of a process (or system) depends solely on its current state and not on its past. Since information about the past is unnecessary, this property is also called memorylessness. The following example demonstrates this behavior.

Example of Exponential Distribution

The documents of a manufacturer of pacemakers show that their lifespan is 5 years on average. A patient who has already been wearing such a pacemaker for 4 years asks whether he can expect another 2 years of functioning. For the probability in the left part of the relationship, according to Eq. 6.24, the following applies.

$$P(X \geq x + y | X \geq x) = \frac{P(X \geq x + y)}{P(X \geq x)} = \frac{1 - P(X \leq x + y)}{1 - P(X \leq x)}$$

$$= \frac{1 - F(6|5)}{1 - F(4|5)} = 0.67.$$

The probability of another 2 years of use can be calculated from this parameter alone and the known lifetime (right side of relation 6.21):

$$P(X \geq y) = 1 - P(X \leq y) = 1 - F(2|5) = 0.67.$$

With an average life span of the pacemaker of 5 years, after 4 years, one can expect a probability of 67% that it will function for another 2 years. However, the information that the pacemaker has already been in use for 4 years is not included in the calculation and is, therefore, irrelevant.

6.2.1.5 Central Limit Theorem

The mathematical formulation of the CLM (Central Limit Theorem) is:

> Let X_i $(i = 1 \ldots, n)$ be n independent, identically distributed random variables with expected value μ and variance σ^2. Then the sum X_i is asymptotically normally distributed with the expected value $n\mu$ and the variance $n\sigma^2$.

If the sum of the random variables is centered and normalized, it converges to the standard normal distribution (Eq. 6.25).

$$\frac{\sum_i X_i - n\mu}{\sigma\sqrt{n}} \overset{n\to\infty}{\longrightarrow} \mathcal{N}(0, 1) \tag{6.25}$$

The relationship Eq. 6.25 can also be formulated using the distribution function (Eq. 6.26).

$$\lim_{n\to\infty} P\left(\frac{\overline{X} - \mu}{\sigma/\sqrt{n}}\right) = F(z) = \mathcal{N}(0, 1) \tag{6.26}$$

In Eqs. 6.22 and 6.23, the location parameter (expected value), the dispersion, n number of variables, X_i is a random variable, z *is* the centered and normalized random variable of the standard normal distribution.

The CWS demonstrates the outstanding importance of normal distribution. In practice, it is often the case that the distribution of the analyzed data is not known for various reasons. The cause may be a too-small data volume ($n < 10^3$) or a combined or nested distribution due to the mixing of several random variables or their influences. In such cases, the CLT offers a way out from a data size of about $n > 30$ if one analyses the sum of the data instead of the data (or other measures that contain the sum, e.g., the arithmetic mean). For the practical analysis, the CGEA can be formulated more pragmatically:

The sum of a large number of random variables distributed at random is normally distributed.

In the analysis, the question arises as to what effects non-compliance with conditions of the CMAA (independence and identical distribution) can have. It is impossible to test the distributions for their equality and independence. Fortunately, the CMAA is relatively robust against violating the two assumptions, as the following example demonstrates.

Example of the CLT

There are two exponentially distributed, centered random variables V_1 and V_2, with parameter = 1, which are mirror-symmetrical, and one equally distributed, centered random variable V_3 (Fig. 6.13). Two new random variables, V_{13} and V_{23}, are formed in which $V_{13} = V_1 + V_3$ and $V_{23} = V_2 + V_3$. On the one hand, this ensures that the random variables V_{13} and V_{23} are *not* identically distributed (mirrored exponential distribution). On the other hand, the summation with V_3 results in a statistical *dependency of* the random variables V_{13} and V_{23} with a Pearson correlation coefficient of = 0.7. For this example, both prerequisites for applying for the CLT are specifically violated. Subsequently, the random variables V_{13} and V_{23} are summed equally over $n = 10, 30, 100$ realizations. The result of the summation is shown in Fig. 6.14. From the sum and order of $n = 10$, one can assume a normal sum distribution from n = 30 at the latest. This example shows that the requirements of the CLT concerning the distribution of the random variables and their independence should not be taken too strictly, as they would not be practically fulfilled. It is rather the case that the summand order from which the CLT is applicable depends in the concrete case on how much one deviates from these requirements. With the distributions and dependencies that frequently occur in practice, one can assume that a summand order of $n = 30$ is sufficient to fulfill the requirements of the CLT.

6.2.2 Statistical Correlation

Most relationships between variables and parameters in biomedicine and biosignal processing are stochastic. On the one hand, the quantities themselves are subject to random fluctuations (to put it more precisely—they cannot be described deterministically). On the other hand, their correlation also has a random character. For example, two non-invasive methods for measuring blood pressure are the Oscillometric Cuff Method (OMM) and the PTT (Pulse Transition Time). Blood pressure is subject to strong fluctuations and is therefore simplified and initially considered a random variable. The two methods mentioned do not measure blood pressure directly but via auxiliary variables (cuff pressure in the OMM, transit time in the PTT), whose relationship to blood pressure is not deterministic but also stochastic. Finally, in the sense of comparing methods, the statistical correlation of the

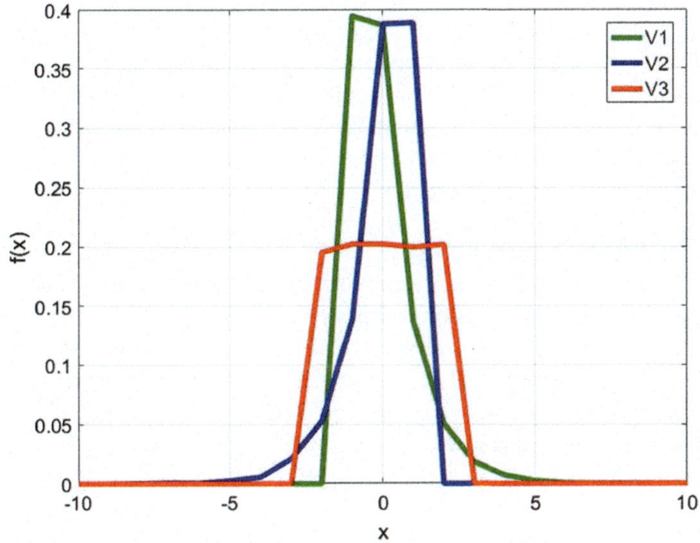

Fig. 6.13 Distributions of independent, centered random variables V1 (exponential, right-skewed), V2 (exponential left-skewed), and V3 (equally distributed)

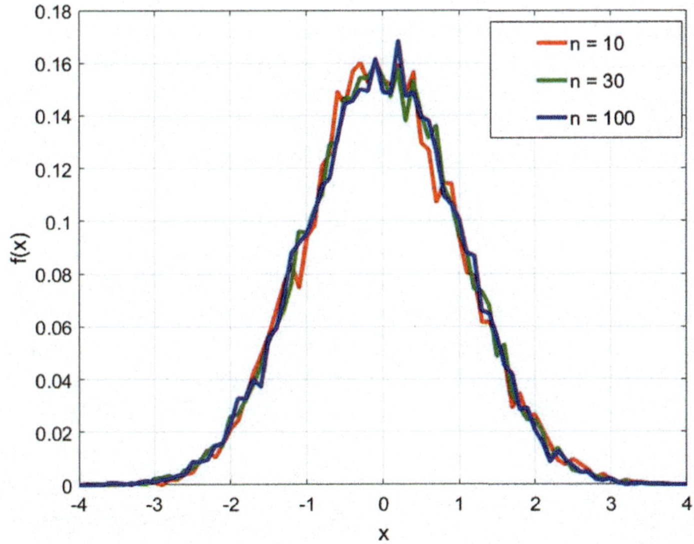

Fig. 6.14 Distribution density (empirical histogram) of the centered and normalized sum of two dependent (= 0.7), not identically distributed random variables from Fig. 6.13 (exponentially distributed, mirrored, with uniform distribution additively superimposed). From a summand order of n = 10, one can assume an approximately normal distribution

results of both methods is determined. The result is a measure of the correlation between two measurement methods (correlation coefficient) to determine which two (or more) stochastic variables and correlations have already been included. It can lead to the fact that, in the end, a *statistical* correlation is verifiably present, but there is no *functional* correlation at all. In the following, we will explain how a statistical correlation can be proven and show that a functional correlation cannot necessarily be derived from it a priori.

6.2.2.1 Population and Sample

Important decisions must be statistically secured in medical diagnostics, therapy, and biosignal processing. Decisions with a given uncertainty (usually 5%) must be justified or substantiated with statistical data. Therefore, enough data are needed to ensure the required statistical certainty (permitted uncertainty) regarding scope. The methodologically simplest way would be to record the entire set of all members of a group who exhibit the characteristic of interest. For example, if the question is how many hypertensives (people with high blood pressure) have been in treatment for more than 10 years, one could ask them all. Such a group of members with the characteristic under investigation is called the *population* (Fig. 6.15).

However, recording the characteristic “10 years of treatment for high blood pressure” would require a logistical peak performance that would not be economically justifiable. One would have to survey about 20 million people in Germany alone.

Therefore, one must be content that one can only draw a *sample from* the population of hypertensives. It must be large enough for statistical purposes. As a general rule, not every sample reflects the distribution of the studied characteristic as it is distributed in the population. If one asks members of a center for competitive sports and an obesity club about their blood pressure or its treatment, one will undoubtedly get highly different results. For a sample to reflect the actual distribution of characteristics in the population, it must be *representative.* Only then can one infer the population from the descriptive measures of the sample. This legitimate requirement can run into a practicality problem: To assess whether a sample is representative, one must know the population’s actual distribution of the characteristic. If one knows the characteristic’s distribution, no more sample is needed. The bottom line is that a sample must be planned (experimental design) in such a way that it is representative a priori. Methods of empirical statistics and planning

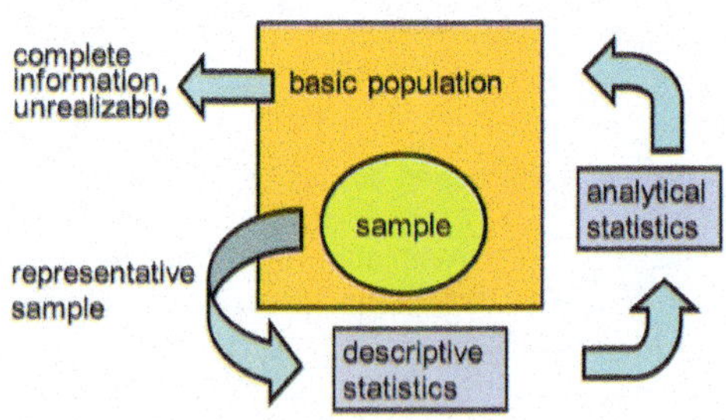

Fig. 6.15 Relationship between complete information on the population and the collection of a representative sample

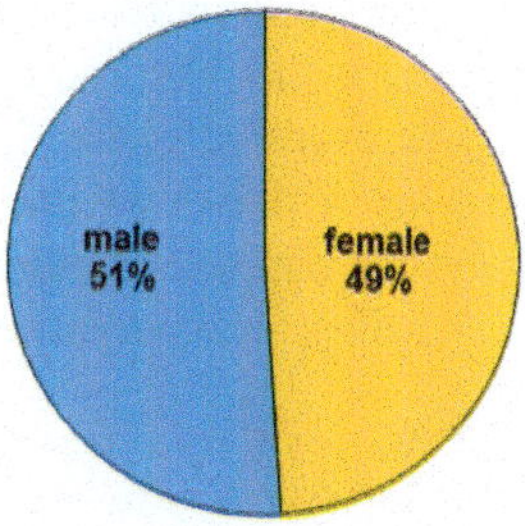

gender	count	part %	angle °
female	54	49	176
male	56	51	184
Σ	110	100	360

Fig. 6.16 Statistics of a maternity ward. This table represents a sample of the population of all births in Germany. It is a representative sample, as the proportions correspond to the national average

are used for this purpose. Once the representative sample is available after the necessary experimental design and survey, its properties are described (descriptive statistics), especially those of the characteristics of interest. The parameters obtained are then examined for correlations and regularities (statistical analysis). Appropriate interventions and measures follow these analyses in the population.

6.2.2.2 Correlation and Representation of Correlations

Histograms (e.g., Figs. 6.13 and 6.14), bar charts, pie charts, and scatter plots are used to visualize distributions and correlations of most two-dimensional random variables. Some display options (pie charts) are suitable for visualizing *standardized random variables.* In particular, probabilities, relative frequencies, and distribution densities are standardized natural quantities (e.g., Eq. 6.21). The following example illustrates the necessity of standardization.

Example of the Representation of Random Variables

The number of male and female newborns is entered in a table (Fig. 6.16). A pie chart can be used to show the gender proportions. The absolute birth numbers (second column of the table in Fig. 6.16) must be normalized (third column) and transferred to the variable angle (fourth column). After normalization, we obtain a 51%/49% proportion for male/female newborns, corresponding to the national average. Due to natural fluctuations, the absolute number of female births may increase slightly in the next month, e.g., to 58. The normalization reverses the ratio in favor of female newborns: female 51%, male 49%. The change in the absolute number of female newborns resulted in a relative change in the proportion of male newborns. This behavior can be generalized; see Fig. 6.16.

Standardization

By normalizing random variables, any change in the absolute number of a characteristic causes a change in all relative measures.

If this effect is not considered when interpreting the results of statistical analyses, it can lead to erroneous conclusions (see formal correlation). Stochastic quantities are also normalized in biosignal processing, e.g., the proportions of spectral power bands of the EEG. Here, a frequently occurring increase in power in the -band can lead to a substantial decrease in the shares of the other bands, although their absolute band power has not changed. The conclusion of these considerations is:

> Normalizing random variables is valuable and necessary to obtain a basis for comparing variables and correlations. The absolute numbers must be included when analyzing and interpreting behavior and correlations to avoid erroneous conclusions.

Visual Inspection

A point cloud is a descriptive means of visually analyzing the distributions and correlations of two or pairs of multiple random variables. Visual inspection is a legitimate means of statistical analysis, which in the first step of the analysis, has to assess whether.

- the sample is homogeneous,
- outliers are present and
- what kind of connection exists.

In biostatistics, especially in biosignal processing, homogeneous samples are a rarity. Therefore, separating distributions with classification methods is often necessary, in the simplest case, by separating lines. A typical example is shown in Fig. 6.17. It is already easy to see visually that the sample (pairs of values of the random variables X and Y) consists of two partial distributions, e.g., height X and weight Y of women (bottom left) and men (top right). Such a sample is inhomogeneous and must be separated into two parts, most simply with a diagonal separating line. Without separation, the correlation coefficient would be $\rho_{xy} = 0.78$, but after separation in the individual point clouds, only $\rho_{xy} = 0.03$ (See Pearson's correlation coefficient). This example shows one of the most common sources of error in the literature and journals: Strong correlation due to the inhomogeneity of the sample.

An important task is to assess a sample to see whether outliers are present. Both the term and its application are handled very differently in the literature. Some sources recommend treating as outliers value two to three times the standard deviation away from the mean. For the evaluation, this would mean that, on average, every 20th to every 500th value would qualify as an outlier and be excluded from the evaluation (Fig. 6.18). This procedure would not be correct from a theoretical and empirical point of view. For biostatistics and biosignal processing, two different terms are used here whose definition and application are based on clear criteria:

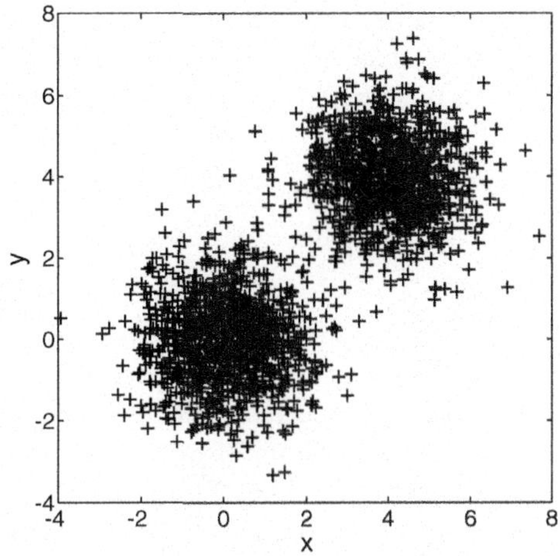

Fig. 6.17 Point cloud of an inhomogeneous distribution. The partial distributions obtained by separation along the main diagonal are homogeneous

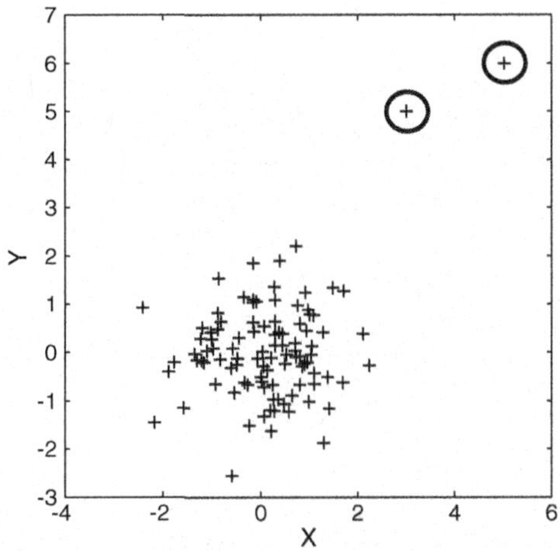

Fig. 6.18 Homogeneous sample of 100 pairs of values of the random variables X and Y with two "outliers" or extreme values marked by circles

Extreme value refers to a value of a sample that is improbable and far from the mean but theoretically possible according to the empirical distribution.

The point cloud shown in Fig. 6.18 shows 100 pairs of values of the normally distributed random variables X and Y, which are independent. The circled values (3; 5) and (5; 6) are improbable but possible and occur on average at least once in correspondingly large samples. Therefore, they belong to the distributions and should not be excluded from the evaluation as supposed outliers; they are considered *extreme values*. A typical example is the measurement of the equivalent noise voltage of a medical measuring amplifier. It is normally distributed and averages about 20 V. However, with a sufficiently long measurement (several minutes), extreme values of 100 V and more also occur. These are part of the sample and must be evaluated. If, however, a measured value of, e.g., 10 V occurs in isolated cases, it must be attributed to a disturbance and excluded as an outlier.

An *outlier* refers to a value in a sample that cannot theoretically or empirically belong to this sample.

Note: In biostatistics/biosignal analysis, a sharp distinction between extreme value and outlier categories *is im*possible since—as is often the case in statistics/stochastics—there is no exact definition for these terms. Therefore, one can quickly fall into the temptation of assigning experimental data to the more favorable category depending on the "desired" result. Such a procedure is methodologically reprehensible and should always be viewed with due skepticism. The above (own) definitions are already very well oriented to reality and can therefore be applied.

In acquiring biomedical data, especially biosignals, interferences sometimes occur that are significantly stronger than the measurement data itself (see Chap. 2). As a rule, values also appear in the measurement data that cannot come from the measured data source with certainty. If, for example, a medium-term (weeks) increase can normally be expected if a patient's muscle strength in rehabilitation is measured daily to observe progress. However, the strength cannot increase from 10^2 to 10^4 N daily. Therefore, this measured value would be excluded from the evaluation as an outlier. This exclusion should be verified for safety with further measurements (on subsequent days).

During visual inspection, one can often recognize the qualitative correlation of the random variables from the point cloud (Fig. 6.19). Ideally, this correlation is already known—at least qualitatively. It is known, for example, that the sensory systems of humans (subjective sensation when seeing, hearing, and touching) are logarithmically dependent on the stimulus intensity (brightness, volume, pressure). Therefore, fitting a corresponding regression curve to the measurement data is relatively simple. In practical analysis, however, the qualitative correlations usually are not known. Therefore, one first tries to find a linear or existing correlation. For the first orientation, the correlation coefficients are helpful.

Correlation Coefficients According to Pearson and Spearman

One needs a measure that reaches the highest value when the variables are equal (most similar) to quantify the similarity of random variables. Following analytical geometry, this problem can be applied to a rectangle that reaches the most considerable area when the two side lengths are equal, thus making the rectangle a square (Eq. 6.27).

$$\kappa_{XY} = X \cdot Y \tag{6.27}$$

Since the random variables X and *Y are* subject to fluctuations (Fig. 6.20), one determines with the help of the expected value. For this purpose, the expectation

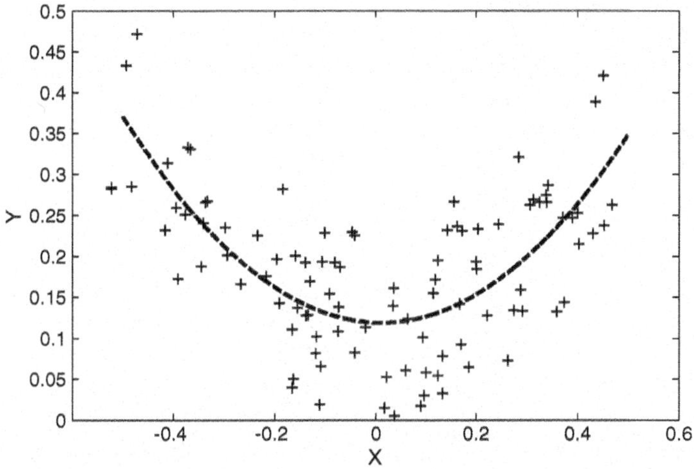

Fig. 6.19 Point cloud of the pairs of values of the random variables Y and X, where Y *is* quadratically dependent on X (the second-order functional relationship is known before the survey sample, e.g., the error functional of an adaptive filter)

Fig. 6.20 Point cloud of pairs of values of two linearly dependent variables, X and Y

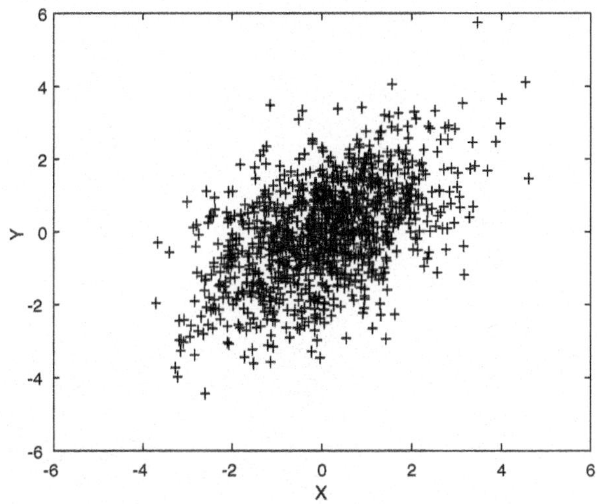

operator is introduced (Eq. 6.28).

$$\kappa = E\{X \cdot Y\} \tag{6.28}$$

The expected value corresponds asymptotically to the arithmetic mean of the measured values. Thus, according to Eq. 6.28, a statistical measure for the linear correlation of the random variables X and Y, which is referred to as *correlation,* is referred to as correlation. The correlation reaches its maximum value when the random variables are completely identical (in terms of value); it is zero when there is no correlation between the centered variables. For the evaluation of the linear correlation based on the point cloud, only the inclination (positive or negative) and the width of the cloud are decisive (Fig. 6.20). The cloud's position above the axes is not a determining factor. In order to achieve independence from the position, the random variables must be centered (Eq. 6.29).

$$\kappa_Z = E\{(X - E\{X\}) \cdot (Y - E\{Y\})\} \tag{6.29}$$

The correlation of centered variables is called *covariance.* According to Eq. 6.29, the relationship follows that if at least one of the mean values is zero, then covariance and correlation are identical. The covariance can reach values in the entire range of real numbers since the product depends on concrete values of the random variables. With k_Z , it is not possible to make a comparison between correlations without normalization. Norming aims at the range of values in the interval [0; 1] or [− 1; 1]. Since the covariance can be positive and negative (positive or negative slope of the point cloud), it is normalized to the range [− 1; 1]. Because the covariance represents the variance of the centered products (Eq. 6.29), it is normalized to the product of the variances. Because of the unit-free normalization (exponents of the same size in the numerator and denominator), the root is also taken from the product (Eq. 6.30).

$$\rho_{XY} = \frac{E\{(X - E\{X\}) \cdot (Y - E\{Y\})\}}{\sqrt{E\{(X - E\{X\})^2\} \cdot E\{(Y - E\{Y\})^2\}}} = \frac{\mathrm{cov}(x, y)}{\sqrt{\mathrm{var}(x) \cdot \mathrm{var}(y)}} \tag{6.30}$$

The normalized covariance, according to Eq. 6.30, is also known as the *correlation coefficient according to Pearson.* The empirical coefficient can be calculated for a sample (Eq. 6.31).

$$r_{xy} = \frac{\sum_{i=1}^{n} (x_i - \overline{x})(y_i - \overline{y})}{\sqrt{\sum_{i=1}^{n} (x_i - \overline{x})^2} \cdot \sqrt{\sum_{i=1}^{n} (y_i - \overline{y})^2}} \tag{6.31}$$

Although the CC (correlation coefficient) is a second-order statistic according to Pearson (the highest exponent sum in the numerator and denominator is 2), it is a measure *only of the* linear correlation or only of its linear component. This fact results from the fact that for a dependence of order *n,* the $n + 1th$ order of a statistic is necessary. A necessary condition is the normal distribution of the examined data. If the conditions are not fulfilled, one needs alternatives to quantify the statistical correlation. *Rank statistics* or rank transformations are particularly well suited for practical analyses (see quartiles, median). The most crucial advantage of rank-based statistics is that the calculation no longer considers the concrete values of the random variables but their rank in the ordered series. This transformation ensures extensive independence from unknown non-linear correlations and extreme values. A disadvantage, however, is that the rank transformation itself has a non-linear

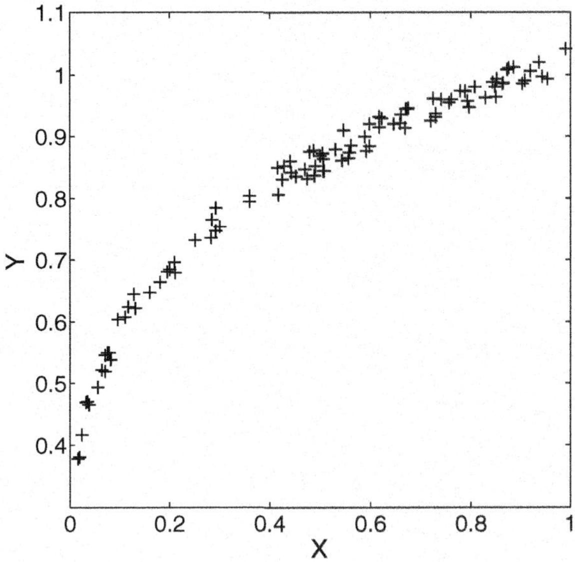

Fig. 6.21 Non-linear dependence of random variable Y on random variable X of the fifth order, monotonic, digressively increasing

stochastic character. Therefore, it is suitable for analyzing correlations but not for inferences from original data (the superposition principle does not apply). The rank-based alternative to Pearson's CC is Spearman's CC (Eq. 6.32).

$$r_{xy} = \frac{\sum_{i=1}^{n}\left(R(x_i) - \overline{R}(X)\right)\cdot\left(R(y_i) - \overline{R}(Y)\right)}{\sqrt{\sum_{i=1}^{n}\left(R(x_i) - \overline{R}(X)\right)^2}\sqrt{\sum_{i=1}^{n}\left(R(y_i) - \overline{R}(Y)\right)^2}} \tag{6.32}$$

In Eq. 6.32 $R(x_i)$ is the rank number of the value x_i in the ordered series of the random variable *X*.

According to Spearman, neither a normal distribution of the data nor a linear relationship is necessary to calculate the CC. However, any non-linearity must be monotonic (falling or rising). An example of correlation coefficients is shown in Fig. 6.21. The random variable *Y* depends on the random variable *X* to the fifth order. Pearson's CC is 0.95, and Spearman's CC is 0.99. Since every non-linearity also has a linear component, this is included in Pearson's CC, although this slightly underestimates the correlation compared to Spearman's CC.

The literature quantifies and presents results using correlation coefficients and (primarily linear) regression. However, when evaluating the correlation, misinterpretations can also occur due to non-compliance with the mentioned prerequisites. The following is an overview of the most common interpretation errors.

Formal Correlation

Due to the normalization of a characteristic explained above, the change in an absolute value influences all other relative values (Weiss, 1999). This relationship is illustrated by the example in Fig. 6.22. An absolute increase in cases of influenza

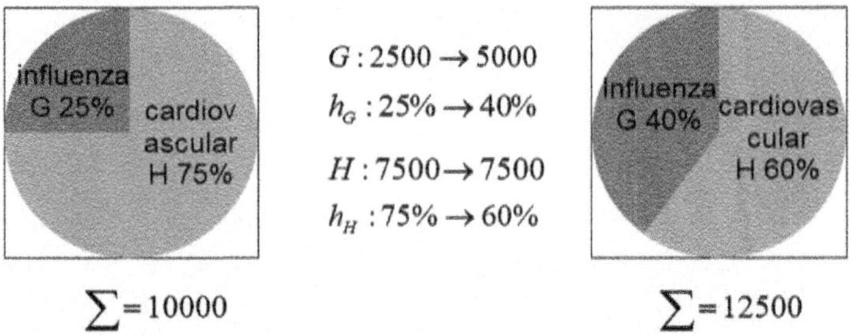

Fig. 6.22 Data example for formal correlation: An absolute increase in influenza cases from $G = 2500–5000$ leads to a relative increase from $h_G = 25–40\%$. At the same time, the proportion of cardiovascular diseases decreases from $h_H = 75–60\%$, although the absolute number $H = 7500$ has remained constant

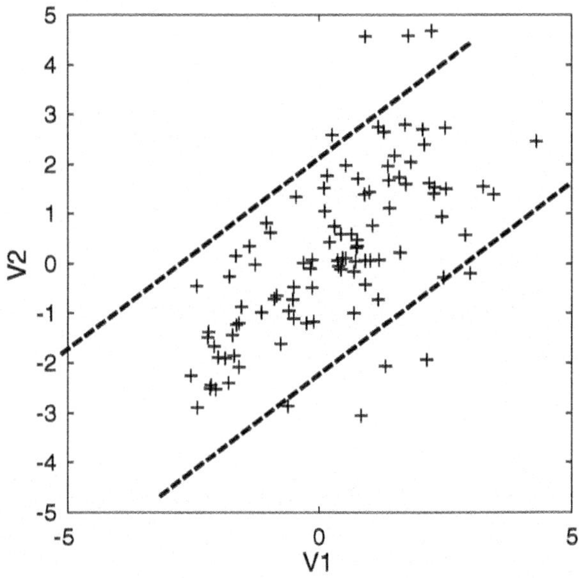

Fig. 6.23 Point cloud of 100 realizations of random variables V1 and V2 pairs. The dashed lines represent limits at 10% trim (trimmed mean)

increases its proportion and thus inevitably reduces the proportion of cardiovascular cases, although their absolute number remained constant. The normalized variables would show a *formal* correlation, although it is clear from the absolute figures that there is no *real* correlation. One could arrive at the completely wrong interpretation that an increase in influenza cases leads to a decrease in cardiovascular diseases. Hence, it is imperative to not only depend on standardized data and instead always consider the absolute data.

Selection Correlation

Selection correlation refers to a statistical relationship from selecting "desired" data from the original sample. On the one hand, selection can occur due to a legitimate demand for a linear but robustified descriptive measure (see Sect. 6.2.1.2, trimmed arithmetic mean). An example of this is shown in Fig. 6.23. From the population of data already used in Fig. 6.14

One hundred pairs of values were randomly selected. According to Pearson, the CC for this sample is 0.65; after a 10% trimming (represented by dashed lines), it rises to a value of 0.88. This example shows that a sensible measure (trimming) of descriptive statistics is not fundamentally suitable for the analysis. The data pairs excluded from the analysis during trimming (left of the upper and right of the lower dashed line) *belong* to the sample. Although they are extreme values, they are not outliers. In analytical practice, *subjectively* motivated (medicine, pharmaceutical industry, technology, and natural science) but *objectively* unjustified selections occur in a methodologically incorrect manner already during the ongoing studies, which are intended to help achieve a "desired result." Apart from the option of excluding clear outliers, such a procedure is (not only) methodologically reprehensible.

Outlier-Induced Correlation

Figure 6.18 shows two (improbable) extreme values as examples. According to Pearson, the CC is $r_{xy} = 0.9$ for all data after excluding the two extreme values $r_{xy} = 0.2$ (a realistic value). Here the extreme values belong to the sample, but according to the homogeneity criterion, one must exclude them from further processing. The problem of distinguishing between extreme values and outliers has already been dealt. At this point, it should be pointed out that—as with all other boundary problems—no threshold one could clearly distinguish. Instead, the transitional area between the two characteristics is always a grey zone with a problem-dependent width. Furthermore, therefore, in analysis practice, not only extreme values but also obvious outliers are deliberately *not* excluded from the evaluation just because they drive the correlation coefficient in the "desired" direction.

Inhomogeneity Correlation

The homogeneity of a sample is the necessary prerequisite for further statistical analysis, as already explained and shown in Fig. 6.17. Taking this concrete example of the weight-body size relationship in men and women as a starting point, one can state: It is plausible and unsurprising that men are, on average taller and heavier than women; hence the CC is relatively high across the sample ($r_{xy} = 0.78$). The sample is inhomogeneous, so the subgroups (women and men) are considered separately. Again, it is plausible that there is no clear correlation between height and weight for either women or men, as individual variability predominates ($r_{xy} = 0.03$). This variant of misinterpreting a correlation is widespread in most cases without attempting to prove a functional or causal relationship.

Commonality Correlation

For years, a strained example of commonality correlation has been the very high correlation coefficient of more than 0.9 between birth rates and stork density. Despite this very high value, it is not difficult to prove (if it is not apparent a priori) that there is no functional or causal relationship. The fact that there is a statistical correlation, although lacking a functional one, is because both characteristics studied depend on a third variable (season).

In today's neurological research, one of the things being studied is correlations between brain regions. A proven statistical correlation between two regions (e.g., between the left and right hemispheres) is often considered functional. For example, it can be statistically proven (certainty > 98%) that the two motor centers (electrode positions C3 and C4 in the EEG) are related. However, it has not yet been proven that there is a functional connection. It is also understandable since both centers are activated by a third variable—e.g., the pupillary reflex in the eye.

In this context, methodological errors can already occur during measurement data acquisition (EEG), which cause a high correlation between brain regions that have no functional connection (see Chaps. 3.1.2 and 5.4.4). Typical examples of such misinterpretations are "paradoxical lateralization" and "mirroring of—waves in the frontal area."

Span Correlation

When developing methods, it is common to compare the results of a new method with a "gold standard" or "ground truth." Where no gold standard exists, a comparison is made with previously established methods. When comparing methods, the working range (span) is crucial. It can be assumed that if the range in the method comparison is only extensive enough or much more extensive than necessary, a high correlation can be brought about. The following example illustrates this effect: S_aO_2 (arterial oxygen saturation) is a vital parameter typically above 95%. Laboratory measurement of the blood gases of an arterial blood sample is considered the gold standard. Since taking blood samples is invasive and discontinuous, non-invasive and continuous methods are being developed. The best known is pulse oximetry which provides a calculated parameter S_pO_2. It is a measurement method alternative to the laboratory measurement of S_aO_2. Comparing measured values of both methods offers an initially good picture (Fig. 6.24); Pearson's CC with $r = 0.96$ shows an almost perfect agreement. However, the controlled measurement range extends between 40 and 100%, which exceeds the 90% to 100% needed in practice many times over. Usually, S_aO_2 is above 90%; when it drops into the 80–90% range, the emergency physician is called. Therefore, readings below 80% have no further practical significance. A 90–100% range should be targeted for comparison under natural conditions (Fig. 6.25).

In this working range, the CC in this example drops to a value of $r = 0.49$, i.e., about half. With some other methods of pulse oximetry, the CC reaches even lower values. These considerations conclude that both values are statistically correct. However, the higher CC presents a "better" result based on a measuring range far removed from practice and, therefore, implausible in measurement methodology.

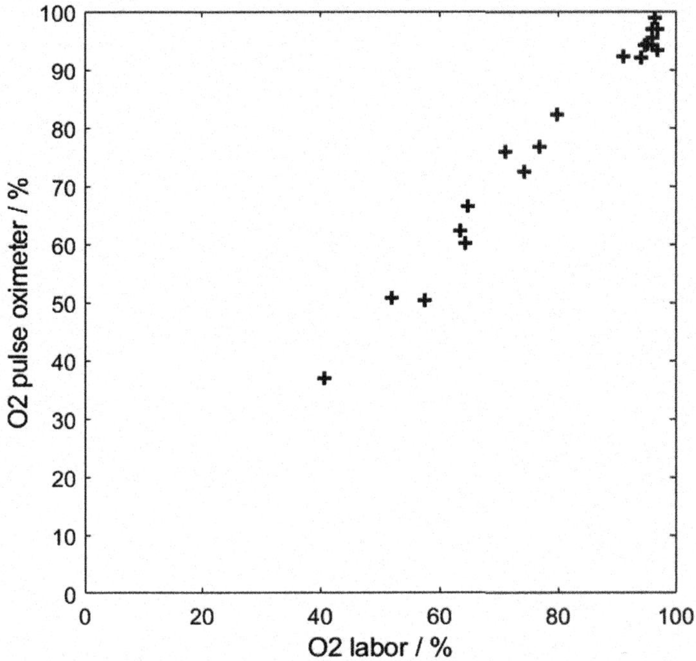

Fig. 6.24 Comparison of measured values between the gold standard (laboratory measurement of arterial blood gases) and a pulse oximetry method in the entire range. According to Pearson, the CC is $r = 0.96$, a highly satisfactory value

6.2.3 Estimation Procedure

6.2.3.1 Basics

The description of statistical data as a basis for the analysis characterizes the data material as comprehensively as possible with a few parameters. Section 6.2.1.2 explained the most critical parameters and their calculation. Therefore, the aim is to determine the true but unknown parameter of the data under investigation. Since only a sample is usually available from a population, the parameter must be *estimated by* suitable means. The task now consists of finding an estimator function (estimator) $g(X)$ in such a way that the true parameter is determined as accurately and reliably as possible (Eq. 6.33).

$$\hat{\vartheta} = g(X) \to \vartheta \tag{6.33}$$

In Eq. 6.33, the true parameter $\hat{\vartheta}$ is the estimate of ϑ, g *is* the estimator, and X is a random variable with a parameter ϑ.

For example, one would like to determine the location parameter of the distribution density of the random variable X, assuming here the distribution of the sample according to Fig. 6.3. As already shown in the figure, there are three possibilities

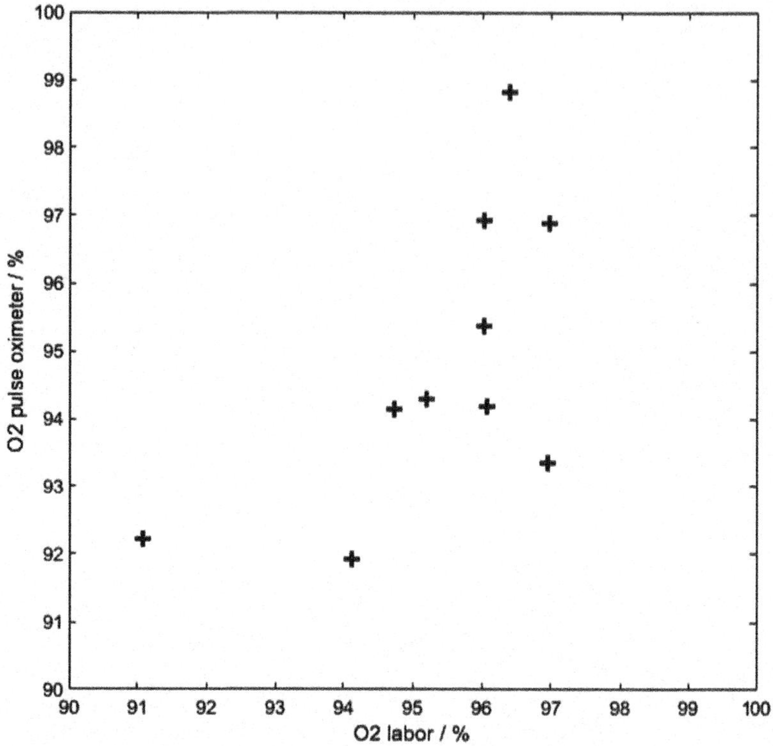

Fig. 6.25 Comparison of measured values between the gold standard (laboratory measurement of arterial blood) and a pulse oximetry method in the natural working range of 90–100%. According to Pearson, the CC is $r = 0.49$, a poor value for the new method, which is unusable

for the estimation function: the mode (Eq. 6.11), the median (Eq. 6.13) and the mean (Eq. 6.5). Since the mode and the median are quantile-based measures and therefore fundamentally non-linear, only the arithmetic mean can be considered for further analysis. Without giving proof, it is true for the mean value (Eq. 6.34) that it corresponds asymptotically (for infinitely large n) to the location parameter.

$$\hat{\mu} = \overline{X} = \frac{1}{n}\sum_{i=1}^{n} X_i \overset{n\to\infty}{\longrightarrow} E\{\overline{X}\} = \mu \tag{6.34}$$

In Eq. 6.34, Xi is the mathematical (theoretical) sample; for an empirical sample, use lower case x_i at this point. $E\{.\}$ is the expectation operator, μ the true location parameter, and $\hat{\mu}$ the estimated value of μ.

For the scatter of a sample (scatter parameter) applies equivalently to the location parameter (Eq. 6.35):

$$S^2 = \frac{1}{n-1}\sum_{i=1}^{n}(X_i - \overline{X})^2 \overset{n\to\infty}{\longrightarrow} E\{S^2\} = \sigma^2. \tag{6.35}$$

Since the estimator of the parameter contains the sum of random variables (Eq. 6.35), one can assume, based on the Central Limit Theorem, that from a sample size of $n > 30$, the estimator $\hat{\mu}$ is normally distributed (Eq. 6.36).

$$\overline{X} = \hat{\mu} \sim \mathcal{N}\left(\mu, \frac{\sigma^2}{n}\right) \tag{6.36}$$

In Eq. 6.36, the variance of the mean is σ^2/n. It is a consequence of averaging n i.i.d (independent, identically distributed) random variables—the variance decreases proportionally with the averaging order. See stochastic processes.

When calculating the dispersion (Eq. 6.35), the sum of the squares of the sample is calculated; therefore, the CLT is not applied here. Assuming that the sample data originate from a standard normally distributed population (Eq. 6.38, Fig. 6.26), the dispersion is χ^2-distributed (Eq. 6.37).

$$S^2 = \hat{\sigma}^2 \sim \chi^2 \tag{6.37}$$

$$\sum_{i=1}^{n} Z_i^2 \sim \chi_n^2, \quad Z_i \sim \mathcal{N}(0,1) \tag{6.38}$$

With a sample size of $n > 100$, the sum of squares approximates a normal distribution (Eq. 6.39).

$$\sum_{i=1}^{n>100} Z_i^2 \sim \mathcal{N}(\mu = n, \sigma^2 = 2n) \tag{6.39}$$

The example of the empirical distribution of the data according to Fig. 6.3 clearly shows that there can be very different estimators for one parameter. In order to assess the quality of the estimators and to compare them with each other, criteria are necessary. The three most important quality criteria are explained below.

Faithful to Expectations

An estimator is said to be expectation-trusted if it fulfills the condition according to Eq. 6.40. Expectation dispersion means that the expected value of the random variable X corresponds to the sought parameter.

$$E\{g(X)\} = \gamma(\vartheta) \tag{6.40}$$

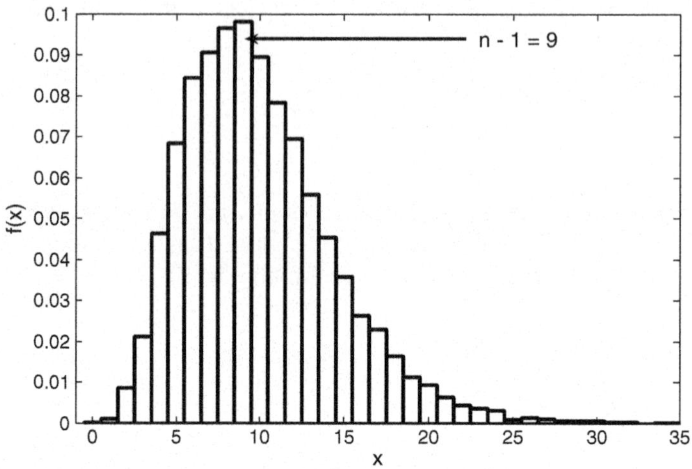

Fig. 6.26 Distribution density of the sum of 10 squared, standard-normally distributed random variables corresponds to a^2 distribution with $n - 1 = 9$ degrees of freedom

In Eq. 6.40, ϑ is the parameter sought of the random variable X, γ (.) is a function of the parameter and g(.) is an estimator. In the simplest case, the functional is the parameter itself. For example, for sensory quantities or spectral estimates, one uses the decadic logarithm of a sensory quantity $\log_{10}(X)$ and specifies it in dB (brightness, loudness).

If the condition for the fidelity of expectations is not fulfilled, an error occurs that is called bias (bias) (Eq. 6.41).

$$b_\vartheta = E\{g(X) - \gamma(\vartheta)\} \tag{6.41}$$

There are several analogies to bias in electronics (bias as quiescent current at the amplifier input, offset voltage as an operational amplifier disturbance variable) or measurement technology (systematic measurement error). The *second central moment* estimates the variance, especially in process analysis (see moments).

$$S_m^2 = \frac{1}{n}\sum_{i=1}^{n}\left(X_i - \overline{X}\right)^2,$$

which differs in the normalization factor compared to the definition according to Eq. 6.35. The comparison of the two equations results in a bias of

$$b_S = E\{S_m^2\} - E\{S^2\} = \frac{n-1}{n}\sigma^2 - \sigma^2 = -\frac{1}{n}\sigma^2.$$

It follows that the second central moment as an estimator of the variance is not expectation-true; it underestimates the true variance. For a huge n, the bias disappears, $\lim_{n\to\infty} b_S = 0$. This property is called asymptotic expectation dispersion. When evaluating estimators, one must distinguish between an a priori and an asymptotic property. Often only small samples are available, so the asymptotic property cannot come into play at all and has a rather theoretical character.

Consistency

An estimator is said to be consistent if the variance ($\mathrm{var}(X) = S^2\,(X)$) of its estimate decreases with increasing sample size (Eq. 6.42). Such behavior is also known as stochastic convergence.

$$n_2 > n_1 \Rightarrow \mathrm{var}_{n2}\{g(X)\} < \mathrm{var}_{n1}\{g(X)\} \tag{6.42}$$

In terms of mathematical statistics, an estimator is either consistent or not. From the perspective of practical analysis, it is also interesting how quickly an estimator converges or how quickly the estimate's variance decreases with increasing sample size. In the Chap. 5.3.3.2, an estimator for the gradient of the error function is described (Eq. 5.49). The variance of this estimator depends only on the adaptation constant and not on the number of estimated values (see Fig. 5.22). Therefore, this estimator is inconsistent. An example of two consistent estimators converging at different rates is shown in Fig. 6.27. On the simulated data from Fig. 6.3, we apply the arithmetic mean and the median as estimators. The median converges faster than the mean. This behavior, applied to the practical analysis of a concrete sample, means that with the median, one obtains a lower variance of the estimate for exponentially distributed data.

Efficiency

An estimator's efficiency increases as its estimate's variance decreases, other things being equal (Eq. 6.43).

$$e\{g(X)\} = \frac{1}{F(\vartheta)\cdot \mathrm{var}(g(X))} \tag{6.43}$$

In Eq. 6.43, X is the random variable, $g(.)$ is the estimator, $F(.)$ is the Fisher information, and $e\{.\}$ is the efficiency.

If two estimators are compared, the one with the lower variance for an identical sample size n is more efficient (Fig. 6.27). For example, it can be shown that for exponentially distributed data, the median is more efficient than the arithmetic mean (Eq. 6.44).

$$n_2 = n_1,\ \mathrm{var}(g_2(X)) < \mathrm{var}(g_1(X)) \Rightarrow e\{g_2(X)\} > e\{g_1(X)\} \tag{6.44}$$

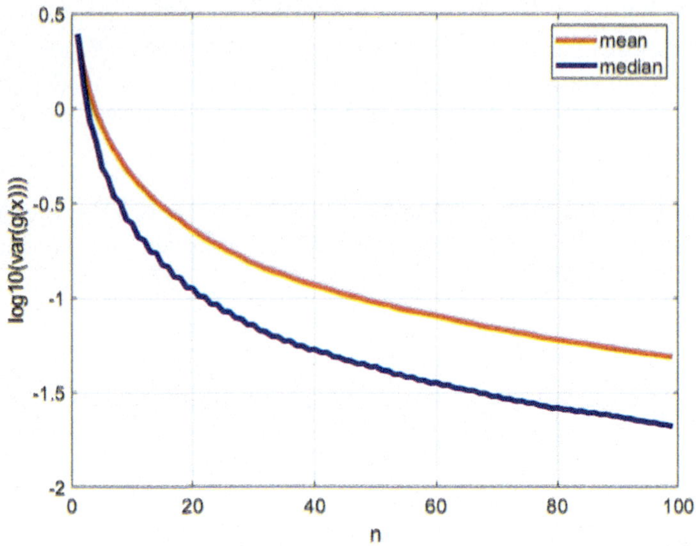

Fig. 6.27 Logarithm of the variance of the estimator's "arithmetic mean" and "median" of double exponentially and right skewed distributed data from Fig. 6.3 as a function of sample size. Both estimators are consistent. The median converges faster and is, therefore, more efficient

6.2.3.2 Methods

To simplify the quality criteria, it makes sense to define a standard measure. In statistics, different measures are used depending on the question. Therefore, a general measure is considered here, which forms a reasonable basis for comparison, especially for biosignal processing—the mean squared error *mse*. It combines the squares of the bias and the scatter or the variance (Eq. 6.45) additively.

$$mse\left(\hat{\vartheta}\right) = E\left\{\left(\hat{\vartheta} - \vartheta\right)^2\right\} = \mathrm{var}\left(\hat{\vartheta}\right) + \left(E\left\{\hat{\vartheta}\right\} - \vartheta\right)^2 = S^2 + b^2 \qquad (6.45)$$

The mse(.) is to be demonstrated using the example of the mean and the dispersion of a normally distributed random variable:

$$\begin{aligned} &X \sim N(\mu, \sigma) \\ &\hat{\mu} = \overline{X} = \tfrac{1}{n}\sum_{i=1}^{n} X_i \\ &mse(\hat{\mu}) = \mathrm{var}(\hat{\mu}) + (E\{\hat{\mu}\} - \mu)^2 = \sigma^2/n \end{aligned}$$

Since the arithmetic mean is an expectation-true estimator, the bias disappears. Therefore, the variance of the mean, also called *sem* (standard error of the mean), depends solely on the dispersion of the original data and the averaging order *n*.

The following applies accordingly for the *mse of* the scatter S (empirical scatter)

$$\sigma^2 = S^2 = \frac{1}{n-1}\sum_{i=1}^{n}\left(X_i - \overline{X}\right)^2$$

$$mse\left(\hat{\sigma}^2\right) = \mathrm{var}\left(\hat{\sigma}^2\right) + \left(E\left\{\hat{\sigma}^2\right\} - \hat{\sigma}^2\right)^2 = \frac{2\hat{\sigma}^4}{n-1}.$$

The estimators of distribution parameters can be developed according to different methods. The three best-known methods are explained below: the method of moments, the maximum likelihood method, and the least squares method.

Method of Moments

A moment of order k is defined as follows (Eq. 6.46):

$$m_k = E\left\{X^k\right\}. \tag{6.46}$$

The moments are always estimated with the arithmetic mean (Eq. 6.47).

$$\hat{m}_k = \frac{1}{n}\sum_{i=1}^{n} x_i^k \tag{6.47}$$

While the first moment directly relates to the position parameter (Eq. 6.48), no direct analogy to the descriptive measures discussed so far can be established from the second moment onwards.

$$\hat{\mu} = \hat{m}_1 = \frac{1}{n}\sum_{i=1}^{n} x_i^1 = \overline{x} \tag{6.48}$$

To establish comparability with the dispersion, a centering of the second moment is necessary (Eq. 6.49).

$$\hat{\sigma}^2 = \hat{m}_2 - \hat{m}_1^2 = \frac{1}{n}\sum_{i=1}^{n} x_i^2 - \left(\frac{1}{n}\sum_{i=1}^{n} x_i^1\right)^2 = \frac{1}{n}\sum_{i=1}^{n}\left(x_i - \overline{x}\right)^2 \tag{6.49}$$

Comparing the moment-based estimate of the variance (Eq. 6.49) with the definition according to Eq. 6.5, we find that the normalization factors $n-1$ and n *are* different. For this reason, the moment-based estimator, according to Eq. 6.49, is not expectation-true but asymptotically expectation-true.

Maximum Likelihood Method (ML Method)

One uses a part (realizations of the random variable X) or the whole sample to estimate a sample parameter. Since the individual realizations are subject to a probability density $f(x_i \mid)$, the probability density of n realizations is the product of the individual densities (Eq. 6.50), i.e., an n-dimensional distribution density.

$$f(x_1, x_2, \ldots, x_n|\vartheta) = \prod_{i=1}^{n} f(x_i|\vartheta) \tag{6.50}$$

According to Eq. 6.50, the function gives the probability density for the occurrence of an *n-dimensional* random variable with a given (known) parameter. In the analysis, however, one needs the opposite—the probability density of the sought parameter with a given realization $\{x_1, \ldots, x_n\}$. According to Eq. 6.50 (Eq. 6.51), it is formulated analogously to the relationship.

$$L(\vartheta|x_1, \ldots, x_n) = \prod_{i=1}^{n} f(\vartheta|x_i) \tag{6.51}$$

The formulation according to Eq. 6.51 is called the ML estimator. Regarding the best estimate, the function *L*(.) must now be maximized. Of course, the estimation function depends on the concrete distribution density of the random variable or its realizations. The ML estimator's derivation for the parameters is demonstrated using the example of a normally distributed variable X.

$$f(x|\mu, \sigma) = \frac{1}{\sigma\sqrt{2\pi}} e^{-\frac{(x-\mu)^2}{2\sigma^2}} \Rightarrow L(\mu, \sigma|x_1, \ldots, x_n) = \prod_{i=1}^{n} \frac{1}{\sigma\sqrt{2\pi}} e^{-\frac{(x_i-\mu)^2}{2\sigma^2}}$$

Now one would have to differentiate the function *L*(.) by and partially in order to determine the conditions for the maximum of *L*(.). However, since the number of realizations n is not known in advance, this task cannot be solved in this way (apart from the computational effort). The following consideration simplifies the optimization problem enormously: If one finds a function *L**(.) *that is* monotonically increasingly related to *L*(.) (a linear function does not solve the problem), then both functions reach their maximum at the identical argument. The function *L**(.) must eliminate the problematic product to suit the logarithm. Since the known distribution corresponds to a normal distribution, the natural logarithm is chosen because of the intrinsic e-function.

$$\ln L(\mu, \sigma|x_1, \ldots, x_n) = -\frac{n}{2}\ln(2\pi\sigma^2) - \frac{1}{2\sigma^2}\sum_{i=1}^{n}(x_i - \mu)^2$$

After the partial derivation of this expression, the two ML estimators are obtained directly:

$$\frac{\partial L}{\partial \mu} = 0 \Rightarrow \hat{\mu} = \overline{x} \quad \frac{\partial L}{\partial \sigma} = 0 \Rightarrow \hat{\sigma}^2 = \frac{1}{n}\sum_{i=1}^{n}(x_i - \overline{x})^2.$$

The ML estimator is expectation-true, but it is not for dispersion, as with the method of moments. ML estimators are usually the most efficient, and if the distribution density assumption is met, they are also consistent. However, the assumption is often unmet in practical analysis so the estimator can become inconsistent. Moreover, the data distribution must be known beforehand, which is rarely the practice case.

LS Method (Least Squares Method)

The principle of the LS method is to try to fit a data model to the empirical data so that the sum (or mean) of the squared errors is minimized (Eq. 6.52).

$$\min_{x} \sum_{i=1}^{n} (y_m - y_i)^2 \tag{6.52}$$

In Eq. 6.52, y_{m} is the modeled quantity, and y_{i} and x_{i} are empirical value pairs whose relationship is to be modeled.

The simplest dependency is the linear correlation, which can be quantified using the well-known linear regression. For the linear case, the optimization problem can be derived from the linear system of equations

$$\mathbf{Ax} = \mathbf{y}_m$$

by minimizing the second norm (sum of the error squares):

$$\min_{x} \|\mathbf{Ax} - \mathbf{y}_m\|_2.$$

Although two points are sufficient to describe a regression line, the system of equations must be overdetermined because of the stochastic fluctuations of the empirical data. It means there must be significantly more than two equations (length of vectors **z** and **b**). The solution to the minimization problem is the parameters of the regression line

$$\hat{y} = y_m = a_{yx} + b_{yx}x$$

with $b_{yx} = \frac{\mathrm{cov}_{xy}}{s_x^2} = r_{xy}\frac{s_y}{s_x}$ and $a_{yx} = \overline{y} - b_{yx}\overline{x}$. ($s_{\mathrm{x,y}}$ are standard deviations, r_{xy} is the Pearson's coefficient, cov_{xy} the covariance of x and y, and the parameters of the straight lines are elements of the vector $\mathbf{z} = (a_{\mathrm{yx}}, b_{\mathrm{yx}})$). Note that the slope of the regression line b_{yx} depends directly on the cov r_{xy}. A regression line can be placed over each point cloud of the value pairs (x_{i}, y_{i}) up to the extreme case $r_{\mathrm{xy}} = 0$. However, when interpreting the results, one should consider the linear correlation and the fact that the linear model becomes less reliable with a decreasing CC. Of course, in practical analysis, the correlations are rarely linear so one can set up a corresponding data model (e.g., with a polynomial) according to the known non-linearity. However, the minimization of the error squares with the system mentioned above of equations is no longer possible because this only applies to the linear case. Iteration methods offer a numerical solution, according to Gauss–Newton, Levenberg–Marquardt, and others.

6.2.3.3 Confidence Interval

Until now, descriptive variables (mean, dispersion, skewness, excess) and dependency parameters (correlation coefficients according to Pearson and Spearman) have been calculated as unique values called *point estimates.* With point estimates, however, one has no information about the calculated value's reliability or in which range one can assume it to be reliable. In order to obtain this critical information, confidence intervale (confidence intervale) (Eq. 6.53).

An interval $I(X_1, \ldots, X_n) = (I_L, I_R)$ is called confidence interval (confidence interval) for a parameter with certainty $1 - \alpha$ if applies

$$P(I_L \leq \vartheta \leq I_R) \geq 1 - \alpha \quad (6.53)$$

In Eq. 6.53, the parameter to be estimated, $0 < < 1$ is the permissible statistical uncertainty, $I_{L,R}$ is the interval limits (left and right), and P *is* the probability.

The confidence interval (KI) is often misinterpreted by claiming that it lies with a probability (PRB) of 1—in the confidence interval. It is correct that the true parameter is placed on the number axis, independent of a calculated interval; it is a fixed quantity. What one can, at best, calculate from a sample are the interval limits $I_{L,R}$, which are also random due to the random character of the data. The correct interpretation is, therefore, that one obtains a confidence interval that includes (covers) the unknown parameter with a PRB of 1.

Confidence Interval for the Mean Value

The derivation and interpretation of the CI are demonstrated in the following example: There is a normally distributed random variable X *for* whose mean value the CI is to be given. According to the following relationships, the centered and normalized mean value usually is distributed:

$$X \sim N(\mu, \sigma) \rightarrow \overline{X} \sim N\left(\mu, \frac{\sigma}{\sqrt{n}}\right) \rightarrow \frac{\overline{X} - \mu}{\sigma/\sqrt{n}} \sim N(\mu = 0, \sigma = 1) \quad (6.54)$$

It means that the AI can be calculated based on the quantiles.

$$P\left(-z_{\alpha/2} \leq Z \leq z_{\alpha/2}\right) = 1 - \alpha \quad (6.55)$$

of the standard normal distribution:

$$I = \left(\overline{X} - z_{\alpha/2}\frac{\sigma}{\sqrt{n}}, \overline{X} + z_{\alpha/2}\frac{\sigma}{\sqrt{n}}\right). \quad (6.56)$$

However, this procedure presupposes that the true dispersion is known. In practical analysis, the true parameters are unknown; otherwise, one would not have to analyze the data. Therefore, the empirical dispersion S must be assumed, which

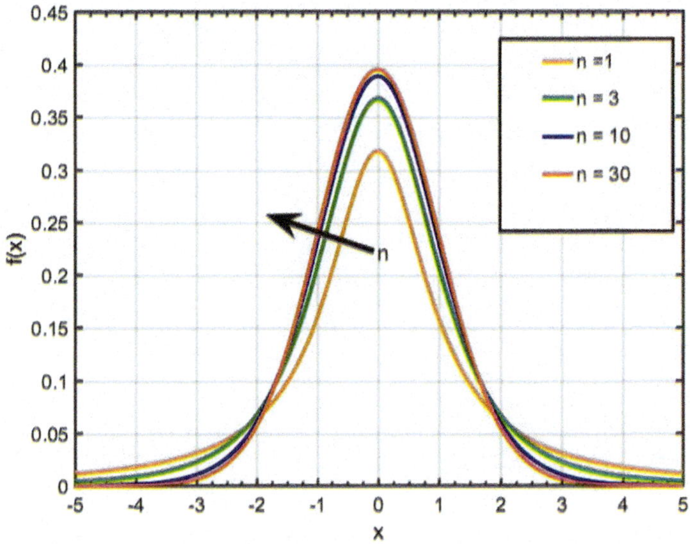

Fig. 6.28 Student distribution (*t-distribution*) for different sample sizes. From $n = 30$ onwards, the CLT comes into play, and the *t-distribution* changes to the normal distribution

means that instead of the standard normal distribution $z(0,1)$, the *t-distribution* (Student distribution) is used instead of the standard normal distribution z(0.1) (Fig. 6.28).

Including the empirical dispersion, the following applies to the distribution of the mean value of the sample of a normally distributed random variable

$$\frac{\overline{X} - \mu}{S/\sqrt{n}} \sim t_{n-1,\alpha/2}. \tag{6.57}$$

The CI of the mean of a standard normally distributed variable for $n = 5$ and $= 0.05$ is shown in Fig. 6.29. The statistical uncertainty is split in half between the two interval boundaries. The graph shows that the CI of $I = [-\ 2.57,\ 2.57]$ includes the true mean of $= 0$ with 95% confidence.

The width of the CI for the mean is more than 2α that of the original data before averaging, so it is a relatively wide range. For the practical analysis, the consequence is that every centered and normalized mean value that lies in the range of 95% certainty for a concrete n can correspond to the true parameter.

Confidence Interval for the Variance

As already discussed, the variance^2 of a normally distributed random variable is^2-distributed:

$$\frac{(n-1) \cdot S^2}{\sigma^2} \sim \chi^2_{n-1}. \tag{6.58}$$

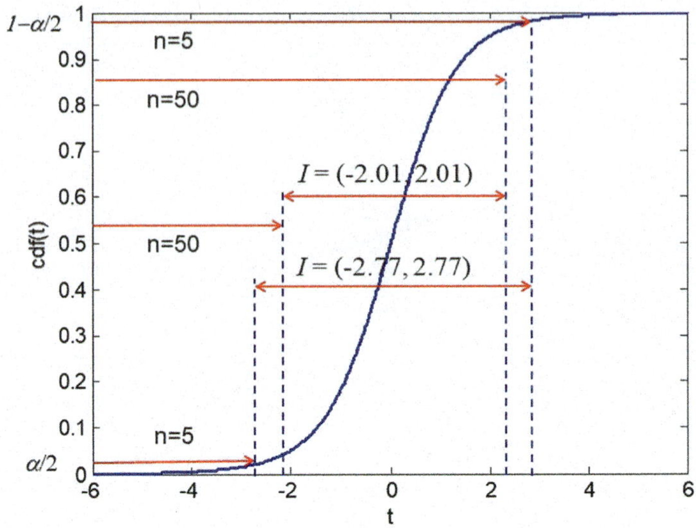

Fig. 6.29 Confidence interval of the mean of a standard normally distributed variable with $n = 5$, 50, and $= 0.05$ from the distribution function of the *t-distribution*

The limits for a desired uncertainty can be determined from the^2 distribution:

$$P\left(\chi^2_{n-1;\alpha/2} \leq \frac{(n-1)\cdot S^2}{\sigma^2} \leq \chi^2_{n-1;1-\alpha/2}\right) = 1 - \alpha. \tag{6.59}$$

By changing over, one obtains the KI

$$I = \left(\frac{(n-1)\cdot s^2}{\chi^2_{n-1;1-\alpha/2}};\frac{(n-1)\cdot s^2}{\chi^2_{n-1;\alpha/2}}\right). \tag{6.60}$$

Properties of Confidence Intervals

For the length of a confidence interval calculated from the *t-distribution, the* following applies

$$L = t_{n-1,\alpha/2}\frac{2s}{\sqrt{n}}. \tag{6.61}$$

This results in the following properties:

- As the sample size n increases, the length L decreases. From a statistical point of view, as the amount of data increases, the certainty or the uncertainty decreases, so the calculated parameter becomes more confident, and therefore its range of variation narrower.

- The length increases as the scatter s *increases. The* more scattered the data or the parameter calculated from them, the greater the range of variation that must be allowed for the AI with the same uncertainty.
- As the uncertainty decreases, the length of L increases. This property can also be read directly from the graph in Fig. 6.29. The more specific an estimate is, the more confident the estimated parameter will not leave the CI by chance, so the CI must correspondingly be longer.

6.3 Statistical Tests

6.3.1 Basics

Based on collected data and a given statistical uncertainty, a statistical test aims to decide which of the pre-test hypotheses to accept and which to reject.

Before a statistical test is carried out, the following methodological steps are necessary:

- With the help of the available means (theory, empiricism, heuristics), one should investigate whether a correlation can exist in the data collected or to be collected and, if so, what the nature of the correlation is. If this first step is omitted, bizarre conclusions can be reached.
- If there is a possibility of a correlation, at least two hypotheses are formulated in order to prepare decision options. Two hypotheses are most common—especially in signal analysis: a null hypothesis (no effect detectable) and an alternative hypothesis (effect detectable).
- One or more samples are planned based on the expected relationship and the hypotheses. A typical example is studies on the effectiveness of medications, in which the comparison before and after (medication intake) is carried out based on two samples. Often, one does not have the choice of planning samples for data whose generation is beyond one's control, e.g., when recording the number of sunspots in daily measurements.
- One or more statistical tests are selected based on the expected relationship, the hypotheses made, and the samples or the characteristics of the data collected. In the test result, the hypotheses are accepted or rejected. In the case of several tests, one must expect that the decisions about the acceptance or rejection of hypotheses will differ. Which of the decisions is accepted is a complex problem that will be dealt with later.

The procedure and the questions to be answered are illustrated in the following example.

6.3.1.1 Example

A new antihypertensive drug is to be tested for its efficacy. Biochemical tests have been carried out to ensure that the new drug could be effective due to its composition. Therefore, it is hypothesized that the drug will work. A clinic receives drugs to test 20 hypertensive patients. For test efficacy, a comparison of the blood pressure before and after taking the new drug is necessary, a typical two-sample problem (paired sample). The measured blood pressure in mmHg is shown in the following table.

Before: 178, 149, 164, 159, 176, 170, 158, 141, 173, 158, 165, 172, 177, 170, 147, 156, 177, 156, 176, mean 165
After: 122, 134, 153, 120, 126, 128, 128, 144, 131, 128, 121, 150, 138, 157, 139, 137, 154, 141, 128, 147, mean 136

The visual inspection of the data and comparing the mean values indicate that the blood pressure decreased after medication. However, this needs to be statistically validated. The clinic's simple statistical program performs a *t-test* with the standard uncertainty (common in medicine) of = 5%. This test confirms the effectiveness of the drug. The question arises regarding how reliable this decision is from a methodological point of view, regardless of the given statistical uncertainty. Compliance with assumptions and prerequisites must be checked. The *t-test assumes* a normal data distribution, as will be shown later.

Consequently, the data must be tested to see if they are normally distributed. It is practically impossible with a sample size of 20, as it is too small for such a test. Therefore, the t-test should not be used at all to be safe. The more difficult problem is that one can only make one null hypothesis here. The test either accepts or rejects this hypothesis. Why this is so is shown below.

6.3.2 Hypotheses for Statistical Tests

In the following, the most common case in signal theory is that of (at least) two hypotheses. The null hypothesis H_0 is the possibility that there is no effect, that there is no change. The alternative hypothesis H_1 (i.e., an alternative to H_0) is the possibility that an effect occurs and that there is a change. The following table shows two possible decisions of a statistical test with two hypotheses resulting in four possibilities. The statistical uncertainties of α (the probability of false positive decision of the test, FPR, false positive rate) and β (probability of false negative decision of the test, FNR, false negative rate) depend on each other so that only one of the two (usually α) is fixed before the test.

The proportion of correct positive decisions $(1 - \beta)$ is called the sensitivity of a statistical test. of statistical test, the proportion of correctly negative decisions $(1 - \alpha)$ specificity. Usually, one expects a statistical test to have a high sensitivity (more than 90%) and a high specificity (more than 90%). In medicine, only a few tests achieve these desired values.

For instance, in ophthalmological diagnostics, intraocular pressure is used to indicate glaucoma, whose sensitivity is about 65% and specificity about 50%. Glaucoma is only detected in two out of three sufferers with the help of eye pressure measurement, but every second healthy person is classified as ill. These are unacceptable values, so further tests must be done (e.g., visual field measurement). The sensitivity and specificity of many tests depend on each other, as shown in the following example.

6.3.2.1 Example

Two large samples (each $n = 10^6$) were taken from a population of adults to determine the body weight of women and men. The distribution of the two characteristics is shown in Fig. 6.30. Based on the display of a patient scale, one is supposed to decide (unrealistically, but statistically impressively) whether a new admission to the hospital is a man or a woman. Therefore, the test variable, in this case, is the measured body weight. Furthermore, a critical value (threshold) of the test size must be determined, which is used to decide whether hypothesis H_0: "New admission is a woman," or hypothesis H_1: "A new admission is a man," is accepted. The threshold was initially set arbitrarily at $x_S = 72$ kg. In order to be able to read off the proportions directly according to Table 6.2, the absolute frequencies from Fig. 6.30 are integrated and normalized to the distribution function (Fig. 6.31). The graph shows that at the critical value of $x_S = 72$ kg, a newcomer with a higher body weight is correctly classified as a man with a PRB of 72%, and a newcomer with a lower weight is correctly classified as a woman with a PRB of 92%.

In contrast, a man who weighs less than 72 kg is misclassified as a woman with a PRB of 18%, and a woman who weighs more than 72 kg is misclassified as a man with a PRB of 8%. From this example, all four proportions (Table 6.2) depend on the decision threshold. One can adjust the threshold of a concrete question depending on the importance of the specificity and the sensitivity of the FPR and the FNR.

6.3.3 The Goodness of Statistical Tests, ROC

A measure and a clear graphical representation of the correlation between the sensitivity and the specificity of the decision threshold are provided by the ROC (Receiver Operating Characteristic). This term originates from message-oriented statistics and has become established in all other fields. The ROC for the above example (Figs. 6.30 and 6.31) is shown in Fig. 6.32. The graph shows that the increase in sensitivity is at the expense of specificity and vice versa. Therefore, the decision on the discriminant threshold always depends on the specific question: Is it essential to find as many positives as possible even if many negatives are mistakenly included in the selection, or is it rather essential to classify negatives as correctly as possible even if many positives are not detected. Typically, such questions arise quite differently in medical diagnostics (identifying as many sick

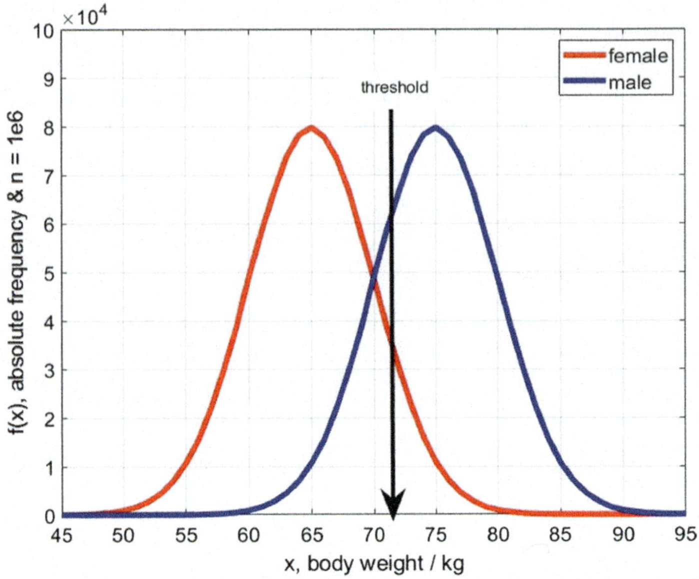

Fig. 6.30 Distribution of body weights of women and men from a sample of $n = 10^6$ each. The mean values are 65 kg for women and 75 kg for men; the dispersion is equal and amounts to 5 kg

Table. 6.2 Possibilities of test results and actual validity of hypotheses, and are the statistical uncertainties in the acceptance of the respective hypothesis

Test result	H_0 true	H_1 true
H_0 adopted	Properly negative 1—Specificity	False negative Error of the 2nd kind
H_1 adopted	False positive Error 1. type	Positive 1—Sensitivity

people as possible as sick, even if healthy people are also included in the selection) than in radical therapy, e.g., in radiation treatment or amputation (wrongly irradiating or amputating as few healthy people as possible). From this point of view, the graph of the ROC in Fig. 6.32 can be concluded from the three marked points:

At a threshold of 65 kg, over 95% of men are correctly identified as men, but almost half of women are incorrectly classified as men. At a threshold of 75 kg, the reverse is true: about 99% of women are correctly classified as women, but less than 50% of men are recognized as such. A balanced success rate is offered by the threshold of 70 kg, where 82% of men and 86% of women are correctly classified.

The quality of a ROC (the statistical certainty of sensitivity and specificity) increases when the graph approaches the upper left corner of the diagram as closely as possible. Looking back at distribution densities (Fig. 6.30), one can state as a general rule that the further apart the two distribution densities are, the better the

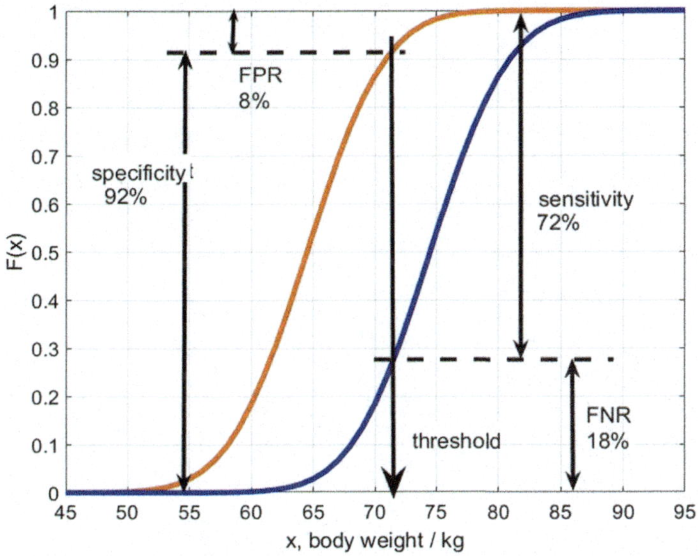

Fig. 6.31 Distribution of probabilities at a decision threshold of 72 kg. The sensitivity, specificity, FPR, and FNR refer to the decision that the patient is a man with a weight greater than 72 kg. One would swap them for whether a patient weighing less than 72 kg is a woman

ROC. However, neither in medical statistics nor in biosignal processing does one have any influence on the collected data. Therefore, instead of using the original data, one can try to use their functionals or other statistical quantities that promise a greater distance of distributions among the hypotheses under consideration. In this concrete example, one could assume that in addition to weight, mean height could also serve as a distinguishing criterion since women are tinier on average than men.

The question arises regarding combining the two statistical variables of body weight and height to achieve better discriminatory power in the ROC. To do this, we first look at the two distributions of body height and weight, as shown in Fig. 6.33. The two distributions for women and men are relatively clearly distinguishable; one can separate them with a linear function (straight line) (dashed line in Fig. 6.33). Since a linear separation is sufficient, a second-order statistical quantity lends itself as a classification criterion: The product of height and body weight, following the cross-correlation. Although products are inherently double-exponentially distributed, the empirical frequencies in Fig. 6.34 show almost normal distributions, primarily due to the large sample. Comparing the distributions in Fig. 6.34 with those in Fig. 6.30, we find they are much more distant from the products in Fig. 6.34. That is what was intended by forming a new statistical quantity: Due to the increase in distance between the distributions, the ROC has also moved significantly closer to the ideal (Fig. 6.35).

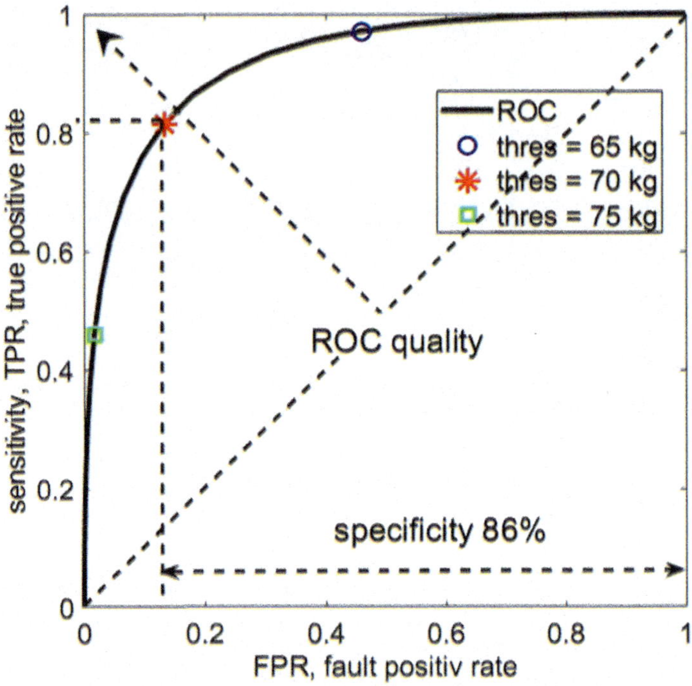

Fig. 6.32 The ROC as a function of the decision threshold thres = 65, 70, 75 kg that a newcomer is a man (see Figs. 6.30 and 6.31). At a threshold of 70 kg, the sensitivity is 82% and the specificity 86% (FPR = 14%)

6.3.4 Parametric Tests

It has already been shown (Sect. 6.2.1) that theoretical or known distributions can be described unambiguously with a few parameters. Usually, two parameters are sufficient to characterize even large data sets, e.g., in the normal distribution case. Parametric tests aim to test the location or the variance of an empirical distribution (one of its parameters) with a given statistical uncertainty concerning the established hypotheses. Typically, one compares the test variables obtained from the empirical data with theoretical distributions, whereby several conditions must be fulfilled. An empirical approach can also be used for practical analysis, as it was treated based on two distributions for the hypotheses H_0 and H_1 (Figs. 6.30, 6.31, 6.32, 6.33, 6.34 and 6.35). The following presents the best-known tests, which are also most frequently used in practice.

Note: For parametric tests with only one sample or one paired sample, only one hypothesis can be made: The null hypothesis H_0, e.g., = 0. Because otherwise, no further empirical data are available for an alternative hypothesis, the alternative hypothesis H_1 is (but methodologically questionable) formulated as a logical negation of the null hypothesis, e.g., H_1: $\mu \neq 0$. This procedure has been

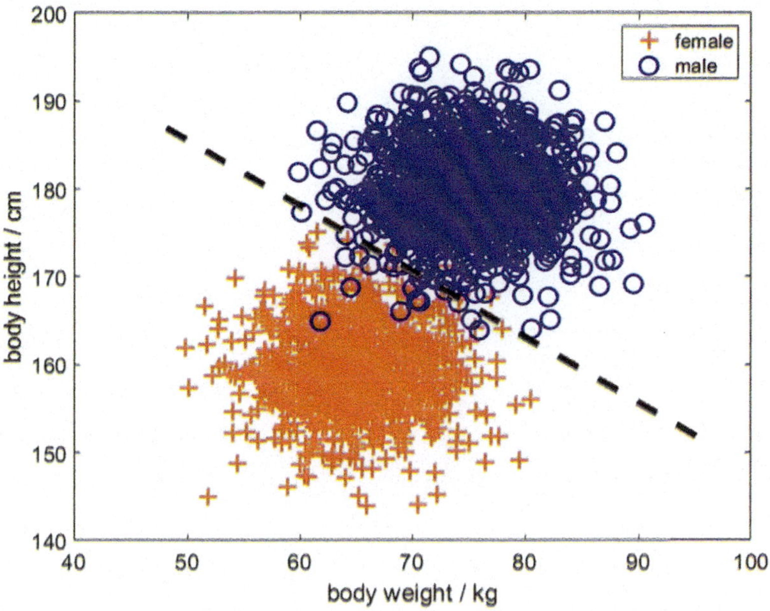

Fig. 6.33 Distributions of the statistical variables body weight and height for women and men

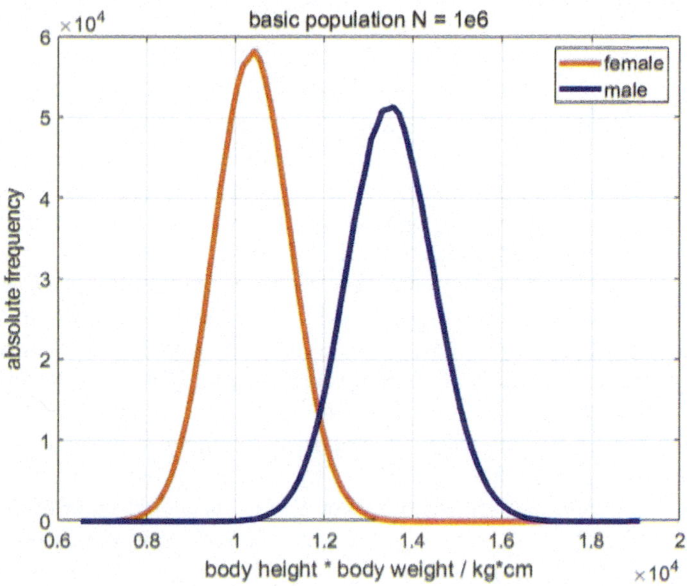

Fig. 6.34 Distributions of the statistical size product of height and body weight in women and men

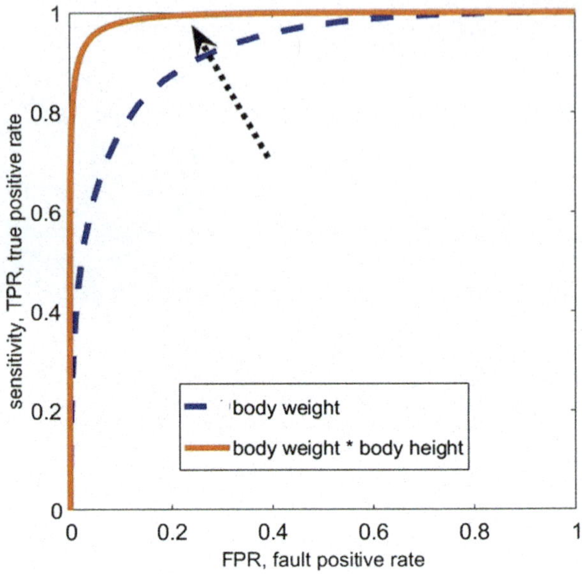

Fig. 6.35 The ROC of the statistical quantity body weight * body height compared to body weight alone to distinguish the classes women and men. The ROC of the composite size is significantly closer to the corner point (FPR = 0, RPR = 1) than the ROC of the body weight alone. Therefore it has a much higher goodness (test power). Numerically, it can be quantified as the area below the ROC: The closer the area converges to 1.0, the higher the quality

recommended, demanded, accepted, and carried out for many years across all disciplines. However, it is clear from the outset that it is methodologically contentious for the following reason: since one has no data for the alternative hypothesis, half of the statistical measures (Table 6.2) cannot be calculated at all (error of the second kind, β; sensitivity, $1 - \beta$). Therefore, with only one sample (also composite sample), one can only set up the null hypothesis H_0 and state the confidence interval for this (methodologically more sensible, as more information about the data). Even rejecting the null hypothesis on uncertainty is problematic. An empirical test statistic (e.g., the arithmetic mean), even if it lies in the rejection range (Figs. 6.38 and 6.39), does not yet provide a plausible reason for rejecting the null hypothesis since the calculated value belongs to the empirical sample at H_0. Why, then, reject the null hypothesis based on an empirical value known to come from the sample for the null hypothesis? The conclusion is: If one has a single (also paired) sample, one cannot logically decide on two hypotheses. One can only show which range (confidence interval), e.g., 95%, the empirical data lie with the desired statistical certainty.

Sample-based statistical tests are required in studies, journals, evaluations, research projects, and qualification papers (according to doctrinal opinion). However, their methodological justification is doubtful; see above. In order to meet the demands of research and teaching in this area, below follows the part of statistical tests as they are demanded and accepted since one will probably have to make do with them for a long time to come. At the end of the chapter, tips are given on methodically avoiding this problem.

An example of the absurdity of this widespread view can be found in the Tasks and Solutions part. It should be investigated (Engelhardt, 2019) whether graduates with a Master's degree (MA) have a higher entry-level income than Bachelor's graduates (BA). Following the doctrinal recommendations, it was formulated in H_0 that MAs have a lower income to prove the opposite statistically in H_1. However, the data situation did not allow this, and H_0 was not rejected. As a (methodologically necessary but mostly ignored) control, the opposite of the first attempt was formulated for H_0. However, here, too, H_0 could not be rejected. The result of this analysis is that one has two unrejected Hull hypotheses, which logically contradict each other. From this, there is a clear conclusion: one of the null hypotheses was wrongly not rejected, and one, therefore, committed an error of the second kind (uncertainty). However, since it is not known, one will not be able to know which of the two null hypotheses should be rejected. This example demonstrates the fundamental dilemma of statistical tests with only one sample: Statistically, confidence at the given uncertainty level rejects the null hypothesis; otherwise, one cannot decide on any of the other three possibilities of the test outcome.

6.3.4.1 The t-Test

The *t-test* is based on the *t-distribution* (Student distribution; Fig. 6.28) and is the same as whether the confidence interval for the mean determined according to Eq. 6.57 includes the value given at H_0. An essential prerequisite for applying the *t-test* is the normal population distribution from which the sample was drawn. Methodically, one must first check whether this condition is fulfilled with a so-called goodness-of-fit test. The standard test used for this is the^2 goodness-of-fit test. However, its application is problematic, as the following example shows.

Example of the χ^2 Adaptation Test

A population of 1000 values was generated, distributed double exponentially, and right skewed (Fig. 6.36). The goodness-of-fit test is performed for different sample sizes (Fig. 6.37). The null hypothesis is that the empirical distribution under investigation is normal. The result of the goodness-of-fit test (Fig. 6.37) states that up to a sample size of 29, the null hypothesis cannot be rejected. In practice, this is wrongly interpreted as confirmation of the null hypothesis and, thus, the assumption of a normal distribution. The correct interpretation of this result is that up to a sample size of 29, the null hypothesis cannot be rejected for the given uncertainty of $= 5\%$, but neither can it be accepted. This problem is because one does not have a distribution for the alternative hypothesis (always), and therefore the error is unknown. That can be very large, especially with small samples. Since there is a high risk of misinterpreting such results, especially in biostatistics, usually only small samples are available, the goodness-of-fit test for small samples ($n < 100$) should be rejected in principle. However, the question then arises as to how one can proceed differently.

As the previous example shows, goodness-of-fit tests are unsuitable for checking whether the assumption about a theoretical distribution is fulfilled, especially with small samples. If one starts from the test size (Eq. 6.62), one can state the

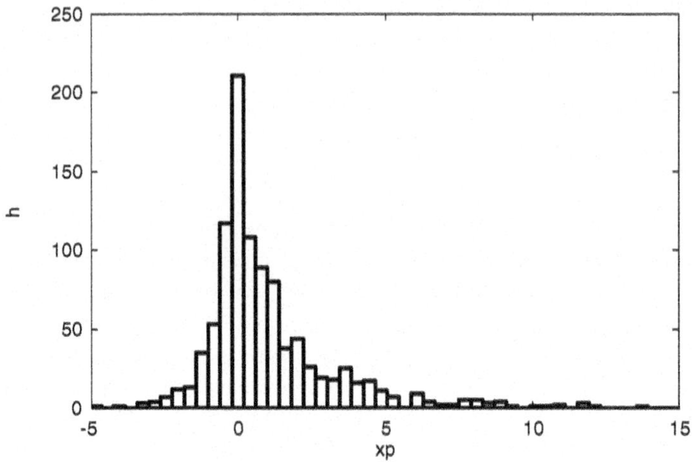

Fig. 6.36 Absolute frequency of 1000 double exponential right-skewed distributed values (skewness is positive, the excess is positive). Considering them as a population for the goodness-of-fit test in the example, it is already evident after the first visual inspection that it cannot be a normal distribution

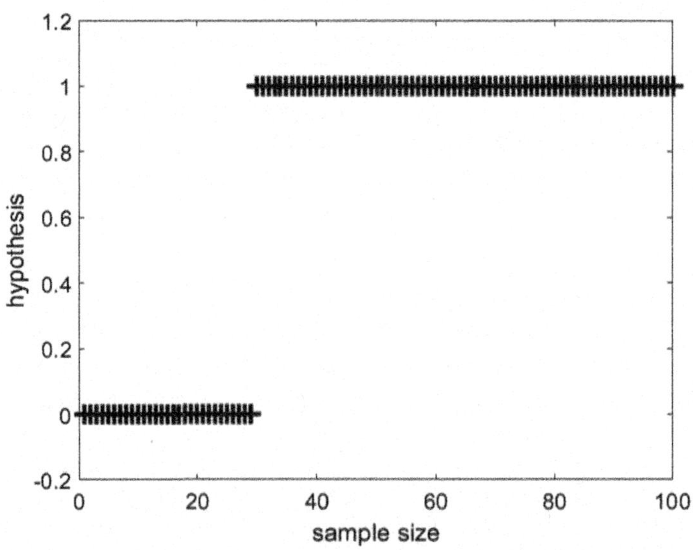

Fig. 6.37 Output of the χ^2 goodness-of-fit test as a function of sample size for data from Fig. 6.36. Note that the null hypothesis is not rejected up to a sample size of 29. Often such a result is wrongly interpreted as meaning that normally distributed data are present since H_0 was not rejected, so it should be correct. However, the fact that the null hypothesis was not rejected merely means that it is not certain that a distribution other than normal is present. Therefore, the non-rejection of the null hypothesis is not yet a confirmation that a normal distribution is present

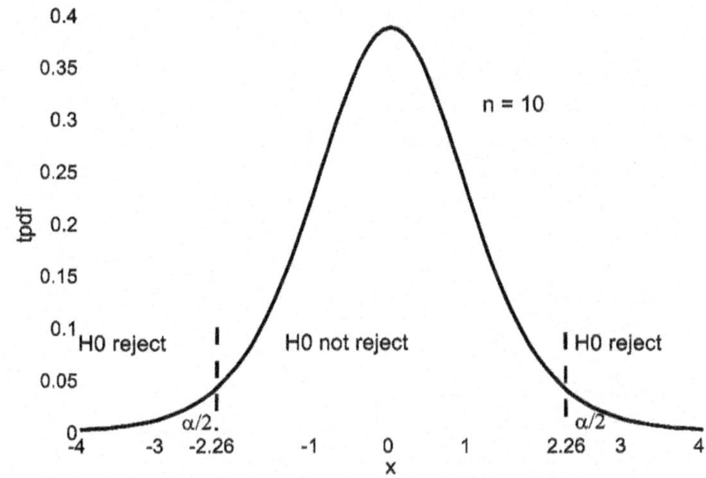

Fig. 6.38 Areas for rejecting and not rejecting the null hypothesis in a two-sided *t-test with a* sample size 10. With such a small sample, it should be sure that the data originate from a normal distribution

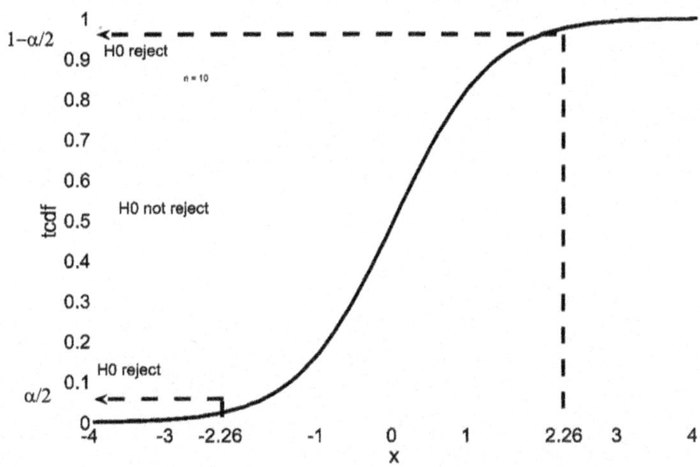

Fig. 6.39 Areas for rejecting and not rejecting the null hypothesis in a two-sided *t-test* with a sample size 10

following:

$$T = \frac{\overline{X} - \mu_0}{S/\sqrt{n}} \tag{6.62}$$

In Eq. 6.62, T is the test size (the parameter to be tested calculated from the sample), $\overline{X}$ is the arithmetic mean of the sample$_0$ is the given parameter to be tested against $\overline{X}$, S is the empirical standard deviation, and n *is* the sample size.

The empirical arithmetic mean is centered (difference formation to$_0$) and standardized to the normalized empirical dispersion. Otherwise, it is a point estimate of the location parameter of the sample under investigation. If one were to calculate more such averages, one could rely on the CLT for a sample size of more than 30: While the population data may not be normally distributed (as in the example above), their mean is taken into the test is, to a good approximation. The analysis concludes that if the sample size is sufficiently large (n > 30), there is no need to check the data distribution, as the CWS ensures it a priori.

Procedure for the t-Test

The following null hypothesis is formulated for a sample (μ_0 is a theoretical, a sufficiently confident empirical, or a hypothetical value):

$$H_0 : \quad \mu = \mu_0$$

If a (theoretical or empirical) distribution for the alternative hypothesis is not available, the negation of the null hypothesis is set up (so far) as an alternative hypothesis:

$$H_1 : \quad \mu \neq \mu_0$$

Corresponding to the test size according to Eq. 6.62, the empirical test size t *is* calculated (Eq. 6.63).

$$t = \frac{\overline{x} - \mu_0}{s/\sqrt{n}} \tag{6.63}$$

For a two-sided test (Figs. 6.38 and 6.39), the critical values for $n - 1$ *degrees of freedom* and the halved statistical uncertainty are determined (from tables or software). For a one-sided test, only the left or only the right is included, depending on the question. Testing of the empirical t-value concerning the critical values is carried out according to the following scheme:

1. $t < t_{n-1,\alpha/2}$, rejection range of H_0 with uncertainty/2, i.e., H_1 is assumed,
2. $t_{n-1,\alpha/2} \leq t \leq t_{n-1,1-\alpha/2}$, H_0 is not rejected, is, therefore, to be accepted,
3. $t > t_{n-1,1-\alpha/2}$, rejection range of H_0 with uncertainty/2, i.e., H_1, is assumed.

In a one-sided test, the left or right side is tested according to the question:

1. $t < t_{n-1,\alpha}$, rejection range of H_0 with uncertainty in the left-sided test
2. $t > t_{n-1,1-\alpha}$, rejection range of H_0 with uncertainty in the right-sided test

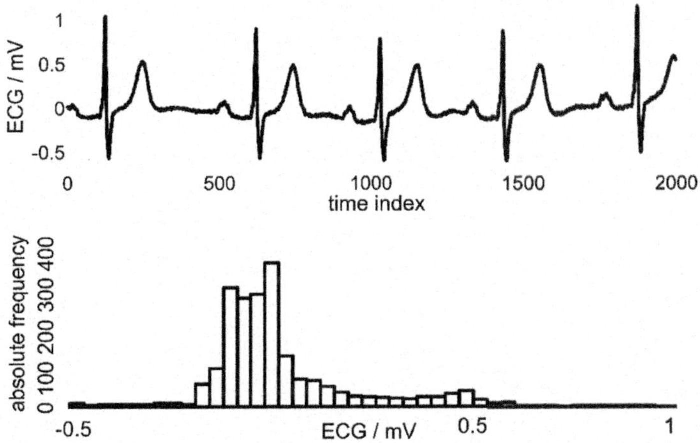

Fig. 6.40 ECG intercept and absolute frequencies of the values. The distribution is multimodal with a high excess of 4.29 and right-skewed; skewness is 1.33. Due to the sample size of N = 2000, it can be assumed that the empirical mean is normally distributed because of the CLT, and therefore H_0 can be tested with the *t-test.* The test rejects H_0 statistically correctly, whereby the proven deviation of 0.02 mV can be classified as unimportant, practically negligible, since it is in the amplifier's inherent noise range

Example

Sections in biosignals in which there is no physiological activity are called the zero line (baseline). It is to be examined whether this designation is justified. The hypotheses are formulated as follows:

H_0: The mean value of an ECG section is zero,
H_1: The mean value of an ECG section is not zero.

An ECG section with 2000 values is available (Fig. 6.40). Despite the inferior distribution properties (excess 4.29, skewness 0.33), due to the large sample of $n = 2000$, it can be assumed that as a result of the CLT, the mean is normally distributed. The *t-value* is calculated from the data according to Eq. 6.63 and compared with the critical value:

$$t = 5.38 > t_{1999\,0.975} = 1.96$$

A two-sided test was carried out since it was unclear whether the mean could become smaller or larger than zero before the test. The empirical *t-value* is larger than the right-hand limit, so the null hypothesis is rejected, and the alternative hypothesis is accepted. From a purely statistical point of view, the designation "null line" is not justified. However, the empirical mean value of 0.02 mV deviates insignificantly from zero and is already in the noise range. For practical analysis, one would assume it to be zero. At this point, it becomes clear that not everything that can be statistically proven also has a practical meaning. In this example, an effect comes into play that must be taken into account in statistical tests:

If the sample is large enough, the null hypothesis is almost always rejected, regardless of the size of the difference, i.e., even in the case of tiny, practically unimportant differences.

The opposite extreme, as shown in the adaptation test example, leads to the following finding:

If the sample is tiny (n < 20), the null hypothesis is seldom rejected, even when there is a clear difference.

Note: The fact that the null hypothesis is not rejected does not necessarily mean it is correct, i.e., it is not necessarily accepted (see note above).

The t-Test for Two Paired Samples

Paired samples are *n* pairs of values $\{x_i; y_i\}$ from populations with expected values μ_1 and μ_2. A typical example is a group of hypertensive patients on whom the effect of a new blood pressure-lowering drug is to be tested. The comparison of the blood pressure "before and after" in the sense of taking the medication is interesting. Differences are formed between the pairs of values, which are then tested for a significant difference of zero, as in a random sample.

$$d_i = x_i - y_i. \tag{6.64}$$

Accordingly, one sets up the hypotheses:

$$H_0 : \mu_1 = \mu_2 \Rightarrow d = 0$$

$$H_1 : \mu_1 \neq \mu_2 \Rightarrow d \neq 0$$

The test variable is calculated analogously to the formula according to Eq. 6.63 (Eq. 6.65):

$$t = \frac{\overline{d}}{s_d/\sqrt{n}} \tag{6.65}$$

The practical advantage of difference formation is that the skewness is partially canceled out in paired samples, and the difference is mainly symmetrical (Fig. 6.41). That means that the requirement regarding the normal distribution of the data is no longer so strict, and one can already calculate with the CDFS from a sample size of n = 30.

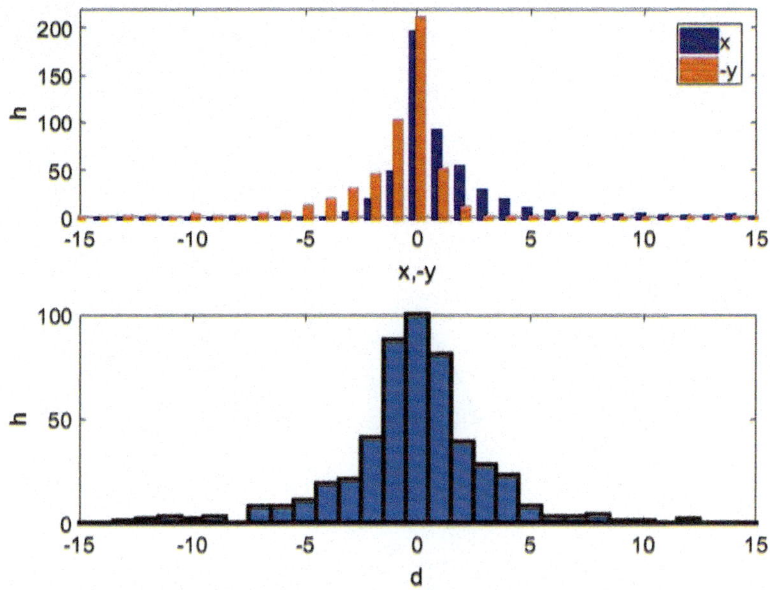

Fig. 6.41 Frequency distributions of two paired samples (top) x and $-$ y as they enter into the difference formation $d = x - y$. Note that x is right-skewed, $-$ y left-skewed. The frequency distribution of the differences (bottom) shows good symmetry

6.3.5 Nonparametric Tests

6.3.5.1 Wilcoxon Test

The Wilcoxon test belongs to the group of rank sum-based tests. The essential feature of this group is the rank transformation of data to rank numbers of a data series ordered by values. There are several rank transformations; the fundamental rank transformation based on the ordered value series will be dealt with here.

Rank Transformation

The basis of every rank transformation is ordering the data into an ascending (or descending) sequence according to the (signed or unsigned) values. The data lose their numerical values and receive rank numbers from the ordered sequence instead (Eq. 6.12). Rank-based statistics and analysis methods are robust and resistant to extreme values and outliers. However, they also have a significant disadvantage:

> The rank transformation is an irreversible stochastic-nonlinear transformation that does not allow any conclusions from linear correlations to original data.

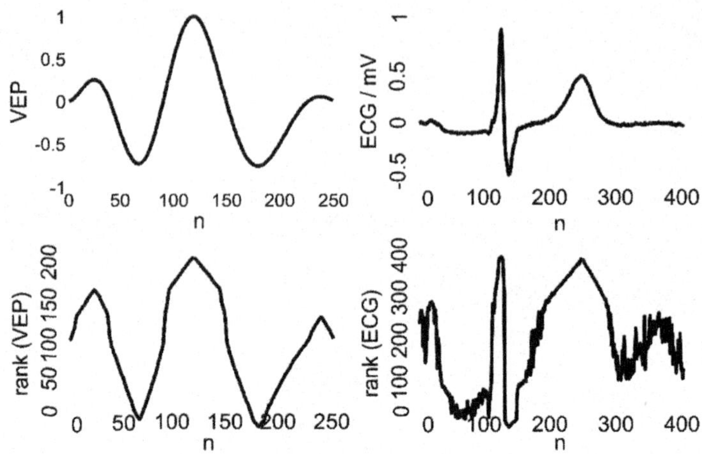

Fig. 6.42 Synthetic (left, VEP) and natural (right, ECG) signals (top) and their rank transformations (bottom)

Stochastic nonlinearity means that the transformed data has a nonlinear relationship to the original data, with the degree of nonlinearity being stochastic. It follows that even established methods of non-linear analysis are not applicable here. The following example demonstrates the effect of the rank transformation.

Example
The effect of the rank transformation on simulated and actual data is shown in Fig. 6.42. These data can be considered decoupled from the time index; the effect is not time-dependent. Both examples show that the rank transformation has a compression character: High levels are attenuated, and low levels are relatively raised. The essential difference to conventional compression methods (e.g., in telecommunications) is that the degree of compression depends on current data and therefore has a random character. In the case of real signals (Fig. 6.42 right), this is aggravated by the fact that the always-present noise is amplified and thus further reduces the already relatively poor SNR.

Procedure for the Wilcoxon Test
When applying the Wilcoxon test, the data must be symmetrically distributed and continuous, not necessarily normally distributed. The continuity condition follows from the fact that the values of the rank numbers should not repeat. This requirement is not critical in the practical analysis since rankings can also be assigned to the same values, e.g., by averaging.

The null hypothesis for paired samples is:

$$H_0: \quad \mu_1 = \mu_2 \quad \text{or } d = 0, \text{ for } d_i = x_i - y_i$$

If a (theoretical or empirical) distribution for the alternative hypothesis is not available, the negation of the null hypothesis is set up as the alternative hypothesis:

$$H_1: \quad \mu_1 \neq \mu_2 \ \text{ resp.}\, d \neq 0$$

The difference formation with paired samples has the positive side effect of having a symmetrizing effect. Therefore, in practical analysis, the symmetry condition is softened in this way and considered to be fulfilled.

The data are sorted according to their amount and are given rank numbers according to their place in the ordered row.

Original data:
4.8399 – 0.0065 – 1.3318 0.0550 0.0142 – 0.1834 0.9504 0.5498 0.2762

Ordered row (by amount):

– 0.0065	0.0142	0.0550	– 0.1834	0.2762	0.5498	0.9504	– 1.3318	4.8399
1	2	3	4	5	6	7	8	9

For$_0$ = 0, the rank sums are determined:

$$R_P = \sum_{x_i > \mu_0} R_i = 32$$

$$R_N = \sum_{x_i < \mu_0} R_i = 13$$

The smaller of the two rank sums is compared with the corresponding critical value:

$$R = \min(R_P, R_N),$$

$$R = 13 > R_{9,0.05} = 10$$

The null hypothesis is not rejected if the determined rank sum exceeds the critical value. As shown in this example, rank sum tests are based on the assumption of a symmetrical distribution. Then the sums under the null hypothesis are approximately equal. If the smaller sum falls below the critical value, one can assume that the assumption of μ_0 does not apply; the empirical distribution lies significantly to the side of μ_0. Because of the rank transformation, the expected value is not tested here (as in the *t-test*) but the median. That is also the reason for the remarkable robustness of the Wilcoxon test. If we look at the ordered series in the example above, the ninth value is an outlier, at least an extreme value. However, it retains its rank whether the original value is corrected accordingly (e.g., by repeated measurement g) or even becomes larger.

6.3.5.2 Wilcoxon-Test Versus t-Test

The robustness of rank sum tests means they are sometimes very conservative in rejecting H_0. They tend not to reject the null hypothesis or to do so late, i.e., they are somewhat skeptical about rejection. The inevitable consequence is that the β-error (false negative decisions, FNR, miss) can rise sharply, although it remains quantitatively unknown. This property is shown by comparing the *t-test* with the Wilcoxon test in Fig. 6.43.

For test data from Fig. 6.41 (double-exponential, right-skewed), the location parameter d was varied between 0.0 and 2.0, with a sample size of $n = 30$. The *t-test* rejected the null hypothesis at $d = 0.8$, and the Wilcoxon test only at $d = 0.97$. Since the data were simulated, it is known that the t-test was decided correctly because of the sample size for applying the CLT is sufficiently large. In practical analysis, however, one does not know the true parameter (d), so one would rather trust the rank sum tests for small sample sizes. The dependence of the test results on the sample size for both tests is shown in Fig. 6.44. As expected, the Wilcoxon test rejects the null hypothesis later than the *t-test*. That the test results vary between the two hypotheses indicates that the default uncertainty = 0.05 is borderline. With a = 0.01, the results would not fluctuate, but a larger sample size would be necessary to reject the null hypothesis.

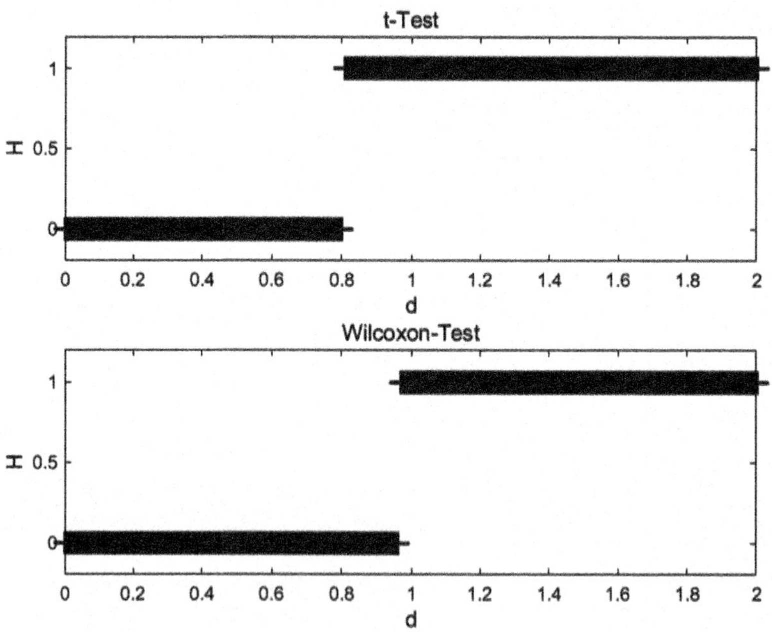

Fig. 6.43 Results (hypotheses) of the *t-test* (top) and the Wilcoxon test (bottom) as a function of $d = 0\ldots2$ (difference) of paired samples with a sample size of $n = 30$. The sample was taken from the data in Fig. 6.41

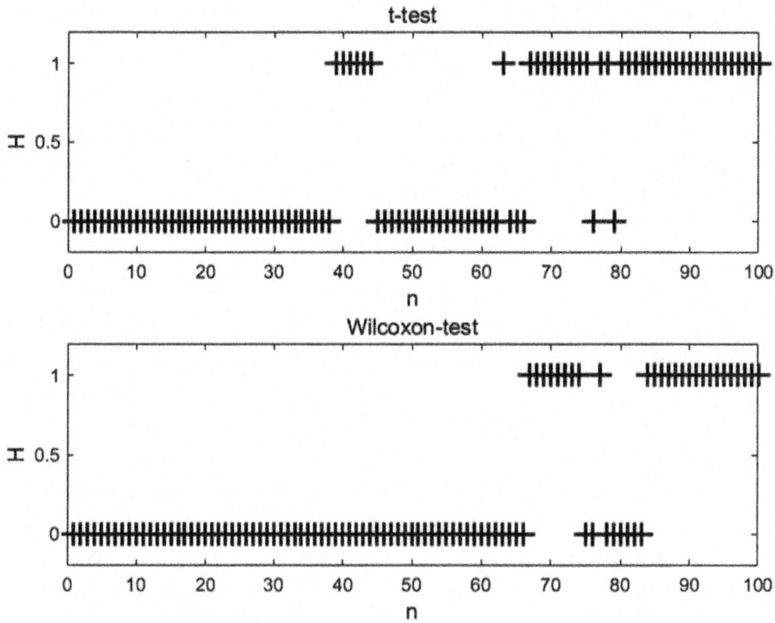

Fig. 6.44 Results (hypotheses) of the *t-test* (top) and the Wilcoxon test (bottom) as a function of sample size *n of* paired samples at $d = 0.6$. The samples were taken from the data in Fig. 6.41 (double exponential, right skewed)

6.3.5.3 Sign Test

The sign test is the most robust statistical test of all. This property results from the approach that there are approximately equal numbers of positive and negative differences in the data distribution under the null hypothesis. Apart from the continuity condition (practically never fulfilled), no other conditions must be met. The continuity condition can be softened in practical analysis by suitable measures, e.g., adding weak noise.

The null hypothesis for paired samples is:

$$H_0 : \mu_1 = \mu_2 \quad \text{or}\, d = 0, \quad \text{for}\, d_i = x_i - y_i$$

If a (theoretical or empirical) distribution for the alternative hypothesis is not available, the negation of the null hypothesis is set up as the alternative hypothesis:

$$H_1 : \mu_1 \neq \mu_2 \quad \text{or} \quad d \neq 0.$$

Similar to the rank sum test, the sums are determined via the signs:

$$R^+ = \sum_{d_i > 0} 1 \quad R^- = \sum_{d_i < 0} 1.$$

The smaller of the two sums is compared with the critical value:

$$R = \min(R^+, R^-) \leftrightarrow R_{n,\alpha}.$$

The null hypothesis is rejected if the calculated sign sum lies outside the critical range. That can be determined for small samples from the binomial distribution,

$$P(X = k) = \binom{n}{k} \cdot 0.5^n,$$

for large samples ($n > 30$), the acceptance range of the null hypothesis can be given based on the standard normal distribution

$$0.5 \cdot n \pm 1.96\sqrt{0.25n} \pm 0.5.$$

Due to its extreme robustness, the sign test is also the most conservative (Fig. 6.45). It follows that if it rejects the null hypothesis, the rejection is also very confident, but at the cost of a very high-uncertainty. Therefore, this test should only be used as a preliminary stage of further investigations in the sense of an initial orientation. If it rejects the null hypothesis, practically no further tests for rejection are necessary. The test robustness is compared to the Wilcoxon test in Fig. 6.45. The sign test rejects the null hypothesis much later than the Wilcoxon test, but this rejection is confident.

6.3.5.4 Test Strategy

Considering the properties of the tests, a recommendation for the procedure can be formulated: If it is not sure that the data are normally distributed (practically the typical case), one starts with the sign test. This decision is confident if the sign test rejects the null hypothesis. If it does not reject the null hypothesis, one continues with rank sum tests for low sample sizes. For sufficiently large samples ($n > 30$, CLT), the more progressive *t-test is* preferred, in which the lowest β-uncertainty is expected.

After a statistical test, one should interpret the test result professionally, especially the quantitative ratios regarding the sample and the demonstrated effect (see the note above on the large sample size).

- Even if the desired result (rejection of the null hypothesis) were achieved, this result would have to be put into perspective concerning the supposed proof of the alternative hypothesis.
- The quantitative evaluation of the confidence interval is beneficial for interpreting the test result. For example, a relatively narrow confidence interval indicates that the random fluctuations are low and that there can only be a systematic error (bias) regarding the true value.

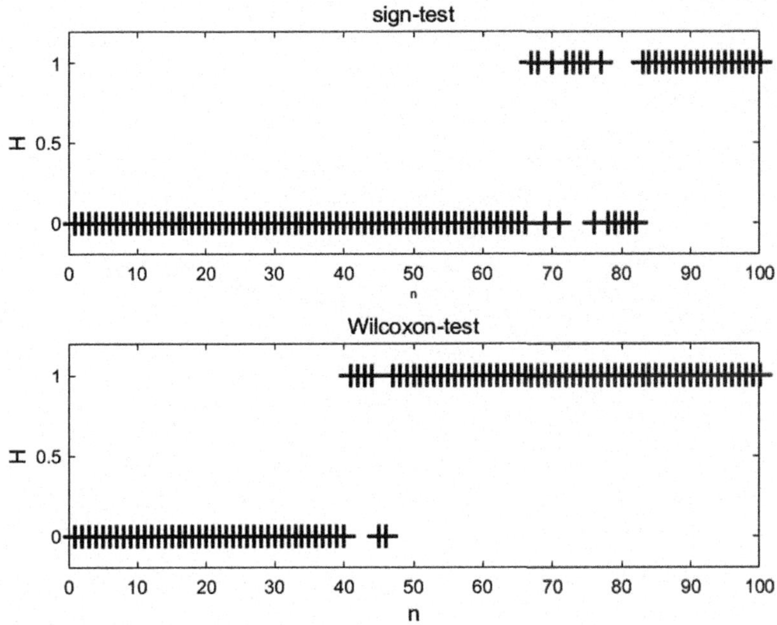

Fig. 6.45 Results (hypotheses) of the sign test (top) and the Wilcoxon test (bottom) depending on the sample size *n of* paired samples at $d = 0.8$. The sample was taken from the data in Fig. 6.45 (double exponential, right skewed). The sign test rejects H_0 much later, only from about twice the sample size of the Wilcoxon test

- One should critically discuss the proven difference (rejection of the null hypothesis) about its meaning. From a medical point of view, the question to be answered is whether the statistically specific difference also means a substantial contribution.

6.4 Statistics and Higher-Order Spectra

6.4.1 Moments and Cumulants

As already explained in Chap. 6.1, biosignals, biological quantities, and their parameters cannot generally be assumed to be normally distributed. The assumption of normal distribution applies exceptionally and only to a few processes, for example, the thermal noise of a medical measuring amplifier. When deriving some statistical measures (arithmetic mean, sum), one can rely on the effect of the CTE leading to the normal distribution. The first two central moments' mean and variance are sufficient for characterizing normally distributed quantities. However, the other 3rd and 4th order moments—skewness and excess—are also needed for all other distributions. In general, one needs measures up to order n, whereby a distinction must be made between two classes—moments and cumulants. From here on, because of the simpler notation, the angular frequency is used instead 2ϕ (Nikias & Petropulu, 1993).

6.4.1.1 Definitions of Moments and Cumulants

For a set of n random variables, $\{x_1^{k1}, x_2^{k2}, \ldots, x_n^{kn}\}$ the moments of order $r = k_1 + k_2 + \cdots + k_n$ are given by (Eq. 6.66)

$$m_r\left[x_1^{k_1}, x_2^{k_2}, \ldots, x_n^{k_n}\right] = E\left\{x_1^{k_1} x_2^{k_2} \ldots x_n^{k_n}\right\} =$$
$$= (-j)^r \frac{\partial^r \Phi(\omega_1, \omega_2, \ldots, \omega_n)}{\partial\omega_1^{k_1} \partial\omega_2^{k_2} \ldots \partial\omega_n^{k_{n1}}}\Big|_{\omega_1=\omega_2=\ldots=\omega_n=0}, \qquad (6.66)$$

where $\Phi(\omega_1, \omega_2, \ldots, \omega_n) = E\{\exp(j(\omega_1 x_1, \omega_2 x_2, \ldots, \omega_n x_n))\}$ is its first characteristic function. For example, for a set of random variables $\{x_1, x_2\}$ and the order $r = 2$, the following moments arise

$$\begin{aligned} m_2(x_1, x_2) &= E\{x_1 \cdot x_2\} \\ m_2(x_1) &= E\{x_1^2\} \\ m_2(x_2) &= E\{x_2^2\} \end{aligned} .$$

The natural logarithm of the first characteristic function yields the second characteristic function $\Psi(\omega_1, \omega_2, \ldots, \omega_n) = \ln(\Phi(\omega_1, \omega_2, \ldots, \omega_n))$, by whose derivative at the origin one obtains the cumulants (Eq. 6.67).

$$c_r\left[x_1^{k_1}, x_2^{k_2}, \ldots, x_n^{k_n}\right] = (-j)^r \frac{\partial^r \Psi(\omega_1, \omega_2, \ldots, \omega_n)}{\partial\omega_1^{k_1} \partial\omega_2^{k_2} \ldots \partial\omega_n^{k_{n1}}}\Big|_{\omega_1=\omega_2=\ldots=\omega_n=0} \qquad (6.67)$$

A determinant relationship exists between the cumulants and moments (Eq. 6.68).

$$c_n = (-1)^{n+1} \begin{vmatrix} m_1 & 1 & \cdots & 0 \\ m_2 & m_1 & & \vdots \\ \vdots & . & . & 1 \\ m_n & \ldots & m_2 & m_1 \end{vmatrix} \qquad (6.68)$$

For the first four orders, the following applies to the relationship between cumulants and moments

$$\begin{aligned} c_1 &= m_1 \\ c_2 &= m_2 - m_1^2 \\ c_3 &= m_3 - 3m_2 m_1 + 2m_1^3 \\ c_4 &= m_4 - 4m_3 m_1 - 3m_2^2 + 12 m_2 m_1^2 - 6m_1^4 \end{aligned} .$$

Typically, one can center the quantities so that $m_1 = 0$. Then the first three cumulants and moments are identical, and for the fourth one, $c_4 = m_4 - 3m_2$, which is already known as an excess. To extend the cumulants and moments to processes, the multidimensional shift$_{1,\ldots,n}$ must be introduced (Eq. 6.69). That turns random variables into temporally shifted variants of one (or more) processes.

$$\begin{aligned} m_n^x(\tau_1, \tau_2, \ldots, \tau_{n-1}) &= E\{X(k) \cdot X(k+\tau_1) \ldots X(k+\tau_{n-1})\} \\ c_n^x(\tau_1, \tau_2, \ldots, \tau_{n-1}) &= c_n\{X(k) \cdot X(k+\tau_1) \ldots X(k+\tau_{n-1})\} \end{aligned} \tag{6.69}$$

Under the normally permissible assumption $m_1 = 0$ (process centering), the following applies for the first four orders (Eq. 6.70)

$$\begin{aligned} & c_1^x = m_1^x = E\{X(k)\} = 0 \\ & c_2^x(\tau_1) = m_2^x(\tau_1) \\ & c_3^x(\tau_1, \tau_2) = m_3^x(\tau_1, \tau_2) \\ & c_4^x(\tau_1, \tau_2, \tau_3) = m_4^x(\tau_1, \tau_2, \tau_3) - m_2^x(\tau_1) \cdot m_2^x(\tau_3 - \tau_2) \\ & - m_2^x(\tau_2) \cdot m_2^x(\tau_3 - \tau_1) - m_2^x(\tau_3) \cdot m_2^x(\tau_2 - \tau_1) \end{aligned} \tag{6.70}$$

6.4.1.2 Properties of Moments and Cumulants

If the sets of random variables $\{x_1, x_2, \ldots, x_n\}$ and $\{y_1, y_2, \ldots, y_n\}$ are independent of each other, the following relationships apply

$$c_n(x_1 + y_1, x_2 + y_2, \ldots, x_n + y_n) = c_n(x_1, x_2, \ldots, x_n) + c_n(y_1, y_2, \ldots, y_n),$$

$$m_n(x_1 + y_1, x_2 + y_2, \ldots, x_n + y_n) \neq m_n(x_1, x_2, \ldots, x_n) + m_n(y_1, y_2, \ldots, y_n).$$

It follows from these relationships that cumulants—in contrast to moments—can be used to either test the independence of processes or to use them to make processes independent (see signal decomposition). Another essential property of cumulants is that they are zero-valued for a normally distributed process from the third order. If the sum $z_i = x_i + y_i$ ($i = 1, 2, \ldots, n$) of two random variables x and y *is* present, where x *is* non-normal, and y is normally distributed and independent of each other, then the following relationships apply under the assumption of zero-mean processes $E\{x\} = E\{y\} = 0$ without the restriction of generality

$$\begin{aligned} c_n(z_1, z_2, \ldots, z_n) &= c_n(x_1, x_2, \ldots, x_n) \\ m_n(z_1, z_2, \ldots, z_n) &= m_n(x_1, x_2, \ldots, x_n) + m_n(y_1, y_2, \ldots, y_n) \end{aligned}.$$

Essential consequences for process analysis result from this property: Independent, normally distributed processes (mostly technical noise and disturbances) disappear in cumulants from order $r = 3$, but not in moments.

For the analysis of correlations, second-order statistics (correlation coefficient, correlation function, coherence) have been used so far, which only provide information about the linear correlation or the linear portion of a correlation. In the case of biosignals, however, one must assume that they and their interdependencies are subject to non-linearities.

The lowest degree of nonlinearity—the second order—is to be treated here. For the analysis of quadratic nonlinearities, third-order moments and cumulants (Eq. 6.71) are needed (assuming $m_1 = 0$).

$$c_3^x(\tau_1, \tau_2) = m_3^x(\tau_1, \tau_2) = E\{X(k)X(k+\tau_1)X(k+\tau_2)\} \tag{6.71}$$

The extension of the dimension of the displacement to calculate the third cumulant is demonstrated by the following example.

Example
A harmonic carrier with $f_T = 50$ Hz is amplitude modulated with a harmonic modulation signal $f_m = 5$Hz, where the modulation index is $m = 1$ (100%). The AM is a non-linear second-order operation (multiplication of the two signals); therefore, there is a quadratic relationship between the carrier and the modulation signal. In addition, the AM signal contains white, normally distributed noise (Fig. 6.46). As expected, the second cumulant (AKF) shows both periodicities (Fig. 6.46 bottom), with reduced noise due to averaging in the cumulant estimation. The second cumulant is an even function, so the APSD is real, and the phase relationship between the two harmonics is lost. The third.

Cumulant (Fig. 6.47) represents the dependence of the third-order product of two independent displacements ω_1 and ω_2. As seen in Fig. 6.47, the third cumulant has three axes of symmetry, so it is not straight in any of the orthogonal directions. It suggests that the two-dimensional spectrum, calculated using a 2D FFT, is complex, and therefore the phase information is obtained.

6.4.2 Higher Order Spectra

Although the term "higher-order spectrum" (HOS) is used in the literature (HOS, Higher-Order Spectrum, polyspectrum) is used in the literature, it refers to the spectral power density of higher order (from order 2). In the following, the term HOS will be used.

6.4.2.1 Definition of Higher-Order Spectra

The order of the moments and cumulants depends on the sum of the exponents r so that the dimension of the independent displacements τ_i is one degree of freedom lower and is $r - 1$. Therefore, the bispectrum can be calculated from the third cumulant. Analogous to the PSD (power spectral density, Wiener-Khinchine-theorem), one first calculates the moments in the time domain and then transforms

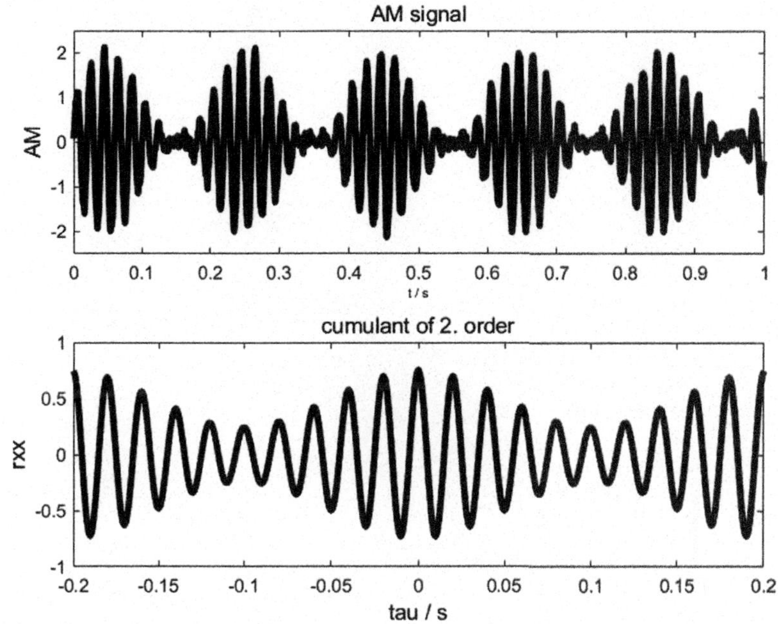

Fig. 6.46 An amplitude-modulated signal ($f_{\mathrm{T}} = 50$ Hz, $f_{\mathrm{m}} = 5$ Hz) with noise (top) and its 2nd-order cumulant (bottom)

them into the frequency domain (Eq. 6.72):

$$\begin{aligned} C_n^x(\omega_1, \ldots, \omega_{n-1}) &= \sum_{\tau_1=-\infty}^{\infty} \cdots \sum_{\tau_{n-1}=-\infty}^{\infty} c_n^x(\tau_1, \ldots, \tau_{n-1}) \\ &\quad \cdot \exp(-j(\omega_1\tau_1 + \cdots + \omega_{n-1}\tau_{n-1})) \\ M_n^x(\omega_1, \ldots, \omega_{n-1}) &= \sum_{\tau_1=-\infty}^{\infty} \cdots \sum_{\tau_{n-1}=-\infty}^{\infty} m_n^x(\tau_1, \ldots, \tau_{n-1}) \\ &\quad \cdot \exp(-j(\omega_1\tau_1 + \cdots + \omega_{n-1}\tau_{n-1})), \end{aligned} \tag{6.72}$$

the following conditions must be met:

$$|\omega_i| \leq \pi, \quad i = 1, 2, \ldots, n-1 \quad \text{and} \quad |\omega_1 + \omega_2 + \ldots + \omega_{n-1}| \leq \pi.$$

6.4.2.2 Properties of Higher-Order Spectra

The HOS are generally complex and can therefore be represented using the magnitude and phase (Eq. 6.73).

$$C_n^x(\omega_1, \ldots, \omega_{n-1}) = \left|C_n^x(\omega_1, \ldots, \omega_{n-1})\right| \exp\left(j\Psi_n^x(\omega_1, \ldots, \omega_{n-1})\right) \tag{6.73}$$

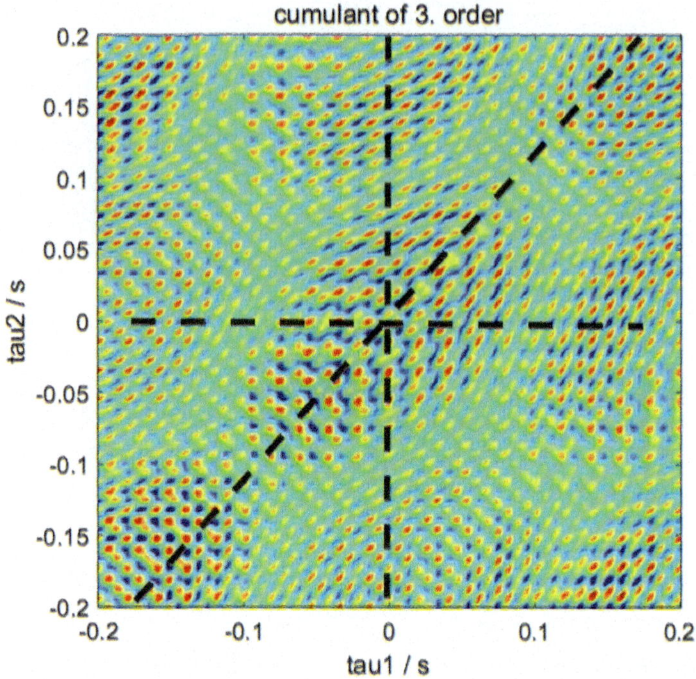

Fig. 6.47 3rd order cumulant of the AM signal from Fig. 6.65. *Note* three symmetry axes at $\omega_1 = 0$, $\omega_2 = 0$, and $\omega_1 = \omega_2$, shown dashed in the graph

The 2nd order cumulant spectrum is the PSD (Eq. 6.74). Since the second cumulant is an even function, the 2nd order HOS is real, and the phase is lost.

$$C_2^x(\omega) = \sum_{\tau=-\infty}^{\infty} c_2^x \cdot \exp(-j(\omega\tau)) \tag{6.74}$$

The 3rd order cumulant spectrum or the *bispectrum* is first calculated from the third cumulant via the two-dimensional FT (Eq. 6.75).

$$C_3^x(\omega_1, \omega_2) = \sum_{\tau_1=-\infty}^{\infty} \sum_{\tau_2=-\infty}^{\infty} c_3^x(\tau_1, \tau_2) \cdot \exp(-j(\omega_1\tau_1 + \omega_2\tau_2)) \tag{6.75}$$

Since the third cumulant (Eq. 6.75) is neither an even nor an odd function of (ω_1, ω_2), the bispectrum is complex, and the first HOS that retains phase information (phase difference). Since the third cumulant has three symmetries (because of the 3rd order products, Fig. 6.47), the 2D-FT (Eq. 6.75) gives twice as many—i.e., six—symmetries (Fig. 6.48).

For the signal analysis of a signal $x(t)$, the first octant or the first quadrant of the bispectrum is sufficient. Figure 6.48 shows the bispectrum of an AM signal

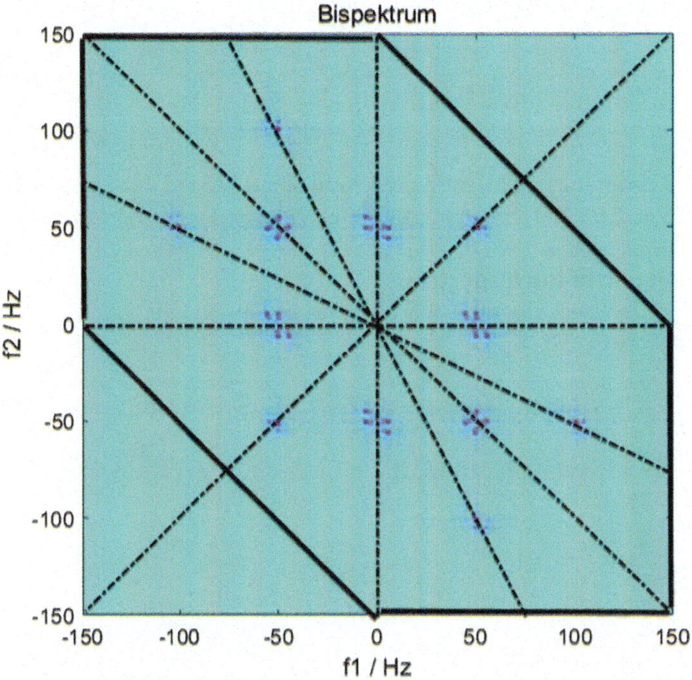

Fig. 6.48 Bispectrum of the AM signal from Fig. 6.46. The bispectrum has 6 symmetry axes, resulting in 12 symmetry regions (thickly framed). The first octant is sufficient for analyzing an auto-bispectrum (a signal with several components)

produced by AM of a harmonic carrier at $f_T = 50$ Hz with a harmonic modulation signal with fm = 5 Hz frequency. The bispectrum shows maxima for the harmonics and the modulation products (mixing frequencies). The influence of the quadratic coupling on the spectral peaks will be investigated in the following.

6.4.2.3 Estimation of the Higher-Order Spectra

The HOS can be estimated indirectly (via the moments) or directly (periodogram), analogous to the PSD. The calculation of the bispectrum exemplifies the procedure.

Indirect Estimation of the Bispectrum

If $x_i(k)$, $k = 0, 1, \ldots, M - 1$ represents the data of the *i-th* segment, the third moment is calculated according to (Eq. 6.76)

$$m_3^{(i)}(\tau_1, \tau_2) = \frac{1}{M} \sum_{k=s_1}^{s_2} x_i(k) \cdot x_i(k + \tau_1) \cdot x_i(k + \tau_2) \tag{6.76}$$

where $i = 1, 2, \ldots, K, = \ldots - 2, - 1, 0, 1, 2, \ldots, s_1 = \max(0_1,_{-2})$ and $s_2 = \min(M - 1, M - 1_{-1}, M - 1_{-2})$. From these moments calculated for each segment, the third moment can be estimated by averaging (Eq. 6.77).

$$\widehat{m}_3^x(\tau_1, \tau_2) = \frac{1}{K}\sum_{i=1}^{K} m_3^{(i)}(\tau_1, \tau_2) \tag{6.77}$$

The third cumulant can be estimated equivalently to the third moment according to Eqs. 6.76 and 6.77. The bispectrum is calculated according to Eq. 6.75.

Direct Estimation of the Bispectrum
Analogous to the periodogram, the bispectrum can be calculated directly via the Fourier transforms (Eq. 6.78).

$$C_3^x(\omega_1, \omega_2) = M_3^x(\omega_1, \omega_2) = X(\omega_1) \cdot X(\omega_2) \cdot X^*(\omega_1 + \omega_2) \tag{6.78}$$

If there are several realizations, the bispectrum is averaged in the polyspectral range (Eq. 6.79).

$$\hat{C}_3^x(\omega_1, \omega_2) = \frac{1}{K}\sum_{i=1}^{K} C_3^{(i)x}(\omega_1, \omega_2) \tag{6.79}$$

Windowing in the direct and indirect estimation of the bispectrum is possible but not helpful. Only a few windows—known from the one-dimensional time domain—apply in two dimensions. In addition, the bispectrum is mainly used to investigate phase couplings between harmonics, i.e., spectrally, to detect harmonic needles in the noise. Since these often lie close together (see Fig. 6.48), they would smear a window and make a distinction impossible. Therefore, windowing is not used in practice, even though the variance in the polyspectral range is significantly higher (due to the third power of the products) than in power density spectra.

6.4.3 Linear and Quadratic Phase Coupling

For most biosignals, the phase is as crucial as the amplitude. It implicitly contains information about a neuronal excitation's latency or running time. If one compares the statistical properties of levels (amplitudes) and the phase of the biosignals, the phase fluctuations are an order of magnitude weaker than the amplitude fluctuations. On the one hand, this is because the signal levels depend on many factors that cannot be influenced, such as the individual physique or the specific measurement conditions, which can never be exactly reproduced. On the other hand, the running times of the neuronal excitations depend largely on the speed of conduction of the neuronal tissue and the qualitatively always the same anatomical structure; they are mechanically fixed and, therefore, relatively stable. The phase information can be used to investigate couplings between neuronal structures, e.g., the cortical control of muscles. Mainly in brain research, one tries to prove connections between brain areas with the help of phase couplings. Since the phase information is already contained in the CPSD (cross PSD), the phase coherence can also be determined in addition to the coherence (Eq. 6.80).

6.4.3.1 Linear Phase Coupling

$$\rho_{xy}(\omega) = \left| E\left\{ \frac{P_{xy}(\omega)}{|P_{xy}(\omega)|} \right\} \right| = \left| E\left\{ e^{j\varphi_{xy}(\omega)} \right\} \right|$$
$$\varphi_{xy}(\omega) = \arctan \frac{\operatorname{Im}\{P_{xy}(\omega)\}}{\operatorname{Re}\{P_{xy}(\omega)\}} \tag{6.80}$$

From a single realization, it is impossible to determine whether a linear phase relationship exists between the signals x(n) and y(n) since the phase of stochastic signals is equally distributed and, on average zero-valued. Therefore, averaging (as with PSD) over several segments is necessary (Eq. 6.81).

$$\hat{\rho}_{xy}(\omega) = \left| \frac{1}{M} \sum_{m=1}^{M} e^{j\varphi_{xy}^{(m)}(\omega)} \right| \tag{6.81}$$

The linear phase coupling (Eq. 6.81) is demonstrated in the following example.

Example
Two harmonics were generated, $s_1(n)$ in each segment m with a constant phase of/4 and a relative frequency of 0.1 and $s_2(n)$ with a phase randomly equally distributed over the segments and a relative frequency of 0.2. The length of the segments is $N = 1000$, and the number of segments for $x(n)$ and $y(n)$ each $M = 1000$. The harmonics are additively embedded in the white noise with $\sigma^2 = 1$. If one calculates the coherence for the signal mixtures $x(n)$ and $y(n)$, it shows, as expected, clear needles of $C_{xy} = 1$ at the relative frequencies of 0.1 and 0.2 (Fig. 6.49). This value is independent of whether or not there is a linear phase relationship. If one calculates the phase coherence for the signal mixture, it shows a deterministic (and correct) value of ρ_{xy} (0.1) = 1 for ϕ_{xy} (0.1) = $\pi//4 = 0.78$ at the relative frequency of 0.1 (Fig. 6.49 bottom). The phase coherence of the signal with random phase has a value close to zero at the relative frequency of 0.2, as do all the other randomly distributed phases.

This example shows that the coherence alone—especially for biosignals with an essentially continuous spectrum—is insufficient for a comprehensive analysis; one should also examine the phase coherence. With a sufficiently high number (M > 100) of segments (realizations), the random phase disappears, and the deterministic (linearly coupled) remains.

6.4.3.2 Quadratic Phase Coupling

While linear phase coupling results exclusively from linear operations (addition, multiplication by a constant), quadratic phase coupling represents a non-linear

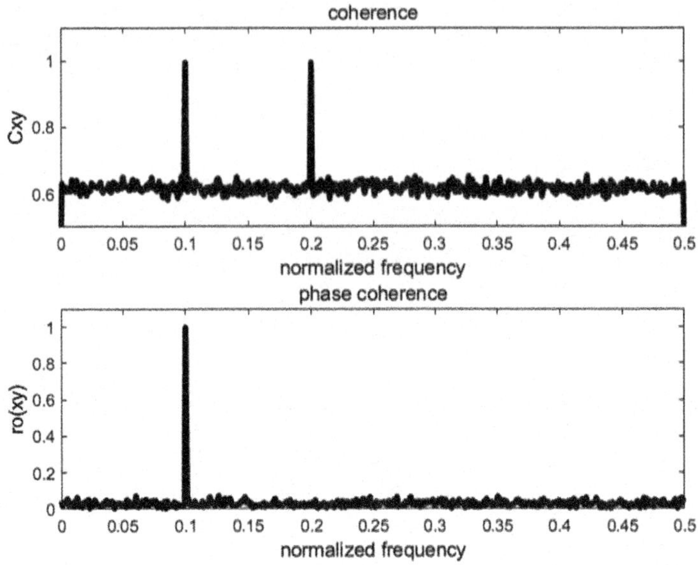

Fig. 6.49 Coherence (top) and phase coherence (bottom) of linearly coupled signals at the relative frequency of 0.1 and with random phase at the relative frequency of 0.2

relationship of second order (e.g., multiplication of two signals in amplitude modulation or frequency modulation). For the analysis of QPC (quadratic phase coupling), cumulants or spectra of (at least) third order are necessary. The signal model can be explained using the example of an AM signal from Sect. 6.4.1.2. In AM, the instantaneous value of a carrier signal is multiplicatively influenced by a modulation signal, which can be generally expressed by Eq. 6.82.

$$u_{AM}(t) = a \cdot u_T(t) + (b + c \cdot u_T(t)) \cdot u_m(t) \tag{6.82}$$

In Eq. 6.82, $u_\mathrm{T}(t) = \cos(2\pi f_\mathrm{T} t + \phi_\mathrm{T})$ is the carrier, $u_\mathrm{m}(t) = \cos(2\pi f_\mathrm{m} t + \phi_\mathrm{m})$ is the modulation signal, where $f_\mathrm{T} > f_\mathrm{m}$; a, b, and c are constants that determine, among other things, the modulation index. According to Eq. 6.82, the relation can be decomposed into frequency components (Eq. 6.83).

$$\begin{aligned} u_{AM}(t) &= a\cos(2\pi f_T + \phi_T) + b\cos(2\pi f_m + \phi_m) \\ &+ \frac{c}{2}\cos(2\pi t(f_T + f_m) + \phi_T + \phi_m) \\ &+ \frac{c}{2}\cos(2\pi t(f_T - f_m) + \phi_T - \phi_m) \end{aligned} \tag{6.83}$$

Through the multiplication of two signals, which corresponds to a second-order non-linearity, second-order products (also called mixed products) arise in addition to the original frequency components (first and second term), whose frequencies result from the sum and the difference of the original frequencies (third and fourth term in Eq. 6.83). Since the sum also forms the phases of the third and fourth terms and the difference of the initial phases, i.e., are coupled via them, this effect is called QPC. If one calculates the bispectrum for the signal according to Eq. 6.83 from only one realization, all frequency combinations of the second order arise. From this picture, one cannot conclude which frequency components are quadratically coupled and which are not. In the bispectrum, such components appear in a realization that consists of three frequencies, whereby one frequency results from the sum of the other two, i.e., $f_3 = f_1 + f_2$. As with the linear phase, one must average several realizations of the bispectrum in order to reduce the influence of the random phase. If there are several realizations, the bispectrum is averaged, analogous to the periodogram (Eq. 6.84).

$$\hat{C}_3^x(\omega_1, \omega_2) = \frac{1}{M} \sum_{m=1}^{M} \hat{C}_3^{x(m)}(\omega_1, \omega_2) \tag{6.84}$$

By averaging the bispectrum, components with a random phase disappear. What remains are components for which, in addition to the condition of the frequency sum, the condition of the phase sum $\varphi_3 = \varphi_1 + \varphi_2$ also applies. It also explains the bispectrum of the AM signal from Figs. 6.48 and 6.50. After averaging the bispectrum over 10 realizations, all components except $\{f_1 = f_T = 50$ Hz; $f2 = f_m = 5$ Hz; $f3 = f_T + f_m = 55$ Hz$\}$ and $\{f_1 = f_T - f_m = 45$ Hz; $f2 = f_m = 5$ Hz; $f3 = 50$ Hz$\}$ disappear visibly in Fig. 6.50 as spectral needles. It follows that the averaged bispectrum can be used to detect QPC. It can also be used effectively with biosignals, as the example of an ECG amplitude modulated with respiration shows (Figs. 6.51 and 6.52). Suppose one designates the components using the terminology of amplitude modulation. In that case, the carrier frequency is $f_T = 1.02$ Hz (primary frequency of the ECG), and the modulation frequency is $f_m = 0.2$ Hz (primary frequency of the respiration) (Fig. 6.53).

The two components of the QPC are marked in Figs. 6.51 and 6.52. In addition to these sought-after components, the bispectrum contains several other relatively strong components that are not quadratically coupled. These components reach extremely high levels in the third order due to the significant amplitude difference between the ECG and the respiration, which could only be suppressed with a high averaging order (M > 1000). In this case, insufficient realizations were available ($M = 8$); therefore, the interfering spectral needles remain relatively strong.

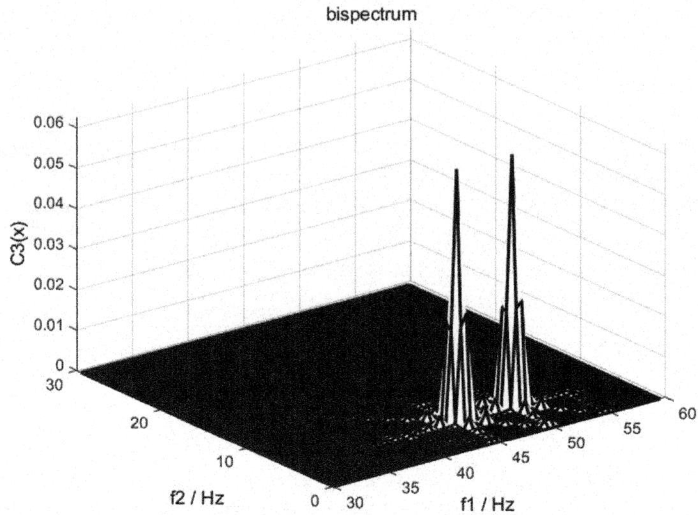

Fig. 6.50 Bispectrum of an AM signal with carrier frequency $f_T = 50$ Hz and modulation frequency $f_m = 5$ Hz

Fig. 6.51 Bispectrum (directly estimated) of an ECG with respiratory components. The parts connected via the quadratic phase coupling are marked in the circled area

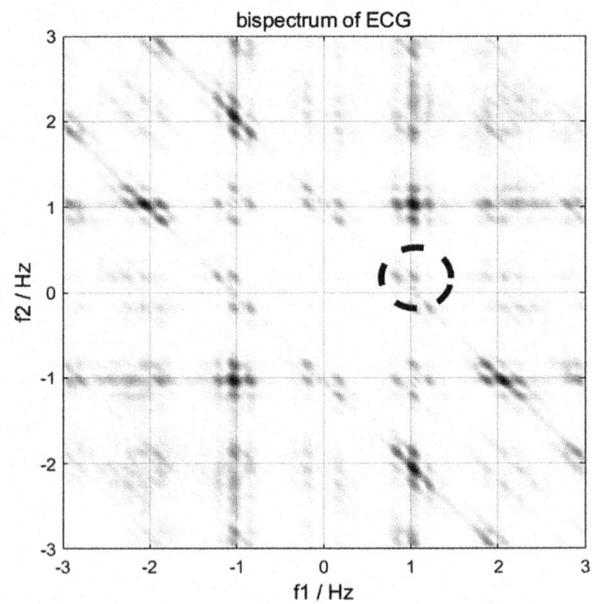

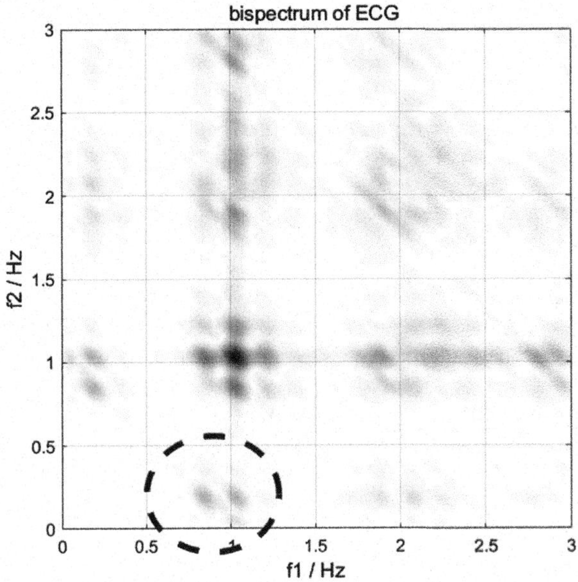

Fig. 6.52 First quadrant of the bispectrum from Fig. 6.51

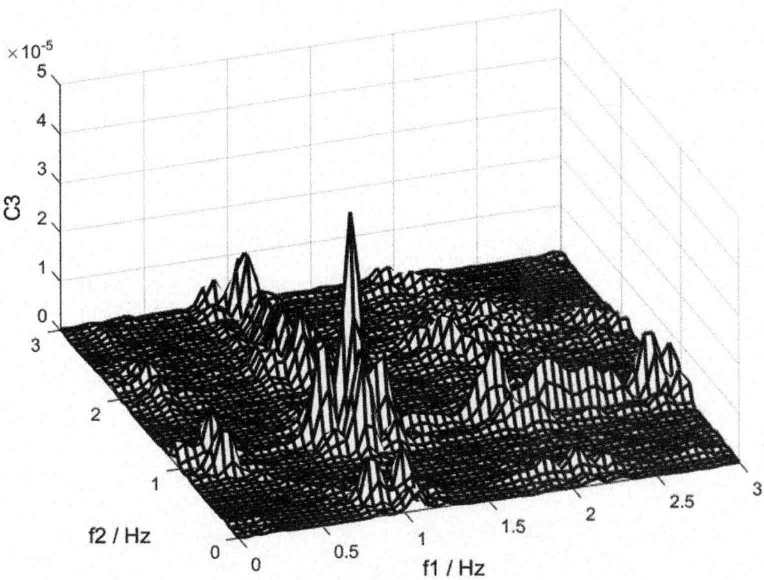

Fig. 6.53 First quadrant of the bispectrum from Fig. 6.52 in 3D representation for size comparison

6.5 Exercises

6.5.1 Tasks

6.5.1.1 Distribution Density and Distribution Function

Calculate the PRB for the occurrence of the value $x = 0$ when the distribution density of the continuous random variable X corresponds to the normal distribution N(0,1). Investigate the influence of the class width of the histogram on the empirical distribution density of a simulated, normally distributed random variable X. The size of the simulated data is $N = 10^6$, and the class width is to be varied between $x = 0.1$ and 0.01. Show how the simulated discrete data can approximately obtain the theoretical distribution density.

6.5.1.2 Arithmetic Mean and Median

The mean and the median are lag estimators. Theoretically, the mean is the best estimator for normally distributed data, which rarely exists, while the median is the most robust. Examine the properties of these estimators for second-order products used to calculate second-order statistical measures (correlation, covariance, coherence, phase coherence). The initial data for calculating the products should be standard normally distributed ($\mu = 0$, $\sigma = 1$), and the data ensemble of the products should be $N = 30$ (sample size) times $M = 10^4$ (number of realizations). Check the applicability of the limit theorem.

6.5.1.3 Statistical Measures and Normalization

In a university hospital, it was investigated (this example is based on a true story) how the shares of personnel and material costs of the medical technique department are distributed. The shares of material costs were further broken down into maintenance, repair, and legally required inspections. The distribution in the interim year comparison is shown in the graph in Fig. 6.54. The external consulting firm appointed for cost minimization submitted these graphs to the management of the university hospital without going into the total costs. During the prescribed cost savings, it was decided to reduce personnel costs. The justification for this decision was that essentially nothing had changed in maintenance and statutory inspections, while personnel expenses had increased by about 4% at the expense of maintenance.

Furthermore, since one could not save on maintenance, personnel costs had to be reduced, i.e., jobs had to be cut. Based on this presentation, the decision would have been justifiable. Try—initially without knowing the total costs—to interpret the two diagrams based on different causes (costs). The solution to this task lists the total costs with corresponding explanations.

6.5.1.4 Statistical Measures and Commonality Correlation

Not surprisingly, medical technology manufacturers have realized that patients' weight will increase in the long term. Their products must become more robust and safer, especially in the mechanical area (beds, trainers, ergometers, couches,

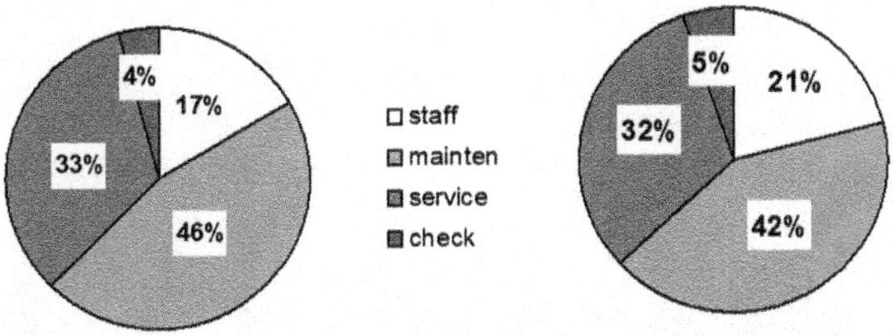

Fig. 6.54 Breakdown of costs of the medical technique department of a university hospital in two consecutive years

CT, MRI). Whereas a maximum patient weight of 100 kg was previously assumed, this must now be set at 120 kg. Since most large companies operate globally, they have investigated how body weight depends on saving unnecessary constructional and technological over-dimensioning and thus costs. After statistical analyses, the following correlations were found, among others:

1. Body weight positively correlates with the respective country's GDP (gross domestic product) related health expenditure.
2. Body weight correlates strongly positively with the geographical longitude of a country in the Northern Hemisphere (from east to west), assuming Japan as the zero reference (ascending Japan, Russia/China, Eastern Europe, Western Europe, USA).

The correlation coefficient between health expenditure and patient weight is so high that strategists decide to conquer the market, starting in Japan with a light and inexpensive design and ending in the USA with elaborate machine-building-like frameworks. The formula is that the higher the health expenditure, the more financially strong the customers and, therefore, the more elaborate and stable technology to sell. From a business point of view, this is a comprehensible plan. Analyze the correlations (better: interrelations) shown according to the causality principle. The questions can be asked: Do the effects show a consequence of direct correlations among themselves or a third influencing variable? If so, which effect is primary and causal?

6.5.1.5 Statistical Tests in Signal Analysis

Detection of a biosignal in noise and interference is a detection problem of message theory. Even with simple parametric tests of statistics, a biosignal can be detected in the noise. Using the example of a synthetic VEP superimposed with a double exponential right-skewed noise, the reliability of the *t-test* and the Wilcoxon test will be examined. To do this, load the file *vep_for_test.mat*. As a check, form the ensemble mean, which must show the course shown in Fig. 6.55.

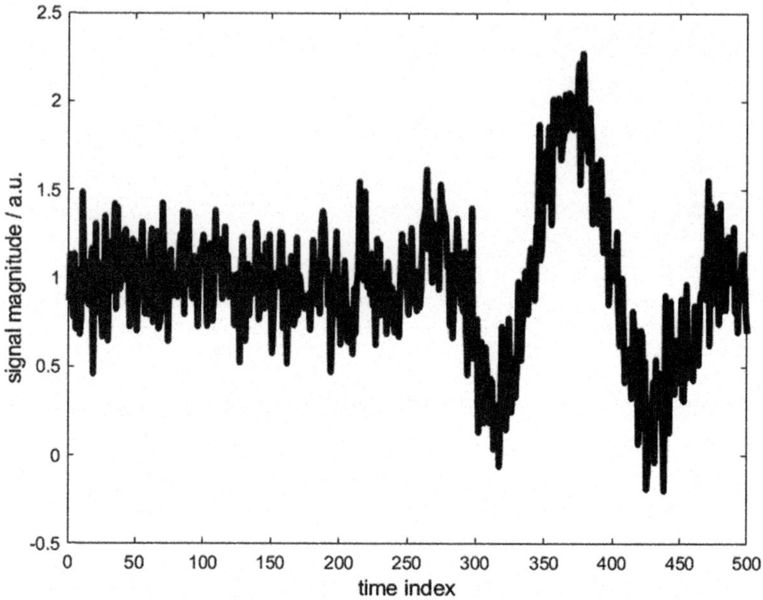

Fig. 6.55 Time course of the synthetic VEP (starts at time index 250) with noise reference (time index less than 250) averaged over 100 realizations. The noise is doubly exponentially distributed and right-skewed

Investigate the reliability of the *t-test* and Wilcoxon test for the sample circumferences $N = 10$, 30, and 100 in the two-sample case. As a time to test for the presence of a VEP, choose one of the waves ($n_{\mathrm{N1}} = 67$, $n_{\mathrm{P1}} = 120$, $n_{\mathrm{N2}} = 180$). Note that the time indices of the waves refer to the middle of the signal at time index 250. The indices N1, P1, and N2 characterize a stimulus response's first, second positive, or negative wave according to standard practice in electrophysiology. The two-sample test must be performed since the zero mean signal cannot necessarily be assumed for real signals.

Any time before the stimulus (time index < 250) is a reference for the null hypothesis. One will get different test statements depending on the time and test chosen. Which result and why will be favored?

6.5.1.6 Test for Income Difference

It is generally assumed that medical technology graduates with Master's degrees receive a higher starting salary than Bachelors's. It is to be statistically examined whether there is such a difference. The following parameters were determined from the sample collected by the survey: With a sample size of 100 masters and bachelors each surveyed, the mean starting salary of the masters was 3050 EUR and of the bachelors 3000 EUR, with a standard deviation of 212 EUR in each case. Is the mean income difference of 50 EUR statistically verifiable with an uncertainty of 5%?

6.5.2 Solutions

6.5.2.1 Distribution Density and Distribution Function

The distribution density of a continuous, normally distributed ZG X with = 0 and = 1 is according to Eq. 6.4 with

$$f(x|\mu = 0, \sigma = 1) = \frac{1}{\sqrt{2\pi}} e^{-\frac{x^2}{2}}$$

is given. If one sets $x = 0$, one obtains $f(x) = 0.4$. It is the value of the distribution density at the point $x = 0$, but not the PRB of its occurrence. With a continuous ZG, an infinite number of values can occur so that the PRB for the occurrence of exactly $x = 0$ is zero-valued according to Eq. 6.2:

$$P(0 \leq X \leq 0) = F(b = 0) - F(a = 0) = 0.$$

Therefore, one can always specify the PRB of a continuous ZG for only one interval, where $a < b$, e.g.

$$P\left(-10^{-3} \leq X \leq 10^{-3}\right) = \left.\frac{dF}{dx}\right|_{x=0} \approx \left.\frac{\Delta F}{\Delta x}\right|_{x=0} = \frac{F\left(10^{-3}\right) - F\left(-10^{-3}\right)}{2 \cdot 10^{-3}} = 0.4.$$

Although this value corresponds to the distribution density for $x = 0$, it applies to the interval $[-\ 10^{-3}, 10^{-3}]$ around $x = 0$.

The following histograms are obtained from the discrete simulated data according to the specifications (Fig. 6.56). Since the histogram indicates the absolute frequencies of the classes, which inevitably deviate from each other for different class widths with a constant amount of data, it is not easy to transfer these to the normalized frequency or the distribution density. To do this, one can exploit that the distribution function $F(x)$ must always reach the value 1 at the end.

One can, therefore, discretely integrate the absolute frequencies of the histograms (add them up cumulatively) and normalize the sum by the number of all values (k is the class index in the histogram, and $h(k)$ is the absolute frequency of a class).

$$F(k) = \frac{1}{N} \sum_{i=1}^{k} h(i).$$

The discrete distribution function is obtained, as shown in Fig. 6.57. To obtain the discrete, normalized distribution density, one differentiates the distribution function:

$$f(k) = \frac{F(k+1) - F(k)}{x(k+1) - x(k)} = \frac{F(k+1) - F(k)}{\Delta x}.$$

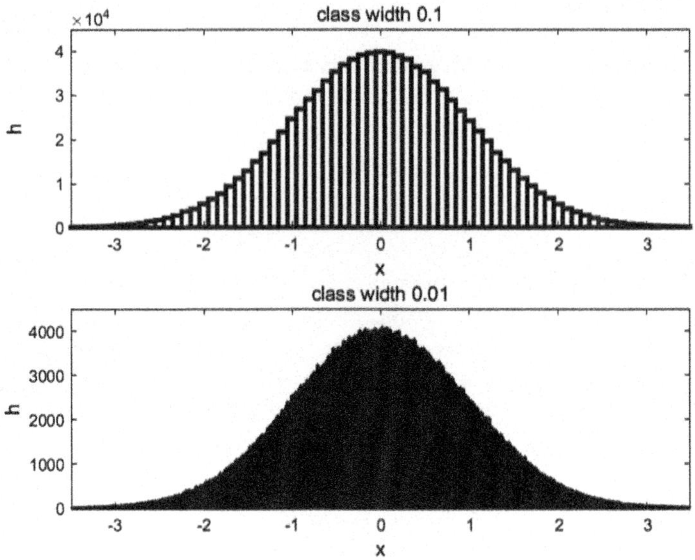

Fig. 6.56 Histograms of a simulated, discrete, normally distributed ZG with = 0 and = 1 for a data size of $N = 10^6$

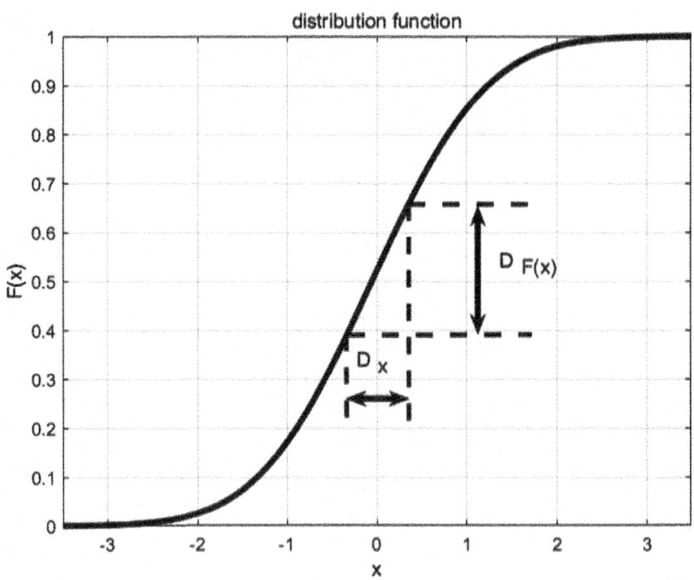

Fig. 6.57 Distribution function of the simulated ZG for calculating the probability density

This way, a normalized distribution density can be obtained from each empirical distribution density via the histogram. The normalization is essential (as mentioned earlier) in order to be able to compare ZGs with each other or their empirical distributions with a theoretical distribution (uebung_6_1.m).

6.5.2.2 Arithmetic Mean and Median

First, generate normally distributed data in the format $x(2, N{*}M)$ and then multiply the rows element by element with each other (uebung_6_2.m)

$$x_2(k) = \prod_{i=1}^{2} x(i,k).$$

The second-order products are rearranged into a data ensemble according to the specifications

$$x_2(n,m)|_{1\le n\le N, 1\le m\le M} := x_2(n + (m-1)N).$$

Then calculate the mean value over the columns

$$\overline{x}_2(m) = \frac{1}{N}\sum_{n=1}^{N} x_2(n,m)$$

and the median

$$\tilde{x}_2(m) = x_2(n,m)|_{P(X\le x(m))=0.5}.$$

If one calculates the variance of the mean and median values ($M = 10^5$), one obtains (in the specific case depending on the simulated data) approximately the following results:

$$\mathrm{var}(\overline{x}_2(m)) = 0.0332, \ \mathrm{var}(\tilde{x}_2(m)) = 0.0084.$$

It follows that the median for the second-order products is about four times more efficient estimator than the mean. Comparing the empirical distributions of the products as well as the mean and the median, we can see the following (Fig. 6.58):

- If the initial data are normally distributed and uncorrelated, their second-order products are doubly exponential and symmetrically distributed. Therefore, the excess is much higher than in the normal distribution (about 6.0), and the mean is no longer the best estimator. Since the median is more robust than the mean, it also shows a lower variance, which justifies its better efficiency.

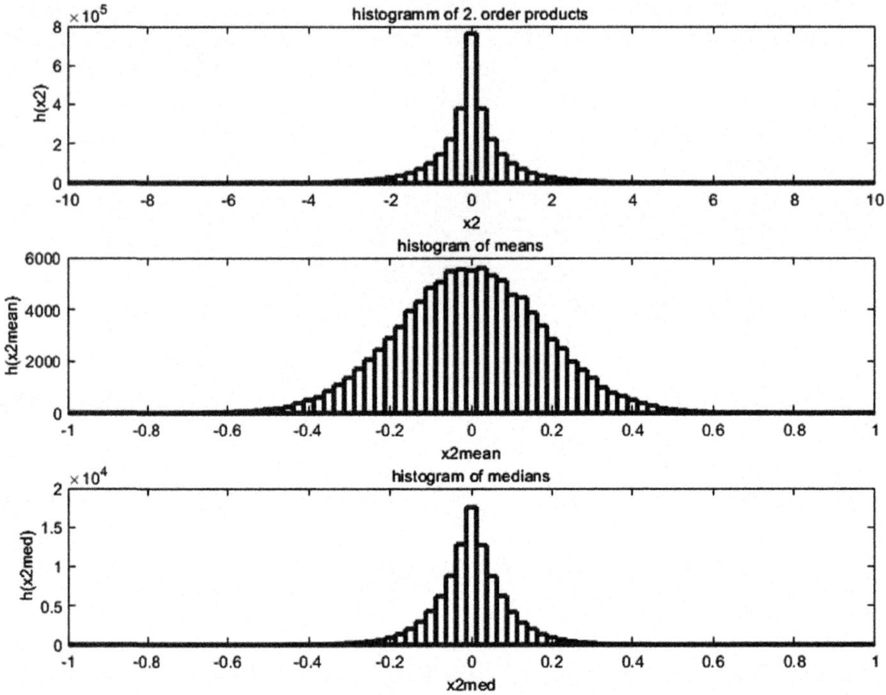

Fig. 6.58 Distribution of 2nd order products (top), double exponential ($N = 30$, $M = 10^4$) for uncorrelated, normally distributed initial data; distribution of means (middle) as well as medians (bottom)

- The mean values are normally distributed (Fig. 6.58 middle) with a variance of 1/30 = 0.033 due to the effect of the CDFS. That is experimental evidence of the CLT. However, the variance is about four times higher than that of the medians (Fig. 6.58 bottom).
- The conclusion is that if the arithmetic mean does not have to be used (linear operations required, limited computing power), the median will be preferred.

6.5.2.3 Statistical Measures and Normalization

In the medical technology department of the university hospital under review, total costs were radically reduced through various measures, especially in maintenance and repair (Table 6.3). The personnel was thus used more effectively by an order of magnitude. Due to the cost reduction of the two items, maintenance and repair, and the normalization of the total sum, the relative share of personnel costs has inevitably increased. Without establishing a reference to the initial data, a formal correlation arises that can be interpreted in different ways:

Table. 6.3 Absolute costs (rounded) in € thousands of the medical technology department of a university hospital

	Staff	Maintenance	Service	Check
1 year	1000	2750	2000	250
2 year	1000	2000	1500	250

- Personnel expenses have increased (e.g., due to new wage settlements) at the expense of maintenance, while maintenance and inspections have remained almost constant (task definition),
- Maintenance costs have been reduced due to improved quality assurance, but personnel costs have increased disproportionately,
- Maintenance must become more effective, and its costs fell by just one percent,
- The exams became light but still more expensive.

If one orients oneself to the initial data in Tab. 6.3, one finds that all the conclusions listed above are wrong:

- Personnel expenses and audit costs have remained constant,
- Maintenance and repair costs decreased significantly by 27% and 25%, respectively.

About 20% of all medical technology costs were saved while personnel costs remained constant. This result speaks for improved economic efficiency and gives no reason for job cuts. This example shows that even correctly calculated statistical measures and collected empirical data can be used for almost any argumentation through economically or politically motivated misinterpretation.

6.5.2.4 Statistical Measures and Commonality Correlation

In the analysis of the correlations, one can proceed as follows (but there are also other ways):

1. Body weight and health expenditure have a strong positive correlation. Can there be a functional correlation? One could hypothesize that obesity increases morbidity, which results in rising costs. However, this effect is insignificant compared to other diagnoses (cardiovascular, cancer, diabetes). Conversely, one could assume that high health expenditure motivates people to gain weight. However, this implication is not demonstrable in real life. So there cannot be a functional relationship. Therefore, there must be a common influencing variable on which both parameters depend. Health expenditure depends on the economic power of a country. Body weight increases on average with wealth, so it also depends on economic power. That is the common influencing factor.
2. Body weight correlates strongly with a country's geographical location. It tends to increase with distance from Japan towards the west. However, the economic power of the countries on this path does not change monotonically. So there

must be another influencing factor because it cannot be the geographical longitude alone. The climatic and geographic conditions, as well as the habitats, are not different enough for that. Even a simple analysis of the possible causes shows that the development of history, mentality, eating habits, and attitudes towards physical exertion is very likely to be the possible explanation.

3. With these two considerations, it initially remained open to how strong the respective influences are. In order to be able to assess this, one would have to subject the factual statistical data to a multivariate analysis, e.g., factor analysis.

6.5.2.5 Statistical Tests in Signal Analysis

Following the procedure for statistical tests, we first examine how the data to be tested are distributed (Fig. 6.59). The empirical distribution of the random variables (data ensemble) corresponds qualitatively to a double exponential, right-skewed distribution. The distribution within the histogram classes is not uniform. From this feature, it can be concluded that the distribution of the ZG is not identical. Therefore, stationarity cannot be assumed, meaning the ergodicity principle cannot be applied. This conclusion is also confirmed by the visual inspection of the data ensemble (Fig. 6.55). The arithmetic mean of the ZG along the time axis is not constant, so neither stationarity nor ergodicity can be assumed.

The statistical test must therefore be applied to a significant difference between pairs of individual times. It makes sense to select the times when the expected waves can occur for testing: N = 317, 370, 430. As a reference for a non-existent signal, any time from the noise can be chosen (N = 1...250), here with N = 100. The statistical tests of the pairwise samples yield the following statements (a 1 corresponds to a rejection of the null hypothesis):

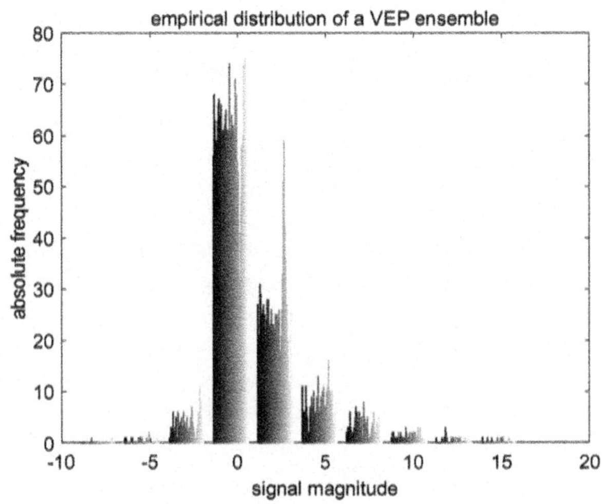

Fig. 6.59 Empirical distribution of the data ensemble from task 6.5.1.5 (*uebung_6_5.m*). Qualitatively, the histogram shows a double exponential, right-skewed distribution. Note that the distribution within the classes is not uniform

- t-Test

H0	N = 317	N = 370	N = 430
M = 10	0	0	0
M = 30	0	1	0
M = 100	0	1	0

- Wilcoxon test

H0	N = 317	N = 370	N = 430
M = 10	0	0	0
M = 30	0	0	0
M = 100	1	1	1

The t-test rejects the null hypothesis at a sample size of M = 30 and the most significant wave (N = 370). However, it is not rejected at the other time points, even with the relatively high sample size of M = 100. The Wilcoxon test rejects the null hypothesis only at M = 100 for all time points tested. Since the prerequisite of normally distributed data for the t-test is massively violated (double exponential, right-skewed), one should trust the Wilcoxon test instead. It is very conservative but robust against the distribution of the data.

6.5.2.6 Test for Income Difference

Following the recommended procedure, one formulates the opposite of what one wants to prove in H_0 (here, too, the question arises as to why someone would want to prove a higher Master's salary and not the opposite, i.e., a pre-trial question).

H_0: MA_Salary ≤ BA_Salary

For the alternative hypothesis (which is no longer necessary), the negation of H_0 remains:

H_1: Ma_Salary > BA_Salary

The right-sided *t-test* does not reject H_0; it is accepted. The *p-value of* 39% shows that one is far from rejection, i.e., the non-rejection is considered reliable. Since H_0 was not rejected, one can (should) swap the hypotheses and test again. However, the left-sided *t-test* for the swapped hypotheses does not reject H_0 either. The *p-value of* 60% shows even more clearly than in the first test that H_0 was rightly and reliably not rejected.

The test results show an apparent contradiction: neither of the logically contradictory hypotheses was rejected. One cannot determine which of the two hypotheses is correct. That is precisely the effect that was and still is ignored in the existing statistical testing methodology: A non-rejection of a hypothesis is not yet their confirmation.

References

Engelhardt, A. (2019): *Crash course in statistics.* http://crashkurs-statistik.de

Nikias, C. L., & Petropulu, A. P. (1993). *Higher-order spectra analysis; a nonlinear signal processing framework.* PTR Prentice-Hall, Inc.

Weiss, C. (1999). *Basic knowledge of medical statistics.* Springer.

Stochastic Processes

7

7.1 Statistical Analysis of Time-Series

7.1.1 From Static Data to Processes

The previous chapters dealt with distributions, statistical measures, and data that can be described as static. It means the simulated or empirical data can change their place or order in the sample without changing the statistical measures. Such data in the sense of expressing a characteristic (e.g., body weight) is called one-dimensional. If two or more random variables are examined (e.g., body weight and height, Fig. 6.35), we use two- or multi-dimensional data. In the case of time series (temporally discrete values, sampled values), an important property is added—the order of the data can no longer be changed, and the data index corresponds to the time index. For statistics, this was an initially simple transition from static data to time series: One interprets a time series as a sequence of random variables ordered by the time index (Eq. 7.1).

$$\{X[n]\}_{n \in Z} = \{X[0], X[1], \ldots, X[n]\} \tag{7.1}$$

In Eq. 7.1, X[n] are random variables, and n is the time index. Square brackets stand for the time sequence.

According to the previous view of multidimensional random variables, statisticians refer to the data analysis of structures according to Eq. 7.1 as an *n-dimensional* problem. This designation method often leads to misunderstandings since, from the point of view of signal analysis, a time series, according to Eq. 7.1 is a one-dimensional signal (e.g., the time course of a voltage). In the further procedure, the sequence according to Eq. 7.1 is a multi-dimensional random variable with several variables ordered according to the time index. Accordingly, the statistical measures are also multidimensional so that they can be formulated for each random variable:

P. Husar and G. Gašpar, *Electrical Biosignals in Biomedical Engineering*,
https://doi.org/10.1007/978-3-662-67998-2_7

- Mean value

$$\mu_x[n] = E\{X[n]\} \tag{7.2}$$

- Auto and cross-correlation function

$$r_{xx}[n_1, n_2] = E\{X[n_1]X[n_2]\} \tag{7.3}$$

$$r_{xy}[n_1, n_2] = E\{X[n_1]Y[n_2]\} \tag{7.4}$$

- The auto and cross-covariance function

$$c_{xx}[n_1, n_2] = E\{(X[n_1] - \mu[n_1]) \cdot (X[n_2] - \mu[n_2])\} \tag{7.5}$$

$$c_{xy}[n_1, n_2] = E\left\{(X[n_1] - \mu[n_1]) \cdot (Y[n_2] - \mu_y[n_2])\right\} \tag{7.6}$$

The following example illustrates the implementation of formulas 7.2–7.6: There are ten realizations (data simulation) of two random variables, X and Y, with a dimension (number of time indices) of $N = 1000$ (Fig. 6.46). The autocorrelation function r_{xx} [100, 300] and the cross-covariance function c_{xy} [900, 700] are to be calculated. The following formulas can be used for this purpose:

$$r_{xx}[100, 300] = \frac{1}{M} \sum_{m=1}^{M=10} x_m[100] \cdot x_m[300]$$

$$c_{xy}[900, 700] = \frac{1}{M} \sum_{m=1}^{M=10} (x_m[900] - \overline{x}[900]) \cdot (y_m[700] - \overline{y}[700])$$

This example makes it clear that, theoretically, for each random variable, several realizations of the random variable must be available at any time n. For practical measurement, it would mean that a locally measured value would have to be recorded simultaneously by several sensors. It would amount to an absurd recording, e.g., of a limb ECG, in which measurements would have to be taken at each limb over an exact point with, for example, ten electrodes. Such a measurement would not be feasible for practical reasons.

7.1.1.1 Sample and Ensemble

In reality, one must make do with a single realization of the ZV (random variable)—with a sampled or a priori discrete-time series. It can be, for example, a discretized biosignal (ECG, EEG) or a discrete-time recording of blood pressure (at a frequency of every 15 min). Of course, there are multisensory measurement set-ups in technology and physics, such as particle physics, designed to detect an event using numerous detectors simultaneously. However, according to probability theory, these multisensory measurement systems increase the probability of detecting an event. Therefore, It is a problem of detection reliability and not of a sufficiently large number of realizations.

Data generated according to the theoretical idea in Fig. 7.1 are referred to as a data set or ensemble. As already explained, such data are hardly ascertainable in the real world. However, since a considerable part of statistical analysis theory is based on this theoretical construct, one tries to come close to it in practical analysis with acceptable restrictions and assumptions. A practicable and practiced approach is to distribute the practically unsolvable problem of parallelism (ensemble) on the time axis—sequencing. The realizations necessary for forming a data set are not recorded simultaneously but one after the other. This procedure is typical, for example, for repetitive stimulation of the visual, acoustic, or somatosensory system, in which stimulus-synchronized EEG sections are arranged one below the other (Fig. 7.2). The problem is that the individual stimulus responses are not identical and thus the theoretical assumptions are violated. The extent of the violations is known and calculable in the result, so this approach is widely used in practical analysis. In a pragmatic assessment, one finds that getting an ensemble that is not entirely correct is better than getting none at all.

In many cases, however, it is impossible to stimulate the system under investigation or measure it synchronously with the stimulus. It is mainly the case with biosignals representing spontaneous activity, e.g., ECG, EEG, EMG, and EOG. In this case, even tolerant ensemble formation is not possible. One must start from a single realization (series of measurements). In such cases, the principle of ergodicity can be applied.

7.1.1.2 Stationarity and Ergodicity

If we look at the data in Figs. 7.2 and 7.3, we see that the individual realizations in the ensemble do not differ from those of a realization along the time axis on *average*. It leads to the approach that one does not have to determine the statistical measures in the ensemble because they are approximately contained in a single realization. Therefore, one interprets a realization along the time axis (realization i in Fig. 7.3) as a realization of a ZV at the index of, for example, $n = 300$. This approach is also called the ergodicity principle. Mathematically it can be expressed as follows (Eq. 7.7).

$$m_x^e[n] = m_x^e(i) = m_x^e \tag{7.7}$$

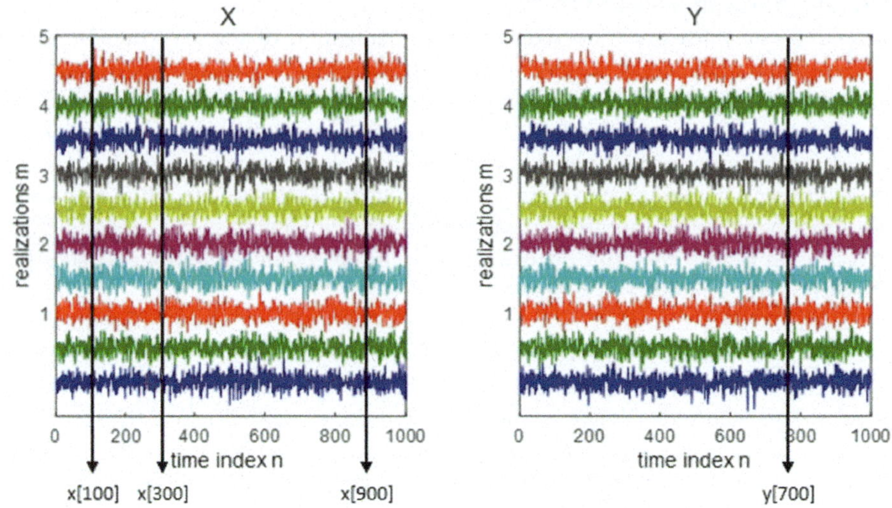

Fig. 7.1 Realisations of the random variables X ($n = 100, 300, 900$) and Y ($n = 700$) with $m =$ 1–10. Theoretically, ten independent measured values (realizations) of the random variables X and Y *are* available at each discrete time point n

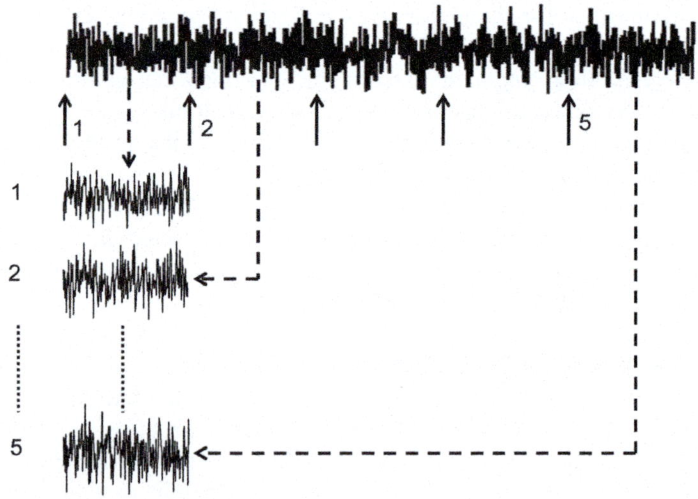

Fig. 7.2 Repetitive stimulation (arrows pointing upwards) and regrouping of consecutive stimulus responses into a data set (empirical ensemble)

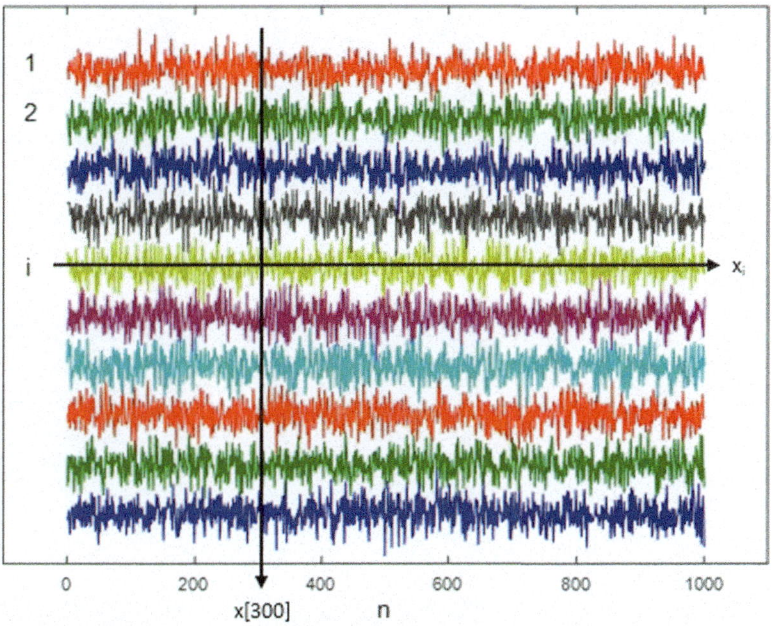

Fig. 7.3 Ensemble of simulated data. The distributions of the random variable X at $n = 300$ and the sample as a time series at $m = i$ *are the* same if stationarity is present

In Eq. 6.64, m^e is a statistical moment of order e, n is the time index in the ensemble, i is the index of a realization in the ensemble, and X *is* the analyzed ZV. Since for a single realization $i = 1$, the index is omitted.

If the condition for ergodicity according to Eq. 7.7 is to be fulfilled; it must necessarily apply that the statistical measures are independent of the time index n. However, this requirement can only be fulfilled if (Eq. 7.8) applies.

$$\begin{aligned} \mu_x[n] &= \mu_x \\ r_{xx}[n, n-m] &= r_{xx}[m] \\ c_{xx}[n, n-m] &= c_{xx}[m]. \end{aligned} \tag{7.8}$$

The formulae, according to Eq. 7.8 are to be interpreted so that the first and second-order statistics are independent of the concrete time n *and* depend solely on the displacement m. Generalized, this means that the statistical process variables are time-independent, and the process must therefore be stationary. In statistics, this property is called weak stationarity in statistics. In contrast, strong stationarity exists if the random variable X[n] distributions are identical. Strong stationarity only exists in theory since this property is not practically verifiable.

Weakly stationary signals exist in nature, technology, and physics. However, biosignals cannot be counted among them. Quite the opposite: according to this

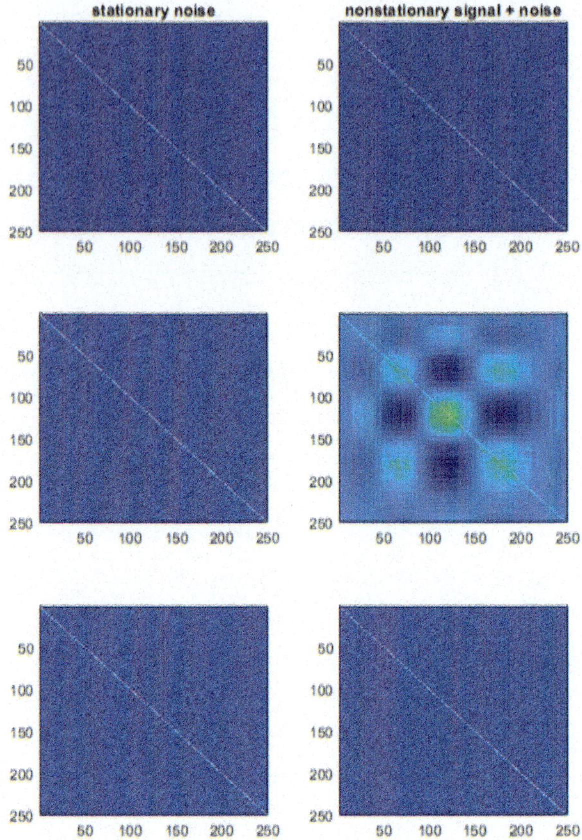

Fig. 7.4 Autocorrelation matrices of a stationary noise (left) and a noisy transient signal (right) in three consecutive segments of equal length $L = 250$ from data in Fig. 6.50 (on axes discrete time shift)

criterion, electrophysiological signals are highly transient. This property is immanent in a biological system, making it challenging to analyze the biosignals regarding the desired ergodicity. According to the condition in Eq. 7.8, the first two moments and the autocorrelation matrix in Eq. 7.9—must be independent of the current time n. Figure 7.4 shows the autocorrelation matrices for three equally long time segments of length L = 250. As expected, the autocorrelation reaches its maximum for $m = 0$, visible in the graph as a maximum (black) level along the main diagonal of the matrices. For stationary noise, the matrices are equally homogeneous, and one can conclude stationarity. If a transient signal is present (right column in Fig. 7.4), the matrices show a different picture; the middle segment differs from the otherwise equal edge segments. It is a solution approach for the analysis of biosignals:

In practical analysis, the ergodicity condition can also be achieved approximately for transient biosignals by forming time segments with approximately constant moments.

Segmentation is not a procedure that can be described precisely, as it is essentially data-dependent and is therefore carried out from an empirical point of view. In the case of simulated data (Fig. 7.5), which were used to calculate the autocorrelation matrices in the example according to Fig. 6.49, the task is relatively simple. Already visually, one can estimate that the middle third ($n = 251$–500) has different levels compared to the first ($n = 1$–250) and the last third ($n = 501{:}750$) of the time series. Therefore, one would form three segments here (the thirds mentioned above) and analyze each separately. However, a generally applicable rule for segmentation cannot be formulated. Automatic segmentation algorithms are usually data-specific and work according to empirical criteria.

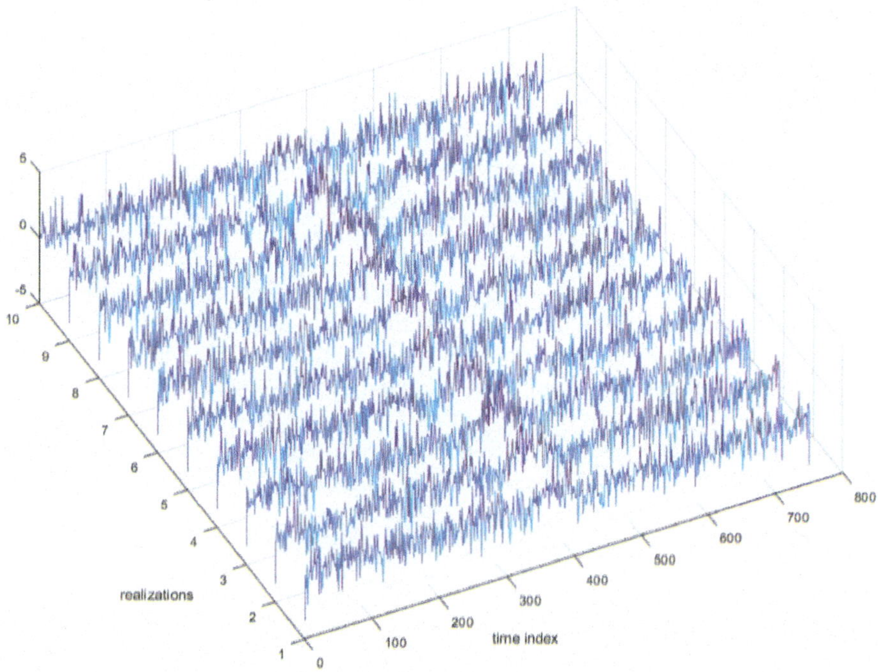

Fig. 7.5 Ensemble of simulated data consisting of a transient signal (simulated VEP between n = 250–500) additively linked with stationary noise. Shown is every tenth realization out of 100, which was used to calculate the autocorrelation matrices in Fig. 7.4 (right column)

Example

A statistic often used in signal analysis is the auto- or cross-correlation matrix (Eq. 7.9).

$$\mathbf{R}_{xx} = E\{x[n] \cdot x[n]\} = \begin{bmatrix} r_{xx}(0) & r_{xx}(1) & \dots r_{xx}(N-1) \\ r_{xx}(1) & r_{xx}(0) & \dots r_{xx}(N-2) \\ \vdots & \vdots & \ddots \vdots \\ r_{xx}(N-1) & r_{xx}(N-2) & \dots r_{xx}(0) \end{bmatrix} \tag{7.9}$$

In Eq. 6.66, r_{xx} (m) is the autocorrelation for the displacement m:

$$r_{xx}(m) = E\{x[n] \cdot x[n+m]\} = r_{xx}(-m).$$

7.1.2 Estimation of the Power Density Spectrum

7.1.2.1 The Parseval Theorem

Both biosignals and noise have continuous spectra, which can be partially overlaid by line spectra (periodic disturbances, stimulus responses). The DFT describes the spectral composition since signal processing today is carried out in the discrete-time domain (and digitized). According to the law of conservation of energy, the amount of energy must not change through the transformation. It must therefore apply that the energy in the time domain is identical to the energy in the frequency domain. Since the signal can also contain periodic components (power signals), it makes more sense to require the identity of the power in both domains (Eq. 7.10). The relationship, according to Eq. 7.10, states that the mean powers in the time and frequency domains are identical; it is known as the *Parseval-theorem* theorem. Regarding the availability of FFT algorithms, the most efficient variant for estimating the power density spectrum (PSD) is the right-hand part of the relationship, according to Eq. 6.67, which is called the *periodogram* (Eq. 7.11).

$$\left.\begin{aligned} X(k) &= \sum_{n=0}^{N-1} x(n) \cdot e^{-j2\pi kn/N} \\ x(n) &= \frac{1}{N} \sum_{k=0}^{N-1} x(k) \cdot e^{j2\pi kn/N} \end{aligned}\right\} \Rightarrow P_x = \sum_{n=0}^{N-1} |x(n)|^2 = \frac{1}{N} \sum_{k=0}^{N-1} |X(k)|^2 \tag{7.10}$$

7.1.2.2 The Periodogram, Direct Spectral Estimation

Since the periodogram (Eq. 7.11) is often used to estimate the PSD due to its practicability, its statistical properties will be examined here.

$$I_x(k) = \frac{1}{N} |X(k)|^2 \tag{7.11}$$

In Eq. 7.11, the factor 1/N is used for normalization. It always depends on the concrete realization of the DFT, according to Eq. 7.10.

When analyzing the signal properties, one can first start from white, normally distributed noise, which is superimposed with three harmonics with the relative frequencies $f_1 = 0.19, f_2 = 0.21$, and $f_3 = 0.25$ ($w(n)$ is white noise, $n = 0–N - 1$, P_w is the power spectral density):

$$w(n) \sim \mathcal{N}(\mu, \sigma),\ P_w(k) = \sigma^2\ s_i = \sin\left(2\pi f_i n/N\right)$$

First, the influence of the window length on the variance of the spectral estimate according to Eq. 7.11 shall be determined (Fig. 7.6). Since the power density of the white noise is constant, the following applies:

$$\overline{I}_w = \frac{1}{N}\sum_{k=0}^{N-1} I_w(k) = \sigma^2 \quad \mathrm{var}(I_w) = \frac{1}{N}\sum_{k=0}^{N-1}\left(I_w(k) - \overline{I}_w\right)^2 = \sigma^4 \tag{7.12}$$

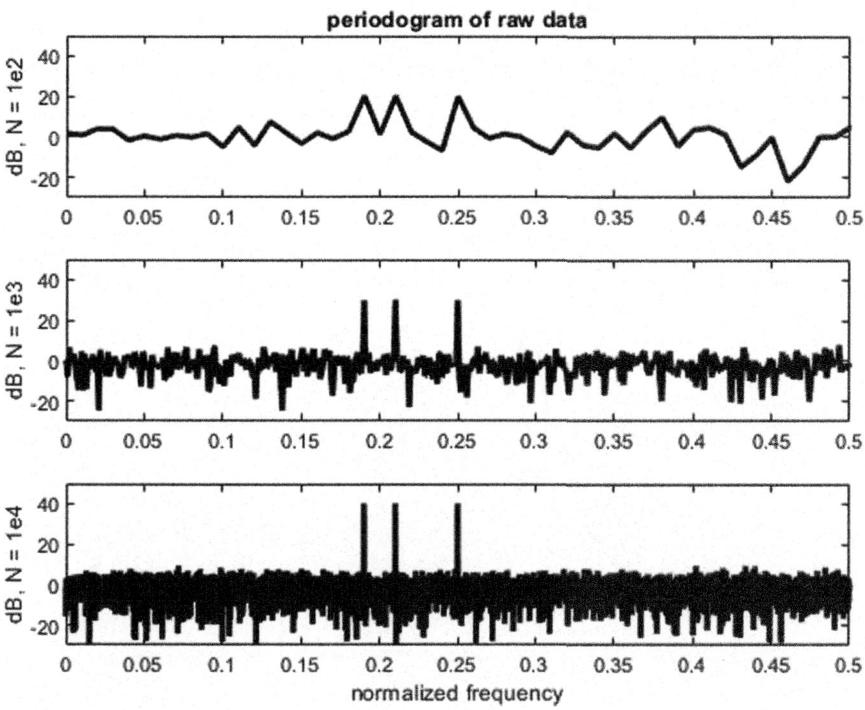

Fig. 7.6 Periodogram of white noise with three harmonics as a function of window length for $N = 10^2$ (top), 10^3 (middle), 10^4 (bottom). The harmonics are at the relative frequencies of 0.19, 0.21, and 0.25. The noise variance remains the same for all windows, regardless of the window length

It follows from Eq. 7.12 that the variance of the periodogram is constant and independent of the length N of the DFT (Fig. 7.6). Therefore, the periodogram is not consistent as an estimator of the power spectral density.

One can reduce the spectral variance with the help of a moving average (Eq. 7.13):

$$\hat{P}_x(\omega_k) = \frac{1}{L} \sum_{l=-(L-1)/2}^{(L-1)/2} (k + l) \tag{7.13}$$

The variance of the periodogram is then lower by a factor of 1/L. However, the spectral resolution deteriorates as a result of the smoothing (Fig. 7.7). While in the example according to Fig. 7.6 and Fig. 7.7, the variance of the periodogram decreases by 21 dB at a window length of $L = 11$, the frequency resolution is no longer sufficient to distinguish the harmonics at f_1 and f_2. At a window length of $L = 31$, the variance is reduced by 30 dB, and the noise is largely eliminated, but a spectral separation of f_2 and f_3 is also no longer possible. For the analysis, this results in the necessity to look for a compromise in the specific case or to use a more suitable window if possible.

Since the Fourier transform is used directly to estimate the spectral power density with the aid of the periodogram and the second moment is only formed in the

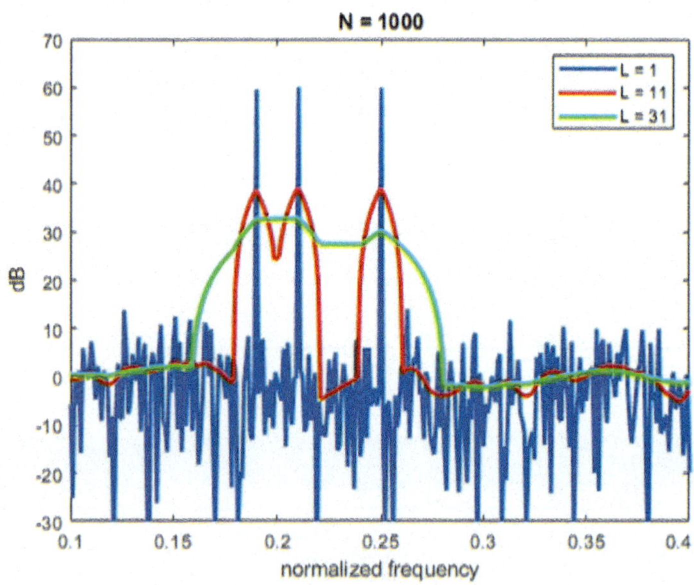

Fig. 7.7 Smoothing of the periodogram with rectangular windows of length $L = 1, 11, 31$. The variance of the spectral estimate decreases with the window length, and the power spectral density of the noise becomes constant, but the spectral resolution deteriorates. At $L = 11$, the harmonics can no longer be distinguished at $f_1 = 0.19$, and $f_2 = 0.21$; at $L = 31$, the discriminatory power of $f_2 = 0.21$ and $f_3 = 0.25$ is also lost

frequency domain, this estimation is referred to here as the direct path. In contrast, estimators that first form the second moment in the time domain are called estimators on the indirect path.

7.1.2.3 Indirect Spectral Estimators

The smoothing with a window, according to Eq. 7.13, is equivalent to the discrete convolution with a rectangle of length L. According to the convolution theorem, it corresponds to multiplication with the window's Fourier transform and the signal's autocorrelation function in the time domain (Eq. 7.14).

$$\hat{P}_x(\omega_k) = \sum_{n=0}^{N-1} w[n] \cdot \hat{r}_{xx}[n] \cdot e^{-j\omega_k n/N} \tag{7.14}$$

In Eq. 7.14, $w[n]$ is a window function (implicitly a rectangle in Eq. 7.13), and $r_{\text{XX}}\ [n]$ is the autocorrelation function.

The relation according to Eq. 7.14 (with rectangular window) is also called the Wiener-Khinchine-The inverse formulation is given by Eq. 7.15.

$$\hat{r}_{xx}[m] = \sum_{k=0}^{N-1} \hat{P}_x(\omega_k) \cdot e^{j\omega_k m/N} \tag{7.15}$$

If one sets $m = 0$ in Eq. 7.15, one contains the mean square power, which according to Eq. 7.16, is another interpretation of Parseval's theorem (Eq. 7.10).

$$\hat{r}_{xx}[0] = \sum_{k=0}^{N-1} \hat{P}_x(\omega_k) \tag{7.16}$$

If a physical interpretation of the spectral power density or the correlation is intended or if these quantities are to be compared for different sampling rates, including the sampling period T_{A} is necessary (Eq. 7.17).

$$\hat{P}_x(\omega_k) = T_A \sum_{n=0}^{N-1} w[n] \cdot \hat{r}_{xx}[n] \cdot e^{-j\omega_k n/N} \tag{7.17}$$

The window functions $w[n]$ were dealt with in Sect. 4.2.1.2, so they are not discussed in detail here. Furthermore, estimating the spectral power density is necessary to estimate the ACF (autocorrelation function) rxx[m]. The ACF results theoretically as the second moment from Eq. 7.18.

$$r_{xx}[m] = E\{X[n].X[n+m]\} \tag{7.18}$$

Normalizing the sum of the products to their actual number in the estimation yields an expectation-true (bias $b = 0$) estimate (Eq. 7.19). This estimator is expectation-true because the sum of the second-order products in Eq. 7.19 is normalized to their actual number.

$$\hat{r}_{xx}[m] = \frac{1}{N-m} \sum_{n=0}^{N-m-1} x[n]x[n+m] \tag{7.19}$$

An alternative possibility is normalizing the total number of values (Eq. 7.20).

$$\hat{r}_{xx}[m] = \frac{1}{N} \sum_{n=0}^{N-m-1} x[n]x[n+m] \tag{7.20}$$

The estimator, according to Eq. 7.20, is unbiased (bias $b \neq 0$). Let us compare the variances of the two estimators according to Eqs. 7.19 and 7.20 (Fig. 7.8). We find that the variance of the expectation-trusted estimate according to Eq. 6.76 increases hyperbolically with increasing m (Eq. 7.21, Fig. 7.8).

$$\mathrm{var}(r_{xx}[m]_{b=0}) = \frac{1}{N-m} \mathrm{var}(x[n]x[n+m]) \tag{7.21}$$

In contrast, the variance of the non-expectation dispersion estimate according to Eq. 7.20 (Fig. 7.8) decreases linearly with increasing m (Eq. 7.22).

$$\mathrm{var}\big(r_{xx}[m]_{b\neq 0}\big) = \frac{N-m}{N^2} \mathrm{var}(x[n]x[n+m]) \tag{7.22}$$

In both relations, according to Eqs. 7.21 and 7.22, the process x[n] was assumed to be stationary, ergodic, and white.

The comparison of these two estimators of the ASF (Eqs. 7.21 and 7.22) demonstrates the fundamental dilemma of parameter estimation:

> Unbiased estimation and variance minimization are opposing demands on statistical measures and parameter estimators.

Therrefore, deciding which estimator is better suited across the board is impossible. It depends on the concrete question. However, since the ACF is often used to estimate the spectral density, one can compare the effects of the different variances on the PSD (Fig. 6.54). Since the variance of the expectation-stratified estimator (Fig. 7.8) is very high compared to the other estimator and shows discontinuities at the edges of the ACF (edge effects), a higher variance of the spectral estimate is also to be expected. Although both PSDs (Fig. 7.9) are qualitatively identical, the variance of the PSD according to the expectation-abundant estimator of the ACF is about five times higher. For the estimation of the power spectral density, this leads to the conclusion that the non-expectation-true estimator of the ACF, according to Eq. 7.20, is better suited for this purpose.

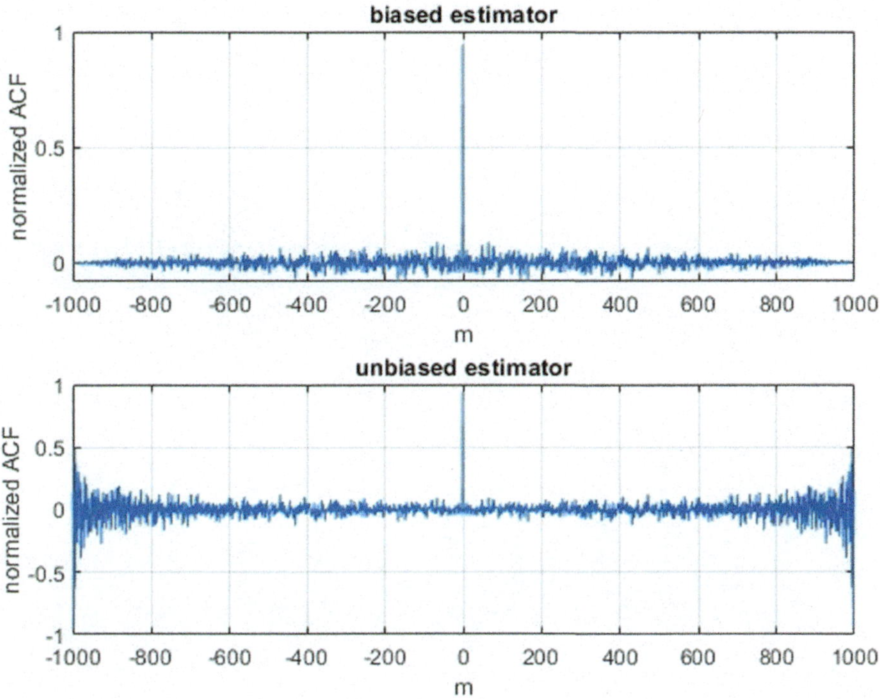

Fig. 7.8 Estimating the ACF biased (top) and unbiased (bottom) sequence of the normally distributed white noise of length $N = 1000$

Since calculating the correlation functions, according to Eqs. 7.19 and 7.20, is very computationally intensive, it can be realized more effectively with the help of the relationship, according to Eq. 7.23.

$$\begin{aligned} x[-n] * y[n] &= \sum_{m=-\infty}^{\infty} x[-m]y[n-m] \\ &= \sum_{m=-\infty}^{\infty} x[m]y[n+m] = r_{xy}[m] \end{aligned} \tag{7.23}$$

In Eq. 7.23, the correlation function $r_{\mathrm{xy}}(m)$ depends on time for formal reasons. It is unimportant for the practical calculation, as it depends solely on the displacement m, independent of the variable designation.

According to Eq. 7.23, the relationship follows that a signal's correlation function and convolution are identical to the time-mirrored signal. According to the convolution theorem, Eq. 7.24 applies to the frequency range.

$$R_{xy}[k] = X[-k] \cdot Y[k] \tag{7.24}$$

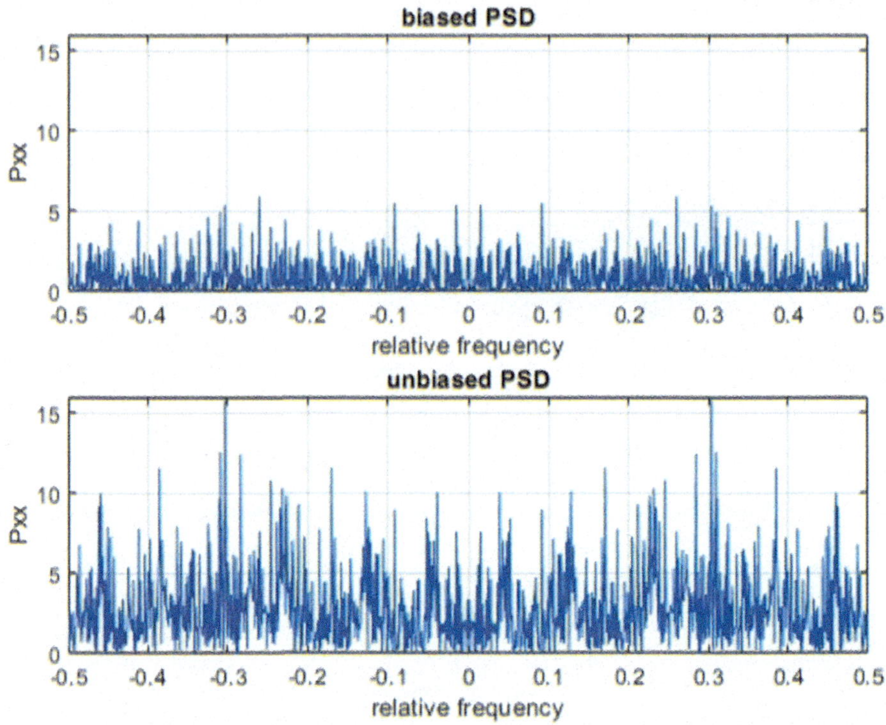

Fig. 7.9 Power density spectra of the noise from Fig. 7.8 for the biased estimate (top) and the unbiased estimate (bottom)

This results in the relationship for the correlation function (Eq. 7.25):

$$r_{xy}[m] = \mathscr{F}^{-1}\{X[-k] \cdot Y[k]\} \tag{7.25}$$

The Fourier transform for the relationship according to Eq. 7.24 and its inverse according to Eq. 7.25 are computationally feasible with fast algorithms of the FFT so that the correlation function can be effectively calculated according to the following algorithm (Eq. 7.26):

$$\begin{aligned} X[k] &= FFT(x[-n]) \\ Y[k] &= FFT(y[n]) \\ \hat{R}_{xy}[k] &= X[k] \cdot Y[k] \\ \hat{r}_{xy}[m] &= IFFT(R_{xy}[k]) \end{aligned} \tag{7.26}$$

7.1.2.4 Methods for Estimating the Power Spectral Density

As already explained, the periodogram, according to Eq. 7.11, is an inconsistent estimator of the PSD; therefore, suitable methods for variance reduction of the

PSD are necessary. If a window function is used for the reduction, this is referred to as a *modified periodogram* or the Blackman-Tukey method. Another approach is to segment the signal and average the periodograms calculated from the segments (Eq. 7.27). In this way, the variance can be reduced by the effect of the arithmetic mean:

$$\hat{P}(\omega) = \frac{1}{M}\sum_{m=1}^{M} I_m(\omega) \rightarrow \hat{P}[k] = \frac{1}{ML}\sum_{m=1}^{M} I_m[k] \tag{7.27}$$

In Eq. 7.27, frequency is in the continuous domain, k *is* the discrete frequency, M *is the* number of segments, and L *is* the segment length.

In the simplest case, the signal is divided into M non-overlapping (and not necessarily adjacent) segments, and the periodograms are calculated and then averaged (Fig. 7.10). This procedure is also called Bartlett's method. The factor M reduces the estimate's variance according to the averaging order (Eq. 7.27). However, the spectral resolution is also reduced since $f \sim 1/t$ (T_A is the sampling period, N *is* the number of all values):

$$\text{var}\left(\hat{P}[k]\right) = \frac{1}{M}\text{var}\left(\frac{1}{L}I_m[k]\right), \quad \Delta f_M = M \cdot \frac{1}{NT_A}$$

The disadvantage of the deteriorated spectral resolution can be relativized by choosing longer segments. However, this would reduce their number M and thus the variance reduction. The possibility of overlapping the segments offers a compromise. This procedure is also called the *Welch* method (Fig. 7.11).

The overlap percentage is, in principle, freely selectable but should be set to 50% for several reasons. Data redundancy occurs if the overlap exceeds 50%; some signal components are overvalued. If less than 50%, locally essential data may be statistically disadvantaged. In order to shed light on this effect, the following explanations are necessary first:

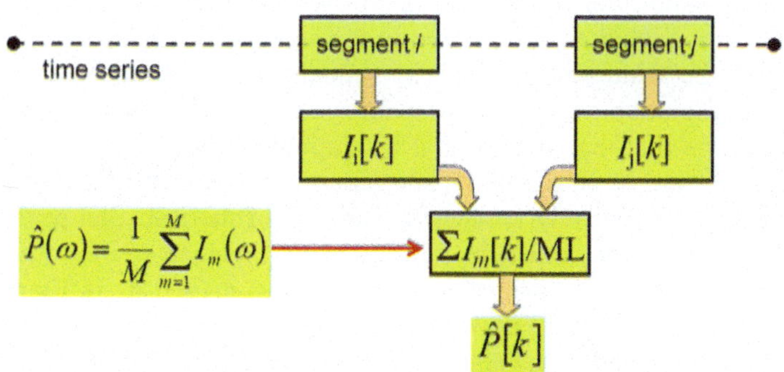

Fig. 7.10 Segmentation of a time series and averaging of the periodograms. The segments do not overlap, their number is M, and their length is L. It is also known as Bartlett's method

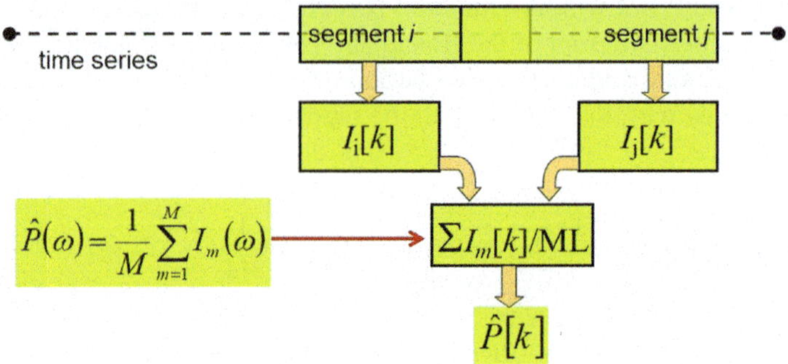

Fig. 7.11 Segmentation of time series and averaging of the periodograms. The segments overlap, their number is M, and their length is L. It is also known as Welch's method

Influence of the Fenestration on the Segment Overlap
For reasons already mentioned, windowing of the data in the time domain is necessary, whereby rectangular windowing is implicit if no window function is chosen. For non-overlapping segments (Fig. 7.10), the choice of window is free; it depends on the known properties of the window functions. From the point of view of signal analysis, however, it can be problematic that the data are weighted less and less toward the edge of the window. Therefore, important process events in these areas may be assigned a reduced significance or no significance. Against this background, it would be crucial to use windowing that assigns the same importance to the data a priori during evaluation. The solution approach is to use overlapping segments, whereby the combined window function (addition of overlapping windows) must always yield one in its result. Of the known window functions, only Bartlett's and Hann's windows come into question for this (see Windowing).

The effect of a 50% overlapping Hann window is shown in Fig. 7.12. The constant weighting condition is fulfilled, meaning all data have the same significance. As mentioned above, the spectral resolution doubles compared to Bartlett's method. A formal but not analytical problem is the double data redundancy. Therefore—as this example shows—the segment overlaps with 50%, and according to Bartlett or Hann, a window is preferred in the analysis.

Of course, the window function also influences the variance of the spectral estimate and resolution. A qualitative comparison of selected window functions (Fig. 7.13) shows that the best spectral resolution can be achieved with the rectangular window but the most significant variance of the PSD. All other windows (Hann, Blackman) effectively reduce the variance of the PSD, but at the expense of the spectral resolution, which is still best with the Hann window for the same window length. With real biosignals, the differences caused by different window functions are not prominent (Fig. 7.14).

Because the simulated signal (Fig. 7.13) forms an extreme combination of white stationary noise and spectral needles (harmonics on grid points of the FFT), any

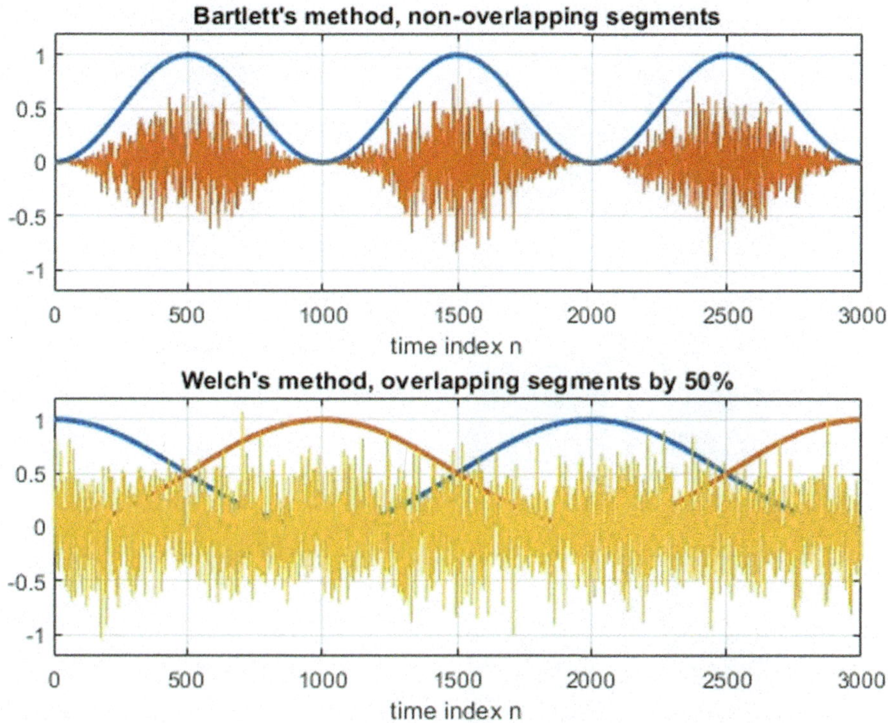

Fig. 7.12 Non-overlapping segmentation with Bartlett's method (top, $L = 1000$) and 50% overlapping segmentation with Welch's method (bottom, $L = 2000$), both with a Hann window. Note the double segment length with Welch compared to Bartlett and the weighting of the data. Original data are normally distributed as white noise, the window functions, according to Hann, are shown in bold

degradation of spectral resolution and estimation variance is immediately effective. Real biosignals have a relatively continuous spectrum in which the noise and interference overlap with the searched signals over wide ranges so that no extreme differences can arise.

Nevertheless, as expected, the rectangular window (Fig. 7.14 top) shows the best spectral resolution but the worst estimation variance. The other window functions Hann (Fig. 7.14 middle) and Blackman (Fig. 7.14 bottom), show no significant differences.

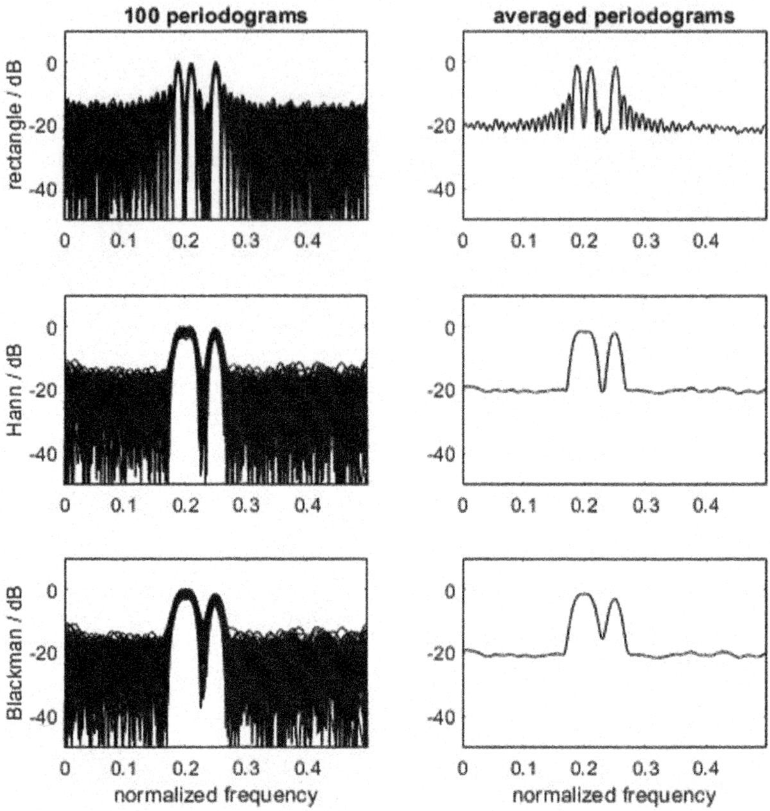

Fig. 7.13 Periodograms of 100 non-overlapping segments (left column) and their averaging (right column) for the rectangular (top), Hann (middle), and Blackman (bottom) window functions. Note the best spectral resolution but the worst variance for the rectangle window (top) and the worst spectral resolution but the best variance for the Blackman window (bottom). The often-used Hann window shows acceptable variance (marginally worse than Blackman) and medium spectral resolution (center). Data were taken from Fig. 7.7 (white, normally distributed noise additively overlaid with harmonics at relative frequencies 0.19, 0.21, and 0.25)

7.1.3 Cross-Power Spectral Density and Coherence

7.1.3.1 Cross Power Spectral Density, the CPSD

The spectral Cross power density, called CPSD (cross-power density spectrum), represents an extension of the relationship according to Eq. 7.14 to two signals, x, and y (Eq. 7.28).

$$\hat{P}_{xy}(\omega_k) = \sum_{n=0}^{N-1} w[n] \cdot \hat{r}_{xy}[n] \cdot e^{-j\omega_k n/N} \tag{7.28}$$

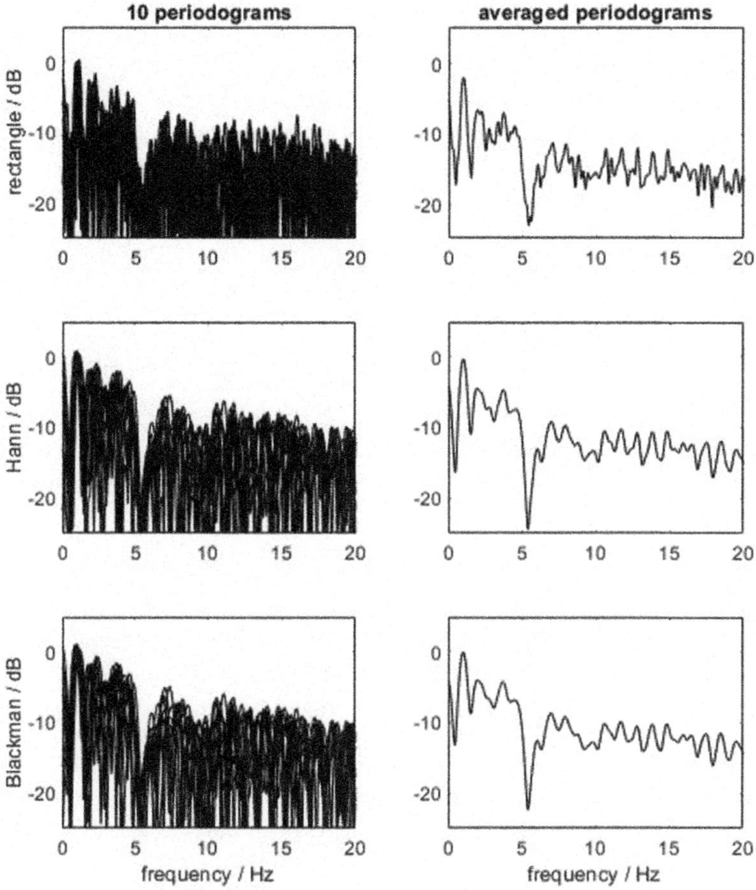

Fig. 7.14 Periodograms of 10 non-overlapping segments of length L = 2000 or TL = 4 s (sampling period $T_A = 2$ ms, spectral resolution $f = 0.25$ Hz) left column and their mean values (right column) for window functions rectangle (top), Hann (middle) and Blackman (bottom). The heart rate is about 1 Hz or 60 bpm (beats per minute, heart actions per minute) and has harmonics up to the 15th order (15 Hz). The spectral notch at about 5 Hz is non-physiological and was caused by implemented signal processing in the amplifier

Since the relations according to Eqs. 7.15–7.22 were first applied to the autocorrelation function (ACF) and the auto power density, they can be regarded as a special case of the CCF and the CPSD. Therefore, they can also be applied without any further restriction, whereby some essential differences result from the extension to two signals: Since the ACF is already an even function, the APSD (auto power density spectrum) is real and non-negative. However, this does not apply to the KLF, which does not have to be even or odd. Therefore, the CPSD is generally a complex quantity (Eq. 7.28). It has far-reaching consequences for the analysis: Despite the formation of second moments, in contrast to the APSD, the phase in the CPSD is preserved (more precisely, the phase difference between $x[n]$ and y[n]). Since the CPSD can be interpreted as a spectral cross-correlation, it provides information about common power components in x and y and their temporal offset. Similar to the estimation of the APSD, the CPSD is determined the direct way (Eq. 7.29, see also Eq. 7.11) or by the indirect way according to Blackman-Tukey (Eq. 7.28) or using the segmenting methods according to Bartlett or Welch (Eq. 7.30, see also Eq. 7.27).

$$\hat{P}_{xy}(\omega_k) = I_{xy}(k) = \frac{1}{N} X(k) \cdot Y^*(k) \tag{7.29}$$

$$\hat{P}_{xy}(\omega_k) = \frac{1}{M} \sum_{m=1}^{M} I_m^{(xy)}(\omega) \rightarrow \hat{P}_{xy}[k] = \frac{1}{M} \sum_{m=1}^{M} I_m^{(xy)}[k] \tag{7.30}$$

The CPSD provides information about the linear relationship (more precisely formulated: about the linear portion of a possible non-linear relationship) between the spectral portions of two signals. The following example will illustrate the possible interpretations.

Example

A patient's ECG and respiration (measured electrically with a strain gauge) were recorded. One would like to eliminate the influence of breathing for diagnostic curve measurement of the ECG.

To do this, one must first know whether respiration is present in the ECG. Therefore, the CPSD is formed between the two signals (Fig. 7.15). Indeed, the spectrum of respiration is already recognizable in the original ECG (Fig. 7.15 center, marked as "linear, additive"). However, this is an unreliable indicator because respiration is not detectable in every ECG. Therefore, the CPSD is calculated for respiration (Fig. 7.15 top) and ECG (Fig. 7.15 middle). The CPSD (Fig. 7.15 bottom) shows a clear spectral correlation between the two signals concerning the common signal component respiration. However, the signal component respiration is also symmetrically distributed around the heart rate (Fig. 7.15 middle), which is caused by the amplitude modulation of the ECG by respiration. It is, therefore, a multiplicative, i.e., a non-linear linkage. The CPSD can be interpreted as an absolute measure of the linear relationship between spectral components of two signals. However, the same question arises as in the case of correlation: How can different CPSDs be compared? The methodologically correct answer, but not completely clear, is normalization. With the help of normalization, one obtains coherence from the CPSD.

7.1.3.2 Coherence

The coherence is defined according to the relationship in Eq. 7.31 and represents the normalized CPSD.

$$C_{xy}(\omega) = \frac{|P_{xy}(\omega)|^2}{P_{xx}(\omega) \cdot P_{yy}(\omega)}, \quad 0 \le C_{xy}(\omega) \le 1 \tag{7.31}$$

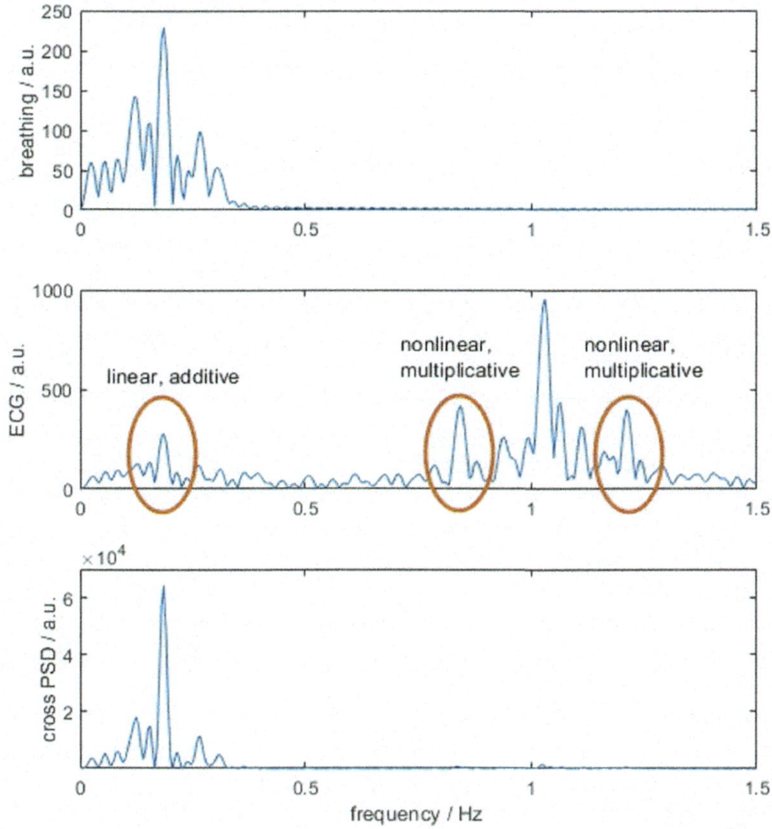

Fig. 7.15 After simultaneous recording, the CPSD between the respiration and the ECG. The respiration alone (top) is spectrally at approximately 0.2 Hz, corresponding to a resting respiration rate of 12 min^{-1}. It is additively (middle) and multiplicatively superimposed on the ECG. The CPSD (bottom) shows a maximum for the respiration's linear components (spectral correlation)

One could interpret the coherence according to Eq. 7.31 as the spectral analog of Pearson's correlation coefficient. However, the coherence interpretation is more complicated and partly misleading, primarily because of the variables' time and frequency. The following example illustrates the fundamental interpretation problem.

Example

Two harmonic oscillations are $x(n)$ and $y(n)$, with identical frequencies of $f_{rel} = 0.1$ and identical phases. If one calculates the coherence for these oscillations according to Eq. 6.88, it has the value $C_{xy}(f) = 1$ over the entire spectrum (Fig. 7.4). It is neither a methodological nor a mathematical error. Admittedly, for all frequencies except for $f_{rel} = 0.1$, C_{xy} results in an indeterminate expression of

$$C_{xy}(f)\big|_{f\neq 0.1} = \frac{0}{0} \to 1.$$

However, it can be shown by limit value calculation that the quotient is always one. Such an initially surprising result can be justified with the help of the analogy to the correlation coefficient: Since coherence is a measure of the linear similarity between two signals at a specific frequency, it is inevitably very high (normalized to one) even if the power densities of the signals at this frequency are zero.

Zero-value power densities are just as similar to high power densities. One approach to solving this problem would be to add uncorrelated noise to the harmonics so that it explicitly violates the similarity outside the frequencies of the harmonics. Intuitively, one will first add a weak noise, not further, to reduce the generally already poor SNR. However, a feeble noise (Fig. 6.61) indicates that the similarity is only slightly violated, and consequently, the coherence deviates slightly from one. Consequently, one must add a strong noise (Fig. 7.16 middle) to the harmonics for a clear coherence relationship. However, this is a contrary requirement for a good SNR.

The above mentioned problem rarely occurs with real biosignals since their spectrum is continuous and overlaid with relatively strong disturbances. If, for example, the CPSD between respiration and ECG (Fig. 7.17 top) is normalized according to Eq. 6.88, a problem typical for real biosignals becomes visible in the coherence diagram (Fig. 7.17 bottom):

While the CPSD correctly represents the common spectral components and is zero-valued above approx. 0.5 Hz, the coherence shows relatively high coherences above this frequency in addition to the correct high values below 0.5 Hz.

This effect occurs when one of the APSD in the denominator of the quotient in Eq. 6.88 is much smaller than the other. Due to the normalization, the weights of the spectra balance out so that the significance of the much weaker signal is increased. It is why the spectral components of respiration, which are about 60 dB below the ECG above 0.5 Hz (Fig. 7.17), nevertheless lead to coherences of up to 0.4.

7.1.3.3 The CPSD in System Identification

Modeling and identification of systems is also an essential field in biosignals. Typical applications are the investigation of the dynamics of the cardiovascular system or the functional testing of sensory systems (visual, acoustic, somatic). From the point of view of system theory, the determination of the system function $h(t)$ with the help of the convolution according to Eq. 7.32 is a methodically simple task: If a Dirac impulse (t) *is* applied to the system input, the impulse response of the system $h(t)$ is obtained directly at the output. It is feasible in most technical systems but not biological ones, in which such an impulse would destroy tissue. A biologically compatible process must therefore be selected as the test signal.

$$y(t) = x(t) * h(t) \tag{7.32}$$

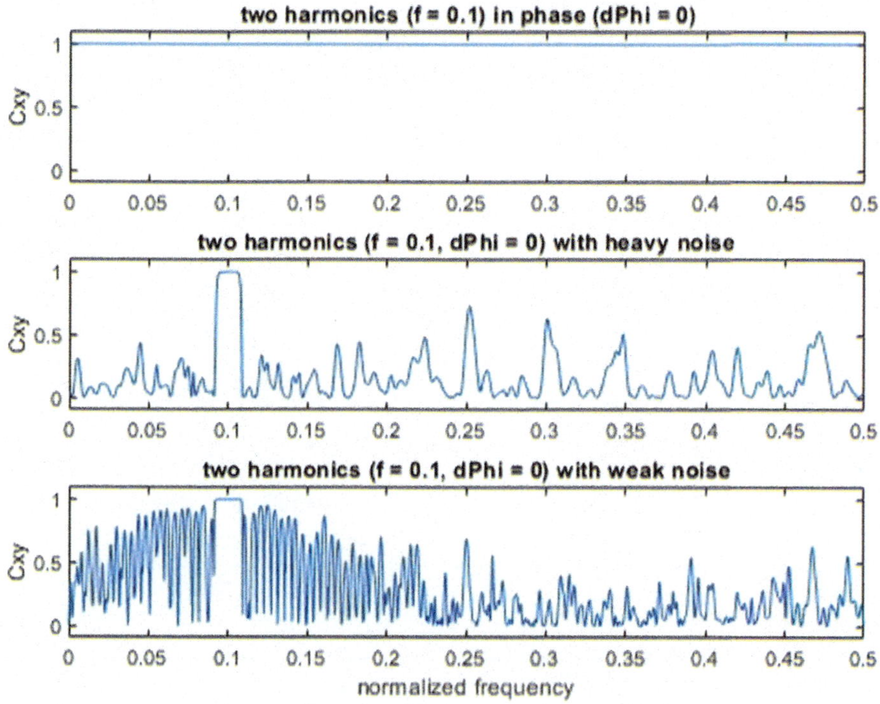

Fig. 7.16 Coherence of two identical harmonic signals at the relative frequency 0.1 (top), superimposed with intense noise (middle) and with weak noise (bottom). The width of the maximum of about $\Delta f = 0.01$ of $C_{xy}(f) = 1$ at $f = 0.1$ results from the length of the analysis window; see windowing

From a signal analysis point of view, the white, normally distributed noise is particularly suitable for the input signal, as it has a constant spectral power density. According to Eq. 7.32, the deterministic relationship must be extended to stochastic processes. First, one folds (Eq. 7.33) both sides of the equation with $x(-t)$,

$$x(-t) * y(t) = x(-t) * x(t) * h(t) \tag{7.33}$$

which enables the transition to statistical moments (Eq. 7.34).

$$r_{xy}(\tau) = r_{xx}(\tau) * h(\tau) \tag{7.34}$$

After the transition into the frequency domain, the system function can be determined from the CPSD and the APSD (Eq. 7.35).

$$P_{xy}(\omega) = P_{xx}(\omega) \cdot H(\omega) \rightarrow H(\omega) = \frac{P_{xy}(\omega)}{P_{xx}(\omega)} \tag{7.35}$$

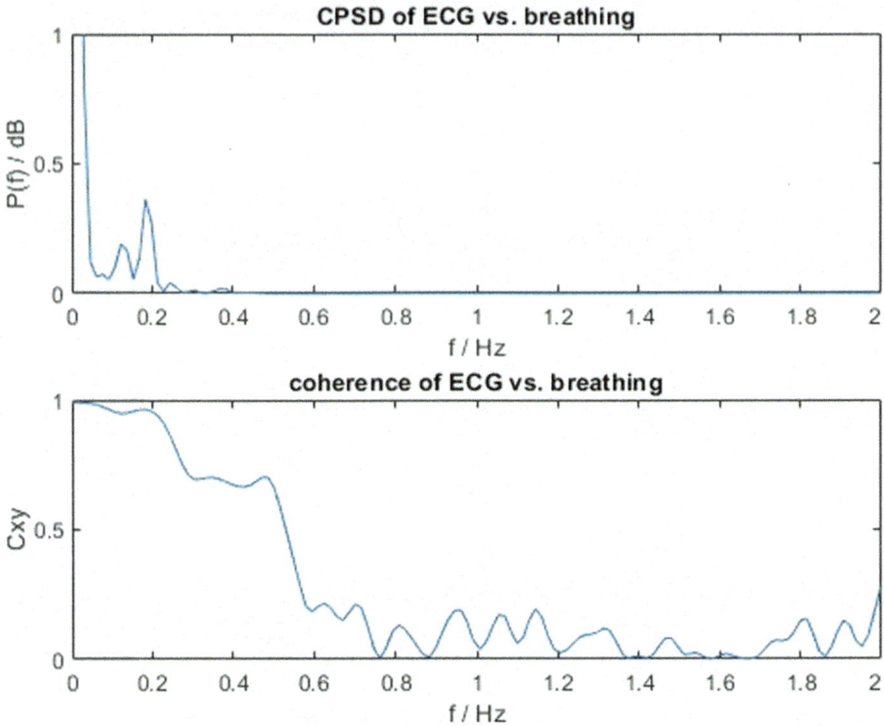

Fig. 7.17 The CPSD between ECG and respiration (top) and the associated coherence (bottom)

System response in the time domain results from the inverse transformation (Eq. 7.36).

$$h(t) = \mathscr{F}^{-1}\{H(\omega)\} \tag{7.36}$$

If white noise with known power is used as the input signal, the system identification is reduced to the determination of the output spectrum (Eq. 7.35):

$$H(\omega) = \frac{P_{yy}(\omega)}{\sigma_w^2}$$

White noise in the visual field could be the picture of a television without a signal (picture noise), and in the acoustic field, for example, the noise of a waterfall. From the point of view of sensory physiology, however, such stimuli are not adequate.

On the one hand, white noise is an analog signal with stochastically distributed levels; on the other hand, with undefined and likewise stochastically distributed edges of the waves. However, since the sensitivity of the senses depends on both the stimulus level and its slope, both must have a known and predetermined value.

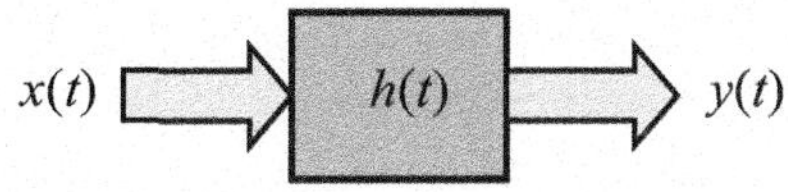

Fig. 7.18 System description in the time domain

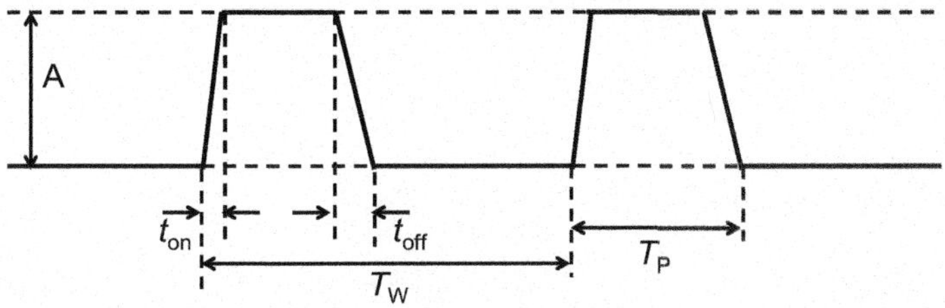

Fig. 7.19 Sensory-physiologically relevant parameters of a stimulus: A—amplitude, t_{on}—rising edge, t_{off}—falling edge, T_P—pulse duration, T_W—repetition period

Therefore, white noise is ruled out as a sensory stimulus. A defined level with a defined slope is provided by binary sequences (Fig. 7.19). Regarding system analysis, it would make sense if these sequences were equal to white noise regarding second statistical moments. Such sequences were developed especially for this purpose, e.g., MLS sequences (Maximum Length Sequences). From the point of view of sensory physiology, they are also particularly well suited for stimulation because they prevent the habituation effect. Periodic stimuli are still used to some extent in the functional control of sensory systems. The periodic stimulation leads very quickly (in the range of seconds) to the sensory system adjusting (adapting) to the periodicity, and the attention (vigilance) and subsequently also the amplitude of the evoked potentials quickly decrease. A pseudo-random binary sequence (PRBS, Pseudo Random Binary Sequence) is generated with an algorithm, so it is known for signal analysis. To the subjects, the sequence nevertheless appears random, preventing adaptation of the sensory system and maintaining vigilance at an almost equal level (Fig. 7.18).

7.2 Signal Detection

Medical diagnostics is mainly based on biosignals and static data and images. In this context, medicine implicitly tries to reduce the sometimes (signal processing) by reducing the sometimes extensive amounts of data to a few parameters (waveform, amplitudes, latencies, time differences, periods, and phases) to classify them differentially into physiological and pathological classes. This procedure is based on empirical knowledge and methodological experience of medicine, which has been accumulated over decades and is mainly used subjectively and time-consuming (diagnosis by a doctor) for current questions. From the point of view of signal analysis, this involves pattern recognition (signal shape) and parameter estimation (curve measurement). Many of these methods can be implemented algorithmically with the help of computer technology, thus reducing routine work to a necessary minimum. However, before the actual pattern recognition and parameter estimation, elaborate and complex SPP (signal pre-processing) must be implemented before the main detection task can be solved. Since SPP is not dealt with sufficiently in the technical literature, the basics will be dealt with in greater methodological depth here.

7.2.1 Signal Detection Using Statistics

Methodically, the statistical proof of an effect (difference concerning H_0) can initially be transferred and applied unchanged to signal detection. However, in this transition from the static domain (an arbitrary sequence of measured values) to the time domain, several methodological extensions must be made, and ways to fulfill the assumptions must be sought.

7.2.1.1 Independence of the Experiments

One of the most critical assumptions of statistics is that experiments are independent. With a few exceptions (manipulated experiments, systematic influence), this assumption is usually fulfilled, so its fulfillment is hardly ever tested in practice. For example, a correlation between blood glucose measurements of different patients is excluded a priori. If this assumption is interpreted in the time domain—a sequence of random variables (Fig. 7.2)—the ACF (autocorrelation function) should be zero-valued everywhere except at $\tau = 0$. The signal's spectrum must be white (constant power spectral density). Biosignals do not have a white spectrum, so they must be whitened before further analytical steps (whitening).

7.2.1.2 Normal Distribution of the Experimental Data

Most parametric tests are based on the normal distribution of the experimental data. Other statistical measures (Pearson's correlation coefficient), distributions (χ^2, Rayleigh distribution), or statistical tests (*t-test*) also require normally distributed data. Biosignals—like all biological quantities—are not normally distributed by nature. Practically, there are two ways to enforce the required normal distribution:

- Use measures or tests where the CLT is applicable and practical, i.e., the sum of experimental data is formed in first order (arithmetic mean, *t-test*).
- Linear transformation containing the linear sum, e.g., the DFT.

Other methods must be used if obtaining an even approximate empirical distribution is impossible—e.g. because the sample size is too small. Well-known are analytically describable transformations (deterministic correlation by logarithm, square root, inverse function) and statistically based approaches (rank or quantile transformations). Besides the intended effect that the data are (approximately) normally distributed after such a transformation, they pose a potentially difficult problem concerning their non-linearity: With deterministic transformations, the degree of non-linearity is known so that its influence on the context under investigation can still be estimated. In the case of rank-based transformations, not even the degree of non-linearity is known due to the stochastic character, so the distortion of the most linear correlation cannot be estimated. It usually leads to the fact that although normally distributed data are obtained for the analysis, the non-linearity distorts a possible correlation to such an extent that it is no longer detectable.

7.2.1.3 Detectability and Effect Size

One of the main problems of statistical effect detection (see Statistical Tests) is that the effect itself (for a given sample size) is too weak for detection. If one does not influence the sample size, the effect under investigation remains unproven. In BSP (Biosignal Processing), to quantify the effect, the quantity SNR (signal-to-noise ratio) can be used to quantify the effect. The normalization of the noise power has the advantage that one is independent of the absolute levels of the signals. However, there is also a detection limit for the SNR below, which makes signal detection impossible. Here, the correct interpretation of the non-rejected hypothesis H_0 becomes even more critical than with the conventional statistical test: If H_0 is not rejected, either no signal is present, or the SNR of an existing signal is too low for detection. Without data for the alternative hypothesis, this question cannot be answered in a differentiated manner.

7.2.1.4 Signal Model and SNR

For further investigations, the straightforward—mathematically statistically favorable—additive signal model of the signal theory is initially assumed:

$$x[k] = s[k] + n[k] \tag{7.37}$$

In Eq. 7.37, x[k] is the measured signal, $s[k]$ is the wanted/searched signal, and $n[k]$ *is* normally distributed white noise.

In the following, the time index from Eq. 7.37 is omitted for better readability. If there is no other index (e.g., for channel realization), it is a time series even without the index $[k]$.

The SNR is conventionally defined as follows:

$$snr = \overline{s^2}/\overline{n^2} \tag{7.38}$$

According to Eq. 7.38, the SNR is well suited for stationary signals as a quotient of mean powers for signal and noise. However, biosignals are partly strongly transient, so a specific definition is used depending on the biosignal.

For signal detection, the conventional formulation of the hypotheses is also used first:

$$H_0: x = \mathrm{n} \quad H_1: x = s + \mathrm{n} \tag{7.39}$$

The task of a signal detector (statistical test) is to decide, according to Eq. 7.39, whether the measurement signal x contains only noise n or the desired signal s.

7.2.1.5 Signal Detection with t-Test

Biosignals are often available as multi-channel (EEG, ECG, EMG) or consecutive stimulus-synchronous recordings (e.g., AEP, auditory evoked potentials in EEG) or as a combination of both (Fig. 7.20).

If it is known at which times the amplitudes ("waves") of the EP occur, one can apply a simple *t-test to the* mean value (point estimate) over all recordings at these times. It can be seen that the most significant amplitude occurs at time 370 ms, so one takes a sample with a size of 200 at 370 ms from the data ensemble of 200 recordings with 500 measured values and tests this against the null hypothesis of $\mu = 0$. The null hypothesis is rejected with a $p = 2.7.10^{-8}$, and the confidence interval is ci $=$ [0.60; 1.22], indicating a reliable rejection of H_0. In an elementary detection problem, as in this example, a single-point estimate would be sufficient for signal detection. However, even in the case of normal (physiological) biosignals, in objective analysis, one is never lucky enough to know the exact time of an amplitude. In addition, in pathologically altered signals, the amplitudes and their latencies are completely different, so one does not know at which point one should test. At this point, one could use a multivariate test, e.g., the multivariate variant of the *t-test*—Hotelling's T^2—or the MANOVA (multivariate ANOVA, Analysis

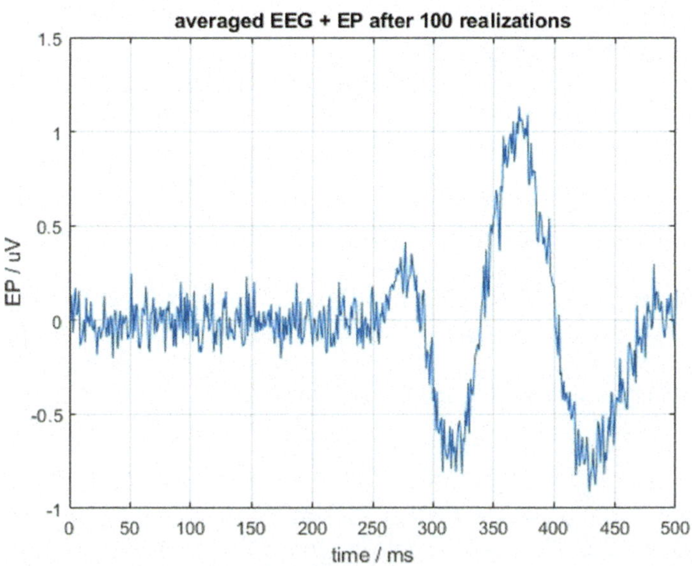

Fig. 7.20 Averaged EEG/EP from 100 individual responses to an acoustic stimulus at the time 250 ms. The simulated noise is white and normally distributed, uncorrelated among the individual recordings. The simulated EP is visible in the right half of the analysis window after the stimulus at 250 ms. The SNR in the original signal (individual recordings) was − 13.2 dB

of Variance), which contains this test in its portfolio. However, the computational effort would be high even for this relatively short signal (quadratic matrices or their inverses), and the assumptions about the signal properties could neither be fulfilled nor tested. Therefore, it makes more sense not to treat the data as a sample (data ensemble) but as a time series, which they form. One thus applies the ergodicity principle, whereby compliance with the necessary stationarity condition remains open.

Since the spectrum of methods for time series analysis is comprehensive and was partly developed for significant problems, selected basics for signal detection and preprocessing are dealt with in the following.

7.2.2 Signal Detection with Energy Detector

The energy detector is based on comparing the energies under the hypotheses H_0 and H_1 according to the signal model in Eq. 6.115. If the energy under H_1 is demonstrably greater than under H_0, the signal's s presence *is* considered confident (with a given statistical uncertainty). Because only the energies are compared, the detection is independent of the signal shape. This is a significant advantage, especially for biosignals with unknown (pathologically changed) signal forms. However, reliable detection is only possible from an SNR of about + 10 dB (Eq. 6.114), so measures to increase the SNR, which is generally low for biosignals (< 0 dB), must normally be connected before the energy detector.

At least approximately several prerequisites must be met before using the energy detector. As explained, the measurement signal (noise and signal) must have a white spectrum. It can be achieved in the time domain, for example, with the method of LPC (Linear Prediction Coefficients), see section Preprocessing Methods. However, one must reckon that whitening (spectral leveling of all components) further worsens the already poor SNR.

The transition from a statistical analysis based on data ensemble to a time series is only possible using the ergodicity principle, which presupposes stationarity. For most biosignals, at least weak and intermittent stationarity can be assumed, especially after whitening. At the same time, the normal distribution of the data must be fulfilled, which, concerning the application to time series, is only possible with the DFT (or another linear integral transformation such as wavelet, STFT). Using the DFT, the normal distribution of the Fourier coefficients is enforced, and a leveling ("washing out") of possible nonstationarity is also achieved through the integration, see above. Usually, the prerequisites for white spectrum and normal distribution would be fulfilled after the previously mentioned measures. However, one should be careful with signals where the instationarities are of diagnostic importance—e.g., with ECG or EMG. Conclusion: It is unnecessary to force the normal distribution by integration (DFT), mainly if the nonstationarity contains essential information.

Procedure for the energy detector using the example of the data ensemble from Fig. 7.3

- According to the formulation of the hypotheses in Eq. 7.39, the question is whether the energy under H_1 is demonstrably higher than under H_0. For this, the *F-test* offers the statistics:

$$\mathrm{H0}: \sigma_2^2 = \sigma_1^2 \quad H_1: \sigma_2^2 > \sigma_1^2$$

For the empirical samples of the two hypotheses:

$$F = \frac{s_1^2}{s_2^2} = \frac{\frac{1}{n_2-1}\sum_{i=1}^{n_2}\left(x_{2,i} - \overline{x}_2\right)^2}{\frac{1}{n_1-1}\sum_{i=1}^{n_2}\left(x_{1,i} - \overline{x}_1\right)^2} \tag{7.40}$$

- According to Eq. 7.40, the test variable is F-distributed, so the critical value F_k can be determined from an appropriate table or software. If the calculated *F value* exceeds the critical value, H_0 is rejected, and H_1 is accepted.
- Since the *F-test* is tailored to random variables—or their variances—it must be adapted for signals on the discrete-time axis. Therefore, Eq. 7.37 can be interpreted as a sample distributed over the time axis (data ensemble) so that the test variable F can be directly equated to the *snr* according to Eq. 7.38. Here, too, one must ensure the independence of the temporally discretized sample values and enforce a white spectrum—at least for the noise n (whitening). The spectrum of the signal s cannot yet be whitened because it is unknown. Practically, one knows the measurement signal x *in the* hope that it is equally effective for both components.
- Since a transient signal is also expected here, a normal distribution forced by the DFT would be useless because of the temporal smearing. Usually, one can reckon with the noise being of technical origin and normally distributed by nature. This assumption is usually valid in practice. The signal s is still unknown; one accepts its non-normal distribution.

In experimental recordings, one obtains only noise (1–250 ms in Fig. 7.21), a practicable noise reference for the energy detector, or a section with noise and possibly present signal (time interval 251–500 ms in Fig. 7.21). Therefore, the SNR must be modified according to Eq. 7.38:

$$snr = \overline{x^2}/\overline{n^2} - 1 \tag{7.41}$$

The *F-test* for a single recording (Fig. 7.21) rightly does not reject H_0. The signal s is present, but with an SNR of − 13.2 dB, it is too weak for detection. It is a typical example that a non-rejection of H_0 does not yet mean its confirmation. The *F-test* confirms H_1 the existence of the signal s only after averaging with

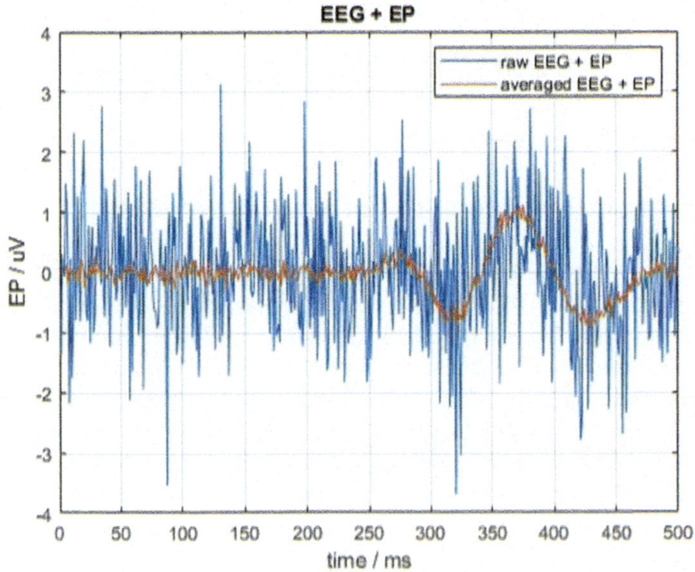

Fig. 7.21 Comparison of two signal characteristics: A single recording (blue) with an SNR of − 13.2 dB. Ensemble average (red) over ten individual recordings with an SNR of − 2.7 dB. An improvement of the SNR by 10 dB was achieved by averaging with an order of ten, sufficient for energy detection in the concrete signal (window length of 250)

the order of 10 after the SNR has been raised by about 10 dB. It can also be interpreted so that in the case of the single recording H_0, it was not rejected, but it was not confirmed either because the effect examined was too weak for proof. This finding leads to the conclusion—contrary to doctrine—that a rejection of H_0 does not yet mean a confirmation of H_1, but a non-rejection of H_0 does not yet mean its confirmation.

> A rejection of H_0 is not yet a confirmation of H_1.
> A non-rejection of H_0 is not yet its confirmation.

After this realization, one can consistently carry the findings (note on hypotheses) to their conclusion: If there is a single (also paired/connected) sample or a single time series, then only one hypothesis can be formulated about this sample/time series. In the case of samples, H_1 is, therefore, the logical negation of H0 at best, and therefore only makes sense for binary hypotheses (component broken? patient pregnant?). For continuous or discretized measured variables in the static range and time series, the formulation of H_1 makes no sense either theoretically or practically since, apart from a reasonably statistically inevitable rejection of H_0, it must leave all three other alternatives between test and truth open.

With a single (even paired) sample/time series of real-valued variables, only one meaningful hypothesis can be made.

7.2.3 Signal Detection with Correlation Detector

The chances of detecting a sought (desired) signal increase considerably compared to the energy detector if the sought signal is known in advance in all its parameters. The signal shape must be available, including all levels/amplitudes and times. Since the correlation only evaluates the linear relationship between two quantities/signals, there must be a shift (translation in level or time) or a multiplication with only one constant (scaling in level). Under these conditions, correlation coefficients (Pearson, Spearman) for static quantities and correlation functions for time series are the best statistical measures. The functioning and the method of the correlation detector are demonstrated with example data from Figs. 6.79 and 6.80, respectively.

In applying the energy detector, it was assumed that the signal sought was unknown. Therefore, it could be detected by its energy alone. It usually requires an SNR of at least 10 dB (10 dB concerning individual measured values). With an analytical window, the window length must be taken into account. In Fig. 7.21, the window length is 250, which brings an SNR improvement of 24 dB through temporal integration). Under ideal conditions (normal distribution, linear relationship), the best measure for the linear relationship is the KK, according to Pearson. The signal searched for is usually a time course (time series) over a limited time (model function, pattern, template). In the search (pattern recognition) for this pattern in the measurement signal, the pattern is shifted over the measurement signal, and the CV is calculated for each shift. This process can be interpreted as calculating the KK in a sliding analysis window. It can be shown that the calculation of the sliding KK corresponds (theoretically and algorithmically) to the convolution of the examined signal with the temporally mirrored pattern. This procedure is identical to filtering the measurement signal with a digital filter whose impulse response corresponds to the time-mirrored pattern signal. This approach is known in signal theory as a correlation or "matched filter." Without showing theoretical derivation and proof here, it can be stated that the correlation filter maximizes the SNR in the time domain. However, this does not yet provide statistical proof of the presence of the sought signal/pattern.

Figure 7.22 shows the basic structure for signal preprocessing and filtering with an MF (matched filter). The biosignal $x\,[n]$ is modeled with an LPC (Linear Prediction Coefficients) in the real-time variant of an LMS algorithm (Least Mean Squares). After the difference between the input signal $x[n]$ and the model, the whitened version of the biosignal $x_{\mathrm{W}}\,[n]$ is obtained. The template (pattern) $d[n]$ is simultaneously whitened with the coefficients of the LPC by an adaptive filter. It provides the filter coefficients for the MF at whose output the signal $y[n]$ with

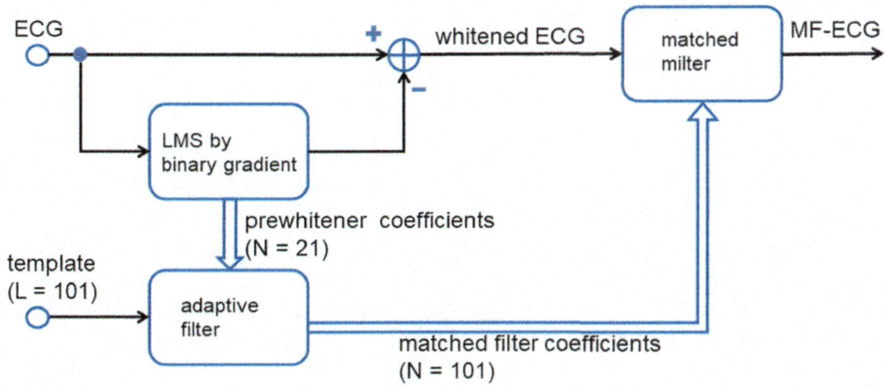

Fig. 7.22 Basic structure for signal preprocessing (whitening) and correlation filter (MF) of the measured biosignal $x[n]$ (ECG) and the pattern (template) $d[n]$

maximized SNR is present:

$$y[n] = \sum_{m=-M/2}^{M/2} x_w[n+m] w_{MF}[-m]. \tag{7.42}$$

In the optimally filtered signal $y[n]$, the exceeding of a detection threshold can be tested in the simplest case in order to decide on the hypotheses according to Eq. 7.43:

$$y[n] \begin{matrix} > \gamma : H_1 \\ < \gamma : H_0 \end{matrix}. \tag{7.43}$$

The determination of the threshold according to Eq. 6.119 is based on existing samples for noise and, if necessary, samples for signal or signal + noise (Fig. 7.21). In general, biosignals have both polarities, so for the detection according to Eq. 7.44, one can use a monopolar signal parameter instead of the signal itself, in the simplest case, the instantaneous power or the signal energy contained in the analysis window:

$$y^2[n] \begin{matrix} > \gamma : H_1 \\ < \gamma : H_0 \end{matrix}. \tag{7.44}$$

If Eq. 7.44 is applied directly, the correlation detector is real-time capable, which is essential for some biosignals (e.g., ECG—R-wave). If the real-time capability is not critical, the instantaneous power in the analysis window can be summed to the energy so that the energy detector, according to Eq. 7.45, is directly applicable to the sliding analysis window:

$$\sum_{1}^{M} y^2[n] \begin{matrix} > \gamma : H_1 \\ < \gamma : H_0 \end{matrix}. \quad (7.45)$$

In all detection variants, a sample of the noise (noise reference) is required in order to be able to establish at least one hypothesis (H_0).

Example of the correlation detector

To demonstrate how the correlation detector works, the same data as the energy detector is used here to allow direct comparison (Figs. 7.20 and 7.21).

The first step applied the original (template) to the raw data. From these, it is known (since simulation) that the noise is white and normally distributed. The output signals of the MF are shown in Fig. 7.24. The MF filtering leads to the fact that the correlation signal of the template itself is present at its output, although pure white noise is fed in at the input (in the simulation). From the system analysis's point of view, the system function's autocorrelation is obtained at the output, not information about the presence of the sought signal. This effect can be observed well in Fig. 7.24 (blue): Although the searched signal (template) is vanishingly weak, one observes its correlation function in the output signal. It leads to the conclusion that a pure shape analysis, according to the MF, is insufficient for detection. Therefore, a discriminator (detector with decision threshold), according to Eqs. 7.43, 7.44, and 7.45 is necessary. The effect of the MF after averaging over 50 realizations is shown in Fig. 7.24 (red): Here, the global maximum of the correlation at time $t = 380$ ms (see also Fig. 7.23) can be correctly detected, e.g., with the detector according to Eq. 7.44.

From Fig. 7.24 and the analyses above, a similar finding as with the energy detector emerges concerning the decision on the hypotheses (Eqs. 7.43, 7.44, 7.45): The hypothesis H_0 can only be rejected above a certain SNR. However, if it is not rejected, it does not mean that it is correct. Concerning the necessary SNR for detecting the sought-after signal, there is an essential difference to the energy detector: The correlation detector is (depending on all signal and method parameters) about 20 dB more sensitive than the energy detector. This difference is usually decisive in many areas of BSP regarding the practical applicability of a method, e.g., the examination duration. Such a difference can decide in clinical examinations for the functional diagnosis of human sensory systems (visual, auditory, somatosensory) whether an examination can take an hour or a minute.

7.2.4 Signal Detection with Combined Detectors

The signal detectors discussed so far (energy and correlation detectors) represent two extremes in the hardly manageable class of detectors:

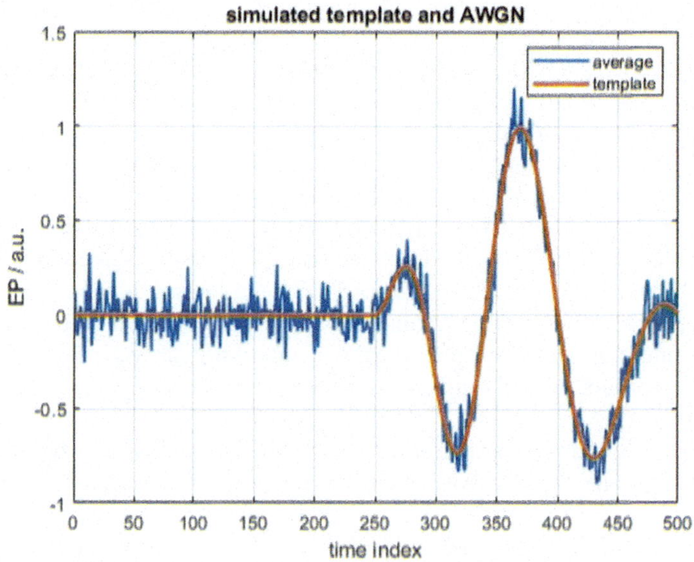

Fig. 7.23 Ensemble average over 200 realizations of a noisy EP (blue) and the template (red). The signal part of the pattern between 250 and 500 ms is used as a template

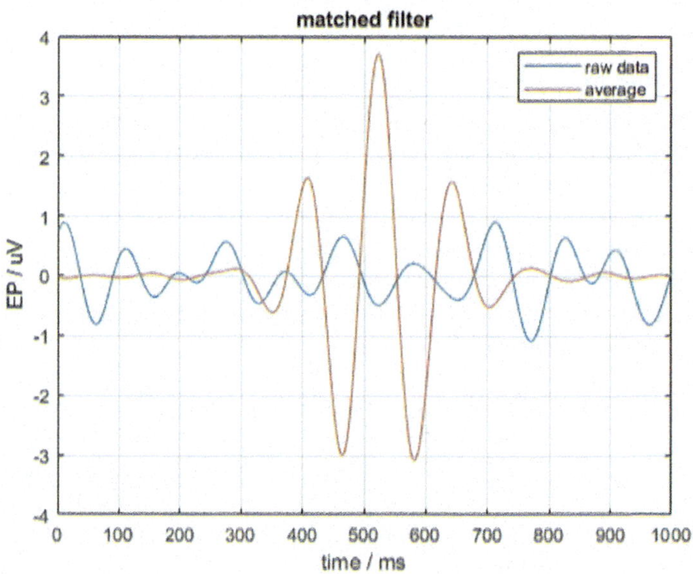

Fig. 7.24 Output signals of the MF with a template from Fig. 7.23 after one realization (blue) and after averaging over 50 realizations (red). Note that after filtering, the result is the moving correlation, not the time signal itself

- For signal detection, the energy detector does not need to know anything about the signal it is looking for, only its energy. This robustness comes at the price of a relatively high SNR, which must be at least 10 dB on average. Such a high SNR is rarely present in biosignals, so pre-processing methods are normally necessary to increase the SNR.
- The correlation detector is the most sensitive detector of all. However, for reliable detection, it needs complete knowledge of the signal it is looking for (signal shape, level, latencies, periods). Such signal knowledge is practically never available for biosignals. However, it is possible to create individual templates with signal preprocessing methods in some cases (e.g., templates from the quiescent signal before an examination) and then use these as patterns during the examination. However, this approach only works in very few cases where at least the qualitative signal form is known (e.g., stress ECG or EP).

Since the above-mentioned extreme detectors can only rarely be realized in practice, a methodically hardly manageable quantity of detectors has been developed, which lie in unspecifiable ranges between the two extremes concerning the required properties of the signal and the detector. As a compromise between the extreme detectors, the oldest, most successful, and still ported detector for the ECG will be discussed here—the R-wave detector, according to Pan-Tompkins (Pan & Tompkins, 1985). Its principle mode of operation and circuit realization were described in the section "Determination of curve parameters." At this point, its statistically essential properties will be analyzed, and problem areas will be highlighted.

The block diagram of the PT detector (Fig. 4.17) shows the methodical procedure: Every detector needs a statistically objective decision criterion for the presence of a signal. In the PT detector, it is the last block of the processing chain—the MA low pass—which, in terms of signal theory, feeds the time-varying signal energy to the decision threshold (Eq. 7.45). Therefore, the PT detector can initially be interpreted as an energy detector. The MA low-pass is preceded by three function blocks that are intended to highlight the most important properties of a QRS complex and therefore make the energy detector more of a waveform-specific detector:

- Bandpass 5–11 Hz: The dominant QRS energy should be transmitted. Today we know that the maximum QRS energy is approximately 10–30 Hz in typical cases. The bandpass is set between 20 and 100 Hz to avoid false positive detections caused by P- and T-waves. Whichever bandpass is used, the prerequisites (white spectrum) for applying an energy detector (MA + threshold) are fulfilled even worse than in the original.
- The differentiator should ensure that the edges in the QRS complex are checked for sufficient steepness. Otherwise, the differentiator positively affects the spectrum—high-frequency components become strong. Thus the narrow bandwidth of the bandpass is partially relativized in the direction of the white spectrum.

- Since the QRS edges should be strong enough regardless of their direction, the differentiator is followed by a squarer, which provides information about this through the instantaneous power formed. Therefore, the energy formed over the mean QRS time in the MA is the measure for recognizing the typical QRS shape.
- It is impossible to determine a generally valid discrimination threshold for the output signal of the PT detector, as none of the essential prerequisites for a statistically reliable detection are fulfilled (white spectrum, normal distribution, stationarity/ergodicity). Therefore, in the practical analysis, a threshold based on empirical data must be determined, which depends on a whole range of parameters: ECG lead system, medical measurement technology, and methods of BSP.
- The PT detector was developed for the physiological ("healthy") QRS progression and, therefore, only functions in this range under the above assumptions and conditions (empirical optimization). The detector fails with slight deviations from the ideal QRS course (extrasystoles) and QRS-like disturbances (movement artifacts in the stress ECG). For these reasons, it is constantly being modified to reduce the frequency of false detections (false positives and false negatives) in clinical practice. For details, please refer to the more detailed literature.

7.2.5 SNR Improvement

Depending on the detector, an SNR of − 10 to + 10 dB is needed for signal detection. For reliable curve measurement, which is clinically standard, an SNR of at least + 40 dB is required. However, compared to publicized presentations in the technical literature and conferences, everyday clinical practice shows a much poorer picture of the actual data: Depending on the biosignal and examination conditions, the usual SNR ranges from + 10 to − 60 dB. The task of BSP is, among other things, to achieve an SNR improvement of at least + 40 dB for functional diagnostics and of at least + 60 dB for course/differential diagnostics. In the following, selected methods of BSP are described, which can also be applied in practice, but some of which have not even arrived in practice yet.

7.2.5.1 Synchronous Averaging

Synchronous averaging is the most effective method for SNR improvement if the assumptions are met. Methodologically, the approach is that event-related recordings by ensemble averaging do not change the signal but reduce the noise by the averaging order. The typical case is shown in Fig. 7.2: Consecutive recordings of a bioelectrical activity are reordered to a synchronous empirical ensemble related to the stimulus. For this, the following conditions must be fulfilled:

- The signal is constant in all recordings regarding signal shape, amplitudes, and latencies.
- The signal is uncorrelated with the noise (CCF zero-valued).
- The noise is uncorrelated in time and realizations/channels (ACF and CCF zero-valued).

If these conditions are fulfilled, one can assume that the SNR is improved by the averaging order N after ensemble averaging:

$$SNR(\overline{x}) = N \cdot SNR(x_i) \tag{7.46}$$

The ensemble average or the achievable SNR reacts very sensitively to a violation of the coincidence of the signal in individual realizations. The effect is demonstrated mathematically for a time difference between two recordings.

- The signal x(t) is time-shifted in two recordings and is otherwise identical

$$x_2(t) = x_1(t + \tau). \tag{7.47}$$

- The ensemble mean contains portions of both time-shifted recordings

$$x(t) = \frac{1}{2}(x_1(t) + x_2(t)) = \frac{1}{2}(x_1(t) + x_1(t + \tau)). \tag{7.48}$$

- If one transforms the ensemble mean into the frequency domain, the time shift becomes a phase shift

$$X(\omega) = X_1(\omega) + X_2(\omega) = \frac{1}{2}X_1(\omega)\left(1 + e^{-j\omega\tau}\right). \tag{7.49}$$

- The spectrum X_1 () is influenced by the phase, whose magnitude can periodically reach two extreme values:

$$\frac{1}{2}\left|1 + e^{-j\omega\tau}\right| = \frac{1}{2}\sqrt{2(1 + cos(\omega\tau))} = \begin{cases} 1, & \tau = 0 \\ 0, & \tau = T/2 \end{cases} \tag{7.50}$$

Equation 7.48 follows (T is the period of a harmonic oscillation of the angular frequency) that a jitter (time/phase fluctuations), depending on the considered frequency, can lead to the preservation of the signal component up to its complete cancellation. Since biosignals generally have a continuous spectrum, the jitter

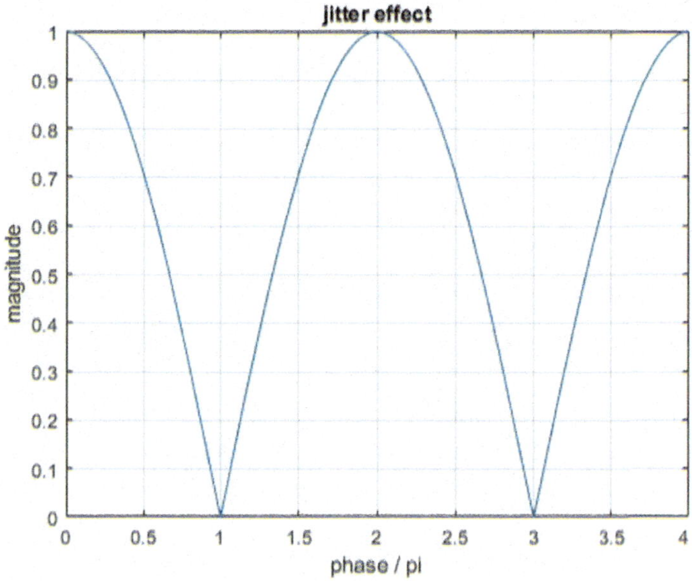

Fig. 7.25 Spectral filter effect of a temporal jitter according to Eq. 7.50 (frequency is normalized to 2π)

leads to frequency-dependent notches, which can be interpreted as time-dependent blocking comb filters (Fig. 7.25).

The following realistic simulation demonstrates the effects of jitter that frequently occur in clinical practice:

- Recordings from eight parietal and occipital channels of the EEG after visual stimulation are available.
- The time shift between the channels is 6 ms.
- According to the actual recordings, the period of the individual waves increases linearly with increasing recording time, i.e., the instantaneous frequency decreases hyperbolically (Fig. 6.85).
- According to the theoretical assumptions, the channel signals are equal; they only show a defined jitter. Concerning the ensemble mean (bold curve), the effects of the jitter can be analyzed and summarised as follows:
  - Since the jitter is constant, higher frequency components are more attenuated than lower frequencies due to its effect, according to Eq. 6.126 and Fig. 6.84.
  - Following the physiological course of the evoked potentials, the jitter has the most substantial attenuation after the stimulus on the higher frequencies in the signal that are diagnostically important (Fig. 6.85).
  - Jitter is one of the most important reasons why the potentials close in time to the stimulus are challenging to detect in stimulus-synchronous recordings of electrophysiological signals.

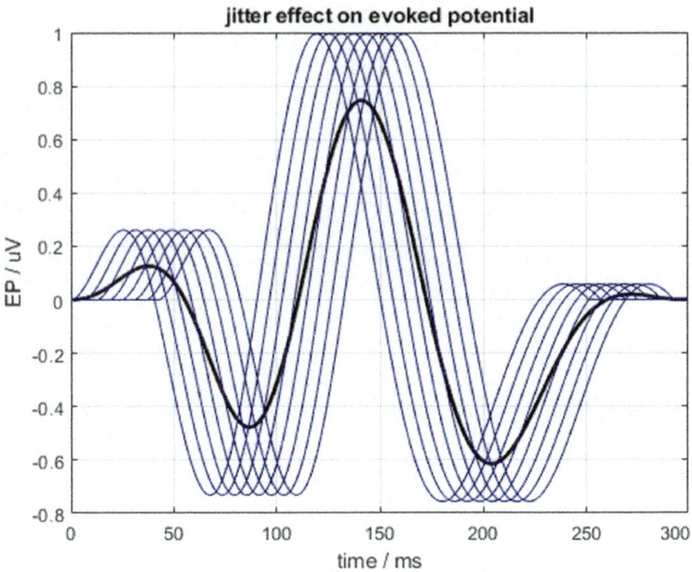

Fig. 7.26 Simulated recordings of a VEP from eight parietal and occipital channels after visual stimulation. The time shift between the channels is 6 ms. The periods of the waves increase linearly with time; thus, the instantaneous frequency decreases hyperbolically. The bold curve shows the simultaneous ensemble mean. Damping decreases with time because of lowering frequency

– Jitter is not only caused by technical signal acquisition. The human being is the much more effective source of jitter: Due to physiological or pathological processes (adaptation, attention, concentration, neurological impairment, or disease), the jitter is variable in time. While the technically induced runtime differences can be primarily reduced (adaptive filter), hardly anything can be done about the individual temporal variability (Fig. 7.26).

7.2.5.2 Spectral Filters

In everyday clinical practice, strong, periodic disturbances from the patient's environment occur most frequently, e.g., disturbances from the 50Hz/60Hz supply network, image frequencies of various monitors (40–200 Hz), the fundamental frequency of the railway (around 16.7 Hz), variable frequencies of asynchronous rotating machines, the fundamental frequency of the air, sea and space transport systems of 400 Hz. The technically generated periodic disturbances are often so strong that they exceed the amplifier's working width (millivolt range), so limitation occurs (supply voltage of the components used). Such a limitation should be monitored and indicated electronically, as it causes non-linear signal distortion and can no longer be corrected by any filter or algorithm. Older measuring systems still have an analog bandstop for the respective mains frequency. It eliminates the mains interference, but because of the analog circuit solution, it distorts the phase frequency response in such a non-linear way that the signal shape suffers and is unacceptable from the BSP point of view (Fig. 7.28). Therefore, only digital band rejection filters that can be constructed with a linear phase, which sometimes causes a considerable delay, are recommended today. A typical bandstop for 50 Hz is shown in Fig. 6.86. The frequency response of the bandstop is almost ideal: The transmission in the passband varies by 1dB at most, and the attenuation in the stopband is better than 60 dB. The phase is linear because the filter has a symmetrical FIR impulse response. However, the running time corresponding to the sampling period of 1 ms and half the filter length is 355 ms (filter length is 710 coefficients). This bandstop is best suited for running (patient monitor) but not for time-critical analysis (pacemaker) (Fig. 7.27).

If a disturbance is known to be periodic but of unknown frequency, adaptive filters (ANC) can be used. Since adaptive filters are normally constructed with the FIR structure (transversal, non-recursive filters), the linearity of the phase frequency response is guaranteed after the adaptation has been completed.

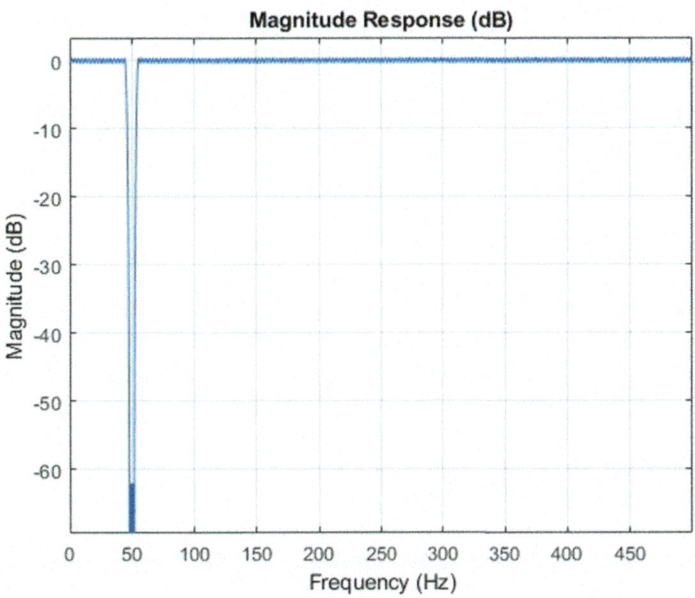

Fig. 7.27 Bandstop for the mains frequency of 50 Hz with a width between 45 and 55 Hz. The attenuation in the stop band is more than 60 dB, and the ripple is 1 dB in the passband. The information loss in the biosignal due to the blocking of the spectral band [45, 55Hz] is negligible, and the phase is linear due to the symmetrical FIR impulse response

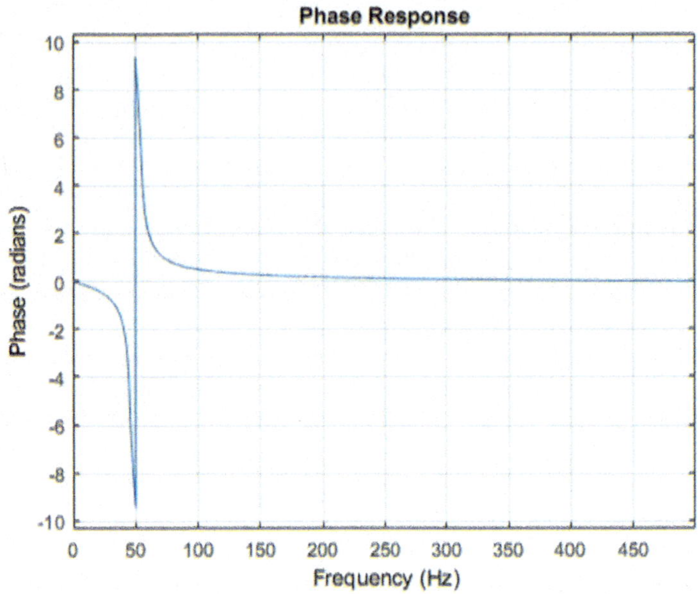

Fig. 7.28 Phase frequency response of a Butterworth-type bandstop with a center frequency of 50 Hz, as used as an analog filter. Note that the strong non-linearity between approximately 30 and 70 Hz leads to distortion of the signal shape

7.2.6 Signal Pre-processing

SPP (signal pre-processing) refers to methods intended to fulfill the assumptions and prerequisites for applying BSP, e.g., white spectrum before signal detection. The methods of SPP can be classified in terms of real-time capability in the domains of time and frequency.

7.2.6.1 Stationarity

Stationarity is a necessary condition from a mathematical point of view but not yet sufficient for the reliable determination of statistical measures and spectral analysis. It is necessary because all integral quantities (correlations, spectra) are based on signal properties not changing in the analysis window. It is insufficient because even with perfectly fulfilled stationarity, it depends on the strength of the investigated effect and whether it can be detected or proven (e.g., energy detector).

With biosignals, one cannot assume stationarity; quite the opposite. A biological system lives from changes; stationarity excludes itself. Since, apart from a few deterministic approaches (state-space model), which are hardly applicable in BSP due to the prerequisites (exact physical model, see Kalman filter), one tries to relativize the demand for stationarity by compromise approaches. Stochastics partly achieve this by partially assuming fulfillment of the ergodicity principle in spectral analysis by temporally shortened and sliding analysis windows. For dynamic spectral analysis, several methods have been developed (STFT/spectrogram, wavelets/scalogram, Wigner/Ville distributions, time-variable filters), which accept or presuppose from the outset instationarities or the dynamic character of the examined signals. The compromise is to use an appropriately short analysis window for the dynamics, in which stationarity is assumed and the relatively poor spectral resolution is accepted.

Stationarity can be established neither methodically nor algorithmically. At best, the examined biosignal can be segmented in such a way that the condition of stationarity is approximately fulfilled within the segments. Of the known methods, wavelets come closest to this view but at the expense of spectral resolution at higher frequencies. Even wavelets or Wigner distributions reach their physical limits at the latest at the Heisenberg uncertainty:

$$\Delta t \cdot \Delta f \geq \frac{1}{4\pi}. \tag{7.51}$$

Relation 7.51 is the Heisenberg uncertainty's signal-theoretical variant, initially formulated in quantum mechanics (Heisenberg, 1927). According to Heisenberg, time–frequency resolutions can be realized purely by impossible calculations, especially with wavelets and Wigner distributions with a two-dimensional kernel. For example, when calculating a Wigner distribution, one can specify a two-dimensional kernel (SPWD) with the following resolutions: In time direction $t = 1$ ms, in frequency direction $f = 1$ Hz. According to Eq. 7.51 with 0.001, the product is clearly below Heisenberg's uncertainty of 0.079, so the desired resolutions cannot be achieved. Although energy distributions can be determined mathematically in the desired time–frequency grid, their content will not fall below the uncertainty principle. Energy distributions that are mathematically above the uncertainty relation appear smoothed or smeared in the result due to the mathematically possible oversampling in the time–frequency composite range.

7.2.6.2 Prewhitening

The term "whitening" is used in the technical literature for methods whose task is to create the prerequisites concerning the signal properties before the application of a primary method (hence the prefix "pre-"). If, for example, the presence of a signal is to be detected using an energy detector, the signal must be "whitened" beforehand. An SPP method must therefore ensure that the global spectrum is constant on average—white in the language of optics—before the actual detection. Prewhitening can be performed in both the time and frequency domains. Temporal methods are algorithmically simple and particularly suitable for signals with high temporal dynamics or highly transient signals, as they are essentially real-time capable (Fig. 7.22). Spectral methods are effective but based on relatively long analysis windows, so they are only suitable for approximately stationary signals for which the temporal dynamics are unnecessary.

Prewhitening in the Time Domain

A relationship between two (and more) random variables is easiest to model linearly:

$$Y = \mathrm{a} + \mathrm{bX}(+\mathrm{a}_1 + \mathrm{b}_1 X_1 + \cdots + \mathrm{a}_\mathrm{n} + \mathrm{b}_\mathrm{n} X_\mathrm{n}). \tag{7.52}$$

The well-known linear regression (regression line) can be calculated using this model, mainly LMS (Least Mean Squares). According to the ergodicity principle, this formulation can be transferred to discrete time:

$$x[\mathrm{n}] = \mathrm{a}_0 + \mathrm{a}_1 x[\mathrm{n}-1] + \cdots + \mathrm{a}_\mathrm{m} x[\mathrm{n}-\mathrm{m}]. \tag{7.53}$$

The model can be interpreted as a linear predictor since the current value x[n], according to Eq. 7.53, results from the linear combination of the m past values. The model coefficients a_i can be calculated for stationary biosignals with the usual methods of AR models (autoregressive models). A computationally effective LMS algorithm is suitable for calculating the coefficients for real-time applications because of the intrinsic dynamics. One can show that the difference between the original signal x[n] and its LPC model is white (Manolakis & Ingle, 2011). Therefore, the filter, according to Eq. 7.53, is also called a *whitening* filter, to SPP a *pre-whitening* filter:

$$y[\mathrm{n}] = x[\mathrm{n}] + \mathrm{a}_1 x[\mathrm{n}-1] + \ldots + \mathrm{a}_m x[\mathrm{n}-m] = x[n] + \mathbf{a}^\mathrm{T}\mathbf{x}[\mathrm{n}-1]. \tag{7.54}$$

According to Eq. 7.54, the algorithm is real-time capable and, therefore, very well suited for time-demanding solutions (pacemaker, insulin pump). Prewhitening is often needed as a preliminary stage for correlation functions (correlation detector, Fig. 7.22) so that both the *whitener* and the MF should, in the simplest case, have a transversal structure (an impulse response, FIR filter). Figure 6.88 shows the effects of whitening on the typical curve features of an ECG recording. An individual template was created from eleven cardiac periods in a relatively poor (disturbed) ECG recording (Pan-Tompkins). The ECG and template were whitened online with the structure in Fig. 7.22 and fed to an energy detector. It enabled detection reliability (sensitivity) of more than 99%, while the FPR (false positive rate) was less than 2%. Without whitening, the FPR was over 25%.

The ECG is also an excellent signal from the point of view of validating methods of BSP since it reveals the performance of diverse solution approaches through its extreme nonstationarity or temporal solid dynamics. Against this background, the effectiveness of the method of an LPC/LMS algorithm applied here can be analyzed vividly in Fig. 6.89: The original ECG has a spectrum typical for this signal, which is composed of a continuous background (typical for all biosignals) and a line spectrum (needles at the harmonics of the fundamental frequency of the ECG). The result after whitening shows remnants of the spectral needles of the heart rate and otherwise an average constant level. Since the LPC implicitly assumes a (linear) stationary process, it is unsurprising that the peaks were not eliminated in the whitened spectrum.

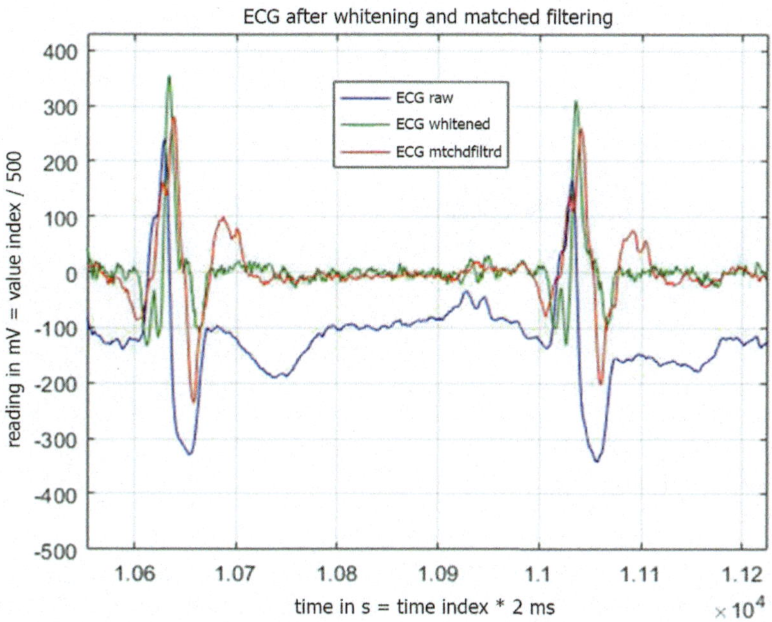

Fig. 7.29 Original ECG (blue) was whitened for the correlation detector (green). The individual ECG template was also whitened (red). After the matched filter, the QRS complex can be detected with a certainty of > 99% (energy detector), even in disturbed recordings

Note to Fig. 7.29: No other biosignal shows the effect of whitening as impressively as the ECG: The raw signal (blue) is reminiscent of the basic course of an ECG, but here it is individually and pathologically altered and affected by "muscle tremor." Whitening levels its spectrum down to some remnants of the ECG spectral needles. It alleviates one of the most challenging problems of BSPthe high variability of signal levels or amplitudes. However, the phase information is preserved even after whitening. Since the phase of all biosignals is more stable than the amplitude or level by one to two decades, a qualitatively almost ideal signal shape is achieved after MF from the electrophysiological point of view. This effect applies in principle to all biosignals, and only it is best observable in the ECG (Fig. 7.30).

Prewhitening in the Frequency Domain

For biosignals, where temporal dynamics are unimportant, one can rely on integral transformations into the frequency domain (or spatial domain), which implicitly allows one to approximate the essential requirement of whitening—stationarity.

Figure 7.31 shows the original spectrum of a verified ECG recording. In the analysis window alone, the level dynamic is over 50dB; up to the Nyquist frequency, it is 80 dB. With such fluctuations in the levels, statistical verification is not possible. Therefore, the spectrum was whitened according to the algorithm

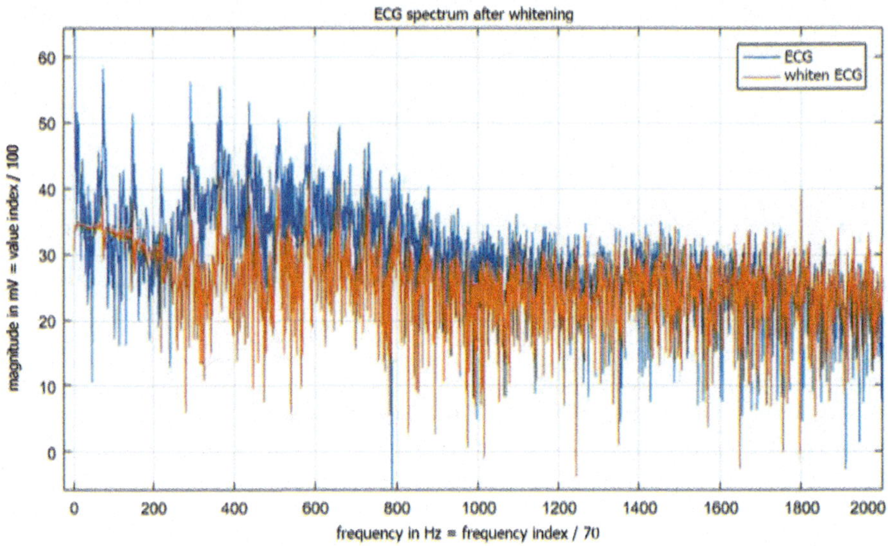

Fig. 7.30 Original spectrum of an ECG (blue) and spectrum after whitening (red) in the time domain (LPC/LMS). The whitened spectrum is almost constant outside the needles of the ECG fundamental frequency

described below, using two different estimators for the trend—the mean and the median. The result is shown in Fig. 6.91: It is easy to see that the mean underestimates the level at the fundamental low frequencies, and the breathing frequency has almost disappeared. It is due to the high-pass character of the moving average. The mean overestimate the spectral level at higher frequencies because of the relatively increasing density of extreme values. In contrast, the median shows a much more robust behavior: t low and high frequencies, it reliably estimates the spectrum.

The following algorithm can be used for whitening in the frequency domain (Fig. 7.32):

$$X[k] = DFT\{x[n]\} \tag{7.55}$$

The Fourier coefficients from Eq. 7.55 are decomposed into magnitude and phase, while the phase is buffered:

$$X_a[k] = |X[k]|,\ P[k] = \arctan(Im(X[k])/\mathrm{Re}(X[k])) \tag{7.56}$$

The amount is logarithmic for the algorithm because of the dynamics, as usual in DSP:

$$X_L[k] = \ln(X_a[k]) \tag{7.57}$$

The spectral background (trend) is estimated by mean or median:

$$X_{Tmean}[k] = \frac{1}{M} \sum_{m=-M2}^{M/2} X_L[k+m], \quad (7.58)$$

$$X_{Tmed}[k] = Q^{0.5}\{X_L[k-u], \ldots, X_L[k+o]\}. \quad (7.59)$$

In Eq. 7.58, M is the odd window length for calculating the mean, and in Eq. 6.135, u is the window's lower, o the upper index for calculating the median.

The trend is de-logarithmized:

$$X_T[k] = e^{X_{Tmean,med}[k]}. \quad (7.60)$$

The real spectrum is normalized to the trend:

$$X_D[k] = \frac{|X_L[k]|}{X_T[k]}. \quad (7.61)$$

The phase is restored:

$$X_W[k] = X_D[k] \cdot \exp(jP[k]). \quad (7.62)$$

The whitened spectrum is transferred into the time domain:

$$x_W[k] = IDFT(X_W[k]). \quad (7.63)$$

The choice of a suitable estimator for the position parameter of the trend in the analysis window—mean (Eq. 7.58) or median (Eq. 7.59)—depends on the properties of a concrete spectrum. For continuous spectra with low dynamics (< 40 dB), the mean is suitable as a signal-theoretically optimal estimator for normally distributed data. One should use the median for spectra with harmonic peaks (Fig. 7.31) because of its robustness against extreme values. However, the median is a stochastic nonlinear estimator, so it should not be used if the whitening is followed by a method explicitly based on measured values (e.g., spectral band power). In this case, one must find a suitable compromise between mean and median-trimmed arithmetic mean.

Prewhitening Using Statistics

In the case of multi-channel signals or repeated biosignal recordings, an ensemble of data, the measurement matrix X, is obtained. Before further analyses or signal processing, it makes sense to transform the matrix X so that individual realizations are linearly independent in pairs. In the terminology of SPP, non-correlation means whitening, i.e., all spectra must be whitened. Methodically, the simplest case would be the eigenvalue decomposition. However, the prerequisites for calculating the eigenvalues/eigenvectors—square matrix, positive values—cannot be

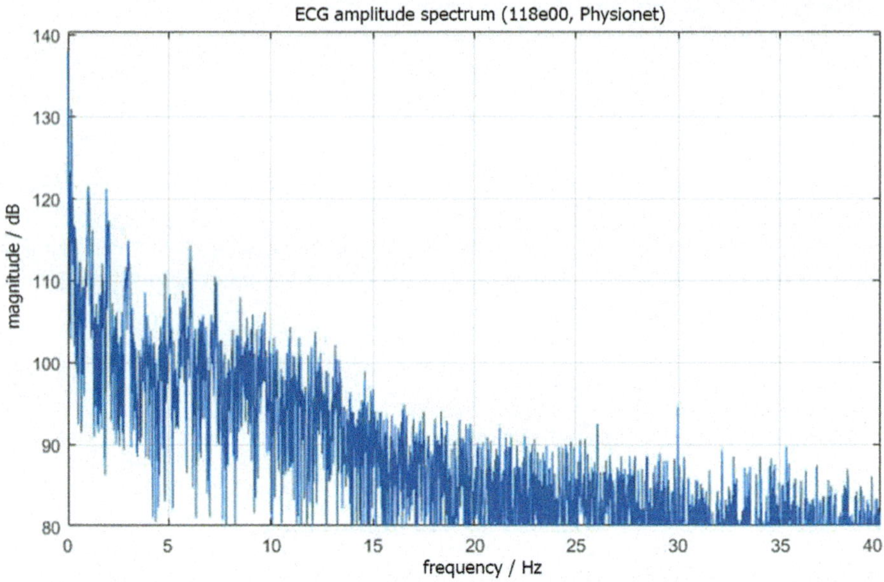

Fig. 7.31 Original spectrum of Physionet recording 118e00. Recognisable is the fundamental frequency of respiration at 0.17 Hz as well as harmonics of the fundamental frequency of the heart rate of 1 Hz

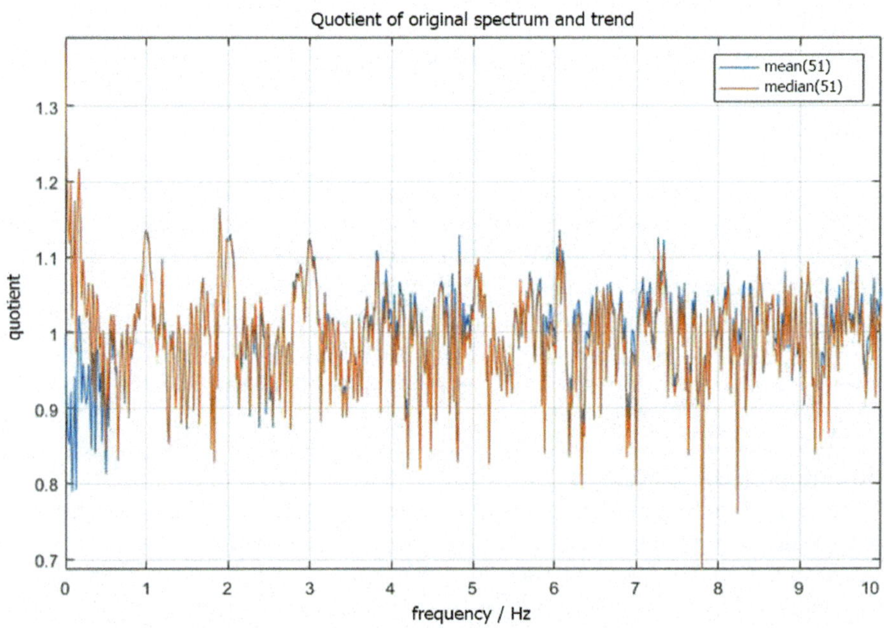

Fig. 7.32 Spectrum of the signal from Fig. 7.31 after whitening. We estimate the trend with mean values (51, blue) and median (51, red). The noise power is one due to the normalization as desired

fulfilled for accurate signals. Therefore, a mathematically and algorithmically similar method is used in practice, also known in the technical literature as Singular Value Decomposition (SVD). The basic signal model of a data ensemble consists of three matrices:

$$\mathbf{X} = \mathbf{U} \cdot \mathbf{S} \cdot \mathbf{V}^{\mathbf{T}}. \tag{7.64}$$

In Eq. 7.64, X is the matrix of the measured values (in columns time series, in rows realizations), U is the left-hand orthonormal matrix of the uncorrelated signal vectors (columns), S is the diagonal matrix of the singular values, V is the right-hand orthonormal matrix ("mixing matrix," weight matrix). Methodological details and applications are dealt with in the next chapter. Therefore, aspects on the topic of whitening or linear independence follow here:

- In principle, stationarity is required in the analysis of stochastic processes. If this requirement is met, the measures and decomposition are considered reliable. However, biosignals are always transient, so the computationally achieved uncorrelatedness of the columns of U does not yet mean the independence of the components.
- After reaching linear independence (white spectra), the components are ordered according to their variance (signal power), independent of desired or undesired signal components. It leads to the fact that there are uncorrelated components, but which desired or undesired components are present is unclear. It is one of the reasons why this group of methods, which ranks the components solely according to their variance/power, is called BSS (Blind Signal Separation).
- The order of the components of U is based on their signal power according to the matrix S. Since this decomposition can also be used to reduce the dimension of the data, it is based on which component has reached sufficient energy from the original signal. Often, for example, 90–95% of the accumulated energy is considered sufficient, so one can omit the remaining components and thus achieve a sometimes considerable reduction in the data. However, decorrelation also leads to the fact that the often very weak biosignals, which are diagnostically important, end up in the energetically rejected part of the decomposition and are entirely omitted from the analysis due to the data reduction. Therefore, SVD should be used solely to secure white spectra to fulfill the subsequent method's requirements.

7.3 Signal Decomposition

7.3.1 Singular Value Decomposition, SVD

In the case of multi-channel or repetitive recordings, one obtains a data ensemble of which one can assume that each realization contains the desired signal and undesired components (Fig. 7.33). (Stewart, 1993).

The following multi-channel EEG recording was simulated to illustrate the mode of action of the SVD:

$$x_i[n] = (i - 1)ep[n] + (16 - i)\sin(2\pi 50n/N) + 5\mathrm{N}(0, 1)[n] \qquad (7.65)$$

In Eq. 7.65, i is the channel index at a total of M channels (column number of matrix X), *ep is* a synthetic evoked potential (Fig. 7.25), N is the length of a realization (row number of matrix X), normal distribution of random data. The channel signals are shown in Fig. 7.33: While the network noise decreases with increasing channel number, the amplitude of the evoked potential increases, the noise power is constant in all channels, and the noise is not correlated in time or space.

With the help of second-order statistics, one can generate orthogonal (decorrelated) components, which are ordered according to their variance (signal power). Such a decomposition can be achieved with the help of PCA, whose basic approach is shown in Fig. 7.34.

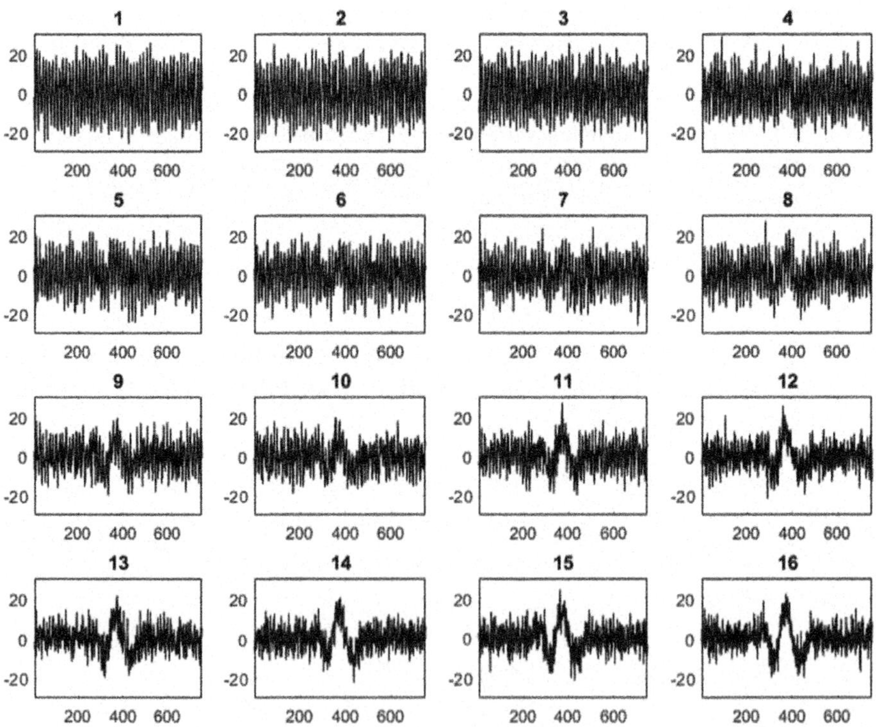

Fig. 7.33 Simulated 16-channel EEG recording with network interference, evoked potential and white noise. The network disturbance decreases with the number of channels, and the amplitude of the evoked potential increases

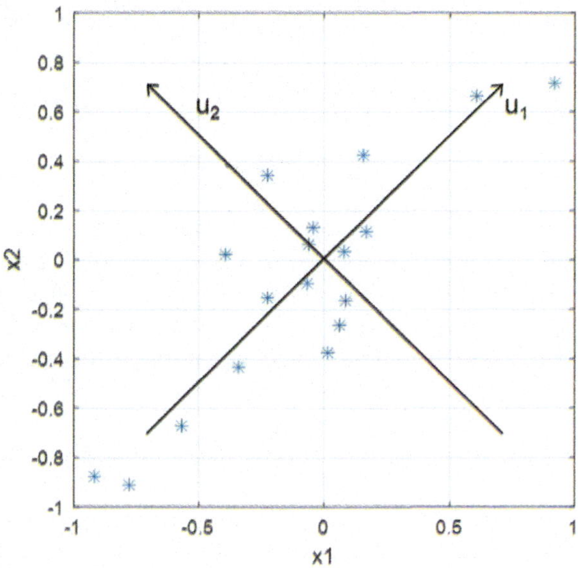

Fig. 7.34 Original data refer to the axes (random variables) x_1 and x_2. After decorrelation, new orthogonal coordinates u_1 and u_2 are created, whose coordinate variance is ordered by size (algorithm SVD)

The decorrelation procedure can be mathematically applied to the EVD (eigenvalue decomposition). Without limiting the generality, one can initially start with two random variables and formulate them as a time series:

$$\mathbf{x} = \begin{bmatrix} x_1 \\ x_2 \end{bmatrix}, \quad \mathbf{X} = [\mathbf{x}(t_1), \dots, \mathbf{x}(t_N)]. \tag{7.66}$$

For the EVD, it is necessary to create the covariance matrix and decompose it:

$$\mathbf{C_{xx}} = E\left\{\mathbf{x}\mathbf{x}^T\right\} \approx \frac{1}{N}\mathbf{X}\mathbf{X}^T = \mathbf{U} \cdot \mathbf{\Sigma} \cdot \mathbf{U}^T, \ E\{\mathbf{x}\} = 0. \tag{7.67}$$

The decomposition is performed according to the eigenvalue equation (**R** is a square, symmetric, real matrix) using the Karhunen-Loève transform (Manolakis & Ingle, 2011):

$$\mathbf{R} \cdot \mathbf{u}_i = \sigma_i \cdot \mathbf{u}_i. \tag{7.68}$$

$$\mathbf{R} = \mathbf{U} \cdot \mathbf{\Sigma} \cdot \mathbf{U}^T \tag{7.69}$$

The matrix **U** is the matrix of the eigenvectors of **R**; it is unitary (orthonormal) and contains a unitary basis for the column space of **R**:

$$\mathbf{U} = [\mathbf{u}_1, \dots, \mathbf{u}_d] \in \mathbb{R}^{d \times d},$$

where d is the number of sensors, **S** is the diagonal matrix of eigenvalues:

$$\Sigma = \begin{pmatrix} \sigma_1 & \cdots & 0 \\ \vdots & \ddots & \vdots \\ 0 & \cdots & \sigma_d \end{pmatrix} \in \mathbb{R}^{d \times d}.$$

In the case of biosignals, one cannot expect a square and symmetrical matrix for the orthogonal decomposition according to Eq. 7.69; quite the contrary: By default, relatively long data matrices are present, i.e., the number of rows (length of the signal) is significantly larger than the number of columns (number of channels or sensors). Therefore, the decomposition was adapted to the matrix dimensions accordingly, and the EVD was converted to SVD (Singular Value Decomposition):

$$\mathbf{R} \in \mathbb{R}^{\mathbf{p} \times \mathbf{d}}, \ \mathbf{R} = \mathbf{U} \cdot \mathbf{S} \cdot \mathbf{V}^T. \tag{7.70}$$

In Eq. 7.70, R is a real data matrix with p rows (samples, time series) and d columns (channels, realizations), where must hold: $p \geq d$.

- **U** is the left-sided unitary matrix that contains the orthogonal components of the data matrix **R** in columns:

$$\mathbf{U} = [\mathbf{u}_1, \ldots, \mathbf{u}_d] \in \mathbb{R}^{p \times p}.$$

It can mathematically reach a large amount of data corresponding to the signal length of p, which would be redundant and unfavorable for signal processing. Therefore, in typical cases, the dimension is reduced to the necessary and sufficient size of $\mathbb{R}^{p \times d}$.

- **V** is the right-sided unitary matrix, which contains the influence of one component on all channels in columns and the influence of all components on one channel in rows:

$$\mathbf{V} = [\mathbf{v}_1, \ldots, \mathbf{v}_d] \in \mathbb{R}^{d \times d}.$$

- **S** is the diagonal matrix of the singular values resulting from the roots of the eigenvalues of **R**. $\mathbf{R}^\mathrm{T}$:

If the SVD is applied to the ensemble in Fig. 7.33, the following numerical solutions for the matrices **U, S,** and **V** *are obtained:*

$$S = \begin{pmatrix} s_1 & \cdots & 0 \\ \vdots & \ddots & \vdots \\ 0 & \cdots & s_{\min(d,p)} \\ 0 & \cdots & 0 \\ \vdots & \ddots & \vdots \\ 0 & \cdots & 0 \end{pmatrix} \in \mathbb{R}^{p \times d}$$

- Left-sided matrix **U**[N, M], orthonormal, contains orthogonal components of **X** in columns (Fig. 7.35).

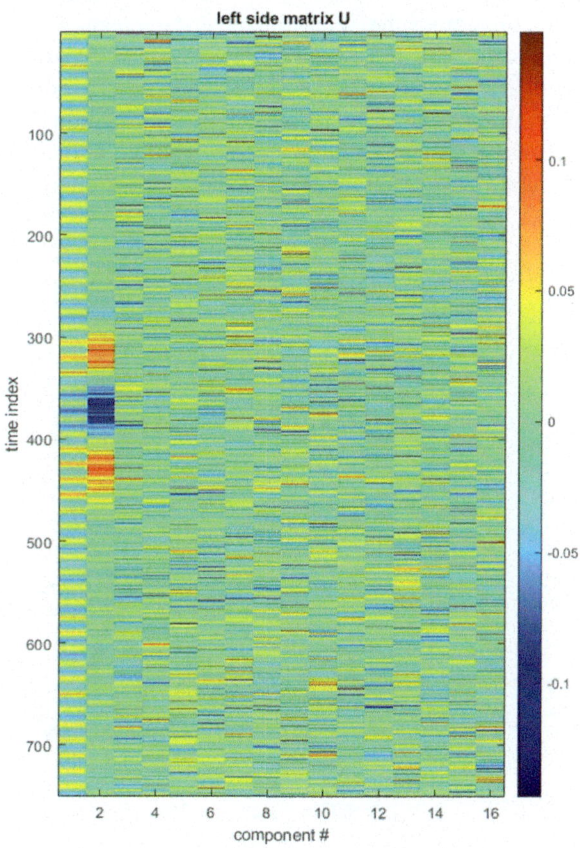

Fig. 7.35 Left-sided matrix **U** after SVD decomposition of the data from Fig. 7.33. The components in column vectors are arranged from left to right in descending order of their signal power

In the graphical representation of the decomposition result in Fig. 7.35, one can already visually recognize important qualitative features: The first component (column) contains a periodic signal, the network disturbance. The second component contains a noisy deterministic signal, the evoked potential. From the third component, one can subjectively recognize neither periodicity nor determinism; it is noise. Therefore, we examine the first two components (Fig. 7.36). The first component clearly shows the periodic disturbance from the network. This example relativizes the assumption often made in signal processing that the most potent component is always the signal we seek. The sought-after signal can be identified only in the second component since its shape is known in advance. Features that should not be present after an orthogonal decomposition can be found in both components. In the first component, residues of the evoked potential (the mean value of the periodic disturbance follows the signal shape of the evoked potential) are recognizable. This effect is a consequence of violating the stationarity assumption—the evoked potential is not stationary. A relatively strong noise is present in the second component, although all noise signals are uncorrelated with each other and concerning the deterministic components. This effect arises as a result of the underdetermination of the linear system of equations according to Eq. 6.140: The simulated system has 18 signal sources (16 noise sources and two deterministic sources), but only 16 measurement equations, the number of sensors is lower than the number of source signals. Therefore, the decomposition components must inevitably also contain extraneous components.

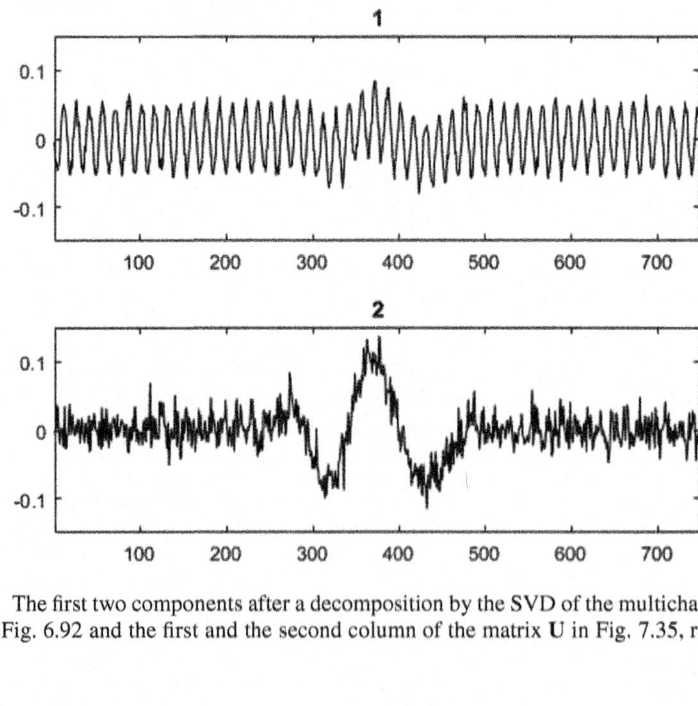

Fig. 7.36 The first two components after a decomposition by the SVD of the multichannel derivative from Fig. 6.92 and the first and the second column of the matrix **U** in Fig. 7.35, respectively

- Diagonal matrix **S** provides information about the components' signal powers (signal energy in the analysis window) (Fig. 7.37). Compared with matrix **U**, it is easy to see that the strongest component is the periodic network disturbance. Only second place follows the sought-after signal component with less than half the signal power. From the third component, one can hypothetically assume pure noise. This result also shows the fundamental problem of the variance-based relevance analysis of the BSS methods: Often, the signal energy alone is used to decide from which component onward all other components are to be irrelevant. It means that if the desired potential level should drop by 40% in this simulation, it falls into the noise range and is routinely excluded from further processing in the sense of data reduction (reduction of the matrix dimension). It can have fatal consequences for the recognition and detection of weak biosignals, also and especially in the EEG. Therefore, SVD should only be used for decorrelation or whitening.
- Right-hand matrix **V** is shown graphically in Fig. 7.38. One can observe the effect of the first two components on the channels: The weight decreases in the first component with the number of channels according to the signal model and increases in the second component. All other weights are randomly distributed according to the noise.

It can be stated that the decomposition of data ensembles (multi-channel recordings, repetitive single-channel recordings) in terms of linear independence using

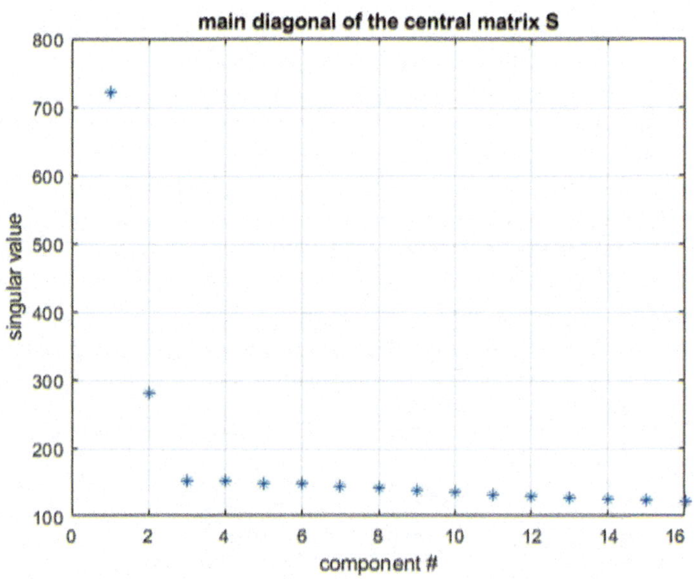

Fig. 7.37 Main diagonal of the singular value matrix **S**. Note that all channels contain the same noise but can only be identified using the level from the third channel onwards

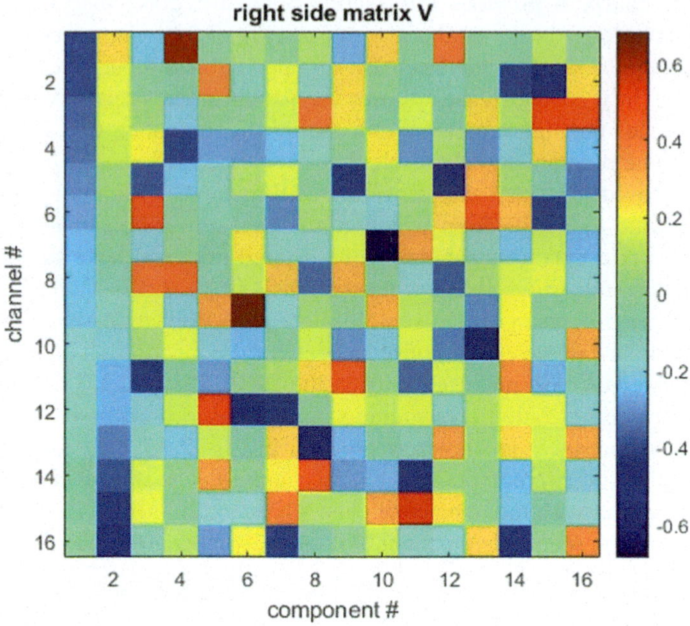

Fig. 7.38 Right-hand side matrix **V** indicating the weight with which the respective component is added to a channel ("mixing matrix")

second-order statistics works correctly if the conditions of stationarity and the exclusively linear correlation, as well as the normal distribution of the data, are fulfilled. The conditions mentioned above in BSP are never fulfilled, so the waveform distortions occur in the practical analysis (Fig. 6.94). The interpretation of the decomposition results, whose signal form mainly contains the matrix **U**, is generally difficult since one does not know before the decomposition and cannot predict in which column vectors the desired and the undesired signal components will appear. It is a fundamental problem of BSS methods. In addition, the decomposition into linearly independent components can also result in entirely new or previously unknown signal characteristics that cannot be assigned to any original biosignal. Therefore, a methodological alternative is needed that accepts both the instationarity and the non-normal distribution and considers non-linear correlations. For this purpose, ICA (Independent Component Analysis) methods have been developed.

7.3.2 Independent Component Analysis

Currently, up to 30 algorithms of ICA are known, which are methodologically based on the fundamental work of Hyvärinen et al. (2001). Unlike previous methods, this methodology claims to find components through the decomposition that are both statistically independent (not only uncorrelated) and non-normally distributed according to reality. It was one of the first approaches to accept the natural signal properties in many areas and develop a new methodology. One of the essential approaches states that "non-Gaussian distributed means independent." At this point, a methodological clarification is necessary: Although the authors use higher order statistics (HOS) to separate the components to achieve independence using maximized non-Gaussianity, these procedures and algorithms result in the non-correlation of non-Gaussian distributed components. As a result, one can still find decorrelation between components after the decomposition, although HOS was used for the separation. This observation follows from the signal model based on addition or linear combination (Fig. 6.97).

7.3.2.1 Signal Model and Assumptions

ICA methods are also based on a simple linear signal model suitable for further processing using linear algebra (Fig. 7.39).

It is realistically assumed that the source signals x_i (t) are not normally distributed. It is a significant difference from previous methodological assumptions. The source signals are weighted with the constants a_{ij} and appear at the sensors, summed up to channel signals y_m (t). One can formulate this model using matrix notation:

$$\mathbf{Y} = \mathbf{A} \cdot \mathbf{X} \tag{7.71}$$

In Eq. 7.71 $\mathbf{Y} \in \mathbb{R}^{p \times N}$ is the matrix of sensor signals, $\mathbf{X} \in \mathbb{R}^{d \times N}$ is the matrix of source signals, $\mathbf{A} \in \mathbb{R}^{p \times d}$ is the mixing matrix, where $p \geq d$, i.e., the number of sensors is at least as large as the number of source signals.

Note: The weights a_{ij} (elements of the mixing matrix **A**) are assumed to be constant in time in order to be able to assume the summed channel signals y_m (t) to

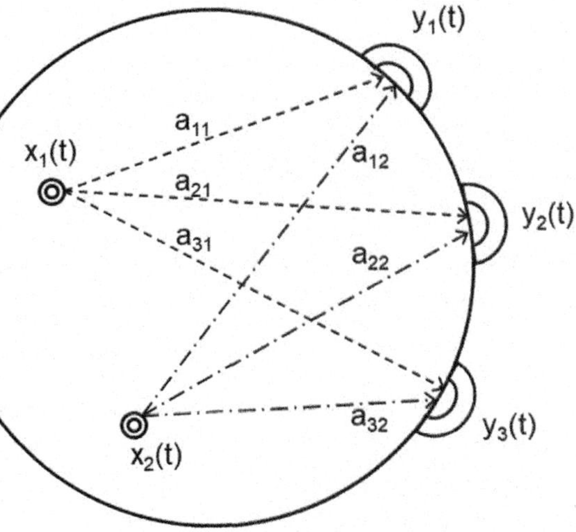

Fig. 7.39 Model for source signals x_i (t) and ym (t) sensors. The transmissions a_{ij} are constant in time

be normally distributed using the ZGWS. This assumption is necessary for further steps but is hardly fulfilled in reality. The actual properties of the mixing matrix **A** are much more complicated: One must assume a non-linear dynamic linkage. Indirectly, the violation of the linearity assumption can be proven by the fact that the channel signals $y_m(t)$ are neither stationary nor normally distributed in practice. For further method development, the linearity assumption is nevertheless accepted.

If the number of channels (second dimension of the matrix **X**) is sufficiently large (> 20), one can assume the effectiveness of the CLT; the channel signals in **Y** are approximately normally distributed. Therefore, one can assume that if one algorithmically trims **X** to maximum non-Gaussianity, one achieves the best separation of the source signals. There are other approaches to separating the components of **X** (minimization of the transinformation, maximization of the differential entropy, tensor decomposition); here, for methodological reasons, the maximization of the non-Gaussianity is dealt with (Hyvärinen, 2001).

If one analyses the decomposition task, one finds that according to Eq. 7.71, the signal model is underdetermined depending on the matrix dimension. In reality, neither the mixing matrix **A** nor the source signal matrix **X** is known. It follows that there are an infinite number of solutions for this. Therefore, the number of degrees of freedom must be reduced appropriately because

- the amplitudes (variances) of the source signals cannot be reconstructed,
- the sequence of the source signals cannot be reconstructed.

One can therefore estimate in advance the expected problems of this BSS method after the decomposition, some of which were already apparent with the PCA/SVD:

- The ICA components are—as with PCA—uncorrelated, in the case of ICA, even independent in the sense of fourth-order statistics. After the decomposition, it is unknown where desired and undesired signal components are located.
- As with PCA, ICA will produce components that have no apparent relation to original biosignals or have new signal shapes.
- In contrast to PCA, there is no order of the components in ICA, nor is the information about the signal power available. It will result in obtaining distinctive waveforms or signal characteristics but no idea about their level, which can have a difference of 100 dB and more.

Fourth-order statistics (kurtosis) based decomposition (FastICA) are applied, and the following assumptions are made regarding the signal properties:

- The source signals **X** are statistically independent (fulfilled in real terms).
- The source signals **X** are centered, i.e., $E\{\mathbf{x}\} = 0$ (well fulfilled in real terms, also by precise centering). The centering is problematic with biosignals where the DC component is essential for signal analysis, e.g., with the plethysmograph.

- Because of the underdetermination of the system of equations, the variances of the source signals must be normalized to one, i.e., $E\{\mathbf{X}\mathbf{X}^T\} = \mathbf{I}$.
- Source signals $\mathbf{X}$ are not normally distributed.
- The mixing matrix $\mathbf{A}$ is quadratic, i.e., the number of sensors p equals the number of source signals d.

7.3.2.2 ICA Through Maximization of Non-Gaussianity

The algorithmic steps (FastICA) are demonstrated using the simulated EEG/EP recording example. Simulated recordings from Fig. 6.92 were used as measurement data; a uniformly distributed noise replaced the original Gaussian-distributed noise. This swap is necessary because the ICA is based on non-Gaussian distributions of the source data.

The following steps are necessary to achieve ICA by maximizing non-Gaussianity (Hyvärinen, 2001):

Considering the data as stochastic processes, Eq. 7.71 can be simplified:

$$\mathbf{y} = \mathbf{A}\mathbf{x}. \tag{7.72}$$

The task of the ICA is to determine the source signals $\mathbf{x}$. Therefore, Eq. 7.72 is rearranged:

$$\mathbf{x} = \mathbf{A}^{-1}\mathbf{y}. \tag{7.73}$$

The chapter "Prewhitening using statistics" explained how a signal could be whitened to be decorrelated (linearly independent) afterward. It can be shown that after whitening, $\mathbf{z}$ applies to a new whitened vector:

$$\mathbf{z} = \mathbf{V}^{\mathrm{T}}\mathbf{x}. \tag{7.74}$$

It transfers the original ICA problem to a problem with an orthonormal mixed matrix: We search for an orthonormal matrix $\mathbf{W}^{\mathrm{T}} = \mathbf{V}$ such that holds:

$$\mathbf{x} = \mathbf{W}^{\mathrm{T}}\mathbf{z}. \tag{7.75}$$

$\mathbf{W}^{\mathrm{T}}$ is a matrix with orthonormal columns:

$$\mathbf{W} = [\mathbf{w}_1, \ldots \mathbf{w}_\mathbf{d}] \in \mathbb{R}^{\mathbf{p}\times\mathbf{d}}. \tag{7.76}$$

As a measure for the non-Gaussianity of the *i-th* component $\mathbf{x}_i = \mathbf{w}_i^{\mathrm{T}}\mathbf{z}$ the kurtosis is used:

$$|Kurt(x_i)| = \left|E\left\{\left(\mathbf{w}_i^{\mathrm{T}}\mathbf{z}\right)^4\right\} - 3\left(\mathbf{w}_i^{\mathrm{T}}\mathbf{w}_i\right)^2\right|. \tag{7.77}$$

For the maximization of the kurtosis using the Lagrange operator and the constraint $\mathbf{w}_i^2 = 1$ holds:

$$\pm \frac{\partial Kurt\left(\mathbf{w}_\mathbf{i}^\mathrm{T}\mathbf{z}\right)}{\partial \mathbf{w}_\mathbf{i}} = 2\lambda \mathbf{w}_i. \tag{7.78}$$

With the help of fixed point iteration, each component can find its local maximum:

$$\mathbf{w}_\mathbf{i}^{(\mathbf{k}+1)} = E\left\{\mathbf{z}\left(\mathbf{w}_\mathbf{i}^{(\mathbf{k})\mathbf{T}}\mathbf{z}\right)^3\right\} - 3\mathbf{w}_\mathbf{i}^{(\mathbf{k})}, \tag{7.79}$$

$$\mathbf{w}_\mathbf{i}^{(\mathbf{k}+1)} \leftarrow \mathbf{w}_\mathbf{i}^{(\mathbf{k}+1)} / \left\|\mathbf{w}_\mathbf{i}^{(\mathbf{k}+1)}\right\|.$$

Since the algorithm, according to Eq. 7.79, does not provide orthogonal result vectors, additional orthogonalization must be carried out after each or after a block of iteration steps, e.g., with the Gram-Schmidt method:

$$\mathbf{w}_\mathbf{i}^\mathbf{k} \leftarrow \mathbf{w}_\mathbf{i}^{(\mathbf{k})} - \mathbf{W}_{\mathbf{i}-1}\mathbf{W}_{\mathbf{i}-1}^\mathrm{T}\mathbf{w}_\mathbf{i}^{(\mathbf{k})}, \tag{7.80}$$

$$\text{with } \mathbf{W}_{\mathbf{i}-1}^\mathrm{T} = \left[\mathbf{w}_1, \mathbf{w}_2, \ldots, \mathbf{w}_{i-1}\right] \text{ and } \mathbf{W}_{\mathbf{i}-1}^\mathbf{T}\mathbf{W}_{\mathbf{i}-1} = \mathbf{I}$$

The ICs after applying FastICA to the modified data from Fig. 7.33 (equally distributed instead of Gaussian distributed noise) are shown in Fig. 7.40. Before the decomposition, it is unclear which column vectors the respective components will appear in. Afterward, there is no information about their order (the constraint with the Lagrange operator is a unit circle); one can judge the components solely by their shape. The interpretation becomes even more difficult if components appear that have no direct relation to the analyzed signals. The two deterministic components, #1 and #8 correspond to the simulated signals EP and net disturbance. No conclusion about the original levels can be drawn from their normalized variance. Both are still affected by the equally distributed noise since the equations system is underdetermined here (18 source signals, 16 channels). An improvement could theoretically be achieved by setting the number of sensors to 18 before the decomposition and applying a suitable algorithm ($p > d$). In practical analysis, however, one does not know how large the source signals are; this can only be estimated after the decomposition (Fig. 7.41).

According to a secondary condition of the FastICA decomposition, the distributions of the source signals must be equal and non-Gaussian. In Fig. 7.42, this condition is very well fulfilled except for the extreme value of the first component (EP). With real signals, however, one cannot check compliance with this condition. At best, one can rely on the fact that if the sensor signals originate from a homogeneous electrically active structure (cortical, cardiac, motor, sensory sources, amplifier noise, technical or biological artifacts), their data distribution is equal. Against this background, one of the fundamental problems of ICA becomes

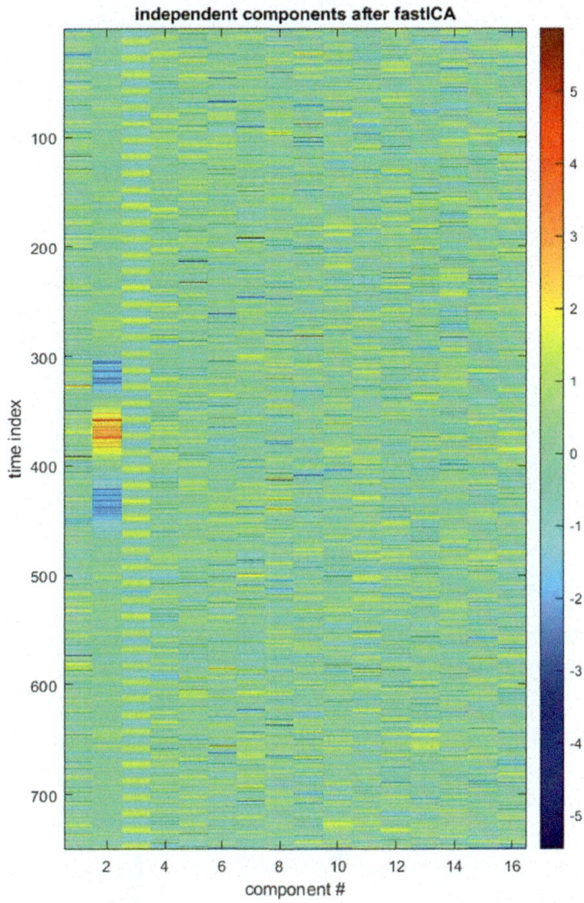

Fig. 7.40 ICs of the data from Fig. 7.33, where the noise is not Gaussian distributed but uniformly distributed. Deterministic components are recognizable in channel 2 (desired EP) and channel 3 (unwanted interference from the main)

clear: extreme values or strong instationarities (component #1 in Fig. 7.43) enter the statistically based decomposition with their mean value in the analysis window; they are statistically "smoothed," so to speak. As a result, the effect of stationary processes (noise, periodic disturbance) is relatively strengthened or raised by the normalization to the unit circle in the Lagrange operator (constraint). However, this is the opposite of what we tried to do in the chapter "Improving the SNR": To improve the already poor SNR of the biosignals. For the practical analysis, it follows that one should (also) ensure a sufficiently large SNR (> + 20 to + 40 dB) before the ICA.

7.3.3 Higher Order Singular Value Decomposition, HOSVD

Although multi-channel biosignals are also primarily one-dimensional (temporal signal course), it became clear at the latest with the introduction of TFA that even

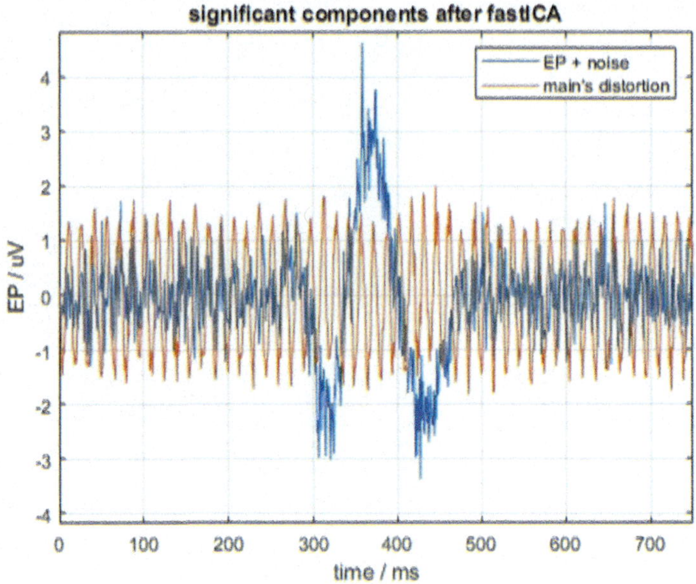

Fig. 7.41 Components #2 (EP) and #3 (network disturbance) from Fig. 7.40. The variance (average signal power) is normalized to one. Conclusions about the original signal are impossible

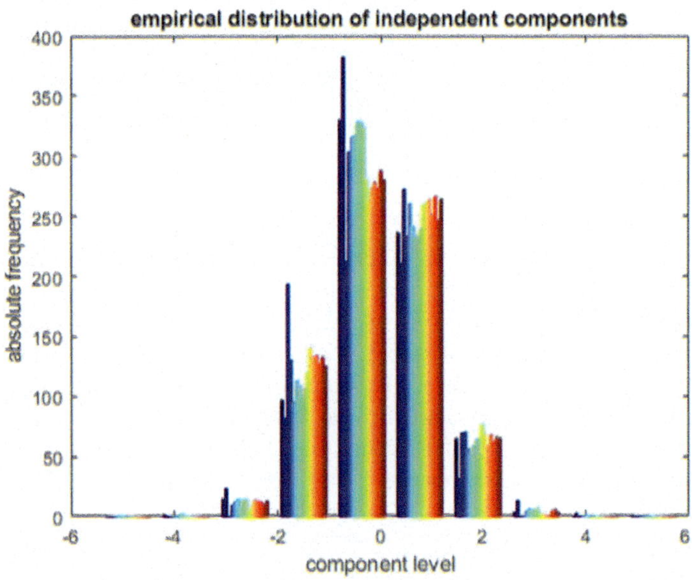

Fig. 7.42 Empirical distribution of the ICs from Fig. 7.40. Except for the extreme value from components #2 (deterministic EP) and #3 (periodical disturbance), the distributions are approximately Gaussian (each color represents one IC). Only source signals, components #2 and #3, are non-Gaussian distributed according to the basic idea of ICA

in linear signal processing, several dimensions are necessary, most frequently time, frequency, and space (sensor position). The temporal data are often accompanied by static data (patient data, examination, measurement conditions), so the number of data sets' dimensions can quickly reach over a decade. For an in-depth (temporal) analysis, the methods used so far (MANOVA) are not sufficient. Multidimensional (multilinear) systems and algorithms offer a theoretically demanding but plausible alternative to solve multidimensional analysis problems efficiently. At this point, the basics are explained (Weis, 2015), and their application is demonstrated by example.

The HOSVD of a tensor $\mathcal{X} \in \mathbb{C}^{I_1 \times \cdots \times I_N}$ of order is defined as follows:

$$\boldsymbol{\mathcal{X}} = \boldsymbol{\mathcal{S}} \times_1 \boldsymbol{U}^{(1)} \times_2 \boldsymbol{U}^{(2)} \ldots \times_N \boldsymbol{U}^{(N)}, \tag{7.81}$$

where $\mathcal{S} \in \mathbb{C}^{I_1 \times \cdots \times I_N}$ is the core tensor of order N. $\boldsymbol{U}^{(n)}$ are matrices of the n-mode singular vectors of $\mathcal{X}$ for n = 1, ..., N. The matrices $\boldsymbol{U}^{(n)} \in \mathbb{C}^{I_n \times I_n}$ are obtained after the SVD of the n-mode deconvolution of $\mathcal{X}$

$$[\boldsymbol{\mathcal{X}}]_{(n)} = \boldsymbol{U}^{(n)} \cdot \boldsymbol{\Sigma}^{(n)} \cdot \boldsymbol{V}^{(n)^{H}}, \tag{7.82}$$

so that they form the unitary basis for the n-mode vector space for $\mathcal{X}$ form. The singular values of the n-mode involutions $[\mathcal{X}]_{(n)}$ yield the n-mode singular values of $\mathcal{X}$ and are denoted by $\sigma_i^{(n)}$ with $i = 1, \ldots, I_n$. After calculating the matrices of the n-mode singular vectors according to Eq. 7.82, the core tensor can be $\mathcal{S}$ can be calculated from Eq. 7.81:

$$\boldsymbol{\mathcal{S}} = \boldsymbol{\mathcal{X}} \times_1 \boldsymbol{U}^{(1)^{H}} \times_2 \boldsymbol{U}^{(2)^{H}} \ldots \times_N \boldsymbol{U}^{(N)^{H}}. \tag{7.83}$$

One does not have to formulate Eq. 7.81 using tensors, although it represents the formally simplest variant. Using the Kronecker product, one can use the conventional notation with matrices, which are linked with each other, or use the most elaborate variant of the scalar decomposition (matrix elements). The tensor formulation, according to Eq. 7.83, is also the most effective algorithmically, as the following algorithm shows:

- for n = 1, 2, ..., N
  - calculate the n-mode unfolding $[\mathcal{X}]_{(n)}$ from $\mathcal{X}$,
  - calculate the SVD: $[\mathcal{X}]_{(n)} = \boldsymbol{U}^{(n)} \cdot \boldsymbol{\Sigma}^{(n)} \cdot \boldsymbol{V}^{(n)^{H}}$,
  - calculate the n-rank of R_n as the number of non-zero elements in $\Sigma^{(n)}$,

  end
- $\boldsymbol{\mathcal{S}} = \boldsymbol{\mathcal{X}} \times_1 \boldsymbol{U}^{(1)^{H}} \times_2 \boldsymbol{U}^{(2)^{H}} \ldots \times_N \boldsymbol{U}^{(N)^{H}}$,
- $\Sigma^{(n)} = diag\,(1^{(n)}, 2^{(n)}, \ldots \mathrm{In}^{(n)})$, n = 1, ..., N.

7.3.3.1 Matrices and Tensors

The data model, which is generally assumed for multi-channel biosignals, is always linear additive in this publication: According to Eq. 7.67, one assumes a linear combination of the source signals $\mathbf{x}$ via the mixing matrix $\mathbf{A}$, whereby for real recordings, an additional AWGN (Additive White Gaussian Noise) $\mathbf{n}$ **is** assumed:

$$\mathbf{y} = \mathbf{A}\mathbf{x} + \mathbf{n}. \tag{7.84}$$

For the linear analysis, the formulation according to Eq. 6.154 is entirely sufficient and, as before, can also be adequately described with matrix algebra. As mentioned above, the complexity of the matrix notation already increases strongly from the third order (e.g., time, frequency, space); at an order of 4, it becomes very confusing and challenging. Therefore, it makes sense to introduce the tensor already from the third order. The tensor concept has existed mathematically since the nineteenth century; M. Grossmann and A. Einstein applied a plausible physical interpretation in 1915 (Renn & Sauer, 1996). For biosignal processing, a tensor can be pragmatically formulated as follows:

> A tensor of order p is a collection of elements (scalars) referenced by p indices.

The difference between matrices and tensors is that although they start from the same data model, tensors offer a much more compact representation of multidimensional data. The following examples demonstrate this compactness (De Lauthauwer et al., 2000).

The compactness of the matrix notation compared to the element-wise formula is already shown by the example of the DFT in Eq. 4.38. For a 3-D data field, we use the tensor notation:

$$\mathcal{X} = \mathbb{C}^{M_1 \times M_2 \times M_3}. \tag{7.85}$$

Like conventional 2-D matrices, the tensor can be represented using modes of order i, see Fig. 7.43. According to this representation, the modes 1, …, n can be decomposed into modal vectors. For the SVD, a two-dimensional matrix representation is needed. Therefore, it is necessary to unfold the respective mode-i vector representation into a two-dimensional matrix. It is shown schematically in Fig. 7.44.

7.3.3.2 Properties of HOSVD

For conventional 2D matrices, according to the SVD:

$$\boldsymbol{X} = \boldsymbol{U} \cdot \boldsymbol{S} \cdot \boldsymbol{V}^{\mathrm{H}} = \boldsymbol{S} \times_1 \boldsymbol{U} \times_2 \boldsymbol{V}^*, \tag{7.86}$$

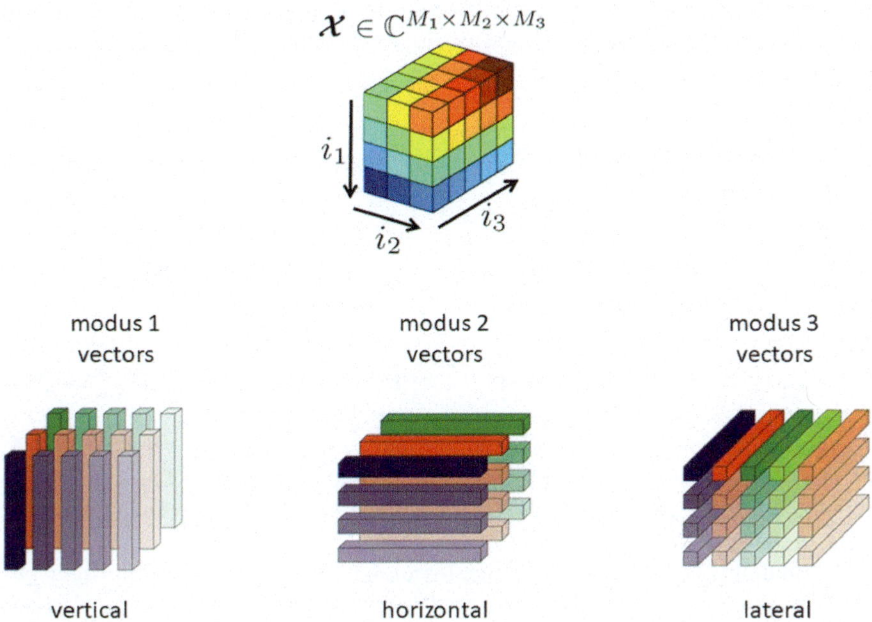

Fig. 7.43 A 3-D data matrix as a 3-D tensor and representation using mode-n vectors (Weis, 2015)

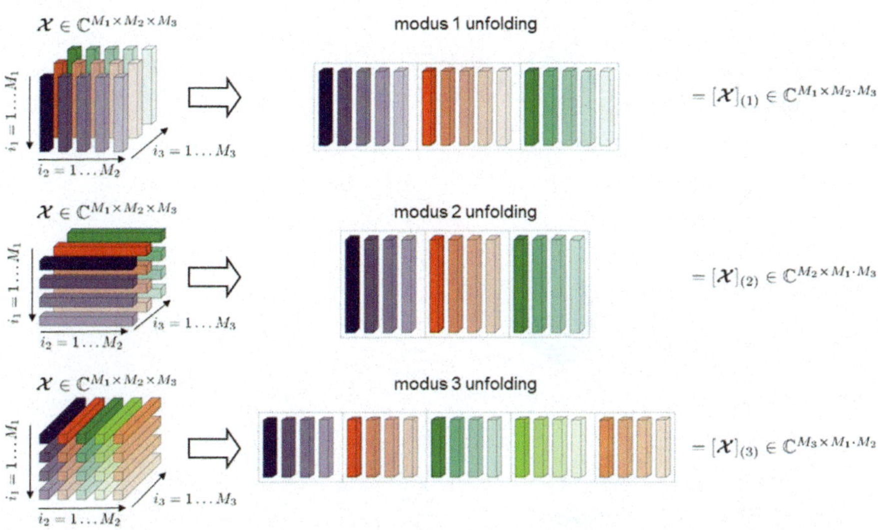

Fig. 7.44 Two-dimensional unfolding of the mode-n vectors from Fig. 6.102 (Weis, 2015)

U is the unitary basis for the column space of X, V^* for the row space. For the more general formulation using tensors, Eq. 7.81 applies, from which the following properties result

$$\begin{aligned}[\mathcal{X}]_{(1)} &= U^{(1)} \cdot S^{(1)} V^{(1)^H} \\ [\mathcal{X}]_{(2)} &= U^{(2)} \cdot S^{(2)} V^{(2)^H}, \\ [\mathcal{X}]_{(3)} &= U^{(3)} \cdot S^{(3)} V^{(3)^H}\end{aligned} \tag{7.87}$$

The matrices $U^{(n)}$ were calculated from the SVD of the unfolded mode-n vector representations. Therefore

$$\mathcal{S} = \mathcal{X} \times_1 U^{(1)^H} \times_2 U^{(2)^H} \times_3 U^{(3)^H}. \tag{7.88}$$

It follows from Eq. 7.87 that S contains three sets of singular values (Fig. 7.45).

The properties of the HOSVD will be demonstrated using the example of a simulated SSVEP (Steady-State VEP), and the analysis of an actual signal will be done in the section "Exercises."

Simulation of a 16-channel parietal-occipital image after periodic light pulse stimulation with 8 pps (pulse-per-second) containing the following signal components:

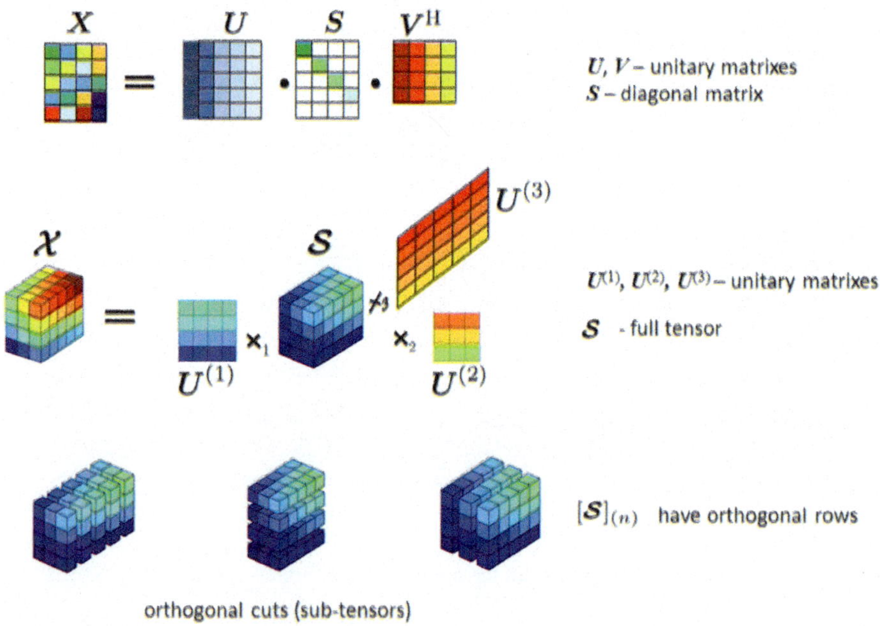

Fig. 7.45 Conventional SVD decomposition of 2D matrices (top), tensor decomposition of a 3D matrix with full tensor S (middle), orthogonal sections of the full tensor S (bottom, full orthogonality) (Haardt et al., 2008; Weis, 2015)

- Harmonics of the stimulation rate of 8 pps at 8, 16, 24, 32, 40 Hz, constant amplitude.
- Harmonics of the mains frequency at 50 Hz and 100 Hz, constant amplitude.
- Transient VEP half a second after the start of periodic stimulation, the amplitude increases with channel number.
- Weak AWGN (additive white Gaussian distributed noise).

Figure 7.47 shows the simulated signal mixture's time–frequency distribution (SPWD) (channel 8) (Fig. 7.46).

After the HOSVD, one can first examine the singular values: The SD in mode 1, in contrast to the other two, shows only two relevant source signals. Since one has simulated the multi-channel signal, one knows that eight deterministic signals are embedded in the noise. Therefore, it is impossible to interpret the singular values in connection with the source signals or to assign them. The SD in Mode-2 and Modu-3 do not contribute to the correct interpretation. In these two, the signal power falls off relatively smoothly, so determining the number of source signals is impossible. The analysis of the singular values concludes that one can distinguish between two to about four uncorrelated source signals, making a direct assignment impossible.

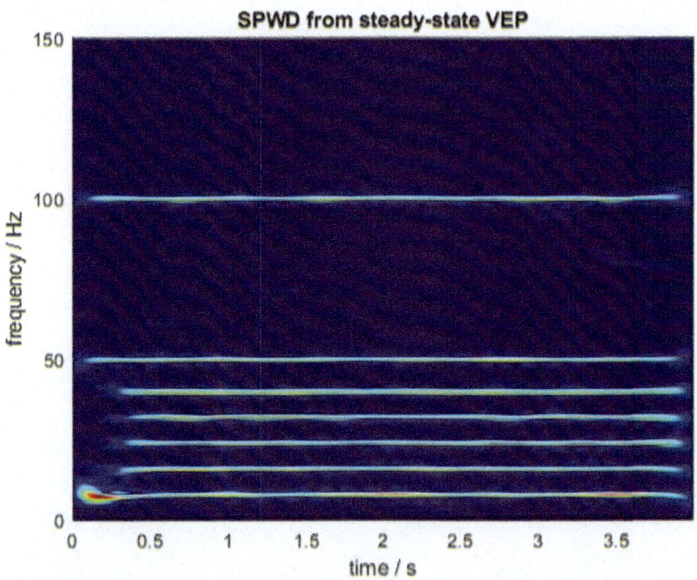

Fig. 7.46 Simulated SSVEP after stimulation with 8 pps with five harmonics of the stimulation rate (8, 16, 24, 32, 40 Hz), the transient component after the start of stimulation (low-frequency energy island in the time range 0–0.5 s), harmonics of the mains frequency at 50 and 100 Hz and a weak AWGN. Edge effects at the beginning and end of the analysis window arise from the relatively long orthogonal windows in frequency direction (1024) and time direction (128) when applying the SPWD

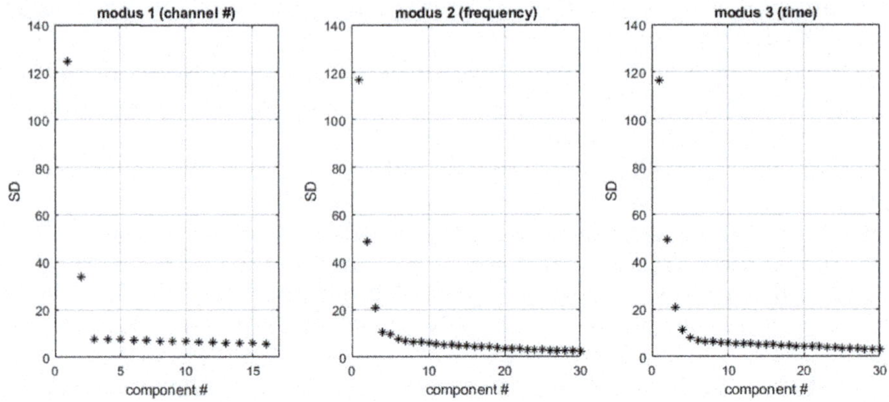

Fig. 7.47 Singular values (SD, standard deviation) of the three sub-tensors (space/channel, frequency, time). The second singular value (frequency) has a total length of 1024, and the third (time) has a length of 4000. The first relevant components are shown

If one analyses the first unitary matrix (Fig. 7.48) according to the HOSVD, one can qualitatively state:

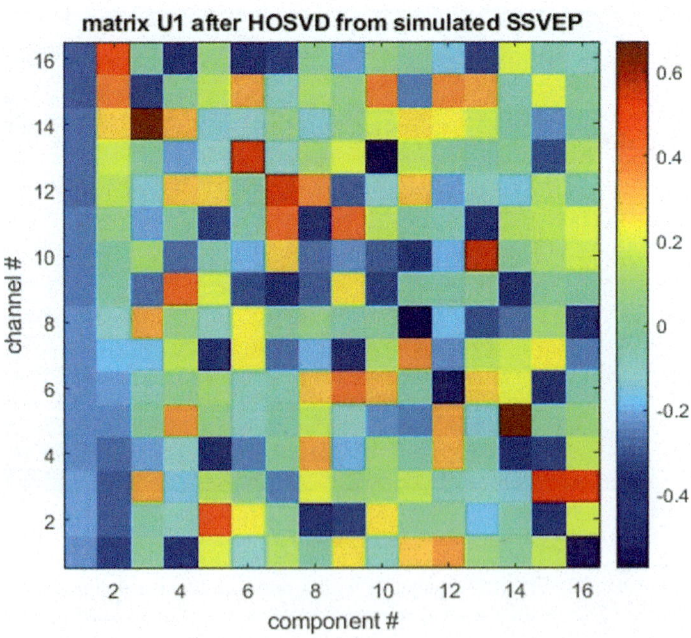

Fig. 7.48 Unitary matrix $U^{(1)}$ after HOSVD of the simulated signal according to Fig. 6.105

- The first two columns seem to have a deterministic character, corresponding to the singular values in Fig. 7.47 on the left.
- According to the qualitative correlation of the channels, all power signals (constant signal power) are assigned in the first column, i.e., all seven harmonics.
- The second column shows an increased signal level with an increasing channel number, so the transient EP is present here. All other columns are stochastic, which is attributed to the noise present.

Without explicitly analyzing the other projections here, it can be stated—assuming multilinear correlations—that the HOSVD leads to the orthogonality of the decomposed components in all directions in any number of dimensions. This result is mathematically and algorithmically justified and assured. However, the same observation applies to the higher dimensions in a figurative sense that was already made for the SVD of two-dimensional matrices:

> Multidimensional orthogonalisation of multi-channel signals or repeated recordings of a single-channel signal yields linearly independent components that are ordered in all dimensions according to their signal power. It does not separate according to signal shape and not according to origin. Emergence of previously unknown waveforms is probable and realistic.

Figure 7.50 shows how difficult it is to interpret the decor-related components even if the composition of the signal to be analyzed is known in advance (Fig. 7.49).

The unitary matrix $\boldsymbol{U}^{(2)}$ (Fig. 6.108 left) shows—in connection with the subtensor $\boldsymbol{S}_{(2)}$ (Fig. 7.47 center)—that there are about three orthogonal components (source signals): While the first component combines all seven harmonic components, the second and third components seem to originate from the transient component. However, such an interpretation is only possible if one precisely knows which signal components are contained. That an interpretation is difficult even with knowledge of the actual signal components is shown by the unitary matrix $\boldsymbol{U}^{(3)}$ (Fig. 6.108 right). Here, only rudimentary transient components can be recognized in the first five components, also attenuated in components #70 to #80, between 0 and 1.2 s. With prior ignorance of the signal components—a practically typical case of signal analysis—a technically competent interpretation of the components after the HOSVD would be impossible. Therefore, the following also applies to the HOSVD:

> For a professional separation of signal components or signals from disturbances, at least the qualitative time courses must be known.

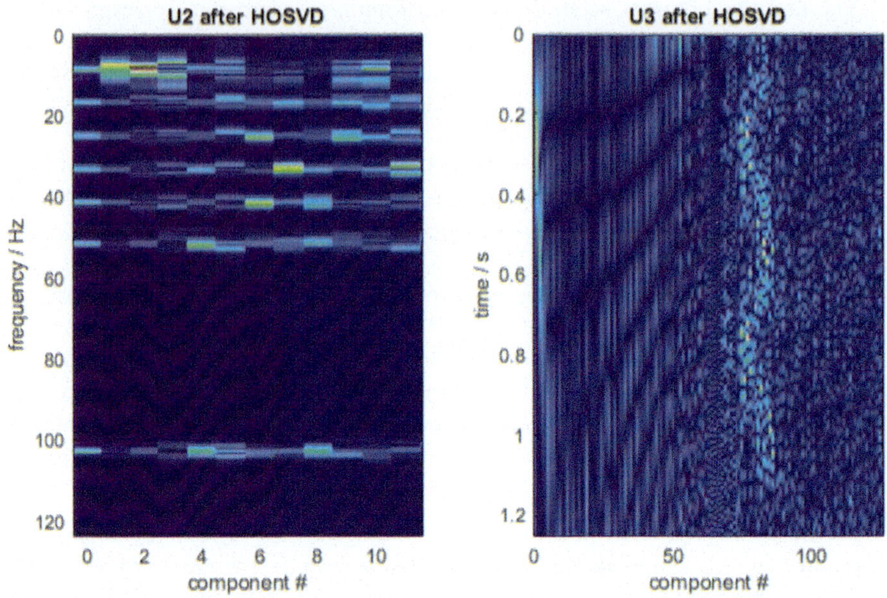

Fig. 7.49 Unitary matrices $\boldsymbol{U}^{(2)}$ and $\boldsymbol{U}^{(3)}$ after HOSVD of the simulated signal from Fig. 7.46

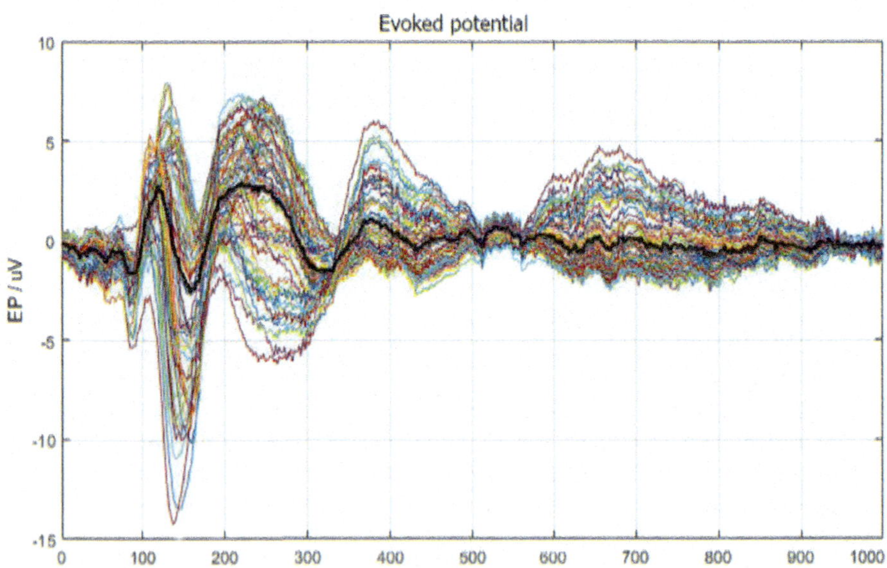

Fig. 7.50 Recording of a 64-channel evoked potential (EEG system 10–10), averaged over 1600 stimulus-synchronous individual responses to adequate optical stimulation. The simultaneous arithmetic mean is marked in bold black

7.4 Exercises

7.4.1 Tasks

7.4.1.1 Signal Detection with t-test

Recordings of an evoked potential are in the file auswtp300_ges.mat (Fig. 7.50). An average recording from 64 EEG channels (10–10 system) over 1600 realizations (synchronous stimulus averaging) is included. These data show that the averaged evoked potential contains almost no spontaneous EEG activity, and the averages are deterministic (stimulus-related) with a high degree of certainty. The task of the BSP is to decide with defined statistical uncertainty whether the recorded and averaged signal occurs randomly or deterministically. To do this, select a suitable statistical test (preferably the *t-test*), check the conditions for its application, and interpret the results.

7.4.1.2 Signal Detection with Energy Detector

Load the file ep_with_nois.mat. It contains 50 consecutive recordings (variable nx) of an artificial EP (evoked potential) embedded in white normally distributed noise with an SNR of ¼ or – 6 dB in the poststimulus part (right half in the time domain) (Fig. 7.51).

In a recording with an SNR of -6 dB, the sought-after signal shape is not identifiable even if well known (Fig. 7.51). For signal detection, one must rely on

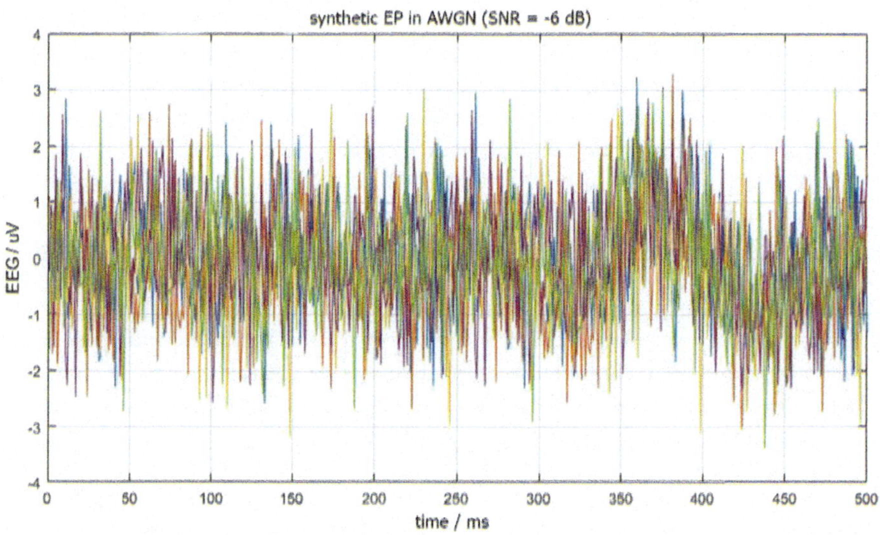

Fig. 7.51 Normally distributed white noise (left half, time 0–249 ms) with a variance of 1.0. Synthetic EP embedded in the same noise (right half, time 250–500 ms) with SNR of 0.25 and – 6dB, respectively. The left half corresponds to the prestimulative EEG (noise reference), and the right half to the poststimulative EEG (noise + signal)

a signal form-independent detector—the energy detector. This one needs a noise reference (signal sections in time or spectrum without signal components) and a measuring section with a possibly existing searched signal. Since the signal form is unknown, one can judge whether a signal is present in the noise based on the signal energy alone.

Based on the data, decide whether a signal with an unknown course is in the right (poststimulative) half of the recording. One can use consecutive (cumulative) stimulus synchronous averaging to increase the SNR to the detection threshold.

7.4.2 Solutions

7.4.2.1 Signal Detection with t-test

With these data (averaging order 64 for the simultaneous mean of the channels), one can rely on the effect of the ZGWS (Central Limit Theorem). Therefore, one can assume that the value averaged at each point in time (printed in bold black in Fig. 7.50) converges to the normal distribution and at least corresponds to the *t-distribution.* Therefore, one can save the adjustment test for a normal distribution here.

Theoretically, 1000 time points are available for the test. However, the requirement in biostatistics for the independence of the experiments would have to be checked or fulfilled here, which in the case of biosignals, amounts to the requirement for a white spectrum. Prewhitening is not a fundamental problem but worsens the already poor SNR. Therefore, one can circumvent this problem by choosing the time points for signal detection "quasi-randomly": Non-randomly, one chooses the electrophysiologically important and known waves: P1 or P100, N1 or N120. These waves and their latencies are known in electroneurophysiology. For example, the time points for the test could be set to t = 120, 160 ms.

The t-test in Matlab yields the following results:

At time $t = 120$: $h = 1$, $p = 1.925e - 12$, $ci = [1.98, 3.16]$. This result can be interpreted according to the school statistics so that the alternative hypothesis H1 can be accepted very confidently. The extremely low value of p suggests a very high reliability of the statistical proof of $H1$. Although the correctness of this result is undoubtedly given (see signal plot), in less convincing cases, one must always remember that the data at hand were evaluated under the null hypothesis $H0$ alone. Although the probability for such a mean value of 2.57 is very low (see p), the value still results from the data for the null hypothesis. From this, the methodologically incorrect conclusion that H1 is to be assumed is still formulated today. However, since no data are available for $H1$, nothing can be said about the reliability of its assumption. In real analysis, the reliability of the $H1$ *assumption* (sensitivity) can even be in the lower percentage range (close), or the uncertainty can reach over 80%.

Methodologically correct would be a control test at a less prominent point of the signal curve, e.g., at time t = 7 ms. No detectable cortical stimulus response can be detected based on the anatomy and electrophysiology. However, the test yields the following results:

At time t = 7 ms: h = 1, p = 3.63e − 4, ci = [− 0.32, − 0.09]. According to the doctrine, there is clear acceptance of hypothesis H1 here, i.e., secure statistical proof of an evoked potential. However, this proof is not electrophysiologically tenable. Therefore, the question arises as to where a signal-analytical problem exists in this test:

- From the point of view of signal processing, there should be no detectable electrical activity at this point—especially with the considerable averaging 1600. Nevertheless, there is a detectable potential due to the signal processing: When recording the EEG, a high-pass filter was used (as is usual in most derivation systems) to eliminate the electrode voltage. After the high pass, the derivative's average time value is zero. However, this also means that if the waves of the stimulus–response are not zero-symmetrical—and they never actually are—signal sections with zero activity (baseline) are shifted into a DC component, in this case in the negative direction. In this way, a high pass feigns electrical activity where none exists.
- Parametric tests are progressive: as the sample size increases (here 1000 × 64), they quickly reject the null hypothesis, even at minor deviations, as here at t = 7 ms with − 0.21 V. In this way, they feign the statistically confident presence of a response where it is not present.

One can methodically devise the idea of detecting simultaneously at several points. From a statistical point of view, several test results are linked. It leads to the so-called accumulation of errors, resulting in the statistical certainty decreasing exponentially with an increasing number of individual tests. Therefore, one would have to introduce the so-called Bonferroni correction. However, one can apply a multivariate test in the time domain from the outset, e.g., the Hotelling test. At this point, we refer to other literature.

The application of the *t-test for the* detection of signal components is relatively simple and also reliable after observing or ensuring the conditions. However, it must be known when the potentials are to be expected or when they cannot occur. It is given for physiological processes but not pathological changes. One must assume that pathological processes change the biosignals unpredictably and thus prevent accurate detection. Therefore, it is necessary to use signal form-independent detectors, especially in diagnostics.

7.4.2.2 Signal Detection with Energy Detector

One can first get a picture of the averaged signal (n = 50) (Fig. 7.52). The analysis of the noise (left half, t = 0–250 ms, noise reference) leads to the conclusion that it is normally distributed white noise according to the prerequisites for the energy detector. The analysis of the mixture noise + signal (right half, t = 251–500 ms) clearly shows that a deterministic component is included, which is not normally distributed and not white. Therefore, the requirement for the signal properties would not be fulfilled. However, one does not have to immediately use the relatively highly averaged signal for detection. One can start with the detection at the first realization with a poor SNR of − 6dB because the requirement for normally distributed data and white spectrum is still fulfilled quite well due to the weak deterministic component here. Afterward, the consecutive mean is always detected iteratively (averaging orders 2, 3, … 50). With each consecutive increase of the averaging order, the SNR improves so that the probability of a successful detection always increases. Commonly, the signal is detected earlier than at the highest averaging order (here at 50). Should all realizations be necessary for averaging (Fig. 7.52), the average must be whitened before detection with one of the whitening algorithms described above.

Corresponding to the relationship for energy detection according to Eq. 7.40, the Matlab test *vartest2.m* can be used. This test decides the null hypothesis for the averaging orders N = 1, 2 during consecutive averaging. From order N = 3, the test favors the alternative hypothesis H1 with p = 0.0248. However, it is already visible in this averaging order that the normal distribution for the poststimulatory part (t = 251–500 ms) is not fulfilled (Fig. 7.53).

While the pre-stimulative data are exactly normally distributed and white (since generated this way), the post-stimulative data (right half) show apparent deviations. Although this detection result can be trusted well (the violation of the preconditions is almost negligible), the data would at least have to be whitened for a correct methodology. A whitening algorithm also improves the empirical data distribution in the direction of normal distribution. However, with increasing averaging

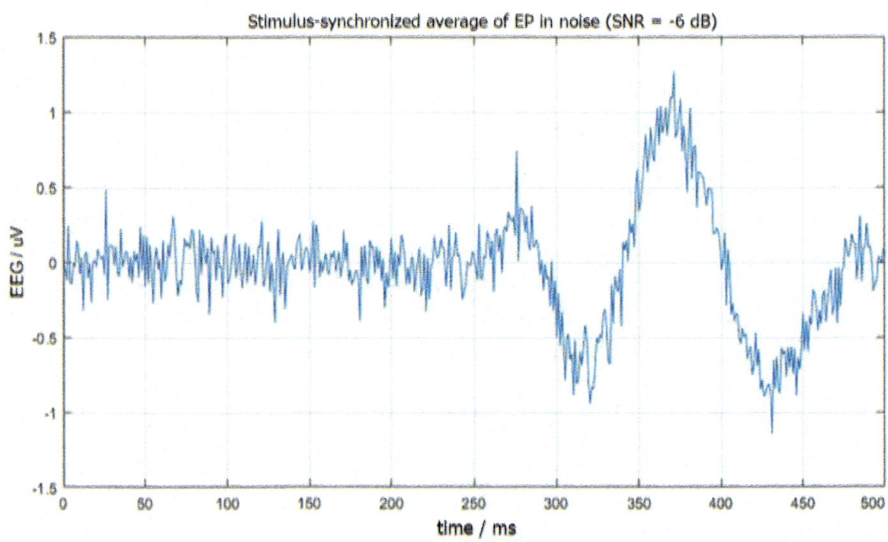

Fig. 7.52 Stimulus-synchronous average (stimulus timing at 250 ms) over 50 individual recordings of a synthetic EP in white normally distributed noise with original SNR of − 6 dB

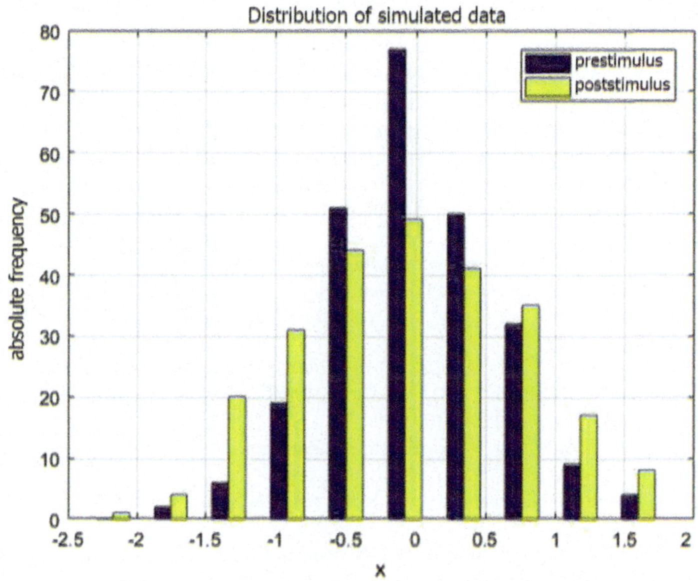

Fig. 7.53 Empirical distributions for prestimulative (left half of the data from Fig. 7.52) and poststimulative (right half of data from Fig. 7.52) after three averages of the original data

order, one must also expect increasing—therefore unacceptable—deviation from the normal distribution. In the course of this deterioration, the assumption of a white spectrum is also less and less fulfilled since the deterministic (non-white) part becomes stronger and stronger. All combined effects lead to an impossible mathematically correct energy detection despite additional measures (whitening, distribution transformation). Already from the low averaging order of three, other methods are necessary to fulfill the prerequisites.

- The normal distribution can be forced with the help of the CLT. In principle, any linear integral transformation is suitable for this. Pragmatically, the application of the DFT is most suitable: according to the DFT, the Fourier coefficients are normally distributed if the signal length is at least 30 measured values long. In practice, this can be fulfilled well.
- In this case, a white spectrum can be created very well with the spectrum-based whitening method.

Thus, both necessary conditions for energy detection are fulfilled, regardless of the noise and signal properties, before the analysis.

For spectral energy detection, please refer to the relevant literature (Liavas et al., 1998).

References

De Lauthauwer, L., De Moor, B., & Vandewalle, J. (April 2000). A multilinear singular value decomposition. *SIAM Journal Matrix Analysis Application, S*, 1253–1278.

Haardt, M., Roemer, F., & Del Galdo, G. (July 2008). Higher-order SVD-based subspace estimation to improve the parameter estimation accuracy in multi-dimensional harmonic retrieval problems. *IEEE Transactions on Signal Processing*, 3198–3213.

Heisenberg, W. (1927). On the descriptive content of quantum-theoretical kinematics and mechanics. *Zeitschrift Für Physik, 43*, 172–198.

Hyvärinen, A. K. (2001). *Independent component analysis*. Wiley.

Liavas, A., Moustakides, G., Henning, G., Psarakis, E., & Husar, P. (1998). Periodogram-based method for the detection of steady-state visually evoked potentials. *IEEE Transactions on Biomedical Engineering*, 242–248.

Manolakis, D., & Ingle, V. (2011). *Applied digital signal processing*. Cambridge University Press.

Pan, J., & Tompkins, W. J. (March 1985). A real-time QRS detection algorithm. *IEEE Transactions on Biomedical Engineering*, 230–236.

Renn, J., & Sauer, T. (September 1996). Einstein's Zurich notebook. *Physikalische Blätter*, 865–872.

Stewart, G. W. (1993). On the early history of the singular value decomposition. *SIAM Review*, 551–566.

Weis, M. (2015). Multi-Dimensional Signal Decomposition Techniques for the Analysis of EEG Data. Ilmenau.

Index

P. Husar and G. Gašpar, *Electrical Biosignals in Biomedical Engineering*,
https://doi.org/10.1007/978-3-662-67998-2

The manufacturer's authorised representative in the EU is Springer Nature Customer Service Centre GmbH, Europaplatz 3, 69115 Heidelberg, Germany. If you have any concerns regarding our products, please contact ProductSafety@springernature.com

Printed and bound by CPI Group (UK) Ltd, Croydon, CR0 4YY

31/07/2026

02183004-0001